High-resolution radio astronomy is burgeoning, with profound contributions coming from recent developments in Very Long Baseline Interferometry and observations with the Hubble Space Telescope in particular. From an exciting international conference in Manchester, these proceedings present the latest results and suggests new ways forward.

Sixteen review articles summarise current knowledge – in fields such as masers, parsec- and kiloparsec-scale radio jets, interstellar turbulence and astrometry – and outline future research. In addition, well over a hundred papers give the latest developments in subjects as diverse as radio stars, the Galactic Centre and gravitational lensing, and present recent advances that will allow the accurate intercomparison of high-resolution radio and optical observations.

This timely volume provides an essential overview of the latest results and points towards future research for graduate students and researchers in the exciting field of sub-arcsecond radio astronomy.

Sub-arcsecond Radio Astronomy

Sub-arcsecond Radio Astronomy

Proceedings of the Nuffield Radio Astronomy Laboratories' conference, held in Manchester, July 20–24, 1992

Edited by
R. J. DAVIS
Nuffield Radio Astronomy Laboratories
University of Manchester
Jodrell Bank, UK

and

R. S. BOOTH
Onsala Space Observatory
Chalmers University of Technology
Gothenburg, Sweden

Published by the Press Syndicate of the University of Cambridge
The Pitt Building, Trumpington Street, Cambridge CB2 1RP
40 West 20th Street, New York, NY 10011–4211, USA
10 Stamford Road, Oakleigh, Melbourne 3166, Australia

© Cambridge University Press 1993

First published 1993

Printed in Great Britain at the University Press, Cambridge

A catalogue record for this book is available from the British Library

ISBN 0 521 43472 6 hardback

Contents

Participants   xix

Preface and acknowledgements   xxv

I. GALACTIC ASTRONOMY

1. STARS

Interferometric observations of radio stars (Review)   1
    A.R. Taylor

Radio jets of SS433 - blobby or continuous   7
    R.C. Vermeulen, R.E. Spencer, R.T. Schilizzi & I. Fejes

Photospheric emission from red giant stars at radio frequencies   10
    M.J. Reid & K.M. Menten

MERLIN observations of radio stars   13
    R.J. Davis

Radio imaging the BE star φ Persei   16
    S.M. Dougherty & A.R. Taylor

High resolution imaging of symbiotic stars   18
    H.T. Kenny, A.R. Taylor, R.J. Davis, P. Pavelin, M.F. Bode
    & M. Bang

The Serpens radio jet   20
    S. Curiel, L.F. Rodríguez, James Moran & Jorge Cantó

The milliarcsec jets of Cygnus X3   22
    C. Schalinski, A. Witzel, K. Johnson, P. Pavelin, R.J. Davis,
    R.E. Spencer & G. Umana

VLBI observations of LSI+61°303   24
    J.M. Paredes, M. Massi, R. Estalella & M. Felli

Cepheus A - an optically obscured Herbig-Haro object?   26
    V.A. Hughes

VLBI Observations of a stellar flare   28
    W. Alef, A.O. Benz, M. Güdel & P. de Vicente

Modelling the strong Cygnus X-3 radio outburst   30
    J.M. Paredes, J. Martí & R. Estalella

Luminosity of radio supernovae   32
    R. Sramek, K. Weiler, S. van Dyk & N. Panagia

LkHα 101: the relation between radio and thermal infrared observations at high spatial resolution 34
    R.M. Danen, C.R. Gwinn & E.E. Bloemhof

Hydrodynamic models of bipolar nova outbursts 36
    H.M. Lloyd, M.F. Bode, T.J. O'Brien & F.D. Kahn

2. THE GALACTIC CENTRE

Compact radio sources in the Galactic Centre (Review) 38
    Jun-Hui Zhao & W.M. Goss

Manifestations of the wind phenomenon at the Galactic Centre 44
    F. Yusef-Zadeh

Could 1E1740.7-2942 be extragalactic? 47
    I.F. Mirabel, L.F. Rodríguez, B. Cordier, J. Paul & F. Lebrun

First 1.3 cm VLBA map of Sgr A 50
    J. Marcaide, A. Alberdi, L. Lara, P. Elósegui, I. Shapiro, W. Cotton, J. Romney & R. Preston

3. MASERS AND MOLECULES

Sub-arcsecond observations of maser stars (Review) 52
    P.F. Bowers

VLBI observations of SiO masers in Europe 59
    F. Colomer

Distance measurements through observations of masers (Review) 62
    J.M. Moran

Maser theories (Review) 68
    D. Field & M. Gray

Preliminary results from VLBI observations of 1720 MHz masers in W3(OH) 74
    M.R.W. Masheder, V. Migenes, M.D. Gray, R.J. Cohen, D.Field & R.S. Booth

Time variations in the flux density of spatially resolved interstellar OH masers 76
    E.E. Bloemhof, M.J. Reid & J.M. Moran

VLBI observations of Methanol masers at 6.7 GHz 78
    K.M. Menten, M.J. Reid, P. Pratap, J.M. Moran & T.L. Wilson

Observations of southern methanol masers 81
    R.P. Norris

Probing the milliarcsecond-scale structure of molecular clouds 84
    A.P. Marscher, T.M. Bania & Z. Wang

4. PLASMA EFFECTS

Interstellar Scattering (Review) B. Dennison	86
A model for extreme scattering events B. Dennison & J.H. Simonetti	92
Galaxy Centre-anticentre asymmetry of low frequency variability radio source distribution: evidence of interstellar matter influence N. Shapirovskaya, O.B. Slee & G.S. Tsarevsky	94
Speckle VLBI C.R. Gwinn, K.M. Desai, J. Reynolds, D. Jauncey, E.A. King, C. Flanagan, G. Nicolson, R.A. Preston & D.L. Jones	96
Highly scattered OH/IR stars at the Galactic Centre P.J. Diamond, D. Frail, H.J. van Langevelde & J. Cordes	99
Spectrum of interstellar turbulence: observations of OH and H_2O masers in W49N K.M. Desai, C.R. Gwinn & P.J. Diamond	102
VLBI structures of low frequency variables: implication for refractive scintillation theory L. Padrielli, M. Bondi, L. Gregorini, F. Mantovani, W. Eastman, N. Shapiroskaya & S. Spangler	104
Three epoch VLBI observations at 1.6 GHz of a sample of low frequency variable sources M. Bondi, L. Padrielli, R. Fanti, L. Gregorini, F. Mantovani, J.D. Romney, N. Bartel, K.W. Weiler & G.D. Nicolson	107
Interferometric study of interstellar turbulence A. Lazarian	109

II. EXTRAGALACTIC ASTRONOMY

1. GRAVITATIONAL LENSES

Observations of gravitational lenses (Review) D. Walsh	111
A (very) few points on the theory of gravitational lenses (Review) C.S. Kochanek	117
Einstein rings and Einstein quads B.F. Burke, S.R. Conner, J.N. Hewitt & J. Lehar	123
High resolution radio observations of 0957+561 M.A. Garrett, R.W. Porcas, L.J. King, R. Calder, P.N. Wilkinson & D. Walsh	128

CONTENTS

Global 6 cm VLBI investigations of 0957+561 R.M. Campbell, B.E. Corey, E.E. Falco, I.I. Shapiro, M.V. Gorenstein, P. Elósegui, J.M. Marcaide & K. Alvi	131
Southern hemisphere observations of PKS1830-211 D.L. Jauncey and the SHEVE team	134
Search for small separation gravitational lens systems A. Patnaik, I. Browne, L. King, T. Muxlow, D. Walsh & P.Wilkinson	137
Multifrequency radio images of MG1131+0456 G.H. Chen & J.N. Hewitt	140
The Jodrell Bank lens survey L.J. King, I.W.A. Browne, A. Patnaik, T.W.B. Muxlow, D. Walsh & P.N. Wilkinson	144
MERLIN observations of 3 gravitational lens candidates M.A. Garrett, A.R. Patnaik, T.W.B. Muxlow, P.N. Wilkinson & D. Walsh	146
Is the BL Lacertae object 1308+326 a microlensed quasar? D.C. Gabuzda & R.I. Kollgaard	148
Northern hemisphere VLBI observations of PKS 1830-211 D.L. Jones and the SHEVE team	150
A VLBI survey of southern hemisphere peaked spectrum sources E.A. King and the SHEVE team	152
Undetected lens systems in the MIT-Green Bank-VLA 5 GHz lens search sample S.R. Conner, A. Fletcher, L. Herold & B.F. Burke	154
The Unusual Radiosource MSH 04-71 J.E. Reynolds & the SHEVE team	156

2.   COMPACT STRUCTURE IN AGN

Intraday variability and high brightness temperatures A. Witzel, S. Wagner, R. Wegner, W. Steffen & T. Krichbaum	159
IDV: polarization and spectral indices R. Wegner & A. Witzel	165
Continuum vs VLBI: a comparison between data derived from flux monitoring and VLBI observations E. Valtaoja & H. Teräsranta	167
Brightness temperatures and sizes of the flaring components in AGN from Metsähovi continuum monitoring H. Teräsranta & E. Valtaoja	170

Parsec-Scale Jets (Review) A.C.S. Readhead	173

New results from VLBI at 43 GHz
  T.P. Krichbaum, A. Witzel, D.A. Graham, C.J. Schalinski &
  J.A. Zensus
	181

The 86 GHz VLBI test with Pico Velat: first detection of quasar
3C454.3 at 3 mm wavelength
  C. Schalinski, A. Greve, M. Grewing, H. Steppe, D. Graham,
  T. Krichbaum, A. Witzel, A. Alberdi, L. Bååth, R.S. Booth
  & F. Colomer
	184

Compact structure in 3C273 Z. Abraham, E.A. Carrara, J.A. Zensus & S.C. Unwin	186
Kinematics of the parsec-scale jet in 3C345 S.C. Unwin	189
Superluminal jet in 3C279 E.A. Carrara, Z. Abraham, S.C. Unwin & J.A. Zensus	191
The search for superluminal motion in a complete sample of lobe-dominated quasars D.H. Hough, R.C. Vermeulen & A.C.S. Readhead	193

Two-epoch VLBI maps of three weak nuclei in lobe-dominated quasars
  D.H. Hough, J.A. Zensus, R.C. Vermeulen, A.C.S. Readhead,
  R.W. Porcas & A. Rius
	195

Extreme superluminal motion in CTA102? F.T. Rantakyrö & L.B. Bååth	197
Geodetic VLBI Monitoring: Parsec-scale structure of Extragalactic Radio Sources C. Schalinski, S. Britzen, A. Witzel, W. Alef, J. Campbell	199
Multi-epoch 8.4 GHz VLBI observations of the nucleus of Centaurus A D.L. Meier and the SHEVE team	201
Parsec-scale properties of FR-1 radio galaxies T. Venturi, L. Feretti, G. Giovannini & A.E. Wehrle	204

VLBI polarization observations of 3C138: preliminary results
  W. Cotton, D. Dallacasa, C. Fanti, R. Fanti, R. Spencer,
  R. Schilizzi & T. Foley
	207

The results of a study of the milliarcsecond scale linear polarization properties of extragalactic radio sources T.V. Cawthorne, J.F.C. Wardle, D.H. Roberts & D.C. Gabuzda	209
3.6 cm VLBI polarization observations of BL Lacertae objects D.C. Gabuzda & T.V. Cawthorne	211

The Caltech-Jodrell Bank VLBI survey
   P.N. Wilkinson, A.G. Polatidis, A.C.S. Readhead, W. Xu
   & T.J. Pearson
	213

Two core-jet sources with large misalignment
   W. Xu, A.G. Polatidis, A.C.S. Readhead, P.N. Wilkinson
   & T.J. Pearson
	216

The spectral indices of the parsec-scale jet components of 3C273 P. Charlot	218
Three-frequency VLBI spectral-index maps of the quasars 1038+528A,B J.C. Guirado, J.M. Marcaide, P. Elósegui & J.L. Gómez	220

Compact structure of 3C286, 3C309.1 and 3C380 CSS quasars
   A.J. Kus, R.S. Booth, A. Marecki, R. Maszkowski, R.W. Porcas
   T.J. Pearson, A.C.S. Readhead & P.N. Wilkinson
	222

3C380: motion down a twisted channel; not a second nucleus A.G. Polatidis, P.N. Wilkinson & C.E. Akujor	225
Core Activities of Compact Steep-Spectrum Radio Sources S. Kameno, M. Inoue, H. Takaba, Rendong Nan, R.T. Schilizzi	227

Compact steep-spectrum radio sources from the Peacock & Wall Catalog
   D. Dallacasa, C. Fanti, R. Fanti, R.T. Schilizzi &
   R.E. Spencer
	229

VLBI observations of compact sources with a limited number of antennas
   D. Dallacasa, M. Bondi, C. Stanghellini, P.L. Cerchiara,
   G. Umana, C. Trigilio & F. Rantakyro
	232

The Quasars 4C39.25 on a scale of tens of mas Shengyin Wu, I.I.K. Pauliny-Toth & R.W. Porcas	234
Twisted Pc-scale structure in the BL Lacertae object Mrk 501 J.M. Wrobel & J.E. Conway	236
A new VLBI image of an archetypal CSS source 3C286 at 5 GHz F.J. Zhang, R.E. Spencer, R.T. Schilizzi & C. Fanti	238

VLBI structure of 3C395 at 2.3 and 8.4 GHz: observations and numerical simulations
   L. Lara, A. Alberdi, J.L. Gómez, J.M. Marcaide &
   T.W.B. Muxlow
	241

2.3 GHz VLBI images of southern hemisphere radio galaxies and quasars D.W. Murphy and the SHEVE team	243

Observations of radio galaxies in the S/X band with the triangle Medicina-Noto-Madrid
   T. Venturi, L. Feretti, G. Giovannini, J. Marcaide,
   L. Lara, M.J. Rioja, C. Trigilio & G. Umana
	245

The structure of the hot spots in 3C295   247
    G.B. Taylor & R.A. Perley

Unusual features in QSR 3C147: a unified explanation of the   249
jets and BLR clouds
    H.S. Chu, F.J. Zhang, R.L. Mutel, L.I. Matveenko &
    R.E. Spencer

3. INTERMEDIATE SCALE STRUCTURE

Observations of jets with MERLIN (Review)   252
    T.W.B. Muxlow

Fine structure in the jets of 3C219   258
    R.A. Perley, A.H. Bridle & D.A. Clarke

The giant radio quasar 4C74.26: pc to Mpc scale structure   261
    K.M. Blundell, P.J. Warner, J.M. Riley, T.J. Pearson
    & P. Alexander

Extended radio jets without hotspots   263
    C.E. Akujor & S.T. Garrington

3C216: a galactic-size powerful radio source?   265
    C.E. Akujor, R.W. Porcas, E. Ludke & D.L. Shone

The sub-arcsecond radio structure of 4C39.25   267
    N. Jackson, I.W.A. Browne, A. Alberdi & J. Marcaide

3C159: a peculiar radio galaxy   269
    P.L. Cerchiara, F. Mantovani, I.W.A. Browne & T.W.B. Muxlow

Superluminal motion at 4 arcseconds in 3C120?   271
    R.C. Walker & J.M. Benson

Polarization observations of CSS sources with the enhanced MERLIN   273
    E. Ludke, T.W.B. Muxlow, S.T. Garrington, R.E. Spencer
    & C.E. Akujor

Polarization observations of compact steep spectrum sources   275
    C.E. Akujor & S.T. Garrington

Radio polarization observations of 807 flat spectrum sources   277
    C.E. Akujor & A.R. Patnaik

Low frequency variability and structure of the quasar 3C345   279
    L.I. Matveenko

On angular structure studies of very-steep spectrum sources   282
found at decametric wavelengths
    K.P. Sokolov

4. AGN AT OTHER WAVELENGTHS

Optical and HST observations of jets (Review) — 284
   W.B. Sparks

The optical counterpart of the east lobe of M87 — 290
   J.I. González-Serrano, I. Pérez-Fournon & W. Junor

Optical observations of compact steep spectrum radio sources — 293
   R. Morganti & C.N. Tadhunter

Optical polarization observations revealing the emergence of new radio components in blazars — 295
   L. Valtaoja

Interpretation of multiwavelength observations of nonthermal extragalactic radio sources (Review) — 297
   A.P. Marscher

New X-ray observations of extragalactic VLBI sources — 303
   D.M. Worrall, M. Birkinshaw & C.R. Gwinn

5. MILDLY ACTIVE GALAXIES

High resolution UV/optical/IR imaging of active galaxies (Review) — 306
   M.J. Ward

MERLIN 5 GHz observations of supernova remnants in M82 — 312
   T.W.B. Muxlow, A. Pedlar, P.N. Wilkinson, D. Axon & A.G. de Bruyn

Subarcsecond observations of the radio jet in NGC4151 — 314
   A. Pedlar, M. Kukula, T.W.B. Muxlow, D.J. Axon, S. Baum, C. O'Dea & S.W. Unger

A VLBI survey of luminous IRAS galaxies — 318
   C.J. Lonsdale, H.E. Smith & C.J Lonsdale

Radio observations of the nuclear environment of Seyfert galaxies — 321
   L. Gregorini, P. Marziani, L. Padrielli & P. Rafanelli

The nuclear ring of the barred spiral galaxy NGC1326 — 323
   J.A. Garcia-Barreto, R.-J. Dettmar, F. Combes, M. Gerin & B. Koribalski

Extragalactic masers — a status report (Review) — 324
   W.A. Baan

Toward weighing a hidden Seyfert nucleus — 331
   C.R. Gwinn, R.J. Antonucci, R. Barvainis, J. Ulvestad & S. Neff

6. THEORY AND JET SIMULATION

Unified Schemes for active galactic nuclei C.M. Urry	333
The effect of turbulence on the large scale structure of radio jets S.A.E.G. Falle	340
Relativistic flow in low luminosity radio jets R.A. Laing	346
Deceleration of relativistic jets in FR-I radio galaxies K.S. Sergey	349
Models of jets with multiscale structures of knots A.C. Raga	352
Interpretation of superluminal radio sources as bent shocked relativistic jets A. Alberdi, J.L. Gómez & J.M. Marcaide	355
Interpreting VLBI sources as twisted jets P.A. Hughes, M.F. Aller, H.D. Aller & A. Rosen	358
Helical jets and the misalignment distribution for core dominated radio source J.E. Conway & D.W. Murphy	360
A magnetized helix in 3C345? W. Steffen, T.P. Krichbaum, A. Witzel & A. Zensus	363
Models of helical jets A.J.Kus	365
CSS sources and projection effects H.S. Sanghera & R.E. Spencer	367
Can the filaments of the radio sources be really due to synchrotron thermal instabilities? E.M. de Gouveia Dal Pino & R. Opher	369
Modelling jets on milli-arcsec scales: polarized structure and flux spectra Y.A. Kovalev & Y. Yu Kovalev	371
A problem of different kinds of asymmetries in extended double radio sources B. Komberg	373
3-D simulations of continuous and recurrent cooling jets E.M. de Gouveia Dal Pino & W. Benz	375
Simulations of relativistic jets M. Bowman	378

III. COSMOLOGY

Milli-arcsecond structures of AGN at different redshifts (Review)   380
    L.I. Gurvits

The angular size – redshift relation for compact radio sources   386
    K.I. Kellermann

IV. ASTROMETRY

Linking the optical and radio reference frames (Review)   390
    R.W. Argyle, L.V. Morrison, J.D.H. Pilkington & A.N. Argue

The current status of HST astrometry for linking the Hipparcos frame to extragalactic objects   394
    P.D. Hemenway & R.L. Duncombe

The Radio/Optical Reference Frame: Progress and Plans   397
    J.L. Russel, A.L. Fey, D.L. Jauncey, K.J. Johnston, N. Kawaguchi, A. Kemball, E.A. King, C. Ma, G. MacLeod, D.F. Malin, P.M. McCulloch, G. Nicolson, J.E. Reynolds, D. Shaffer, Y. Takahashi, C. de Vegt, G.L. White & N. Zacharias

Astrometry with optical interferometers   403
    C.A. Hummel

Astrometry and structure of SiO Masers   406
    A. Baudry

VLBA phase referencing tests   409
    J.E. Conway & R.C. Vermeulen

VLBI phase referencing at 5 degrees separation   412
    P. Elósegui, J.M. Marcaide, A. Alberdi, M.I. Ratner, I.I. Shapiro, A. Quirrenbach, A. Witzel, F. Mantovani & A. Rius

High precision, VLBI trigonometric parallaxes of three radio-emitting stars σ CrB, UX Ari and HR5110   415
    J-F. Lestrade, R.B. Phillips, D.L. Jones, RA. Preston & D.C. Gabuzda

High-accuracy proper motions of millisecond pulsars from the Nancay timing program   418
    I. Cognard, G. Bourgois, D. Aubry, B. Darchy, J-P. Drouhin & J-F Lestrade

Towards VLBI determinations of pulsar parallaxes and proper motions   420
    R.M. Campbell, N. Bartel, I.I. Shapiro, M.I. Ratner, R.J. Cappallo, A.R. Whitney, W.H. Cannon, W.T. Petrachenko, J. Popelar & T.M. Eubanks

V. INSTRUMENTS AND TECHNIQUES

MERLIN – Phase 2   422
    P.N. Wilkinson

VLBA capability and status   425
    R.C. Walker

The southern hemisphere VLBI experiment (SHEVE)   428
    R.A. Preston and the SHEVE team

Millimetre VLBI capability status   431
    L.B. Bååth

Space VLBI Project VSOP   434
    M. Inoue

Preliminary space VLBI requirements for observing time on ground radio telescopes   437
    D.L. Meier, D.W. Murphy & R.A. Preston

32 meter antenna for Torun radio astronomy observatory   439
    S. Gorgolewski, A. Kus & B. Krygier

Portuguese radio interferometry the MAGRIÇO project   441
    M. João Martins & A.A. da Costa

AUTHOR INDEX   443

OBJECT INDEX   446

SUBJECT INDEX   449

LIST OF PARTICIPANTS

Zulema Abraham, IAG/USP Brazil
Chidi Akujor, Onsala Space Observatory, Sweden
Antonio Alberdi, Instituto de Astrofisica de Andalucia, Spain
Walter Alef, MPIfR, Bonn, Germany, Germany
Bryan Anderson, University of Manchester, NRAL, Jodrell Bank, UK
Bob Argyle, Stellar Reference Frame Group, RGO, Cambrdige, UK
Vadim Artyukh, Lebedev Physical Institute, Moscow, Russia
David Axon, University of Manchester, NRAL, Jodrell Bank, UK
Willem Baan, Arecibo Observatory, Puerto Rico
Lars Bååth, Onsala Space Observatory, Sweden
Alain Baudry, Observatoire de Bordeaux, France
Arnold Benz, MPIfR, Bonn, Germany
John Biretta, NRAO VLA, Socorro, NM, USA
Adam Black, Cavendish Laboratory, Cambridge, UK
Eric Bloemhof, Harvard-Smithsonian, CfA, Cambridge, MA USA
Katherine Blundell, Cavendish Laboratory, Cambridge, UK
Mike Bode, Liverpool Polytechnic, Liverpool, UK
Walter Bogers, Kapteyn Astronomical Institute, Groningen, Netherlands
Roy Booth, Onsala Space Observatory, Sweden
Phillip Bowers, SFA Inc, Landover, Maryland, USA
Mark Bowman, University of Manchester, NRAL, Jodrell Bank, UK
Ian Browne, University of Manchester, NRAL, Jodrell Bank, UK
Bernard Burke, MIT, Cambridge, MA USA
Harvey Butcher, NFRA, Dwingeloo, Netherlands
Robert Campbell, Center for Astrophysics, Cambridge, MA USA
Everi Antonio Carrara, Universidade de Sao Paulo, Brazil
Tim Cawthorne, Harvard-Smithsonian, CfA, Cambridge, MA USA
Patrick Charlot, Observatoire de Paris, Paris, France
Grace Chen, Dept of Phys, MIT, Cambridge, MA USA
Ismael Cognard, Obs. de Meudon, France
Jim Cohen, University of Manchester, NRAL, Jodrell Bank, UK
Francisco Colomer, Onsala Space Observatory, Sweden
Samuel Conner, MIT, Cambridge, MA USA
John Conway, NRAO VLA, Socorro, NM USA
Robin Conway, University of Manchester, NRAL, Jodrell Bank, UK
Alan Cooper, Open University, Milton Keynes, UK
Antonio da Costa, Astronomy Dept, University of Manchester, UK
William Cotton, Jr. NRAO, Charlottesville, VA USA
Salvador Curiel, Harvard-Smithsonian, CfA, Cambridge, MA USA
Rustam Dagkesamansky, Astro Space Centre, Moscow, Russia
Daniele Dallacasa, Istituto di Radioastronomia, Bologna, Italy
Rod Davies, University of Manchester, NRAL, Jodrell Bank, UK
Richard Davis, University of Manchester, NRAL, Jodrell Bank, UK
Brian Dennison, Virginia Tech, VA USA
Ketan Desai, NRAO VLA, Socorro, NM USA
Peter Dewdney, DRAO, Penticton, BC Canada
Philip Diamond, NRAO VLA, Socorro, NM USA
Sean Dougherty, Phys & Astr. Calgary, Canada
John Dyson, Astronomy Dept, University of Manchester, UK
John Ellithorpe, Dept of Physics, MIT, Cambridge, MA USA

LIST OF PARTICIPANTS

Pedro Elósegui, Harvard-Smithsonian, CfA, Cambridge, MA USA
Sam Falle, Applied Mathematics, Leeds, UK
Carla Fanti, Istituto di Radioastronomia, Bologna, Italy
Roberto Fanti, Istituto di Radioastronomia, Bologna, Italy
David Field, School of Chemistry, Bristol, UK
Andre Basil Fletcher, MIT, Cambridge, MA USA
Tony Foley, NFRA Dwingeloo, Netherlands
Ed Fomalont, NRAO, Charlottesville, VA USA
Antonio Frasca, Istituto di Radioastronomia, Bologna, Italy
Denise Gabuzda, University of Calgary, Canada
J. Antonio Garcia-Berreto, Instituto de Astronomia, UNAM, Mexico
Michael Garrett, University of Manchester, NRAL, Jodrell Bank, UK
Simon Garrington, University of Manchester, NRAL, Jodrell Bank, UK
Jingping Ge, Brandeis University, Waltham, MA USA
Jose Gomez, IAC, Spain
Jesus Gomez, Centro Astronomico de Yebes, Spain
Ignacio González-Serrano, Dept Fisica Moderna, Cantabria, Spain
Stanislaw Gorgolewski, Torun Radio Astronomy Observatory, Poland
Miller Goss, NRAO VLA, Socorro, NM USA
Ann Gower, University of Victoria, Canada
Loretta Gregorini, Istituto di Radioastronomia, Bologna, Italy
Jose Carlos Guirado, IAA, Granada, Spain
Leonid Gurvits, Arecibo Observatory, Puerto Rico
Carl Gwinn, University of California, Santa Barbara, CA USA
Paul Harrison, University of Manchester, NRAL, Jodrell Bank, UK
Cyril Hazard, Institute of Astronomy, Cambridge, UK
Paul Hemenway, Austin, TX USA
David Henstock, University of Manchester, NRAL, Jodrell Bank, UK
Lori Herold, MIT, Cambridge, MA USA
Jaqueline Hewitt, MIT, Cambridge, MA USA
David Hough, Dept of Phys. Trinity Univ, San Antonio, TX USA
John Howard, Penny & Giles, UK
Philip Hughes, Astron.Dept Univ of Michigan, Ann Arbor, MI USA
Vic Hughes, Queens University, Ontario, Canada
Christian Hummel, USNO/NRL, Washington, DC USA
Markoto Inoue, Nobeyama Radio Observatory, Japan
Yakov Istomin, Lebedev Physical Institute, Moscow, Russia
Neal Jackson, Sterrewacht Leiden, Leiden, Netherlands
David Jauncey, ATNF, COSSA, Australia
Adrian Jenkins, Dept Physics & Astronomy, Cardiff, UK
Rachel Johnson, University of Manchester, NRAL, Jodrell Bank, UK
Ken Johnston, Naval Research Laboratory, Washington, DC USA
Dayton Jones, Jet Propulsion Laboratory, Pasadena, CA USA
Seij Kameno, Nobeyama Radio Observatory, Japan
Dr. Hans Jürgen Kärcher, MAN GHH, Germany
Ken Kellermann, NRAO, Charlottesville, VA USA
Neil Killeen, AAO, Epping, NSW Australia
Edward King, Phys Dept, Univ of Tasmania, Australia
Lindsey King, University of Manchester, NRAL, Jodrell Bank, UK
Chris Kochanek, Harvard-Smithsonian, CfA, Cambridge, MA USA
Boris Komberg, Astro Space Centre, Moscow, Russia
Sergi Komissarov, Astro Space Centre, Moscow, Russia
Yuri Kovalev, Astro Space Centre, Moscow, Russia
Thomas Krichbaum, MPIfR, Bonn, Germany
Marek Kukula, University of Manchester, NRAL, Jodrell Bank, UK
Andrej Kus, Torun Radio Astronomy Observatory, Poland

LIST OF PARTICIPANTS

Robert Laing, RGO, Cambridge, UK
Glen Langston, NRAO, Charlottesville, VA USA
Lucas Lara, Instituto de Astrofisica de Andalucia, Spain
Anthony Lasenby, Cavendish Laboratory, Cambridge, UK
Alex Lazarian, DAMTP, Cambridge, UK
Patrick Leahy, University of Manchester, NRAL, Jodrell Bank, UK
Joseph Lehar, Institute of Astronomy, Cambridge, UK
Mikael Lerner, Onsala Space Observatory, Sweden
Jean-Francois Lestrade, Observatoire de Meudon, Paris, France
Huw Lloyd, Liverpool John Moores University, UK
Colin Lonsdale, Haystack Observatory, Westford, MA USA
Sir Bernard Lovell, University of Manchester, NRAL, Jodrell Bank, UK
Everton Lüdke, University of Manchester, NRAL, Jodrell Bank, UK
Peter McCulloch, Dept of Phys, University of Tasmania, Australia
Jonathan McDowell, Space Science Lab, Alabama, USA
Franco Mantovani, Istituto di Radioastronomia, Bologna, Italy
John Marcaide, Instituto de Astrofisica de Andalucia, Spain
Maria Marcha, University of Manchester, NRAL, Jodrell Bank, UK
Andrej Marecki, Torun Radio Astronomy Observatory, Poland
Alan Marscher, Boston University, Boston, MA USA
Jennifer Marshall, Cavendish Laboratory, Cambridge, UK
Mike Masheder, Dept of Physics, University of Bristol, UK
Jean-Michel Martin, Observatoire de Paris, France
Leonid Matveenko, Space Research Institute, Moscow, Russia
David Meier, Jet Propulsion Laboratory, Pasadena, CA USA
Karl Menten, Harvard-Smithsonian, CfA, Cambridge, MA USA
Felix Mirabel, Service d'Astrophysic, Yvette, France
Alice Monet, US Naval Observatory, Washington, DC USA
Nicholas Moore, Dept of Physics & Astronomy, Cardiff, UK
James Moran, Harvard-Smithsonian, CfA, Cambridge, MA USA
Raffaella Morganti, Istituto di Radioastronomia, Bologna, Italy
Ian Morison, University of Manchester, NRAL, Jodrell Bank, UK
David W. Murphy, Jet Propulsion Laboratory, Pasadena, CA USA
Tom Muxlow, University of Manchester, NRAL, Jodrell Bank, UK
Alistair Nelson, Univ of Wales College, Cardiff, UK
George Nicolson, Hartebeesthoek Radio Astr.Obs. S.Africa
Colin Norman, Space Telescope Science Institute, Baltimore, MD USA
Jan Noordam, NFRA, Dwingeloo, Netherlands
Ray Norris, Australia Telescope National Facility, Australia
E. Otobe, Waseda University, Tokyo, Japan
Lucia Padrielli, Istituto di Radioastronomia, Bologna, Italy
J.M. Paredes, Dpt Astronomia i Meteorologia, Barcelona, Spain
Paola Parma, Istituto di Radioastronomia, Bologna, Italy
Alok Patnaik, University of Manchester, NRAL, Jodrell Bank, UK
Tim Pearson, Caltech, Pasadena, CA USA
Alan Pedlar, University of Manchester, NRAL, Jodrell Bank, UK
Ismael Perez-Fournon, IAC La Laguna, Tenerife
Rick Perley, NRAO VLA, Socorro, NM USA
Marta Peracavla, Dept of Astronomy, University of Barcelona, Spain
Elisabete de Gouveia Dal Pino, Harvard-Smithsonian, CfA, Cambridge, MA USA
T.B. Pjatunina, Institute of Applied Astronomy St. Petersburg, Russia
Antonis Polatidis, University of Manchester, NRAL, Jodrell Bank, UK
Richard Porcas, NRAO VLA, Socorro, NM USA
Robert Preston, Jet Propulsion Laboratory, Pasadena, CA USA
Eugen Preuss, MPIfR, Bonn, Germany
Alex Raga, Astronomy Dept, University of Manchester, UK

LIST OF PARTICIPANTS

Fredrik Rantakyrö, Onsala Space Observatory, Sweden
Tony Readhead, Caltech, Pasadena, CA USA
Mark Reid, Harvard-Smithsonian, CfA, Cambridge, MA USA
Nan Rendong, Beijing Astronomical Observatory, China
John Reynolds, Australia Telescope National Facility, Australia
Maria Jose Rioja, Instituto de Astrofisica de Andalucia, Spain
E.F. Rizov, Astro Space Centre, Moscow, Russia
David Roberts, Brandeis University, Waltham, MA USA
Clive Robinson, University of Manchester, NRAL, Jodrell Bank, UK
Paola Rossi, Osservatorio Astronomico di Torino, Italy
Jane Russell, ARC/NRL, Washington, DC USA
Hardip Sanghera, University of Manchester, NRAL, Jodrell Bank, UK
Cornelius Schalinski, IRAM, Grenoble, France
Peter Scheuer, Cavendish Laboratory, Cambridge, UK
Richard Schilizzi, NFRA Dwingeloo, Netherlands
Rolf Schwartz, MPIfR, Bonn, Germany
Giancarlo Setti, Istituto di Radioastronomia, Bologna, Italy
N. Shapirovskaya, Astro Space Centre, Moscow, Russia
William Sherwood, MPIfR, Bonn, Germany
Chu Han-Shu, Purple Mountain Observatory, China
Richard Shubert, Calif. State University, Fullerton, CA USA
Ashok Singal, NFRA Dwingeloo, Netherlands
C. Slottje, NFRA Dwingeloo, Netherlands
Slava Slysh, Astro Space Centre, Moscow, Russia
Kostia Sokolov, Institute of Radio Astronomy, Kharkov, Russia
Barkat Sorathia, Dept of Physics, Newcastle-upon-Tyne, UK
William Sparks, Space Telescope Science Institute, Baltimore, MD USA
Ralph Spencer, University of Manchester, NRAL, Jodrell Bank, UK
Herr D. Spickermann, MANN GHH Germany
Richard Sramek, NRAO VLA, Socorro, NM USA
David Stannard, University of Manchester, NRAL, Jodrell Bank, UK
Wolfgang Steffen, MPIfR, Bonn, Germany
Naoki Tanaka, Japan
Russ Taylor, Dept of Physics, Calgary, Canada
Greg Taylor, Osservatorio Astrofisico di Arcetri, Italy
Harri Terasranta, Finland
Peter Thomasson, University of Manchester, NRAL, Jodrell Bank, UK
Bob Thomson, Institute of Astronomy, Cambridge, UK
Wan Tongshan, Shanghai Observatory, China
Corrado Trigilio, Istituto di Radioastronomia, Noto, Italy
Gregory Tsarevsky, Astro Space Centre, Moscow, Russia
Stephen Turner, Cavendish Laboratory, Cambridge, UK
Tasso Tzioumis, Australia Telescope National Facility, Australia
Grazia Umana, Istituto di Radioastronomia, Noto, Italy
Stephen Unwin, Caltech, Pasadena, CA USA
Meg Urry, Space Telescope Science Institute, Baltimore, MD USA
Esko Valtaoja, Tuorla Observatory, Finland
Leena Valtaoja, Tuorla Observatory, Finland
Huib Jan van Langevelde, Sterrewacht Leiden, Netherlands
Tiziana Venturi Istituto de Radioastronomia, Bologna, Italy
Rene C. Vermeulen, Caltech, Pasadena, CA USA
Pablo de Vicente, Centro Astronomico de Yebes, Guadalajara, Spain
Stefan Wagner, Heidelberg, Germany and Observatoir de Meudon, Paris, France
Craig Walker, NRAO VLA, Socorro, NM USA
Jasper Wall, RGO, Cambridge, UK
Johan van der Walt, Potchefstroom Univ. S. Africa

LIST OF PARTICIPANTS

Dennis Walsh, University of Manchester, NRAL, Jodrell Bank, UK
Martin Ward, Department of Astrophysics, Oxford, UK
Ralf Wegner, MPIfR, Bonn, Germany
R. Wietfeldt, ISTS/York Univ, Toronto, Canada
Peter Wilkinson, University of Manchester, NRAL, Jodrell Bank, UK
Arno Witzel, MPIfR, Bonn, Germany
Diana Worrall, Center for Astrophysics, Cambridge, MA USA
Joan Wrobel, NRAO VLA, Socorro, NM USA
Shengyin Wu, Beijing Astronomical Observatory, China
Wenge Xu, Caltech, Pasadena, CA USA
Jeremy Yates, University of Manchester, NRAL, Jodrell Bank, UK
Farhad Yusef-Zadeh, Northwestern University, IL USA
Fujun Zhang Shanghai Observatory, China
Herr Th. Zimmerer, MAN GHH Germany

Preface and Acknowledgements

The past decade has seen a major advance not only in resolution but also in imaging techniques in radio astronomy. High quality images are now being produced on all resolution scales down to milliseconds of arc and recently some reliable images have been produced at a resolution of 50 microseconds. Corresponding linear scales in the nearer extragalactic objects are as small as light months (0.1 pc) and in stars they may correspond to light seconds (10^9 m).

The main aims of the conference on Sub-arcsecond Radio Astronomy were to present an overview of the astrophysical results achieved with todays high resolution imaging instruments. The most powerful of these, the VLA, MERLIN and the Australia Telescope are producing deep images with a resolution that is now matched in optical astronomy both with the Hubble Space Telescope and with modern ground based telescopes. Some first comparisons across the spectrum are presented here, with reference in particular to Active Galactic Nuclei and Quasars.

Another exciting step achieved through improvements in the radio technique is the possibility to detect and image the radio emission from stars. Although only one session was devoted to stars in the conference, the number of poster presentations in this area shows how it is gaining in maturity. New images of gravitationally lensed extragalactic sources are presented which impose tight constraints on the distribution of mass in galaxies. Other topics covered include unified schemes, astrometry, extragalactic supernovae, line observations of molecular clouds, the Galactic centre and interstellar scattering.

The conference, Sub-arcsecond radio astronomy, held 20-24th July 1992, at the University of Manchester gathered an assembly of about two hundred participants. Conferences on Sub-arcsecond radio astronomy and in particular VLBI have been held at four year intervals since 1970. The conference was supported by the University of Manchester, the Royal Society, URSI, the Royal Astronomical Society, Cambridge University Press, Interferometics Inc. Penny & Giles, MAN GHH and the EuropeanVLBI Network. The Scientific Organizing Committee included R.S. Booth (Chair), R.D. Davies, W.M. Goss, A.C.S. Readhead, D.L. Jauncey, M. Inoue, C.A. Norman, M.J. Reid, R. Fanti and A.R. Taylor and the Local Organizing Committee included J.P. Leahy, A. Pedlar, I.W.A. Browne, P.N. Wilkinson, S.T. Garrington, J. Dyson and T.W.B. Muxlow. We would like to thank the Director, R.D.Davies, all committee members and all the staff of Jodrell Bank and especially R.E. Spencer of the Local Organizing committee and Mrs J. Eaton and Mrs J. Warren for all their efforts to make the conference and the publication a success.

Richard Davis
Roy Booth

NRAL, Jodrell Bank
November 1992

Interferometric Observations of Radio Stars

A.R. Taylor [*]

Abstract

The past decade has seen tremendous progress in the study of radio continuum emission from stars. Most of this progress can be attributed to high resolution, interferometric observations. Conventional VLBI observations have traced the structure and variations of non-thermal emitting systems, such as x-ray binaries and RS CVn stars. High sensitivity, connected element arrays have revealed jets, shells and bipolar structures of thermal emitting circumstellar gas at sub-arcseconds scales in a number of stellar systems.

Over the next few years the technique of phase referencing applied to VLBI should allow milli-arcsecond scale imaging of non-thermal radio stars at the level of a few mJy or less. In addition, as the sensitivity and resolving power of connected element arrays such as the VLA and MERLIN continue to improve, imaging of faint, thermal sources at resolution of 10's of milli-arcseconds, or better, will allow the possibility of imaging a number of new stellar sources and phenomena.

1. Stellar Radio Emission

As the sensitivity of radio telescopes have improved, more and more stars have been detected as radio emitters. However, radio emission from stars is still indicative of "abnormal" stellar phenomena. To illustrate this it is instructive to compare the normal emission from stars with the capabilities of the most powerful of current radio telescope arrays. At a radiation frequency ν, the flux density, S_ν, brightness temperature, T_B and angular radius θ of a radio source are related by the simple expression

$$S_\nu = 5.7 \times 10^{-4} \left(\frac{\nu}{\text{GHz}}\right)^2 \left(\frac{\theta}{''}\right)^2 T_B \quad \text{mJy}.$$

At a frequency of 5 GHz, the brightness temperature and angular size are related to the minimum detectable flux density of an instrument, S_{min} by $T_B \theta^2 = 70 S_{min}$. This relationship defines straight lines in a $\log(T_B)$ - $\log(\theta)$ plot, as shown in figure 1. Also

---

[*]Department of Physics and Astronomy, University of Calgary, Calgary, Alberta, T2N 1N4, CANADA.

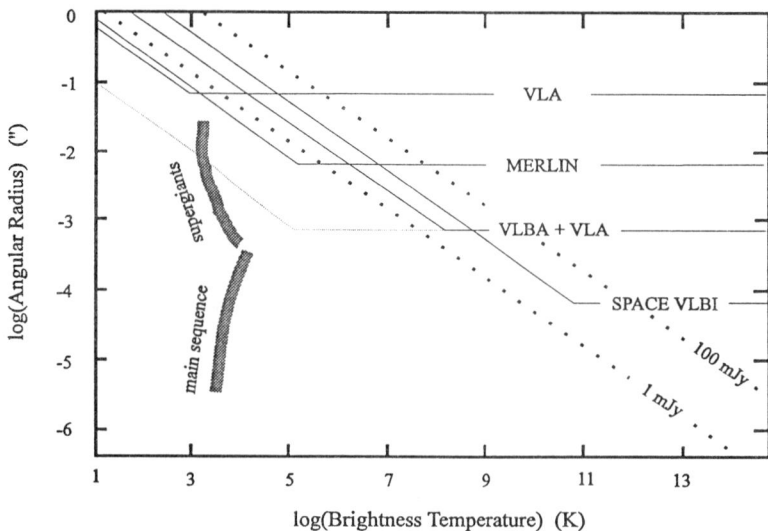

Figure 1. The sensitivity and resolving power of interferometric arrays compared to the temperature and angular radii of normal photospheric emission from stars. For most stars, detectable radio emission implies either an angular size much larger than stellar radii or a brightness temperature much in excess of stellar effective temperatures.

shown on figure 1 are the brightness temperature and angular radii of normal stars along the main sequence and supergiant branches of the HR diagram. Large, cool red stars at the tip of the supergiant branch lie very close to the detection limits of connected element arrays such as the VLA and MERLIN. Most stars, however, lie well below the detection capabilities of even the most powerful arrays.

In order to become detectable, radio emission associated with stars much either have brightness temperature much in excess of stellar effective temperatures, or dimensions much larger than typical stellar radii. This fact allows us to broadly classify radio emitting stars into two catagories: non-thermal emitters, which have high brightness temperatures and potentially very small angular radii, and thermal emitters, which have low brightness temperatures and dimensions much larger than the associated star. Because of the generally smaller dimensions of the emitting region, non-thermal radio stars are characterized by highly variable radio emission, with outbursts or flares with time-scales, ranging from seconds (dMe stars) to hours or days (x-ray binaries, RS CVn stars and others). Thermal radio emission from stars is also often associated with outburst phonenema, although in this case the time scales are much longer – months to years (novae, symbiotic stars). A larger fraction of thermal stars are quasi-steady state sources, for example, mass losing stars such as O-B stars, Be stars and planetary nebulae.

Although there is an ever increasing variety of stars that exhibit radio emission, the phenomenon of outflow in either a steady-state condition or as result of an outburst appears to be a common feature of almost all radio stars. Interferometric observations have played a central role in investigating these objects. At the mJy level, interferometric observations are required to unambigously detect radio emission from stars, both because of the precise postions required for identification and because mJy flux densities are below the confusion level of single dish antennas at cm wavelengths. Beyond this, interferometric observations provide measurements of the structures and/or angular dimensions of the radio sources and their time evolution. It is impossible to review the progess that has been made in this field in a few short pages. A few highlights in the non-thermal and thermal regimes are outlined below.

2. Non-thermal Radio Stars

The most luminous non-thermal radio stars are the x-ray binaries. This otherwise heterogeneous group share the common characteristics of strong, highly variable radio emission, ranging in luminosity from a few 10^{35} erg-s^{-1} (Cyg X-3) to 10^{30} erg-s^{-1} (Sco X-1). The high luminosities, reflect the presence of highly energetic phenomena. One x-ray binary, SS433, is distinguished by continuing eposides of ejection of radio emitting plasmons with a bulk velocity of $0.26c$. The mean trajectory of the plasmons follows a kinematic model in which the ejection axes of two oppositely directed beams precess with a period of 164 days [AM79]. Multi-epoch, interferometric observations show that the ballistic trajectories remain unmodified from milli-arcsecond scales [Ve87] through sub-arcsecond and arcsecond scales [Sp79], [HJ81] corresponding to linear dimensions from 5 to 5000 Astronomical Units. VLBI images also show general agreement with the Doppler boosting due to the relativistic motion of the opposed jets [Fe86].

Although all strong radio emitting x-ray binaries exhibit an expanding emitting region during outburst, no other shows evidence for precessing ejection axes and emission covering the the same range of linear scales. Angular size measurements and multi-epoch studies of other radio emitting x-ray binaries show a wide range of expansion velocities. Molnar et al. [MR88] observed two flares of Cyg X-3. Modelling the time dependence of the visibilities with a double-sided jet, implied a source expansion primarily in North-South direction with velocity between $0.16c$ and $0.31c$. VLBI imaging of LSI+61°303, on the other hand, indicate expansion velocities of only a few hundred km-s^{-1} [TK92], [MP92]. VLA images of the recently detected Galactic Center Transient [Zh92] imply expansion at a velocity of 6000 km-s^{-1} ($0.02c$). X-ray binaries are thought to be systems in which a gravitationally collapsed object accretes matter from a companion star. The wide range of expansion velocities suggests a range of properties of the compact components or, perhaps, a difference in the mechanism that converts gravitational potential energy to bulk motion during outbursts.

As the sensitivity of interferometers has improved a number of other types of star have been detected as radio sources and have become the object of interferometric study. In addition to the well studied magnetic binary systems such as RS CVn and Algol type binaries (see, for example, [Lj88], [MM88]), the list of detected classes has grown to include single stars, including synchrotron emission from the winds of luminous early-type stars [PT90], from peculiar B stars [PL88] and from flares in late-type pre-main sequence stars [FL89]. Because of the low flux density level of these type of stars, VLBI observations have been generally limited to a few large antennas and the non-thermal nature of the emission is inferred by high brightness temperature that are estimated from simple fits to a few visibility data points.

3. Thermal Radio Stars

Thermal emission from stars is characterized by brightness temperatures of order 10^5 K or lower. Hence interferometric studies have been almost exclusively limited to sensitive, connected element arrays such as the VLA which has maximum resolution of about 0.1".

Sub-arcsecond imaging of outbursting type thermal sources has allowed dynamic imaging of ejecta. Observations of this type test models that have been based on the evolution of radio light curves. For example, early models of radio outbursts of novae were involked spherically symmetric ejection of hot plasma with the time dependence of the radio flux determined largely by the velocity distribution of the ejecta [SP77], [HW79]. Recent high resolution images suggest a range of behaviour that is not consistent with these simple models. Radio images of the recurrent nova RS Oph revealed a bipolar radio "jet" with a combination of low-frequency non-thermal and high-frequency thermal emission [TD89]. Nova QU Vul underwent a change from a bipolar shape about 200 days past optical maximum to a more classical spherically symmetric shell at 900 days [TH88].

A class of objects related to novae are the symbiotic stars. Some of these objects undergo radio outburst that have properties similar to radio detected novae. One such object of note is CH Cygni which underwent a series of outburts beginning in September of 1984. Coincident with the radio outburst was the production of a double-sided jet of thermal emitting gas with an angular expansion velocity of 1.1 $" - yr^{-1}$ [TS86]. Subsequent monitoring of the radio structure has revealed a precession of the ejection axis. This effect is illustrated in figure 2, which shows images of the radio jet at 1.5 and 2.9 years after onset of the radio outbursts. The trace in the figure represents the loci of ballistic trajectories for an SS433-type kinematic model with a precession cone angle of 34° and a precession period of 7.2 years. The dots are separated by 200 days and the outermost dots represent material ejected at onset of the outbursts. For a distance of 400 pc, the ejection velocity is 1050 km-s^{-1}.

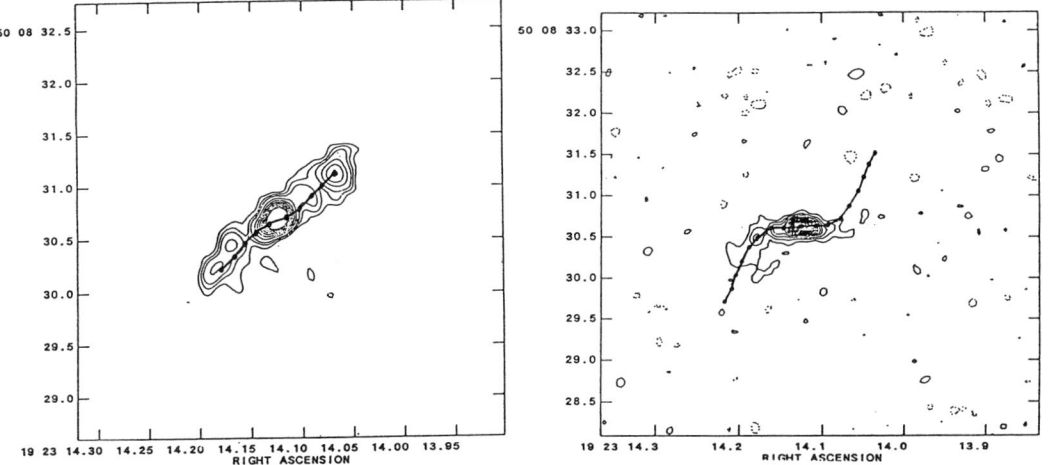

Figure 2. VLA images of the symbiotic star CH Cygni showing the precession of the ejection axis. The mages were obtained in February 1986 (left) and July 1987 (right).

4. The Future

Figure 3 shows the number density of known radio stars as a function of flux density as of 1987. The number of stars per mJy decreases with flux density as $S_\nu^{-2.1}$. Since the sample of known radio stars is essentially complete for flux densities greater than about 1 Jy but vastly undersampled at low flux densities, the actual slope of the $\log(N)$-$\log(S)$ curve for radio stars must be much steeper. Improvements in sensitivity will allow many more stars to be studies with interferometric techniques. Phase referencing with the VLBA and other VLBI arrays should allow milli-arcsecond imaging of non-thermal stars down to flux densities of $\sim$1 mJy. The recent enhancements to MERLIN will allow imaging with resolution of about 10 milli-arcseconds down to brightness temperatures of order 10^4 K. This potential opens up thermal radio sources to high resolution studies. Looking to the future beyond this, one might imagine a global array of 100 meter class telescopes capable of phase referencing interferometry. As shown by the dotted line in figure 1, such an array would open up a large region of the (T_B, θ) diagram to interferometric imaging.

References

[AM79] Abell, G.O. and Margon, B. *Nature*, 279, (1979) 701.

[FL89] Feigelson, E.D., *et al. Bull. A.A.S.*, 21, (1989) 1003.

[Fe86] Fejes, I. 1986, *Astronomy & Astrophysics*, 166, (1986) L23.

[HW79] Hjellming, R.M., *et al. Astronomical Journal* 84 (1979) 1619.

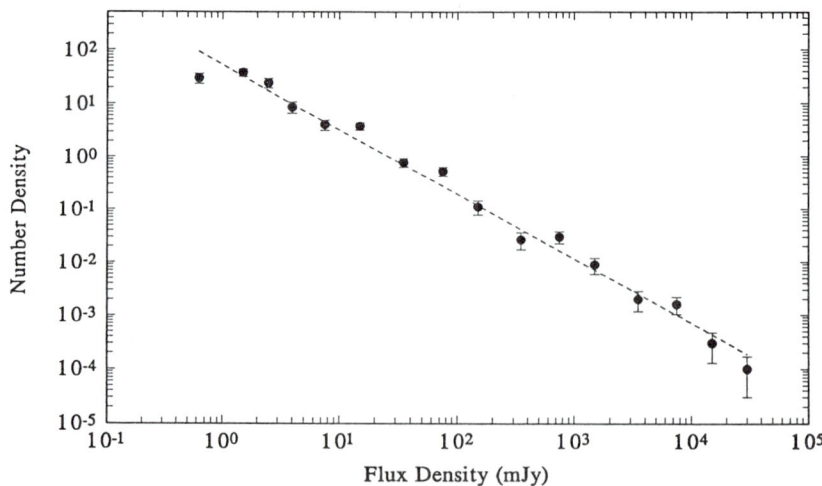

Figure 3. log(N) - log(S) curve for radio detected stars. Based on data from [We87]

[HJ81] Hjellming, R.M. and Johnston, K.J. 1981, *Nature, 290 (1981) 100.*

[Lj88] Lestrade, J-F. et al. *1988, Astrophysical Journal 328 (1988) 232.*

[MM88] Massi, M., et al. *Astronomy & Astrophysics 197 (1988) 200.*

[MP92] Massi, M., *et al. Astronomy & Astrophysics, (1992) in press.*

[MR88] Molnar, L.A., *et al. Astophysical Journal 331 (1988) 494.*

[PL88] Phillips. R.B. and Lestrade, J-F. *1988, Nature 334 (1988) 329.*

[PT90] Phillips, R.B. and Titus, M.A. *Astrophysical Journal 359 (1990) L15.*

[SP77] Seaquist, E.R. and Palimaka, J. *Astrophysical Journal 217 (1977) 781.*

[Sp79] Spencer, R.E. *1979, Nature 282 (1979) 483.*

[TS86] Taylor, A.R., Seaquist, E.R. and Mattei, J.A. *Nature 319 (1986) 38.*

[TH88] Taylor, A.R., *et al. Nature 335 (1988) 235.*

[TD89] Taylor, A.R.,*et al. 1989, Mon. Not. roy. Astr. Soc. 237 (1989) 81.*

[TK92] Taylor, A.R., *et al. Astrophysical Journal 395 (1992) 268.*

[Ve87] Vermeulen, R.C., et al. *Nature 328 (1987) 309.*

[We87] Wendker, H.J. *Astronomy and Astrophysics Supplement 69 (1987) 87.*

[Zh92] Zhao, J.H., et al. *1992, Science 255 (1992) 1538.*

The jets of SS 433; blobby or continuous ?

René C. Vermeulen * *Ralph E. Spencer* [†]
Richard T. Schilizzi [‡] *Istvan Fejes* [§]

Abstract

Ballistically moving expanding VLBI blobs in the jets of SS 433 can be associated with radio flares; relativistic particle generation is sustained for several days. Conglomerations of such knots, ejected over the course of a flaring episode, may form MERLIN knots. There is also radio emission from a more continuous underlying jet. The spectral index steepens along the jets.

A sequence of six 5 GHz European VLBI Network (EVN) images of the inner jets of SS 433 was obtained at two-day intervals in 1987 by Vermeulen et al. (1993a); see also Fig. 1, discussed below. The morphology is consistent with the standard kinematic model; nodding motion has now been discovered in radio images. The distance to SS 433 is found to be 4.85 ± 0.2 kpc. The relative alignment of this sequence is fixed unambiguously by requiring that the major VLBI features have a constant apparent proper motion, and do not slow down and speed up at random. In the most plausible "absolute" registration of the core, the global jet morphology is symmetric. Features in the approaching (eastern) jet are initially brighter than their receding counterparts due to Doppler favouritism, but then fade faster as a result of light travel time differences. The core region always has wing-like extensions, spanning ~ 100 AU each. Their flux density ratio indicates continuously ejected matter moving at $0.26c$. Further out, the jets are dominated by individual knots. The bright pair formed during the 1987 sequence was connected with two distinct flares, which differed in their spectral evolution (Vermeulen et al. 1993b). Relativistic particle generation was probably sustained for a few days, with a softening of the particle spectrum. The first flare took place in the core wing-region (Image 2). There was then a drop in total flux density, before the onset of the second flare, which was due to the brightening of the knots as they moved out to a distance of $\sim 4 \times 10^{15}$ cm, or 250 AU (Images

---

*Radio Astronomy 105-24, California Institute of Technology, Pasadena, CA 91125, USA.
 This author was supported by the NSF under grant number AST-9117100.
[†]University of Manchester, Jodrell Bank, Macclesfield, Cheshire, SK11 9DL, UK.
[‡]NFRA, P.O. Box 2, NL-7990 AA, Dwingeloo, The Netherlands.
[§]Satellite Geodetic Observatory, Penc, H-1373, Budapest, Pf 546, Hungary.

3–5). This region seems to be a "brightening zone". Similar brightening and fading patterns apply to all of the jet material, sometimes leading to phase effects and seemingly stationary features at ~ 250 AU, even when there are no bright knots.

Beyond the brightening zone, the VLBI knots fade rapidly, presumably due to adiabatic expansion. However, this cannot continue unchecked for long, since MERLIN images, with a resolution of $\sim 10^{16}$ cm, typically show substantial compact emission. The images are again consistent with the kinematic model, without evidence for slowing down of the ejecta. A series of seven MERLIN 18cm images, spaced by 2 to 3 weeks, was obtained in 1988 (Spencer et al. 1992). Two pairs of knots were formed, which were linked to flaring episodes. They decayed in brightness with a 1/e time of ~ 20 days, rather more slowly than VLBI knots. The brightness temperature of MERLIN knots is $\sim 10^5$ K, compared to $\sim 10^8$ K for VLBI knots. Whereas the latter can be related to individual flares, MERLIN knots may consist of a conglomeration of such VLBI features, ejected over an outburst period; the details of this process are unclear. In MERLIN maps, there is substantial continuous emission in between the brighter knots. It is unclear whether this fill-in emission also exists, at a low level, on VLBI scales of $\sim 10^{15}$ cm, or whether it is formed at larger distances from the core. It is known that there is little compact flux on scales $\leq 10^{14}$ cm (Walker et al. 1981, Vermeulen et al. 1993a).

Simultaneous MERLIN and EVN data is available for the 1987 monitoring sequence. However, their combination is complicated by the lack of overlapping uv-coverage, by further gaps due to technical failures, by the proper motion of features in the jets (9 mas/day), and by the occurrence of major flares. The final combined images, shown in Fig. 1, should be interpreted with caution, since they may differ in dynamic range and image fidelity. Nevertheless, when compared to the EVN-only images of Vermeulen et al. (1993a), there are intriguing hints of more extended features. Image 1, taken when SS 433 was still quiescent, shows relatively much extended emission, indicating, perhaps, that similar possibly continuous emission could not be mapped properly in the later images, when SS 433 was flaring. The EVN-only images did not show the emission immediately ahead of the two brightest knots, seen especially to the east in Images 3 and 4, and to the west in Image 5; these features would correspond to the core wings in Image 2. There is marginal evidence in Image 5 for the continued presence of the outermost western knot; from the EVN data alone, it is visible only in Images 1–4. Furthermore, those data did not indicate a corresponding feature in the eastern jet. While this could be ascribed to different fading rates resulting from light travel time differences, the combined data may have revealed the counterpart, though only in Images 1 and 3. Image 5 shows yet another pair of features, at ~ 100 mas, which may not correspond to anything in the earlier images. It is possible that we are witnessing the recurrent brightening and fading of ejecta, but the data quality does not allow an unequivocal conclusion. New combined EVN and MERLIN observations will be conducted in 1993.

In March 1990, SS 433 was observed on the same day with the EVN at 5 GHz and with the US VLBI Network at 1.6 GHz. Unfortunately, technical failures have seriously compromised the similarity of the uv-coverage between the two networks, have made their relative amplitude calibration rather uncertain, and have left efforts to secure the relative registration of the two images through phase referencing unsuccessful. The resultant spectral index map is therefore of limited significance. Nevertheless, the data firmly indicate that the core region has a rather flat (possibly even inverted) spectrum, and that the spectral index then steepens along the jets. Interestingly, the jets can be traced out to $\sim 10^{16}$ cm at 5 GHz, further than in most earlier images. Furthermore, the jet morphology, while still resembling a chain of knots, is rather more continuous than in most VLBI images made to date. Spectral index imaging of SS 433 will be pursued on the VLBA.

References

Spencer R.E., Vermeulen R.C., Schilizzi R.T., 1992, *VLBI and MERLIN Observations of the Moving Knots in SS433*, in "Stellar Jets" (Capri September 1991), in the press

Vermeulen R.C., Schilizzi R.T., Spencer R.E., Romney J.D., Fejes I., 1993a, A&A in the press

Vermeulen R.C., McAdam W.B., Trushkin S.A., Facondi S.R., Fiedler R.L., Hjellming R.M., Johnston K.J., Corbin J., 1993b, A&A in the press

Walker R.C., Readhead A.C.S., Seielstad G.A., Preston R.A., Niell A.E., Resch G.M., Crane P.C., Shaffer D.B., Geldzahler B.J., Neff S.G., Shapiro I.I., Jauncey D.L., Nicolson G.D., 1981, ApJ 243, 589

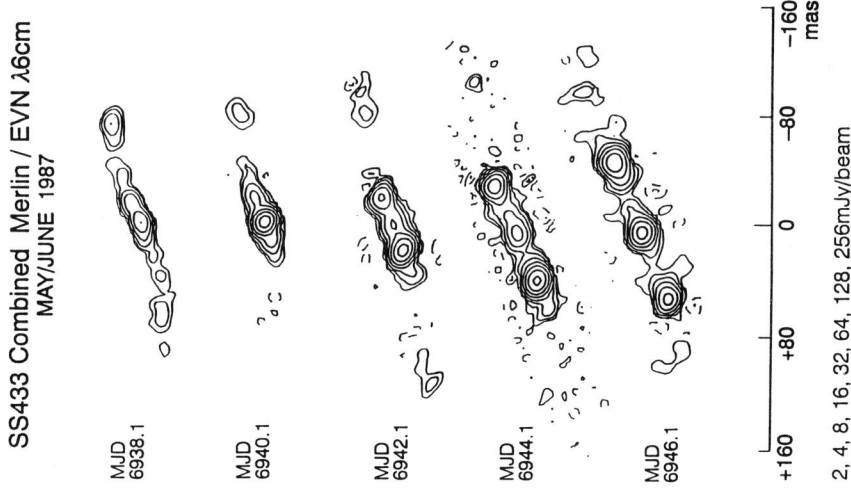

Figure 1 Combined EVN/MERLIN images of SS 433 taken in 1987.

PHOTOSPHERIC EMISSION FROM RED GIANT STARS AT RADIO FREQUENCIES

Mark J. Reid and Karl M. Menten [*]

Abstract

We report the detection of cm-wavelength emission from a sample of nearby Mira and semi-regular variables. The continuum emission in the radio band has a spectral index of 2 for all stars detected, as expected for optically thick black-body emission. The flux densities are consistent with emission from the outer portion of the optical photospheres of these stars. We partially resolve the stellar disk for the star W Hya and find a diameter of about $0\rlap{.}''08$.

1. Introduction

Long Period Variables (e.g., Miras) vary dramatically in optical light and are thought to change temperature and size by about 30% over periods of 100 to 1000 days. We do not have a good theoretical understanding of the stellar pulsation mechanism or even of the observational properties of the pulsations. Non-radial pulsations have been suggested, and there have been reports of significant departures from spherical symmetry for *o* Ceti (e.g., KNP91).

The sizes of a few nearby Mira variables have been determined directly by speckle techniques (e.g., WeW80), and by lunar occultation (NaW73) to be between 0.02 and 0.10 arcsec. However, there is a significant variation of the measured size with observing wavelength, probably caused by strong molecular opacity effects. Perhaps because of the difficulty of the observations and of modelling the opacity changes over a stellar cycle, a complete series of measurements of size as a function of time (stellar phase) is lacking. Mira variables have photospheric temperatures of ≈2000 K and sizes of 10 or more astronomical units. The Rayleigh-Jeans tail of the black-body emission from such a star at a distance of 100 pc would be of the order of 1 mJy at a wavelength of 1 cm. This level of emission is easily detectable with the Very Large Array (VLA) and motivated a survey of these red giants for radio emission from the stellar photosphere.

---

[*] Harvard–Smithsonian Center for Astrophysics

2. Observations

We observed 8 stars between 1989 December 2 and 1991 February 27, using the VLA in its smallest (D) configuration. We used a standard continuum observing technique and observed at 8, 15 and 22 GHz for most stars. The stars were chosen to be nearby (< 250 pc) and distributed uniformly in right ascension. All stars were detected in at least one of the observing bands, and the measured flux densities are plotted in Figure 1 for a few of the stars.

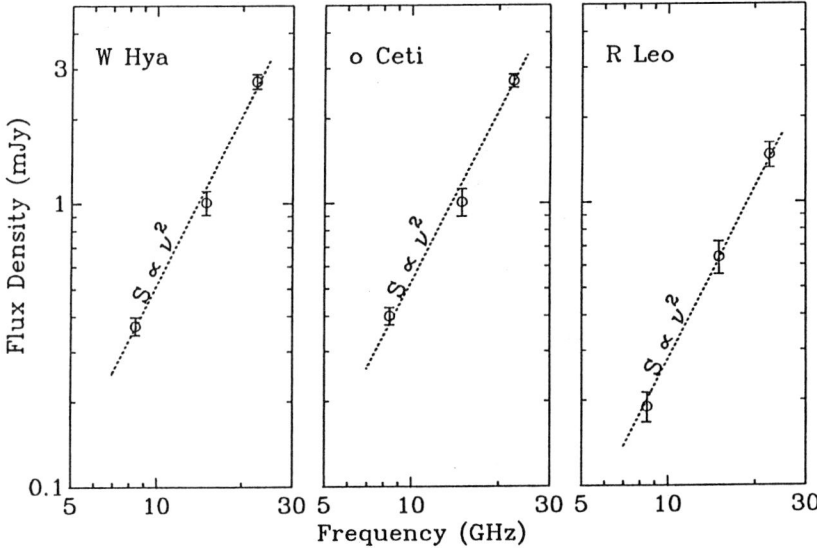

Fig. 1: Radio frequency spectra of three Mira variables. The fitted lines have a slope of 2, characteristic of optically thick black-body emission.

We found that all stars had 1) power law radio spectra (X, U, and K band) consistent with optically-thick black-body emissions, 2) flux densities close to those expected from optically determined temperatures and sizes, and 3) positions coincident to better than $0\rlap{.}''5$ with the optical positions of the stars. Clearly we detected photospheric (or near photospheric) emission from these red giant stars.

We attempted to measure directly the angular size of the semi-regular variable W Hya when the VLA was in its largest (A) configuration. At 22 GHz the VLA has a synthesized beam of of $0\rlap{.}''08$, which is comparable to the angular size, measured optically, of this star. Since, radio frequency "seeing" (determined mostly by fluctuations of the water vapor in the atmosphere) is rarely better than $0\rlap{.}''1$, we developed a novel observing procedure in which H_2O masers associated with the stars served as a phase reference for the continuum observations. This calibration allowed nearly perfect "seeing" to be achieved.

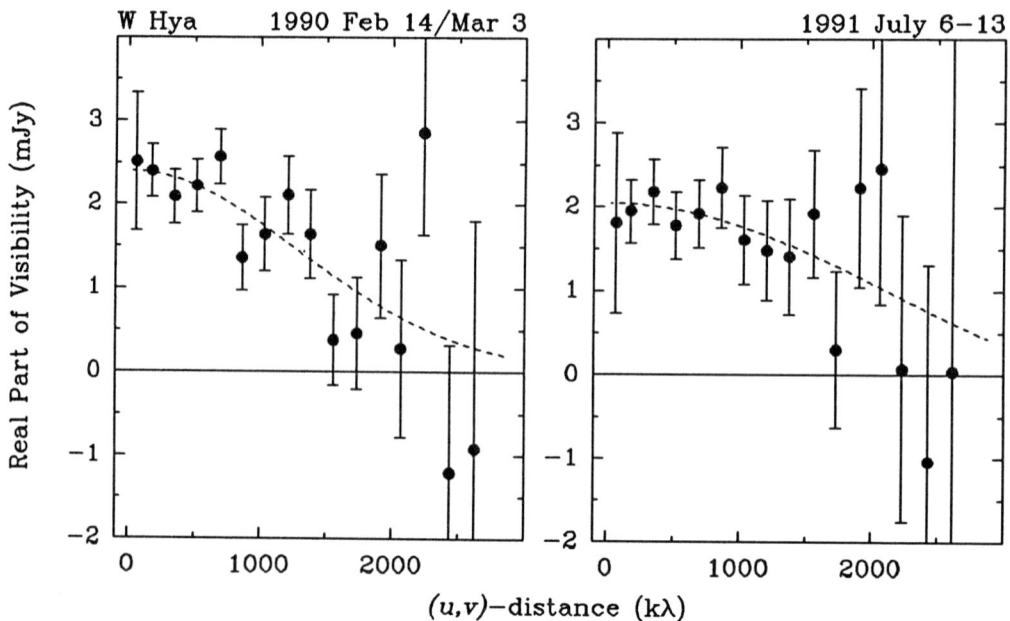

Fig. 2: Fringe visibility versus interferometer baseline length for W Hya at 22 GHz measured near optical phase (*left*) 0.7 and (*right*) 0.0.

The broad-band observations at 22 GHz of W Hya reveal a 2.2 mJy continuum source. This emission is probably from the outer photosphere of W Hya. Figure 2 shows the (real part of the) fringe visibility versus interferometer spacing. These visibilities indicate a partially resolved object. Modelling the variation of fringe visibility with baseline length for the 1990 observations yields a (uniform disk) diameter of 0.″09 ±0.″02. The 1991 observations yield a size of 0.″07 ±0.″02. For a distance of 125 pc to W Hya, a 0.″07 diameter of the star corresponds to about 1.3×10^{14} cm ($\approx$ 9 AU) and a radio brightness temperature of $\approx$ 1600 K, close to that expected from the outer portions of the optical photosphere for this variable star.

References

[KNP91] Karovska, M., Nisenson, P., Papaliolios, & Boyle, R. P. *Asymmetries in the Atmosphere of Mira.* Ap. J. (Lett), Vol. 374 (1991), pp. 51-54.

[NaW73] Nather, R. E. and Wild, P. A. T. *The Angular Diameter of R Leonis,* A. J., Vol. 78 (1973), pp. 628-631.

[WeW80] Welter, G. L. and Worden, S. P. *The Angular Diameters of Supergiant Stars from Speckle Interferometry,* Ap. J, Vol. 242 (1980) pp. 673-683.

Merlin Observations of Radio Stars

Richard J. Davis *

Abstract

The advent of Merlin phase 2 enables significant developments in the study of radio stars. The new sensitivity achieved allows detection of non-thermal and thermal radio emission, thus increasing the types of radio star that can be studied with the increased resolution of the instrument. Examples are given involving the early detection and resolution of a classical nova, the imaging of an early type and also a Wolf-Rayet star. The radio stars play an important part in astrometry particularly in linking the radio and optical frames. Measurements indicate that a precision of 1-2 mas can be achieved with relative phase results from Merlin.

1. Introduction

The new interest in radio stars stems from the performance of Merlin phase 2 ([W92]). The essential two parameters are the sensitivity and resolution. At 5GHz the resolution is 50 mas which is 50 a.u. at 1 kpc and the sensitivity after a 12 hr track corresponds to a brightness temperature of 1000 K. Thus the instrument can be used to study 10,000 K thermal material as well as any non-thermal emission. Objects of particular interest are stars with mass loss envelopes. Examples include Late type stars such as Betelgeuse where Merlin will probe down to the photosphere. This will enable a mapping of the wind which should show if the flow is isotropic or are there density enhancements and also comparison with high resolution optical images. Early type stars will also be of considerable interest. In particular Emission line stars, T Tauri stars, Young stellar objects, Wolf-Rayet stars. Here there is evidence of thermal and non-thermal emission and bipolar outflow. For these objects the astrophysical mechanisms are unclear.

2. WR147 and P Cygni

The particular Wolf-Rayet star WR147 has been studied before with the VLA and Merlin ([MDB89]) Multi-frequency mapping of this Wolf-Rayet star reveals two re-

---

*Nuffield Radio Astronomy Laboratories, Jodrell Bank, Macclesfield, Cheshire SK11 9DL, UK.

solved components separated by 0.6 arcsec ; a Southerly, thermal component corresponding in position with the optical star and a Northerly, non-thermally emitting component thought to result either from interaction with a binary companion or shock processes in the extended atmosphere of the star.

Preliminary new observations with Merlin show essentially the same double features but the increased performance now produces a highly complex image showing evidence for bipolar outflows leading in the Northerly component to a highly flattened region which was identified as the shocked region producing the non-thermal emission.

White and Becker ([WB82]) model fit a diameter of .12 arcsec to P Cygni at 5 GHz using the VLA. They find that the visibility function declines linearly with baseline length consistent with a spherically symmetric wind. New Merlin results show a shell-like structure which to our surprise is not symmetric.

3. Astrometry

The long term aim of astrometry is to link the VLBI, FK5 and Hipparcos reference frames. Without going into the astrometric arguments it appears that the radio stars have a very important part to play. An important result of such a linkage is then to align the optical and radio images for astrophysical purposes. Both Hipparcos and Merlin should give accuracies of 2 mas so there is a good match here. It is important to bear in mind that stars are not radio points, and that there is often thermal and non-thermal emission.

Broadly speaking VLBI arrays only see non-thermal emission. This may in some cases be dangerous since in the case of WR147 the non-thermal emission is spatially displaced from the star which appears only to produce thermal emmision. Thus the importance of Merlin observations is their resolution to spatially separate while at the same time retain the sensitivity to see both types of radiation.

To achieve astrometric performance on Merlin a number of refinements have been commissioned. The barometric pressures at each site have been accurately calibrated and are in use in the tropospheric phase correction. Polar motion and UT1-UTC are now inserted on-line from the IERRS predictions at weekly intervals. This gives sufficient accuracy for 2 mas relative phase measurements over a 2 degree throw between sources. An error in the implimentation of the diurnal aberration was also corrected.

4. Nova Cygni 92

The radio discovery of Nova Cygni 1992 was announced by Pavelin and the author in IAUC 5534 [PD92]. The optical outburst ocurred on Feb 19 and Merlin observations were made on May 9 centred on the reported position and phase referenced on the

VLBI point source 2031+549. Emission was detected with a peak flux density of 280+/-50 microjansky at roughly 2 arcsec from the reported optical position.

Subsequent astrometry with the Carlsberg AMC at La Palma shows the optical position to be within 50 mas of the Merlin position which is estimated to have an uncertainty of +/-5 mas. Our current thinking is that the suggested optical projenitor 2 arcsec away is a separate star seen on the Palomar sky survey and that the nova has come from beneath the plate limit of the 1952 survey. The proper motion required to move the suggested projenitor to the nova position is too great.

The radio source is extended with an integrated flux density of 1360+/-50 microjansky. This is the first time that the radio emission has been mapped at such an early phase in the evolution of a classical nova. At this stage four images have been made and the centimetre radio spectrum still indicates that the region is optically thick and the flux density is still rising. The peak brightness temperature is only 5000 K. This, and the inverted radio spectrum between 5GHz and the millimetre measurements made with the JCMT, indicates that the emission is thermal.

From the angular expansion rate deduced from the maps, and a speed deduced from the opical line widths a distance of 1kpc is found. The maps show asymmertric expansion possibly indicating injection of energy by jets and also the existence of different temperatures in the expanding envelope.

The images suggest that the evolution of the ejecta is more than a simple linear or spherical expansion. We appear to be witnessing a transition from an early (possibly magnetically contrained) linear bipolar-type of outflow to a more spherically-symmetric, expanding shell. Work is currently in hand to model the nova in detail with expanding clouds of different temperature and density and thus different optical depths with time.

References

[W92] Wilkinson P. N., *Merlin Phase 2 . This volume*

[MDB89] Moran J. P.,Davis R. J.,Bode M. F.,Taylor A. R., Spencer R. E.,Argue A. N.,Irwin M. J.and Shanklin J. D., *Merlin observations of the Wolf-Rayet star AS431 Nature 340 (1989) 449-450*

[WB82] White R. L.and Becker R. H., *P-Cygni observations Astrophys. J. 262 (1982) 657*

[PD92] Pavelin P. and Davis R. J., *Nova Cygni 1992 IAUC 5534*

Radio Imaging the Be star ψ Persei

Sean M. Dougherty and A.R. Taylor [*]

Abstract

Using the 'A' configuration of the VLA we have successfully resolved the circumstellar envelope of ψ Persei at 15 GHz and find that the radio emission arises from a non-spherically symmetric distribution of thermally radiating gas. A Gaussian fit to the visibility data indicates a radio emitting region that has a major axis size of 111 ± 16 milliarcseconds (mas) and is unresolved in the minor axis direction, with a 3σ upper limit of 68 mas.

The geometry of the circumstellar plasma around Be stars has been a subject of considerable debate. Interpretation of observations at different wavelength regions indicate apparently conflicting characteristics of the plasma. UV absorption line profiles of highly ionized metals *e.g.* Si IV and C V, suggest the existence of very high velocity (~ 1000km s^{-1}), low density plasma. However, optical emission line strengths and the level of excess continuum emission from near-IR to radio wavelengths indicate the presence of low velocity (a few tens of km s^{-1}), high density plasma. Both spherically symmetric and non-spherically symmetric models of the circumstellar gas distribution have been suggested to account for these observations.

We observed ψ Persei in August 1991 at 15 GHz with the 'A' configuration of the VLA. The synthesized beam full width, half maximum was 176×144 mas. The complex gains of the antennae were monitored by frequent observation of the nearby bright source 3C84. The total integration time was approximately 11 hours, giving a 1σ noise level on the synthesized image of 40μJy.

To determine if ψ Persei is resolved, we examined the complex visibility data. After phase-centering the source, the real visibility amplitudes of ψ Persei were calculated as a function of baseline length, independent of hour angle (figure 1b). The real visibility amplitude of ψ Persei decreases by $\sim 75\%$ at the longest baslines. The emission is well resolved. A circular Gaussian fit gives a mean angular size of 74 ± 13 mas. The real visibility amplitude of 3C84 decreases by less than 1% over the baseline range. We conclude, therefore, that the decrease in visibility amplitude of ψ Persei at

---

[*]Department of Physics and Astronomy, University of Calgary, Calgary, AB, T2N 1N4, CANADA.

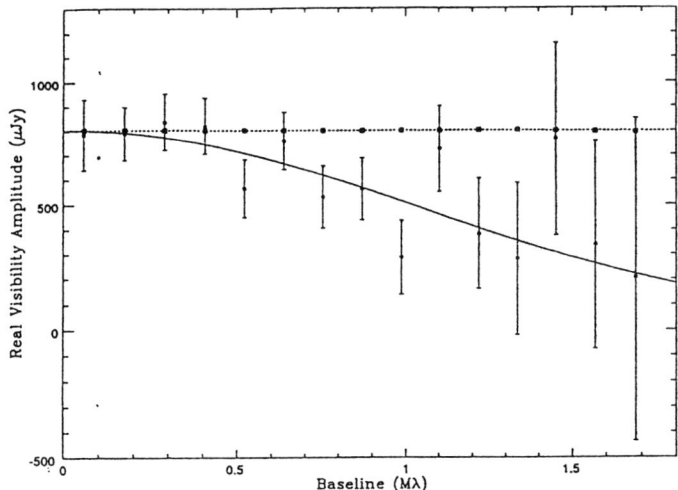

Figure 1. The real visibility amplitude of ψ Persei (solid circles) and 3C 84 (open squares) at 15 GHz, as a function of baseline length. The data for 3C 84 have been normalized to the zero baseline amplitude of ψ Persei. The lines represent circular Gaussian fits to the visibilities.

long baselines is a resolution effect and not due to de-correlation due to atmospheric effects.

To determine the shape of the radio brightness distribution of ψ Persei on the plane of the sky, the visibility data were binned into a 30 × 30 grid in the u,v plane. A 2D Gaussian fit to the gridded data gives a source that has a total flux density of $837 \pm 60 \mu$Jy, and a major axis of 111 ± 16 mas at a position angle of $158 \pm 10°$. The source is unresolved in the minor axis direction. A 3σ upper limit of 68 mas was established by Monte Carlo simulation.

Clearly the radio emitting region around ψ Persei is not symmetric on the plane of the sky, demonstrating conclusively that a spherically symmetric circumstellar gas distribution is not applicable to this star. This has been previously inferred from linear polarization observations of ψ Persei that suggest an equatorially enhanced circumstellar electron distribution with a major axis at a position angle of $\sim 135°$ (Poeckert et al., 1979, Astron. J., **84**, 812). This is in good agreement with the position angle derived from this analysis

This observation is the first time the circumstellar material around a Be star has been resolved and is the first direct confirmation of equatorially enhanced circumstellar plasma distributions for Be stars originally proposed in 1931 by Struve (Struve, 1931, Ap. J., **73**, 94).

High Resolution Imaging of Symbiotic Stars

H.T. Kenny, A.R. Taylor, *R.J. Davis, P. Pavelin †
M.F. Bode, and M. Bang ‡

Abstract

New observations are reported for six symbiotic stars at 5 GHz, made in A configuration of the VLA. Each source was observed for one hour in August 1991, and AG Pegasi was also observed in an earlier two hour session in March 1990. The observations reveal new detail concerning the quiescent environment of symbiotic novae (HM Sge and V1016 Cyg), outburst characteristics (CH Cyg and AG Peg) and the angular extent of fainter systems (Z And and V1329 Cyg).

Of the six objects observed in this program, significant structure was observed only in AG Peg (Figure 1), HM Sge and V1016 Cyg (Figure 2). The radio jet source, CH Cyg, was unresolved with an angular size upper limit of 140 mas. Its total flux density (3.2 ±0.2 mJy) was significantly higher than the quiescent level (~ 1 mJy) resumed following the most recent outburst. Z And (1.3 mJy) was also unresolved to a limit of $\sim$100 mas, while V1329 Cyg (0.9 mJy) had an angular size of 0."53 × 0."44.

The image of AG Peg, from March 1990, is shown in Figure 1 below. The total flux of 9.9 mJy is concentrated in two features: a central unresolved object (1.0 mJy) and a more extended feature on the scale of 2". The major lobe enhancement of the extended feature seems to be linked to the central object. The position angle of lobe enhancements is variable (see Kenny, *et al.* 1990 *Ap. J.*, **366**, 549, for observations in 1984 and 1987), however, position angles of the enhancements do not correlate simply with orbital phase (orbital period = 820 days).

The nebular region associated with AG Peg is expanding. Angular sizes for each epoch were measured from the visibility data using a spherical shell model. Comparison of the new data with earlier epoches (Kenny *et al.* 1990) indicates an expansion velocity of 53 ± 4 km s^{-1} and implies a kinematic age of 49 years (*cf.* 150 years after the onset of the most recent nova-like outburst). This finding is consistent with the interpretation of the nebula as an interaction feature between the red giant wind

---

*Department of Physics and Astronomy, University of Calgary, Calgary, AB, T2N 1N4, Canda
†Nuffield Radio Astronomy Laboratories, Jodrell Bank, Macclesfield, Cheshire, SK11 9DL, UK
‡School of Physics and Astronomy, Lancashire Polytechnic, Preston Lancashire, PR1 2TQ, UK

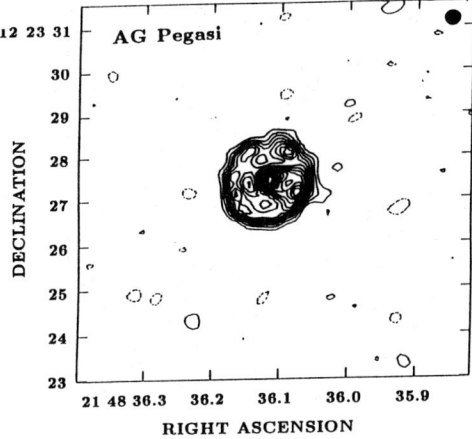

Figure 1. Radio Image of AG Pegasi at 5 GHz, March 1990.

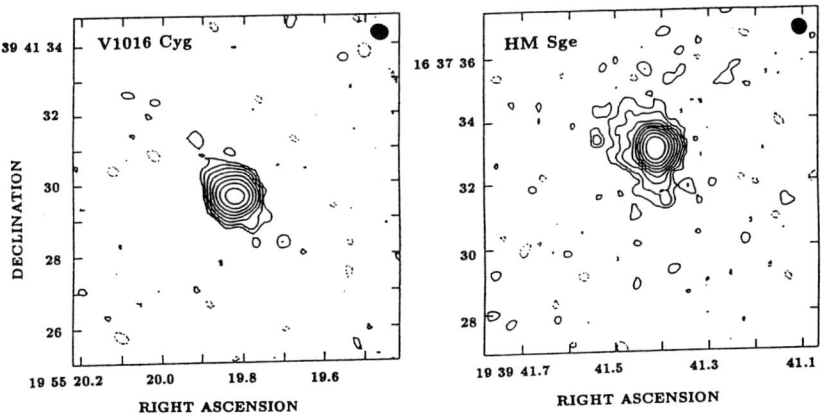

Figure 2. Radio Images of V1016 Cyg and HM Sge at 5 GHz, August 1991.

(remnant, previous to outburst), and a more recently initiated wind from the compact object. The mass of the nebula indicated by the model is $4.0 \pm 0.4 \times 10^{-5} M_\odot$.

The structure observed in both V1016 Cyg (44 mJy) and HM Sge (56 mJy) aligns approximately with the inner, bipolar morphology revealed in the 15 GHz images of Taylor (1988, in *The Symbiotic Phenomenon*, ed. J. Mikolajewska, M. Friedjung, S. J. Kenyon, and R. Viotti, Dordrecht: Reidel, p. 77.). The images at 5 and 15 GHz apparently present different phenomena. The 5 GHz images show material at larger distances from the inner binary star systems and may be associated with mass lost previous to the the recent outbursts. The 15 GHz images show more complex structure, at smaller linear distance from the inner systems. This structure shows evidence of the outbursts themselves.

THE SERPENS RADIO JET

Salvador Curiel *   *Luis F. Rodríguez* [†]
James M. Moran [‡] *Jorge Cantó* [§]

Abstract

The triple radio continuum source in Serpens has been identified as a radio jet associated with a star forming region. It exhibits a *one-sided radio jet morphology*, being in many respects very similar to Herbig-Haro optical jets, but having smaller dimensions. It exhibits an extended and knotty structure connecting the central source with one of the outer components. We propose that the Serpens radio jet could be a *proto-Herbig-Haro System*, in an early stage of evolution.

Several optical jets have been observed in association to young stellar objects, which are mainly observed in the infrared. On the other hand, there are only a few galactic outflows driven by central radio continuum jets. Probably the most interesting of these radio jets is the triple radio continuum source in Serpens. While this source is associated with a star forming region, its outer components exhibit non-thermal spectra (characteristic of optically thin synchrotron emission) and large proper motions of about $0.12''$ per year [RCMMRG89].

Recent high angular resolution VLA observations, at 3.6 cm wavelength with the A-array, have shown that this source exhibits a jet-like morphology at two different scales. Its main body has an extended and knotty structure connecting the central source with the NW outer component while no emission is detected between the central source and the SE components (see Figure 1; [CRMC92]). The knotty structure in the NW lobe is very similar to that observed in optical jets such as HH34, and HH111, except that it has a physical scale ($\sim 6''$ or 1800 AU at a distance of 300 pc) of about one order of magnitude smaller than the optical jets ($\sim 26''$ [12000 AU] and $54''$ [25000 AU] for the jets in HH 34 and HH 111). Likewise, the central source presents a jet-like structure similar in size and morphology to that observed in the

---

[*]Harvard-Smithsonian Center For Astrophysics, 60 Garden Street, Cambridge, MA 02138, USA.
[†]Instituto de Astronomía, UNAM, Apartado Postal 70-264, 04510 México, D.F., México.
[‡]Harvard-Smithsonian Center For Astrophysics, 60 Garden Street, Cambridge, MA 02138, USA.
[§]Instituto de Astronomía, UNAM, Apartado Postal 70-264, 04510 México, D.F., México.

energy sources of Herbig-Haro objects (see [CRMC92]), such as VLA1 and L1551 IRS5, the exciting sources of HH 1-2 and HH 28-29, respectively.

We identify the triple radio continuum source in Serpens as a young radio jet emanating from the central component, having a dynamical age of only ~ 50 years. Furthermore, we believe that the knotty structure in the NW radio jet and the SE components are the result of discrete ejection of material (or "bullets") from the central source, with ejection intervals of about 10 years (CRMC92). There is also some evidence of periodicity in the order of 10^3 years in the ejection of molecular clumps in the bipolar outflow of L1448 and IRAS 3283 (e.g., [BMP91]). The time separation between ejections in Serpens is then about 10^2 times smaller. Finally, based on the morphological and kinematical similarities between this radio jet and optical Herbig-Haro jets, and their different length scales and dynamical ages, we have proposed that the Serpens radio jet could be a *proto-Herbig-Haro System*, in an early stage of evolution (CRMC92).

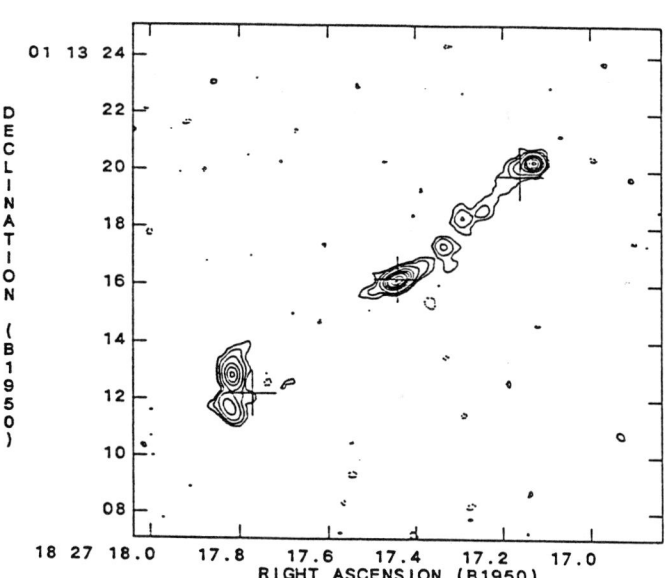

Figure 1. —Natural-weighted VLA map of the Serpens radio jet made at 3.6 cm in the A configuration. The crosses mark the position of the peak emission at 6 cm in 1984. Contours are -6,-3,3,6,12,24,48,70, and 100 times the 1-σ rms value of 13 μJy per beam. It is important to notice the knotty structure observed in this radio jet.

References

[BMP91] Bachiller R., Martin-Pintado J., Planesas P., *Astron. Astrophys.*, Vol. 251 (1991), pp. 639-648.

[CRMC92] Curiel S., Rodríguez L.F., Moran J.M., Cantó J., *To be submitted to Ap. J. (1992)*.

[RCMMRG89] Rodríguez L.F., Curiel S., Moran J.M., Mirabel I.F., Roth M., and Garay G., *Ap. J. (Letters)*, Vol. 346 (1989), pp. L85-L88.

The Milliarcsecond Jets of Cygnus X-3

C.J. Schalinski, * A. Witzel, † K.J. Johnston,‡
P.E. Pavelin, R.E. Spencer, R.J. Davis, § G. Umana¶

Observations of the X-ray binary system Cygnus X-3 using the Very Long Baseline Interferometry technique allow simultanously investigations of the structure of the radio emission, especially during large outbursts, and in the presence of strong scattering allow the study of properties of the interstellar medium. Immediately following the report of a radio flare detected by the NRL-Greenbank Interferometer (Johnston, priv. com.) we have organized adhoc experiments using telescopes of the European VLBI network.

The first successful experiment was conducted during the 1985 October flare with the telescopes at Effelsberg, Westerbork, Medicina, Onsala and Jodrell Bank using the Mk-II system (2 MHz bandwidth) at a wavelength of 6 cm at the time of a regular network session. The radio emission reached a peak flux density of ~ 15 Jy at 6 cm wavelength on October 10, and was followed by flares with smaller amplitudes on October 17, 20 and 24. We detected compact radio emission from Cygnus X-3 at 6 epochs between October 7 and 16 with VLBI. These data indicate that the source is heavily resolved even on the shortest baseline from Effelsberg to Westerbork (~ 50 mas fringe spacing at 6 cm), with peak correlated flux densities between 0.5 and 1 Jy. On October 10 and 15/16, when the 100m telescope could participate, sufficient data were obtained to allow a modelfit (Schalinski, Johnston and Witzel, 1990): the best representation of the data is a three component model, consisting of a compact "core" and two components on either side separated by about 30 mas each. The position angle of this *linear* structure, $180^\circ \pm 20^\circ$, and $205^\circ \pm 20^\circ$, respectively, is in agreement with the position angle of the structure found on larger scales (Spencer et al., 1986; Strom, van Paradijs, and van der Klis, 1989).

The angular size of the central compact component, 15 ± 5 mas, is significantly broadened compared to the 7 mas-beam of the VLBI-network used. By adding published data at lower frequencies we find a $\theta \propto \nu^{-2.07 \pm 0.04}$-dependence. This can be explained by interstellar scattering due to electron inhomogeneities along the line of sight with a powerlaw spectrum $C_n^2 q^{-\alpha}$ with scattering amplitude C_n^2, wavenumber q, and spectral index $\alpha = 3.87 \pm 0.07$, to be compared to the theoretical value of 11/3 for Kolmogorov turbulence. In good agreement with our data Wilkinson, Spencer and Nelson (1988) obtain $\theta \propto \nu^{-2.06 \pm 0.03}$ on the basis of MERLIN observations at 73 cm wavelength. The

---

*Institut de Radio Astronomie Millimétrique, Grenoble, France.
†Max-Planck-Institut für Radioastronomie, Bonn, Germany.
‡E.O. Hulbert Center for Space Research, Naval Research Laboratory, Washington D.C., U.S.A.
§Nuffield Radio Laboratory, Jodrell Bank, U.K.
¶Istituto di Radioastronomia, Bologna, Italy.

image broadening may be attributed to Refractive Interstellar Scattering (Schalinski, Johnston and Witzel, 1990), and further data should allow to check on an ellipticity of the compact component as expected from scattering anisotropies.

The identification of the flare on October 10 as onset of a linear expansion on either side of the central source gives $0.29 \pm 0.06\,c$ at an assumed distance of 10 kpc. On the basis of the sizes of the secondary components a lower limit of about 50% of the above value is estimated for a transverse source expansion (Schalinski et al, 1993). Although, together with previous velocity estimates by Geldzahler et al. (1983) (5 mas/day), Spencer et al. (1986) (6 mas/day), and Molnar, Reid and Grindlay (1988) (0.16-0.31c) there is now good evidence for jets emanating from Cygnus X-3 with $\sim 0.3\,c$. Obviously higher sensitivity is required to reveal details of the kinematics of the jets.

During the January 1991 radio outburst we were able to observe Cygnus X-3 with EVN-telescopes at 6 cm wavelength on four consecutive days - including the 100m telescope at two epochs - with the MK-III system and a bandwidth of 56 MHz. The substructure seen in the visibility amplitudes clearly indicates the presence of a complex radio jet. Modelfitting of these data requires at least 5 components, attributed to 2 jets on either side of a central component. There is evidence that the multiple outbursts visible in the total flux density curve (Waltman, priv. com.) is correlated with the creation and motion of compact components on the milliarcsecond-scale (Schalinski et al., in prep.). Further analysis of the data may show, if the significant bending of the jet structure can be attributed to a precession of the binary system.

References

Geldzahler, B.J., Johnston, K.J., Spencer, J.H., Klepczynski, W.J., Josties, F.J., Angerhofer, P.E., Florkowski, D.R., McCarthy, D.D., Matsakis, D.N., and Hjellming, R.M., 1983: *Ap.J. (Letters)*, **273**, L65
Molnar, L.A., Reid, M.J., and Grindlay, J.E., 1988: *Ap.J.*, **331**, 494
Schalinski, C.J., Johnston, K.J., and Witzel, A., 1990: in: *Parsec-scale radio jets*, ed. J. A. Zensus and T. J. Pearson (Cambridge University Press), p. 140.
Schalinski, C.J., Johnston, K.J., Witzel, A., Spencer, R.E., Fiedler, R.L., Waltman, E.B., Pooley, G.G., Hjellming, R.M., and Molnar, L.A., 1993: *Ap.J.*, accepted
Spencer, R.E, Swinney, R.W., Johnston, K.J., and Hjellming, R.M., 1986: *Ap.J.*, **309**, 694
Strom, R.G., van Paradijs, J., and van der Klis, M., 1989: *Nature*, **337**, 234
Wilkinson, P.N., Spencer, R.E., and Nelson, R.F., 1988: in: *IAU Symposium 129, The Impact of VLBI on Astrophysics and Geophysics*, ed. M. J. Reid and J. M. Moran (Dordrecht: Kluwer), p. 305.

VLBI observations of LSI+61°303

Josep M. Paredes [*]   *Maria Massi* [†]   *Robert Estalella* [*]

Marcello Felli [†]

Abstract

An hybrid map of the X-ray binary LSI+61°303 has been obtained at 6 cm wavelength by using the European VLBI Network (EVN) and the VLA. The map of the source reveals two components, separated 0.9 mas. The overall size of the source, at a level of half peak intensity, is about 1.6 × 1.0 mas. The values derived from the data for the magnetic field and energy content are consistent with an adiabatic expansion model of a synchrotron emitting source with prolonged injection of energetic particles.

1. Introduction

LSI+61°303 is the optical counterpart of the variable radio source GT0236+610 and it is located at a distance of 2.3 kpc. It presents periodic radio outbursts, of variable amplitude, with a period of 26.5 days [Tay82]. In addition, the peak flux density of the radio outbursts varies with a 4-yr cycle [Gre89] [Par90].

LSI+61°303 also presents optical variability. The values obtained from radial velocity observations are consistent with the radio period and give support to the presence of a companion star [Hut81]. The system is a weak (10^{33} erg s^{-1}) X-ray source and probably a γ-ray source.

2. Observations and results

The observations were made with the MkIII VLBI system in the standard A mode on 6 June 1990 at 6 cm wavelength. The antennas used were Effelsberg, Westerbork,

---

[*]Departament d'Astronomia i Meteorologia, Universitat de Barcelona, Av. Diagonal 647, E-08028 Barcelona, Spain; and Laboratori d'Astrofísica, Societat Catalana de Física, IEC, Spain. This work has been supported by CICYT (Spain) under contract ESP88-0731.

[†]Osservatorio Astrofisico di Arcetri. Largo E. Fermi 5. 50125 Firenze. Italy.

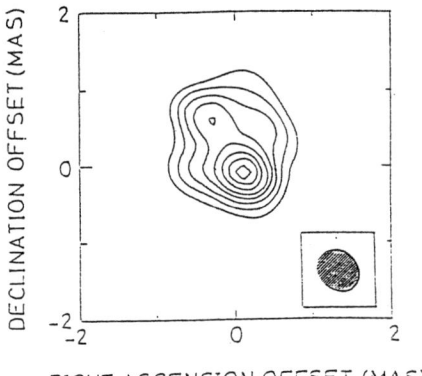

Figure 1. Map of LSI+61°303. The restoring beam is shown as a hatched ellipse.

Medicina, Onsala and the VLA. The observation, with a duration of 14 hours, was carried out two days after the maximum of the 26.5 d periodicity.

Figure 1 shows the final map obtained with a HPBW of 0.6×0.5 mas (p.a.= 45°). The map shows two components separated 0.9 mas, the southern component being the strongest one. The overall size of the source, at a contour level of half peak intensity, is $\sim 1.6 \times 1.0$ mas, corresponding to a linear size of $\sim 5.5 \times 3.4\,10^{13}$ cm. The total flux density in the VLBI map is 200 mJy. The estimated brightness temperature is $T_B = 8.8\,10^9$ K. This value confirms the suggested non-thermal nature of the radio star LSI+61°303.

Assuming that both components expand away from a common center and that the observed radio outburst started 4 days before our observation, a velocity of $4.4\,10^7$ cm s^{-1} results. Assuming equipartition of high-energy particles and magnetic energy a minimum energy of $4.2\,10^{39}$ erg, and a magnetic equipartition field of 0.7 G have been derived. The data are consistent with an adiabatic expansion model which assumes a prolonged injection of energetic particles [Par91].

References

[Gre89] Gregory P.C., Huang-Jian Xu, Backhouse C.J., Reid A., 1989, ApJ 339, 1054

[Hut81] Hutchings J.B., Crampton D., 1981, PASP 93, 486

[Par90] Paredes J.M., Estalella R., Rius A., 1990, A&A 232, 377

[Par91] Paredes J.M., Marti J., Estalella R., Sarrate J., 1991, A&A 248, 124

[Tay82] Taylor A.R., Gregory P.C., 1982, ApJ 255, 210

CEPHEUS A - An Optically Obscured Herbig - Haro Object?

V. A. Hughes
Queen's University, Kingston, Ontario

Cepheus A has been monitored at the VLA since 1981. It was seen originally to consist of lines containing about 13 radio objects, which appeared to be thermal, the regions central to the area being consistent with compact optically thick HII regions, while those further out were more diffuse and optically thin. If the objects were HII regions, then each could be produced by a B3 star, and the total IR emission was that expected from 13 B3 stars (Hughes, 1985). However, this interpretation in terms of classical HII regions was somewhat suspect, since their linear size was only about 1,000 AU. This means that either there must be some mechanism which contains the regions and prevents them from expanding, or there must be some trigger mechanism which causes all 13 stars to be in the same stage of evolution the very narrow time interval of about 1,000 years. It is clear that the central objects are compact, $< 0\rlap{.}''1$ in diameter (< 70 AU at the distance of 725 pc), and are believed to be stellar in nature, while the others appear to be much more diffuse ionized regions. More recently, two highly variable radio objects have been detected, which can appear and disappear in under 50 days. The spectrum of the latter is consistent with gyrosynchrotron radiation from a region of size 1 AU, temperature of 10^8 K, magnetic field of ~ 100 G, and density of 10^6 cm^{-3} (Hughes, 1991). It is believed that these may be low mass stars shedding their magnetic fields as they go through the Hyashi phase. In addition, in the line of diffuse objects referred to as Source 7, one of them, namely Source 7(c)(ii), is moving with respect to its neighbours at a speed of 300 km s^{-1}. It is compact, being the only object in that group which is detected at the wavelength of 2 cm, and appears to be a "radio bullet" (Hughes, 1993). It is moving in a direction away from the highly variable Source 9, as shown in the Figure, which may be the powerhouse for the "bullet". It appears that in these cases, the powerhouse could be a T-Tauri star. The overall appearance of the region is very similar to that of the few Herbig-Haro objects known at radio wavelengths, such as HH-1 and -2 (Herbig & Jones, 1981, Rodriguez et al. 1990), and also to the triple radio source in Serpens (Rodriguez et al. 1989, Torrelles et al. 1992). A comparison of images in the radio, in 2μm continuum, and in 2μm line from H$_2$ (Nadeau, Doyon, & Rowlands, 1992), shows no correlation between any of the features, at any of the wavelengths.

REFERENCES

Herbig, G.H., & Jones, B.F. 1981, AJ, 86, 1232.

Hughes, V.A. 1985, ApJ, 298, 830.

Hughes, V.A. 1991, ApJ, 383, 280.

Hughes, V.A. 1993, AJ, January.

Nadeau, D., Doyon, R., & Rowlands, N. 1992, Meeting of the Canadian Astronomical Society, Halifax, Nova Scotia, 1992 June 27.

Rodriguez, L.F., Curiel, S., Moran, J.M., Mirabel, I.F., Roth, F., & Garay, G. 1989, ApJ, 346, L85.

Rodriguez, L.F., Ho, P.T.P., Torrelles, J.M., Curiel, F., & Canto, J. 1990, ApJ, 352, 645.

Torrelles, J.M., Gomez, J.F., Curiel, S., Eiroa, C., Rodriguez, L.F., & Ho, P.T.P. 1992, ApJ, 384, L59.

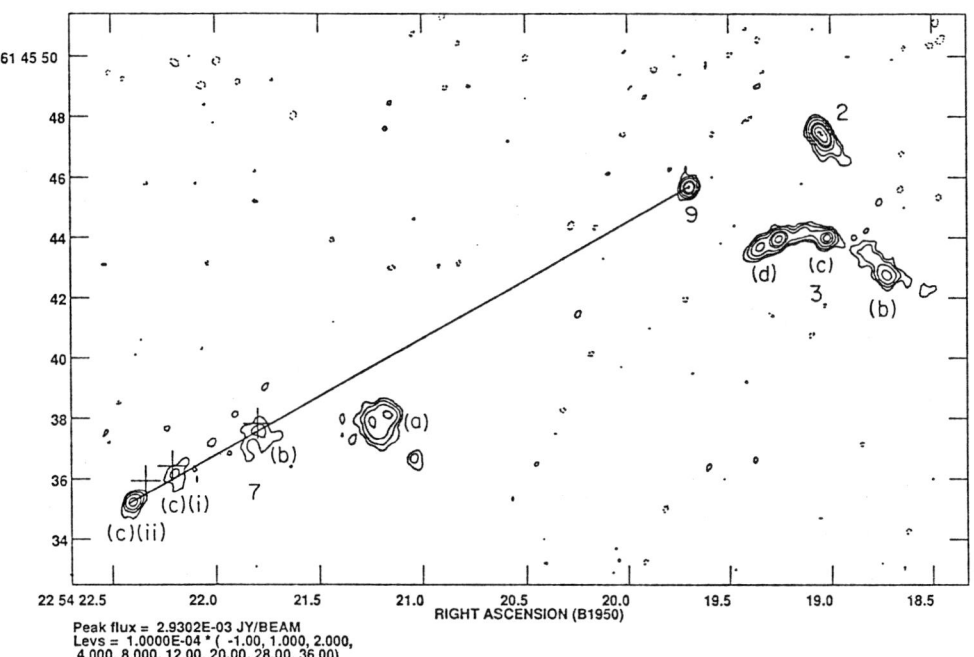

Figure 1. The Cepheus A region.

VLBI Observations of a Stellar Flare

Walter Alef [*]   *Arnold O. Benz* [†]   *Manuel Güdel* [‡]
Pablo de Vicente [§]

Several nearby dwarf M-stars have been detected in an intercontinental VLBI experiment at 18 cm including the Arecibo, Effelsberg, Jodrell Bank, Owens Valley, and VLA telescopes [1]. We used the Mark III VLBI recording system with a bandwidth of 28 MHz in each hand of circular polarization and achieved a threshold sensitivity of about 0.9 mJy ($\approx 7\sigma$) on the most sensitive baseline, Arecibo-Effelsberg.

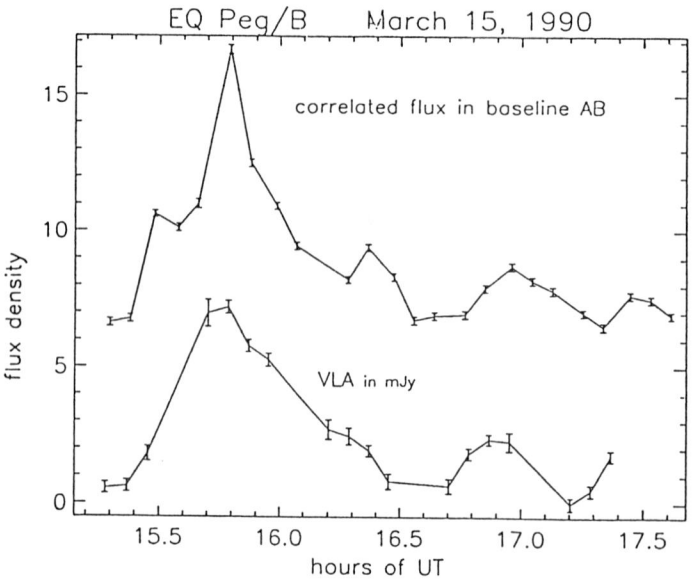

Fig.1: Time profile of full event. The correlated flux in the Arecibo-Bonn baseline correlates roughly with the total flux observed by the VLA.

By coincidence, a flare occurred during the observation of EQ Peg/B. The correlation of the VLA flux density time profile with the correlated flux of the longer baselines

---

[*] Max-Planck-Institut für Radioastronomie, Auf dem Hügel 69, D-5300 Bonn 1, Germany.
[†] Institute of Astronomy, ETH, CH-8092 Zürich, Switzerland.
[‡] JILA, University of Colorado, Boulder, USA.
[§] Centro Astronomico de Yebes, Spain.

indicates that the source was largely unresolved (Figure 1). The observed decrease of the correlated flux with baseline length suggests a formal source size of 1.2 ± 0.1 mas (or 3 stellar diameters), limited, however, by the unknown calibration errors. We estimate an upper limit of 1.3 mas, corresponding to a peak brightness temperature of at least $6.5 \cdot 10^9$ K. A high degree of circular polarization confirms the operation of a coherent emission process.

The closure phase shows significant deviations from zero during the peak of the flare (Figure 2). It indicates that the source is slightly resolved and has a time variable structure.

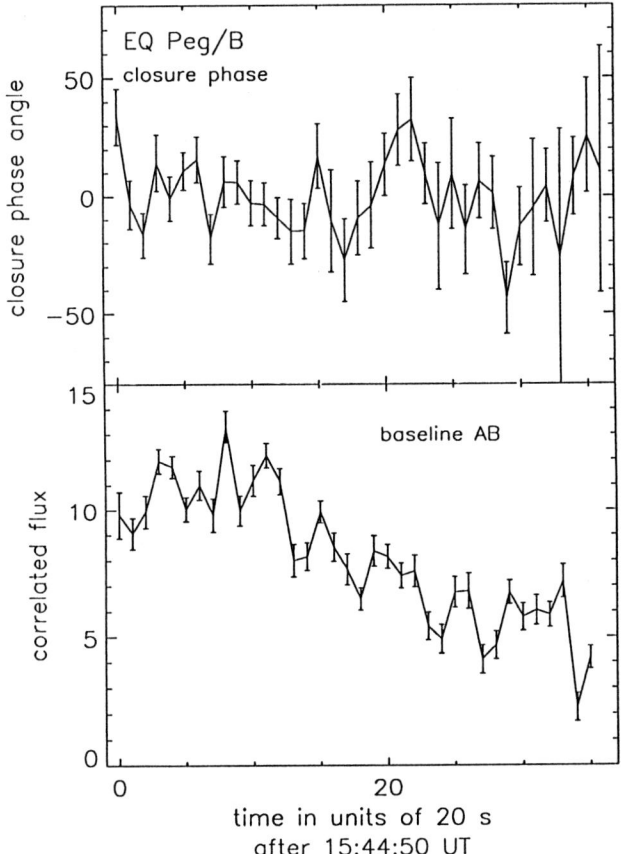

Fig.2: The closure phase of the most sensitive triangle (Arecibo - Bonn - Green Bank) during peak flux.

References

[1] Benz, A.O., Alef, W., *VLBI observations of YZ CMi - a single dMe Star.* Astron. Astrophys. Vol. 252 (1991), pp. L19-L22.

Modelling the strong Cygnus X-3 radio outbursts

Josep M. Paredes * *Josep Martí* * *Robert Estalella* *

Abstract

A jet model, consisting of synchrotron emitting plasma and undergoing lateral adiabatic expansion, is proposed as an explanation for the strong radio outbursts of the low mass X-ray binary Cyg X-3. The model is applied to the historical 1972 first observed giant radio outburst. Satisfactory agreement is obtained over more than two decades in frequency, from 0.4 GHz to 90.0 GHz.

1. Introduction

Cyg X-3 is assumed to be a low mass X-ray binary whose emission is powered by mass accretion from a red dwarf onto a white darwf. In 1972 September this object was reported to exhibit strong radio outbursts reaching a peak flux density of $\sim$ 20 Jy in a time scale of a few days [Gre72]. In the recent times, VLBI observations of Cyg X-3 have revealed that this system actually presents a N-S elongated structure [Gel83] [Spe86] [Sch89]. This suggests that Cyg X-3 contains a twin jet similar to SS 433. Bulk motions between 0.16 and 0.31 c have been estimated by Molnar et al. [Mol88]. These authors also obtain an expansion rate transverse to the jet axis of $\sim$ 0.13 c.

2. The model and its application

We describe the Cyg X-3 strong radio outbursts as time evolving injection events of synchrotron emitting particles into twin jets undergoing lateral adiabatic expansion. Initially, this is done at an exponential rate, and the magnetic flux is always conserved. We apply this model to the multifrequency data ([Gre72] and papers following this reference) obtained during the first giant radio outburst of 1972 September.

---

*Departament d'Astronomia i Meteorologia, Universitat de Barcelona, Av. Diagonal 647, E-08028 Barcelona, Spain; and Laboratori d'Astrofísica, Societat Catalana de Física, IEC, Spain. This work has been supported by CICYT (Spain) under contract ESP88-0731.

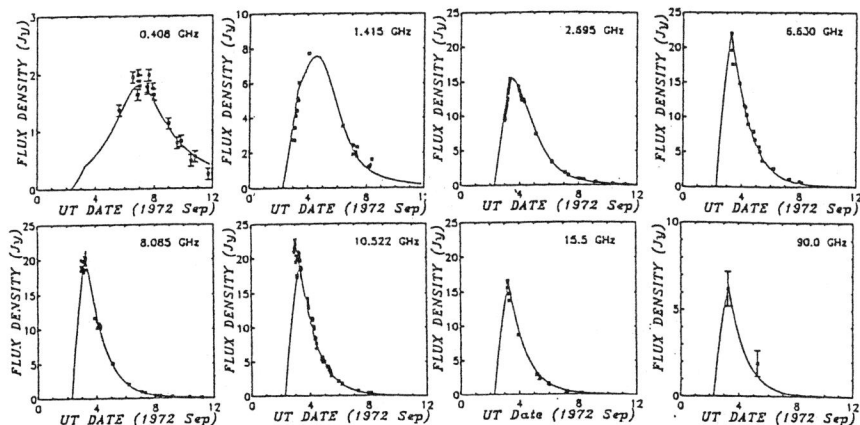

Figure 1. Outburst peak flux density computed from the present model together with the observed data points at eight frequencies.

The best fit values are: initial magnetic field $B_0 = 0.1$ G, initial lateral expansion velocity $v_0 = 0.068\ c$, injection time interval $t_f = 1.0$ d, injection rate of relativistic electrons $\dot{M}_{\rm rel} = 8.5\ 10^{-12}\ M_\odot {\rm d}^{-1}$, initial thermal electron density $n_{\rm th}^0 = 4.4\ 10^5\ {\rm cm}^{-3}$, and power law index $p = 2.1$. Fig. 1 contains the computed radio light curves at the eight frequencies observed. Details of the present model are discussed in Martí et al. [Mar92].

References

[Gel83] Geldzahler B. J., Johnston K. J., Spencer J. H., Kepczynski W. J., Josties F. J., Angerhofer P. E., Florkowski D. R., McCarthy D. D., Matsakis D. N., Hjellming R. M., 1983, APJ 273, L65

[Gre72] Gregory P. C., Kronberg P. P., Seaquist E. R., Hughes V. A., Woodsworth A., Viner M. R., Retallack D., 1972, Nature 239, 440

[Mar92] Martí J., Paredes J.M., Estalella R., 1992, A&A 258, 309

[Mol88] Molnar L. A., Reid M. J., Grindlay, J. E., 1988, APJ 331, 494

[Sch89] Schalinski C. J., Johnston K. J., Witzel A., 1989, in Parsec-scale Radio Jets, Ed. J. A. Sensus, T. J. Pearson, Cambridge Univ. Press, p. 141

[Spe86] Spencer R. E., Swinney R. W., Johnston K. J., Hjellming R. M., 1986, APJ 309, 604

Luminosity of Radio Supernovae

Richard Sramek, *Kurt Weiler, Schuyler Van Dyk* [†]
Nino Panagia [‡]

Abstract

Preliminary results from a VLA program to detect radio emission from recent bright optically discovered supernovae have yielded no new detections.

Starting in January 1985 until mid-1989, any new supernova with an optical magnitude brighter than 14.5 was observed with the VLA in an attempt to establish the luminosity distribution for radio supernovae. The new supernovae were observed near the time of discovery and then one week, two weeks, one month, two months, four months, eight months, 16 months, and 32 months later. In this way radio emission on a wide range of time scales could be detected.

Preliminary results for 15 supernovae are given in the table. No new supernovae were detected in this program; three sigma upper limits are shown in the table. Upper limits on the radio luminosity of some of these supernovae (e.g. 7×10^{25} erg/sec/Hz for SN1985H) are a factor of 500 below the luminosity of the powerful radio supernovae like SN 1986J (luminosity of 2×10^{28} erg/sec/Hz at 6 cm). Simply looking at our optical flux limited sample, the detection rate of supernovae with a radio flux of > .2 mJy seems to be less than a few percent. This is not very encouraging for hunting radio supernovae.

---

[*]National Radio Astronomy Observatory, Socorro, New Mexico, U.S.A. The National Radio Astronomy Observatory is operated by Associated Universities, Inc. under cooperative agreement with the National Science Foundation.
[†]Center for Advanced Space Sensing, Naval Research Laboratory, Washington, D.C., U.S.A
[‡]Space Telescope Science Institute, Baltimore, MD., U.S.A.

Table 1. List of supernovae observed.

SN	Type	Max mag	Gal	Dis Mpc	Flux Limit (mJy)		Lumin. Limit (erg/s/Hz)	
					20 cm	6cm	20cm	6cm
SN1983K	II	B 12.4	NGC 4699	26		< 0.24		< 1.9×10^{26}
SN1984E	II	B 15.2	NGC 3169	20		< 0.23		< 1.1×10^{26}
SN1984R		V 13.0	NGC 3675	13	< 0.28	< 0.17	< 5.5×10^{25}	< 3.3×10^{25}
SN1985A	Ia	14.5	NGC 2748	24	< 0.27	< 0.20	< 1.8×10^{26}	< 1.4×10^{26}
SN1985B	Ia	V 13.0	NGC 4045	32	< 0.50	< 0.17	< 6.0×10^{26}	< 2.0×10^{26}
SN1985F	Ib	B 12.1	NGC 4618	7	< 0.18	< 0.18	< 1.2×10^{25}	< 1.2×10^{25}
SN1985G	II	B 15.0	NGC 4451	17	< 0.54	< 0.20	< 1.8×10^{26}	< 6.8×10^{25}
SN1985H	II	V 16.4	NGC 3359	19	< 0.54	< 0.15	< 2.4×10^{26}	< 6.6×10^{25}
SN1985L	II	12.5	NGC 5033	19		< 0.63		< 2.6×10^{26}
SN1986A	Ia	B 14.4	NGC 3367	44	< 1.10	< 0.15	< 2.5×10^{27}	< 3.4×10^{26}
SN1986E	II	B 13.4	NGC 4302	17	< 1.00	< 0.30	< 3.4×10^{26}	< 1.0×10^{26}
SN1986G	Ia	B 12.5	NGC 5128	5		< 9.00		< 2.6×10^{26}
SN1986I	II	V 14.2	NGC 4254	17	< 0.25	< 0.27	< 8.5×10^{25}	< 9.1×10^{25}
SN1986O	Ia	V 14.0	NGC 2227	28		< 0.30		< 2.7×10^{26}
SN1987D	Ia	B 13.7	MGC+0-32	30		< 0.18		< 1.9×10^{26}

LkHα 101: The Relation between Radio and Thermal Infrared Observations at High Spatial Resolution

R. M. Danen [*]   *C. R. Gwinn* [†]   *E. E. Bloemhof* [‡]

Abstract

We have obtained diffraction-limited images of the pre-main-sequence emission-line star LkHα 101 at $\lambda = 10$ μm using a 3 m telescope. The point-spread function has FWHM $\sim 0.7"$; images of LkHα 101 are consistent, within errors, with images of the unresolved star α Tau. Deconvolving the point-spread function from LkHα 101 places an upper limit of $\sim 0.3"$ (240 AU) on the diameter of LkHα 101. We measure the 10 μm flux density to be ≈ 325 Jy. This emission probably arises from spatially compact circumstellar dust. Radio observations reveal that LkHα 101 is the source of a strong ionized wind. Our observations place the dust at radial scales between about 50 and 120 AU, comparable to the inner scales considered in source models derived from radio data.

LkHα 101 is a massive pre-main-sequence emission-line star 800 pc distant. It radiates most of its luminosity in the infrared, and is also a bright radio source. From IRAS data, the luminosity of LkHα 101 is 4.8×10^4 $L_\odot$ [BSS90], which corresponds to an O9 ZAMS star. The radio emission from LkHα 101 is divided into two components. One is an optically thin H II region (S222) about 1' in extent; there is evidence that a hole with a diameter of about 30" exists in this region [BeW88]. The other component is a radio source smaller than 1" in diameter, with a spectrum that suggests the radio emission arises in an ionized wind.

The infrared observations reported here were made on 1991 November 24, using a scanned linear array on the 3 m Shane telescope at Lick Observatory. Thirteen contiguous pixels of the array, each subtending 0.34", are oriented N-S in the focal

---

[*] Physics Department, University of California, Santa Barbara, CA, USA 93106. This author is supported by a Predoctoral Fellowship from the Smithsonian Astrophysical Observatory.
[†] Physics Department, University of California, Santa Barbara, CA, USA 93106.
[‡] Harvard-Smithsonian Center for Astrophysics, 60 Garden Street, Cambridge, MA, USA 02138.

plane. Earth rotation then provides a smooth and accurate west-to-east scan over the source. The array is read out rapidly, and a complete scan across a field of a few arc seconds is completed in a few hundred ms, less than the characteristic timescale of atmospheric fluctuations at $\lambda = 10$ μm. Thus, the resolution of images we obtain is set by the diffraction limit of the telescope. The theoretically expected point-spread function of a 3 meter telescope is an Airy pattern with FWHM $\sim 0.7''$.

Our 10 μm images of LkHα 101 show no significant spatial extent in comparison to images of the unresolved star α Tau. Numerical deconvolution, using α Tau as the experimental point-spread function, places an upper limit of $\sim 0.3''$ on the diameter of LkHα 101. We measure a flux density of 325 ± 27 Jy from this spatially unresolved core.

The 10 μm emission from this compact core is not photospheric emission; an optically thick source with the radius and effective temperature derived for an O9 ZAMS star at 800 pc emits less than one Jansky at $\lambda = 10$ μm. Also, the 10 μm emission is not free-free radiation from the ionized wind. [BSS90] derive a spectral index of a=+0.67 from 2.7 to 100 GHz for the compact radio component, which corresponds to what is predicted for a constant-velocity, isotropic, ionized stellar wind. Extrapolation of this spectrum into the infrared yields a flux density less than 10 Jy at $\lambda = 10$ μm. Hence the compact core that we observe is probably due to circumstellar dust.

Assuming reasonable properties of graphite dust, we calculate from considerations of grain heating and cooling that grains of radius 0.03 μm cannot exist any closer than ~ 50 AU (0.06'') to the central star. Actual dust condensation may occur substantially further out than this. We are thus able to constrain the dust to lie in a relatively small spread of radial distances (50 AU $\lesssim$ r $\lesssim$ 120 AU) in which it coexists with an energetic ionized wind. We are exploring the implications of this coexistence [DGB93] for models of LkHα 101 based on high-resolution radio observations [BSS90].

References

[BSS90] Barsony, M., Scoville, N.Z., Schombert, J.M., Claussen, M.J., *The Circumstellar Environment of the Emission-Line Star, LkHα 101.* Ap.J., 362 (1990), pp. 674-690.

[BeW88] Becker, R.H., White, R.L., *LkHα 101: The Stellar Wind, the Surrounding Nebula, and an Associated Radio Star Cluster.* Ap.J., 324 (1988), pp. 893-898.

[DGB93] Danen, R.M., Gwinn, C.R., Bloemhof, E.E., *Subarcsecond* 10 μm *Imaging of LkHα 101.* Ap.J., (in preparation).

Hydrodynamic Models of Bipolar Nova Outbursts

H.M. Lloyd [*] *M.F. Bode* [†] *T.J. O'Brien* [‡] *F.D. Kahn* [§]

Abstract

RS Ophiuchi is a recurrent nova which last underwent outburst in 1985. The nova binary system contains a red giant with a slow wind and the ejection therefore occurs in a dense circumstellar medium. The nova was mapped by EVN at 1.7 GHz, 77 days after outburst, showing non-thermal emission and a bipolar morphology; the outburst was also detected in X-rays by EXOSAT. We have performed two–dimensional numerical hydrodynamic calculations of bipolar recurrent nova outbursts with particular reference to RS Oph (1985). Two general cases have been investigated – point injection of energy at the centre of an anisotropic wind, and the ejection of material along preferred directions into a spherically symmetric wind. The calculations must take account of the recurrent nature of the outbursts, which results in the wind being of finite extent ($\sim 10^{15}$ cm) having been cleared by the last outburst (although some material may remain in the equatorial plane). For the anisotropic wind models it is found that for no reasonable parameter values can a sufficient fraction of the outburst energy be removed to the required distances to supply synchrotron emission at the observed level. However, this can be achieved in models where material is ejected in the polar direction. Model X-ray spectra then agree with the EXOSAT data.

Previous spherically symmetric hydrodynamic models of RS Oph (1985) [OBK92] give rise to X-ray emission in reasonable agreement with EXOSAT data, however the EVN map [TDPB85] shows clear evidence for a bipolar structure. We have therefore performed 2-D axisymmetric hydrodynamic calculations of recurrent nova outbursts using a second-order Godunov scheme [F91] in an attempt to model both the X-ray spectra and radio morphology. Figure 1 shows gas density and velocity in the

---

[*]Chemical and Physical Sciences, Liverpool John Moores University, Liverpool L3 3AF. This author is supported by an SERC research assistantship.
[†]Chemical and Physical Sciences, Liverpool John Moores University, Liverpool L3 3AF.
[‡]Computing and Mathematical Sciences, Liverpool John Moores University, Liverpool L3 3AF.
[§]Department of Astronomy, The University, Manchester M13 9PL.

(r, z) plane for a bipolar ejection model with $\dot{M}/u = 7 \times 10^{12}$g cm^{-1} (see [OBK92]) and an ejection velocity of 3500km s^{-1} (in line with optical and UV spectroscopic measurements). Note that the remnant shock has broken out of the confining wind, resulting in material streaming out into the surrounding low density medium. The corrugated appearance of the leading edge of this material is due to the Rayleigh–Taylor instability. The blobs of ejecta remain relatively undecelerated and have a thermal energy content greater than the equipartition energy derived by [TDPB89]. X-ray spectra calculated from the models agree with those observed by EXOSAT. For the anisotropic wind models (using the density distribution of [KW85]) it was found that for no reasonable parameter values could a sufficient fraction of the outburst energy be removed to the required distances for agreement with the 1.7 GHz data. In a further paper, we will discuss our models in more detail, calculating radio maps, X-ray spectra and light curves.

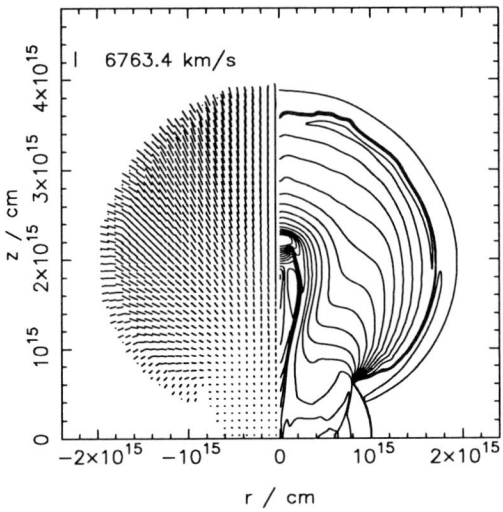

Figure 1. Logarithmic density contours and velocity vectors for a bipolar ejection model at 77 days after outburst. See text for details.

References

[F91]   Falle, S.A.E.G., *Mon. Not. R. astr. Soc.*, Vol. 250 (1991), p 581

[KW85]   Kahn, F.D., West, K., *Mon. Not. R. astr. Soc.*, Vol. 212 (1985), p 837

[OBK92]   O'Brien, T.J., Bode, M.F., Kahn, F.D., *Mon. Not. R. astr. Soc.*, Vol. 255 (1992), p 683

[TDPB89] Taylor, A.R., Davis, R.J., Porcas, R.W., Bode, M.F., *Mon. Not. R. astr. Soc.*, Vol. 237 (1989), p 81

Compact Radio Sources in the Galactic Center

Jun-Hui Zhao and W. M. Goss [*]

Abstract

We review new observational results of the compact radio sources SgrA* and the galactic center transient (GCT) within the central 3 pc of the galactic center. The immediate environs of these compact radio components are also discussed.

1. Introduction

Since the discovery of the compact radio source Sgr A* in 1974 by Balick and Brown [BB74], numerous investigations of this source have been carried out. Sgr A* is embedded in the HII region Sgr A West which consists of a spiral shaped structure of size 1' or 2.5 pc at 8.5 kpc (see [EGSG83] & [LC83]).

The properties of Sgr A* have been summarized by Lo ([LO87] & [LO89]). The compact non-thermal source has a spectral index of ~ 0.25 ($S \propto \nu^\alpha$) with a brightness temperature of $> 7 \times 10^8$ K at λ 1.35 cm. The source is scatter-broadened with a size dependence of λ^2 ($\lambda > 1.35$ cm). At this wavelength, the inferred linear size is < 13 AU. The radio luminosity of Sgr A* of $\sim 10^{34}$ erg s^{-1} is three orders of magnitude smaller than the nuclear source in the nearby spiral galaxy M81. The accurate proper motions determined with the VLA have been summarized by Backer and Sramek [BS87]; any peculiar velocity is < 25 km s^{-1} setting a lower limit to the black hole mass of 100 M$_\odot$ within 10 AU of Sgr A* (Backer and Sramek, private communication).

In this review, we will discuss recent VLA observations of the H92α line in Sgr A West (§2), the galactic center transient of 1990 - 1992 (§3), the nearby environs of Sgr A* (§4), recent observations of the time variations of Sgr A* (§5) and recent VLBI images of Sgr A* (§6). We will not discuss VLA observations of the mass-losing supergiant IRS 7 ($\sim 5''$ north of Sgr A*). Yusef-Zadeh and Morris [YM91] and Yusef-Zadeh and Melia [YM92] have summarized the radio "cometary tail" of ionized gas from IRS 7.

---

[*]NRAO, P. O. Box O, Socorro, NM87801, USA. The National Radio Astronomy Observatory is operated by Associated Universities, Inc., under a cooperative agreement with the National Science Foundation.

2. H92α Line and Images of Sgr A West

Roberts and Goss [RG93] have carried out a VLA survey of Sgr A West in the H92α line at 3.6 cm. The angular resolution is 1 - 2″, the velocity resolution is 14 km s^{-1} and the rms noise is $\sim$ 0.15 mJy beam^{-1}. The resolution and sensitivity represent an order of magnitude improvement compared with the earlier VLA images in the H76α line described by van Gorkom et al. [GSB85] and Schwarz et al. [SBG89]. In agreement with the earlier observations, three major kinematic features have been identified: the western arc, the northern arm and the extended bar. The western arc can be modelled as a portion of a ring with an inner radius of 25″ (1 pc). The interior mass of $\sim 3.5 \times 10^6$ M$_\odot$ is based on a simple model of circular rotation. The previously reported high electron temperatures south of Sgr A* can now be explained as a superposition of the bar (v $\sim$ 0 km s^{-1}) and the northern arm (v ~ -250 km s^{-1}); a uniform electron temperature of 7000 K is observed throughout Sgr A West.

3. The Galactic Center Transient

The transient radio source (the galactic center transient or GCT), located 36″ or 1.5 pc southeast of Sgr A*, was discovered by Zhao et al. [ZRGFLE91] using the VLA. Following the radio detection, the location, nature and origin of the GCT have been investigated with intensive observations at radio and infrared wavelengths [ZRGFLSKEABS92]. Many characteristics, such as high radio luminosity (1×10^{33} erg s^{-1}), steep spectrum ($S_\nu \propto \nu^{-1.2}$), elongated radio morphology, lower expansion rate ($\sim$ 6000 km s^{-1}) (compared to a young supernova remnant) along with lack of infrared counterpart, suggest that the transient radio emission of the GCT is likely caused by an ejection of a relativistic plasma from an X-ray binary system.

Radio observations of the GCT have continued using the VLA. Fig. 1 shows the latest image observed at λ 3.6 cm, on 19 January 1992, using the VLA. At a resolution of $\sim$ 0.5″, the GCT was not resolved. Fig. 2 shows the radio light curves at λ 20, 6, 3.6, 2 and 1.3 cm during 11 December 1990 to 3 April 1992. The radio light curves of the GCT are unique in several aspects. At 20 cm, a number of maximum and minimum values of flux density have been observed during this period, showing

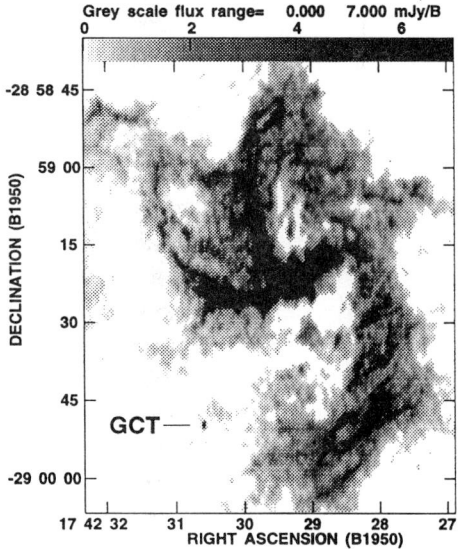

Fig. 1: A recent VLA image of Sgr A West, showing the GCT, observed at λ 3.6 cm on 19 January 1992 with a beam size (FWHM) = 1.0″ × 0.5″ ($-6.3°$).

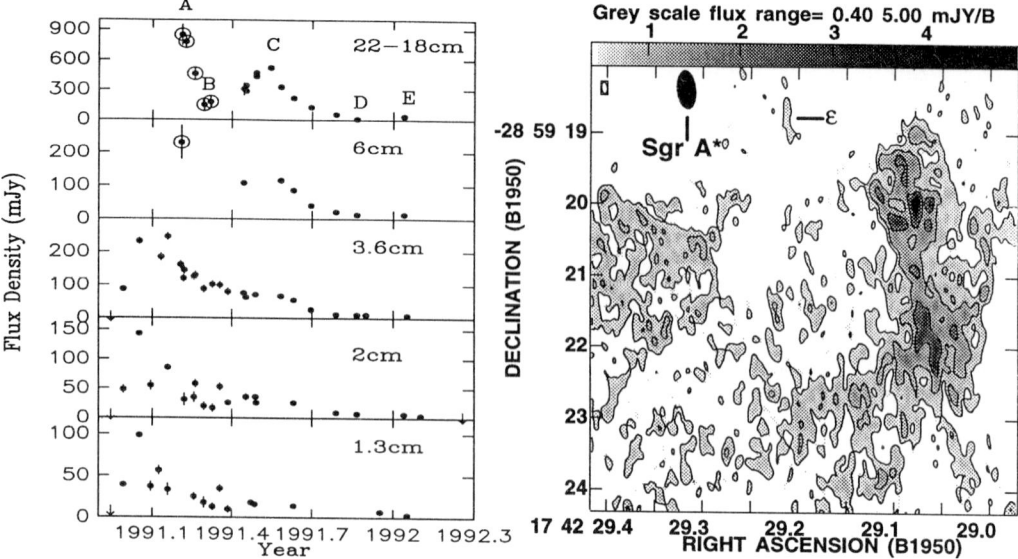

Fig. 2: Radio flux density observations of the GCT as a function of time at λ 1.3, 2, 3.6, 6 and 18-20 cm. The VLA measurements are indicated by dots. The circuled dots denote the AT observations.

Fig. 3: The VLA image of the immediate environs of Sgr A*, observed at λ 1.3 cm with a beam size (FWHM) = $0.15'' \times 0.08''$ (6.2°) indicated in the upper-left corner.

large amplitude variations with time scales of $t_{\rm FWHM} = 1\text{-}3$ months. These distinctive features are prominent at λ 20 cm and are less prominent as the wavelength decreases. The variations of the GCT at long wavelengths can also be recognized from its changing spectrum. During the maxima (March 1991, July 1991 and January 1992), steep spectra were observed with a typical spectral index $\alpha = -1$ between λ 20 and 1.3 cm. The spectrum flattened at λ > 4 cm during the minima in April 1991 and November 1991. This variability at long wavelengths is likely caused by varying optical depth. It is not clear if this optical depth effect is internal or external to the source.

The global evolution of the flux density in the decaying phase appears complex. No significant variations have been observed on a time scale of a day. The rapidly rising phase characterized by a steep spectrum ($S_\nu \propto \nu^{-0.5}$) is distinct from the synchrotron bubble model [HJ88]. During an initial decaying phase between late January 1991 and mid July 1991, the evolution of flux density can be described by $S_\nu \propto (t-t_0)^{-0.67}$ at wavelengths between 6 and 1.3 cm, with the reference time t_0 the end of 1990. A more rapid evolution of flux density can be fitted with $S_\nu \propto (t-t_0)^{-2.2}$ following this initial decay phase.

In addition, in agreement with the VLA results, a linear radio morphology of the GCT has been confirmed using MERLIN at 6 cm in June 1991 with a resolution of 50 mas × 150 mas (Pedlar, private communication).

4. The Environs of Sgr A*

The surroundings of Sgr A* have been observed using the VLA, obtaining images at λ 2 and 1.3 cm with a resolution of 0.1″ or 0.005 pc ([YME89], [YME90] & [ZGLE91]). The faint extended radio structures in the immediate vicinity of Sgr A* ($\alpha - \gamma$ in [YME90]) are not confirmed in the new observations [ZGLE91] at λ 2 and 1.3 cm and are likely low-level artifacts [ZGLE91]. The feature ϵ, slightly elongated and $\sim$ 1″ southwest of Sgr A*, is confirmed.

The details of the radio emission surrounding Sgr A* revealed in the earlier VLA images at λ 2 cm ([YME89], [YME90]), such as a small cavity of size $\sim$2.5″ southwest of Sgr A*, are also observed at λ 1.3 cm with a FWHM beam of 0.15″× 0.08″ (see Fig. 3). At this resolution, a few bright emission components along the western rim of the cavity are observed. The typical surface brightness temperature of these components is $\sim$1000 K. This brightness temperature gives an emission measure of 2×10^8 cm^{-6} pc (T$_e$ =10000 K). The typical angular size $\sim$ 0.2″(0.01 pc) of these components indicates a large electron density in this region (n$_e$ $\sim 10^5$ cm^{-3}).

The asymmetric distribution of the radiation intensity along the rim of the cavity suggest that the cavity is a result of an impact of a wind on the ionized gas in Sgr A West. The large excess of radio emission on the western edge of the cavity may indicate that the ionized gas has been compressed by the wind. A high temperature emitting gas bubble observed at λ 2.117 μm possibly arising from [Fe III] or [Fe XII] [EGKHWD92], coincident with the radio cavity, is also evidence for the strong interaction occurring in this region. The major problem discussed by several authors ([EGKHWD92], [WY92]) is the source (IRS 16 or Sgr A*) of the wind.

5. Time Variability of Sgr A*

The time variability of Sgr A* was investigated by Brown and Lo [BL82]. They found Sgr A* to vary significantly at λ 3.7 and 11 cm on time scales ranging from years to days. In particular, the flux density at λ 11 cm showed a steady secular increase. Long-term variations of the source at λ 6 and 20 cm were also observed from a collection of available flux density observations from 1974 to 1987 [ZEGLN89]. The time scale of the variation at λ 20, 11 and 6 cm scales as λ^2, suggesting that the radiation is substantially scattered by the ISM along the line of sight.

A preliminary result of the VLA observations of Sgr A* at λ 20, 6, 3.6, 2, and 1.3 cm during 1990/1991 was reported by Zhao et al. [ZGLE92]. Flux density variations of Sgr A* are prominent at the shorter wavelengths (see Fig. 4). At least three radio events can be identified during 1990/1991. The amplitude of the events appear to diminish with increasing wavelength. The typical length of the outbursts is about one month. Both the amplitude of the flux density variation ($\Delta S/2S \sim 50\%$) and the time scale ($t_{\rm FWHM} \sim 1$ month) at λ 1.3 cm are too large to be explained by scattering in the interstellar medium. The steepening of spectrum ($\alpha \sim 1$) during the events suggest that the large amplitude variations at the shorter wavelengths are caused by optically thick synchrotron bursts which are likely intrinsic to Sgr A*.

6. VLBI Images of Sgr A*

Using the partially completed VLBA in conjunction with the phased VLA, the Goldstone 70-m, the Haystack 36-m and NRAO 43-m telescope, Sgr A* has been imaged at λ 3.6 and 1.3 cm by Lo et al. [LO92]. Fig. 5 shows a VLBI image at λ 3.6 cm. At this wavelength the source is well represented by an elliptical gaussian model, with a major axis size (FWHM) of 17.5±0.5 mas and an axial ratio of 0.5±0.05. The position angle of the major axis is 87±5 degrees. At λ 1.3 cm, the source has a size (FWHM) of 2.5±0.3 mas. Both the elliptical gaussian brightness distribution and the angular size dependence of λ^2 agree with previous observations ([LO85], [JAUNCEY89] & [MARCAIDE92]). No compact substructure (< 10 mas) has been detected at a flux density limit of 4 mJy at 3.6 cm. This lack of small source structure supports the hypothesis that the radio radiation from Sgr A* is strongly scattered by the interstellar medium.

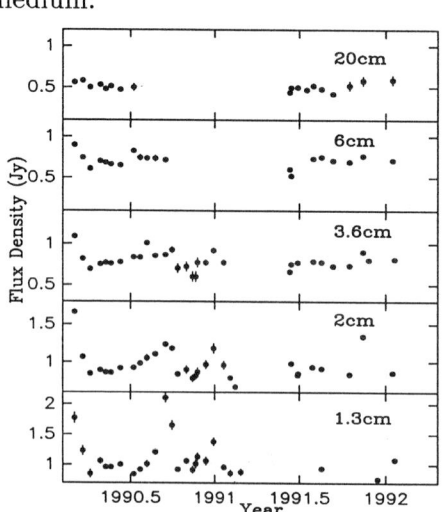

Fig. 4: Radio flux density observations of Sgr A* as a function of time at λ 1.3, 2, 3.6, 6 and 20 cm using the VLA.

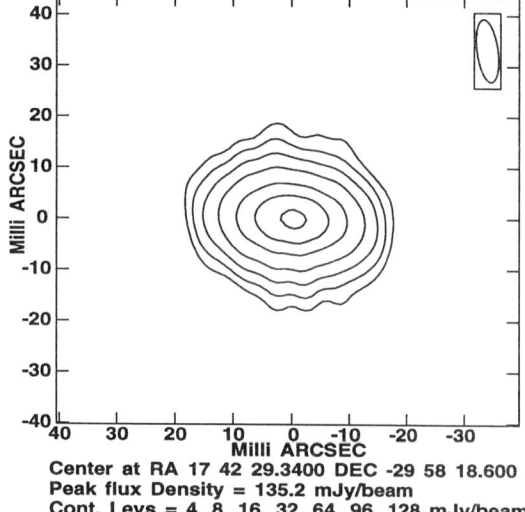

Fig. 5: VLBI image of Sgr A* at λ 3.6 cm with a beam size (FWHM) = 12.4mas × 3.6mas (6.6°) indicated in the upper-right corner.

References

[BB74] Balick, B., Brown, R. L., 1974, *Astrophy. J.*, **194**, *265*.

[BS87] Backer, D. C., Sramek, R. A., 1987, *The Galactic Center*, ed. D. C. Backer

(AIP, New York), p.163.

[BL82] Brown, R. L., Lo, K. Y., 1982, ApJ., **253**, 108.

[EGSG83] Ekers, R. D., et al., 1983, A.A., **122**, 143.

[EGKHWD92] Eckart, A., et al., 1992, Nature, **355**, 526.

[GSB85] van Gorkom, J. H., Schwarz, U. J., Bregman, J. D., 1985, *The Milky Way Galaxy (ed. von Woerden et al.)*, (Dordrecht), p.371.

[HJ88] Hjellming, R. M., Johnston, K. J., 1988, ApJ., **328**, 600.

[JAUNCEY89] Jauncey, D. L. et al., 1989, A. J., **98**, 44.

[LO92] Lo, K. Y. et al., 1992, Nature, submitted.

[LO89] Lo, K. Y., 1989, *The Center of The Galaxy (ed. Morris, M.)*, p.527.

[LO87] Lo, K. Y., *The Galactic Center*, ed. D. C. Backer (AIP New York), p.30.

[LO85] Lo, K. Y. et al., 1985, Nature, **315**, 124.

[LC83] Lo, K. Y., Claussen, M. J., 1983, Nature, **306**, 647.

[MARCAIDE92] Marcaide, J. M. et al., 1992, A. A. **258**, 295.

[RG93] Roberts, D. A., Goss, W. M., 1993, ApJ Suppl., in press.

[SBG89] Schwarz, U. J., Bregman, J. D., van Gorkom, J. H., 1989, A.A., **215**, 33.

[WY92] Wardle, M. and Yusef-Zadeh, F. 1992, Nature, **357**, 308.

[YM92] Yusef-Zadeh, F., Melia, F., 1991, Ap.J.(Lett.), **385**, L41.

[YM91] Yusef-Zadeh, F., Morris, M., 1991, Ap.J.(Lett.), **371**, L59.

[YME90] Yusef-Zadeh, F., Morris, M., Ekers, R. D., 1990, Nature, **348**, 45.

[YME89] Yusef-Zadeh, F., Morris, M., Ekers, R. D., 1989, *The Center of The Galaxy (ed. Morris, M.)*, p.443.

[ZRGFLSKEABS92] Zhao, J.-H., et al., 1992, Science, **255**, 1538.

[ZGLE92] Zhao, J.-H., Goss, W. M., Lo, K. Y., Ekers, R. D., 1992, *Relationships between AGNs and Starburst Galaxies (ed. Filippenko, A.) (ASP)*, p.295.

[ZGLE91] Zhao, J.-H., Goss, W. M., Lo, K. Y., Ekers, R. D., 1991, Nature, **354**, 46.

[ZRGFLE91] Zhao, J.-H., et al., 1991, IAU Cir. No. 5210.

[ZEGLN89] Zhao, J.-H., et al., 1989, *The Center of The Galaxy (ed. Morris, M.)*, p.535.

Manifestations of the Wind Phenomenon at the Galactic Center

Farhad Yusef-Zadeh *

Abstract

Recent study of the inner two pcs of the Galaxy suggest compelling morphological and, in some cases, kinematic evidence for the interaction of stellar winds from the IRS 16 cluster with i) the black hole candidate which is thought to be coincident with a nonthermal radio source, Sgr A*, ii) the mass-losing supergiant IRS 7, located within a light year of the Galactic center, and iii) the ionized gas associated with Sgr A West. Highlights of a range of arguments are given that seek to unify a number of observations of Galactic center components within a framework that employs wind phenomena. While the presence of a wind blowing from the Galactic center is not a new result, the idea that it manifests itself in different ways as it interacts dramatically with different components is new. It is the latter that must be taken seriously when interpreting the detailed morphology and kinematics of the Galactic center ISM.

The center of our Galaxy is known to be a complex and puzzling region consisting of numerous components whose roles and whose physical associations with each other have been a subject of much debate. In particular, the components in the inner couple of pcs of the dynamical center of the Galaxy have raised questions as to what extent they are responsible for outflow and/or infall activity in this region. The main idea that we intend to use to explain a number of Galactic center features is centered on two primary objects (see [YuW92] for more details). One is the cluster of mass-losing blue and luminous objects having outflow velocities $\approx$ 500-700 km/s and an inferred mass-loss of 10^{-3} $M_\odot$ yr^{-1} ([GKBW91]). The other is the compact radio source, Sgr A* ([Lo89]), which is generally *assumed* to have $\approx 10^6$ $M_\odot$. The centroids of these two objects lie to within 1" (0.04 pc) of each other. IRS 16C and Sgr A* lie on an east-west line with Sgr A* to west. It is quite natural to consider consequences of the physical interaction of the IRS 16 winds with the gravitational potential of the scattering center, Sgr A*. A portion of the matter could provide the energy to power the Galactic center. Recent work by [Mel91] has been particularly enlightening. He derives a reasonable theoretical fit to the spectrum of Sgr A* if it has $\approx 10^6$ $M_\odot$. An

---

*Department of Physics and Astronomy, Northwestern University, Evanston, Illinois. This author thanks Mark Wardle for useful discussions. This work was supported by NASA grant NAGW-2518.

implication of this idea is that an accretion disk around Sgr A* may not be needed, thereby explaining the lack of IR emission from Sgr A*.

What happens to the outflowing winds that are not captured by Sgr A*? The portion of the wind passing at a distance of 0.1pc with a velocity of 500 km/s from Sgr A* will be able to escape from the strong gravitational potential of Sgr A*. However, Sgr A* will be able to change the trajectory of the wind particles and act as a lens to focus the winds on the side opposite to IRS 16. This focusing mechanism will, in principle, allow the diffuse supersonic wind particles to collide with each other and form a denser and hotter gas to the west of Sgr A*. The dense gas will then escape from the gravitational potential of Sgr A* in the direction away from Sgr A* as well as IRS 16. [WaY92] have recently applied this idea to explain a number of blobs that are seen beyond 1" to the west of Sgr A*. Possible detection of four satellite sources within 1" of Sgr A* were reported by [YME90] but [ZGLE91] did not confirm these sources. On a smaller scale, however, both VLBI ([Lo89]) and VLBA measurements by Brown and Benson (private communication) have shown an east-west elongation of Sgr A* and weak features in the east-west direction. All these features in the immediate vicinity of Sgr A*, perhaps, indicate that the blobs are outflowing materials induced as a result of the interaction of the sources of winds and the source of concentrated mass. Future proper motion studies of these blobs would be valuable to test the above hypothesis.

An extension of the above idea may explain the origin of a hole and its 2.217μm emission ([Eck92]). Collision of fast moving blobs with the orbiting gas associated with the bar of Sgr A West ([WaY92]) produce a hole and a surrounding shell of hot shocked gas in the continuous flow of ionized gas. This hole is known as a "mini-cavity", and is characterized by high temperatures and a velocity structure which deviates from a circular rotation ([Eck92]; [RoG92]). If the 2.217μm emission is produced by FeXII at temperature of 10^6K, then the dissipation of the kinetic energy of the blobs every 25 yrs or so is quite consistent with the energy and continuity requirements to explain the luminosity of the hot gas. However, if the line is identified as FeIII, then a temperature of few times 10^4K is implied, and the formation of the high velocity ionized gas has to be continuous. Furthermore, the dissipation of the continuous flow of ionized gas will not explain the energetics involved unless uv radiation contributes ([Eck92]). The difference in accounting for FeXII and FeIII line emission arises from the fact that while the temperature needed to excite FeIII is lower by more than an order of magnitude, the cooling time of FeIII is faster than that of FeXII by more than a factor of 60 ([WaY92]).

While the outflowing gas from IRS 16 becomes collimated as it travels toward Sgr A*, it is oblivious to the strong gravitational potential of Sgr A* as the fast moving particles follow a linear trajectory and interact with other prominent components of the Galactic center. In particular, interaction with a mass-losing red supergiant, IRS 7, which lies only a light year in projection to the north of IRS 16C takes place when it is exposed to the 700 km/s wind. The interaction of the IRS 16 winds and the IRS 7 wind have manifested themselves beautifully by forming a bow shock structure in the atmosphere of the supergiant as well as forming a tail of ionized gas. The apex of the bow shock faces IRS 16C whereas the ionized tail is seen swept away from the supergiant in the direction away from the IRS 16 cluster ([SLA91]; [YuMo91]; [YuMe91]).

Another fascinating aspect of the interaction of the IRS 16 winds with nearby components involves the orbiting (or perhaps with infalling) ionized gas that surrounds the Galactic center. This interaction may explain the waviness noted along the Northern arm of Sgr A West ([YuW92]). This waviness is most prominent near IRS 16 cluster, where the ram pressure of the wind is expected to be strong enough to compress the gas, induce Raleigh-Taylor instability and produce the sinusoidal-shaped pattern. Another consequence of the compression of the gas and dust is enhancement of the magnetic field resulting in enhanced polarized dust emission, as has been observed by [AGSMR91].

References

[AGSMR91] Aitken, D.K., Gezari, D., Smith, C.H., McCaughrean, M. and Roche, P.F. 1991, *Ap.J.,* **419**, 419.

[Eck92] Eckart *et al.* 1992, *Nature,* **355**, 526.

[GKBW91] Geballe, T.R., Krisciunas, K., Bailey, J.A., Wade, R. 1991, *Ap.J.* **370**, L73.

[Lo89] Lo, K.Y. 1989, *IAU Symposium No. 136, The Center of the Galaxy*, ed. M. Morris, p527.

[Mel91] Melia, F. 1991, *Ap.J.,* **387**, L25.

[RoG92] Roberts, D. and Goss, M. 1992, *Ap.J.*, in press.

[SLA91] Serabyn, E., Lacy, J.H., and Achtermann, J.M. 1991, *Ap.J.,* **378**, 557.

[WaY92] Wardle, M. and Yusef-Zadeh, F. 1992, *Nature,* **357**, 308.

[YuMe91] Yusef-Zadeh, F., and Melia, F. 1991, *Ap.J..* **385**, L41.

[YuMo91] Yusef-Zadeh, F. and Morris, M., 1991, *Ap.J.,* **371**, L59.

[YME90] Yusef-Zadeh, F., Morris, M., and Ekers, R. 1990, *Nature,* **348**, 45.

[YuW92] Yusef-Zadeh, F. and Wardle, M. 1992, *Ap.J.*, in press.

[ZGLE91] Zhao, J.H., Goss, W.M., Lo, K.Y., and Ekers, R.D. 1991, *Nature,* **354**, 46.

Could 1E1740.7-2942 be Extragalactic ?

I.F. Mirabel [*] *L.F. Rodríguez* [†]
B. Cordier [*], *J. Paul* [*], *and F. Lebrun* [*]

Abstract

The double radio jet associated with the compact e^+e^- annihilator 1E1740.7-2942 may resemble, at first glance, a radio galaxy. However, when its morphology is considered together with its radio power and time variability, the properties of this radio source are not consistent with the usual properties of radio galaxies. Besides, the X-ray properties of 1E1740.7-2942 indicate that if this object were at a large distance beyond our Galaxy, it would be the most luminous hard X-ray AGN so far detected.

1. Introduction

The galactic coordinates of 1E1740.7-2942 are $l \sim -0.9°$, $b \sim -0.1°$. Since no optical or infrared counterpart has been found its distance is still unknown. The radio measurements to detect proper motions or flux variations in the lobes which would prove that it resides in our Galaxy may take years. Since the morphology of the radio jet is reminiscent of a radio galaxy, several experienced radio astronomers suggested at the Manchester meeting that it is an extragalactic object. Here we discuss this hypothesis.

The radio jet was reported in [1] and [2]. The core central source is inside the ROSAT error circle determined by Prince [3]. The integrated flux at $\lambda 6$ cm is ~ 2 mJy; its total angular size $\sim 1'$. Two important observational facts are the following: 1) the central, unresolved radio source varies in a correlated way with the hard X-ray photon counts from 1E1740.7-2942, and 2) the radio source does not consist only of three point sources aligned along a straight line, but of two resolved hot spots located at the end of extended (jet-like) emission that is aligned with the central source.

---

[*]Service d'Astrophysique. Centre d'Etudes de Saclay. 91191 Gif/Yvette. FRANCE.
[†]Instituto de Astronomía. UNAM, Apartado Postal 70-264, 04510 México, DF, México.

2. An unrelated background radio galaxy ?

Someone playing the role of Descartes' evil genius would certainly ignore the correlated variability between the X-ray source and the core radio source [1,2] to propose that the radio jet is an unrelated double-lobed radio galaxy that happens to be seen through the galactic plane. This is unlikely because:

1) From radio source counts the probability of a coincidental superposition of the compact radio source inside the ROSAT error box is $\leq 10^{-3}$.

2) The angular size of the radio source is extraordinarily large for its integrated flux. It is known since almost two decades [4] that among radio galaxies there is a definite relationship between morphology, size, and luminosity. Double-lobed radio galaxies are luminous sources with radio powers in the range of 10^{35-38} W; the typical distances between hot spots are in the range of 100-500 kpc. Double-lobed radio galaxies with fluxes ≤ 2 mJy and sizes $\geq 1'$ are a rather small fraction ($\leq 5\%$) of the extragalactic population of faint radio sources [5]. This brings the probability of coincidental superposition to $\leq 5\ 10^{-5}$.

3) To our knowledge, the variability by a factor of 5 that we observed [1,2] in the central component has never been seen in an extragalactic source. Strongly variable extragalactic sources appear to be core-dominated.

4) Most double-lobed radio galaxies show one-sided extended emission ("jet") instead of two-sided jets. In the handful of double-lobed sources where a weak counter-jet is seen, the flux of the lobe is ≥ 100 times the flux of the counter-jet.

5) Unlike the case of Scorpius X-1, where one of the outer lobes is unresolved and there are no visible jets connecting the hot spots to the central source [6], the lobes in the radio counterpart of 1E1740.7-2942 are resolved and we do see extended emission aligned with the central source.

6) Another triple radio source with similar properties has recently been found [7] associated to GRS1758-258, which is the second persistent hard X-ray source near the centre of the Milky Way. The probability of coincidental superposition of two weak extragalactic triple radio sources with the two persistent hard x-ray sources of the galactic centre region must be extremely low.

3. An extragalactic X-ray source ?

Since the variations between the core radio source and the X-ray source are correlated, we now ask if the X-ray source could be extragalactic.

The burst detected by SIGMA in the 300-600 keV energy range has been interpreted as 511 keV line emission from e^+e^- annihilation [8]. A Gaussian fit of this feature gives a centre at 410 keV [9]. If the $z \sim 0.20$ redshift were cosmological, the source would be at a distance of ~ 800 Mpc (for $H_o = 75$ km s^{-1} Mpc^{-1}). It is interesting that at such distance, the 1' angular size of the radio source would correspond to ~ 230 kpc, which is typical of the sizes of extragalactic double radio sources. 1E1740.7-2942 would then be an extraordinary source because:

1) Its **hard X-ray** (40-150 keV) luminosity would be $\sim 4\ 10^{47}$ erg s^{-1}. This is ~ 10 times the hard X-ray luminosity of 3C 273 and $\sim 10^4$ times that of Cen A. SIGMA has observed several nearby radio galaxies and quasars and no such **persistent** X-ray ultraluminous AGN's have been observed. 1E1740.7-2942 would be the most luminous hard X-ray source detected so far, with a rather unusual hard X-ray to radio luminosity ratio.

2) SIGMA has been observing numerous fields of $5^o \times 5^o$ outside the galactic plane and no seredipitous persistent extragalactic source with comparable photon counts has been found. Besides, from the results of the HEAO 1 full sky survey [10] we know that only 20% of the X-ray sources detected above 80 keV are extragalactic. In this context, it would be ironic if the most luminous hard X-ray AGN found so far happened to be opticaly hidden at $\leq 1^o$ from the galactic centre.

3) Since 1E1740.7-2942 was detected at soft X-rays by ROSAT [3] it is unlikely to be an extragalactic object on the galactic plane. The absorption at soft X-rays indicates a column density to the source of $\sim 1.5\ 10^{23}$ cm^{-2} [11]. However, observations of molecular transitions [12] indicate that the total column density of molecular gas through the whole Galaxy in that direction is $\sim 4\ 10^{23}$ cm^{-2}.

Acknowledgements: I.F. Mirabel thanks J. Ballet, A.S. Wilson, J. Dickey and colleages from Jodrell Bank for enlightening discussions.

References

[1] Mirabel, I.F., Rodríguez, L.F. et al. *Nature 358, (1992) 215-217.*
[2] Mirabel, I.F., Rodríguez, L.F. et al. *Astr. Astrophys. Supp. Ser. in press*
[3] Prince, T.A. et al. *Private communication to be submitted to ApJ (1992).*
[4] Faranoff, B.L., Riley, F.M. *MNRAS 167, (1974), 31p-35p*
[5] Fomalont, E.B., et al. *Astron. J. 102, (1991), 1258-1277*
[6] Fomalont, E.B., Geldzahler, B.J. *Astrophys. J. 383, (1991), 289-294*
[7] Rodríguez, L.F., Mirabel, I.F., and Martí, J. *Astrophys. J. in press*
[8] Bouchet, L. et al., *Astrophys. J. 383, (1991). L45-L48.*
[9] Sunyaev, R.A. et al. *Astrophys. J. 383, (1991). L49-L52*
[10] Levine, A.M. et al. *Astrophys. J. Supp. Ser. 54, (1984), 581-617*
[11] Kawai, N. et al. *Astrophys. J. 330, (1988), 130-141*
[12] Mirabel, I.F. et al. *Astron. Ap. 251, (1991), L43-46*

First 1.3cm VLBA Map of SgrA*

J. Marcaide *   A. Alberdi †   L. Lara †   P. Elósegui ‡
I. Shapiro ‡   W. Cotton §   J. Romney §   R. Preston ¶

Abstract

1.3cm VLBI observations of SgrA* (5 VLBA antennas plus phased-VLA) have allowed us to make a low resolution hybrid map and to model the source as an elliptical gaussian of major axis 2.6±0.2 mas at position angle 75°±10°, with axis ratio 0.5±0.2.

On 23 June, 1991 we observed SgrA* at 1.3cm and LCP with a VLBI array of 13 telescopes, 9 in the U.S.A., and 4 in Europe. Various technical problems at 2 American and 2 European stations, coupled with the relatively large size of SgrA*, resulted, for mapping purposes, in an effective array of 6 antennas (in parentheses are the symbols for the antennas used hereafter): phased-VLA (Y) and 5 VLBA antennas: Pie Town (P), Kitt Peak (T), Los Alamos (X), OVRO (O), and Fort Davis (F). Of these, the OVRO station suffered severe pointing problems. This array did not provide the resolution needed to study SgrA* in detail. We also obtained 3 (marginal) detections for the Effelsberg-Madrid baseline which will not be further considered here.

To try to ease the always severe calibration problems in observations of SgrA*, we observed the maser 1744-280 for 1 min after each 12 min scan of SgrA*. We recorded all data in a manner compatible with MkIII mode B and we correlated them at the Max Planck Institut für Radioastronomie, Bonn. Initially, we calibrated the data using measured system temperatures and known antenna sensitivities. Starting from a point-source model we autocalibrated the data iteratively using the program AMPHI of the Caltech Package, and a sky window of 4mas×4mas (and 6mas×6mas in the final iteration) centered in the source for CLEAN-ing, to produce the hybrid map shown in Figure 1. The residual map contains no significant structure. Recurrent technical (read-out) problems have prevented us, so far, from using the maser information for calibration purposes.

---

*Dpto. de Matemática Aplicada y Astronomía, Universitat de València, 46100 Burjassot, Spain.
†Instituto de Astrofísica de Andalucía, Apdo. 3004, 18080 Granada, Spain.
‡Center for Astrophysics, 60 Garden St., Cambridge, MA 02138, U.S.A.
§N.R.A.O., Socorro, NM, U.S.A.
¶Jet Propulsion Laboratory, Pasadena, CA, U.S.A.

To obtain more quantitative information about SgrA*, we also fit an elliptical gaussian model to the autocalibrated data using the program MODELFIT. We found that an elliptical gaussian with flux density 1.1±0.1 Jy, major axis 2.6±0.2, axis ratio 0.5±0.2 and position angle 75°±10° fits the data well. We show in Figure 2 a representative sample of correlated amplitudes and closure phases corresponding to two antenna loops; the continuous lines are the predictions from the elliptical gaussian model. The orientation of its major axis is consistent with that found by Jauncey et al. (1989) at 3.6cm (82°±6°) and Lo et al. (1985) at 1.3cm (87°±30°).

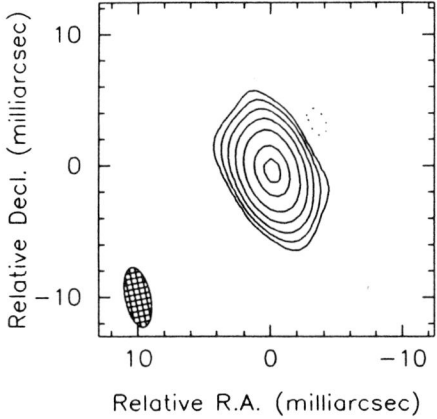

Figure 1: Hybrid map of SgrA* at 1.3cm. Lowest contour is 2% of peak. The restoring beam is shown at the lower left.

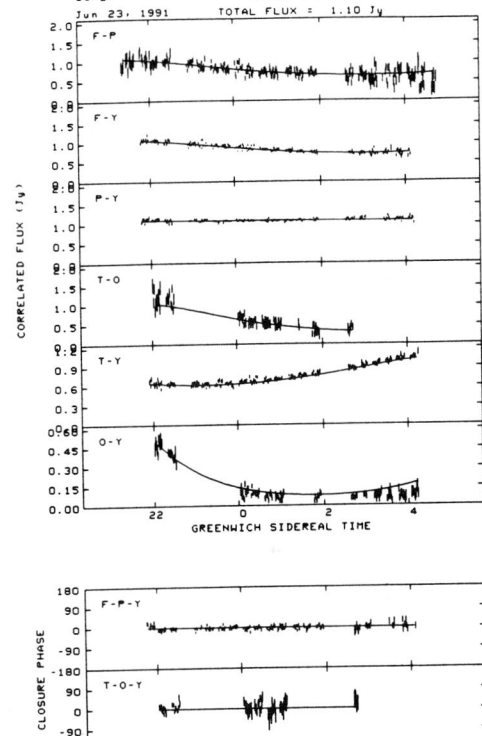

Figure 2: Correlated amplitudes and closure phases of SgrA* at 1.3cm. The continuous lines correspond to the gaussian model given in the text.

References

[1] Jauncey D.L., Tzioumis A.K., Preston R.A., et al., *A.J.*, 98 (1989), p44.

[2] Lo K.Y., Backer D.C., Ekers R.D., et al., *Nature*, 315 (1985), p124.

Sub-arcsecond Observations of Maser Stars

P. F. Bowers *

Abstract

A review is given of recent results pertaining to interferometric observations of SiO, H_2O, and OH masers found in the envelopes of evolved, mass-losing stars. For most AGB OH/IR stars, the brightness distributions of OH masers indicate that at large radii ($R > 10^3$ AU) the gas is distributed in virtually all directions over a timescale of 10^3 yr. To a first-order approximation, the data can be modeled by a spherical shell of gas with a constant outflow velocity. Evidence of axially symmetric structure is common for the large OH shells of massive supergiants and also is seen for post-AGB stars. At smaller radii, data become more complex. For classical Mira variables, OH masers ($50 < R < 500$ AU) and H_2O masers ($15 < R < 50$ AU) exhibit indications of outflow, but there are significant departures from the standard expanding shell model; SiO ($R < 10$ AU) and H_2O distributions can change strongly with time.

1. Introduction

In the course of evolution from the main sequence phase to white dwarfs, stars with masses < 8 $M_\odot$ pass through the asymptotic giant branch of the HR diagram, developing large ($R_\star \approx$ 1–3 AU), cool (2000–3000 K), convective photospheres and producing extensive envelopes of dust and molecular gas which eventually dissipate into the interstellar medium. Neither the the mass loss mechanism nor the structure of the outflow is well understood, but once dust grains form, radiation pressure from the luminous star is likely to play a critical role for accelerating the dust which, in turn, accelerates the gas to a typical terminal velocity of 10–20 km s^{-1}. If the gas is oxygen-rich (O > C) and if the rate of mass loss exceeds about 10^{-7} $M_\odot$ yr^{-1}, there is a high probability of maser emission from various transitions of SiO, H_2O, and OH. The majority of maser stars are AGB (Mira or OH/IR) stars, but masers also are found from some protoplanetary (PPN) and planetary (PN) nebulae and from a few massive supergiants.

---

*SFA, Inc., 1401 McCormick Dr., Landover, MD 20785, USA. Work performed at the Center for Advanced Space Sensing, Naval Research Laboratory, Washington, DC, under contract # N00014-89-C-2398.

Interferometric observations of circumstellar masers reveal information about the properties of the masers and of the envelope structure. Such data can be used to establish constraints on models of the mass loss mechanism, to determine the interaction of the outflowing material with the ambient medium, and to explore the evolutionary connection between AGB stars, PPN's, and PN's. However, the interpretive link between the observed brightness distribution and the envelope structure is difficult to establish uniquely.

Maser data typically are interpreted in terms of an expanding shell model (e.g., Chapman & Cohen 1985). The outflowing gas (expansion velocity V_e) is assumed to be uniformly distributed in a thin shell. Amplification of the emission occurs as along as the line-of-sight velocity gradient is less than the line width. If the shell is spherical, if V_e is constant throughout the shell, and if the shell is not extremely thin (Bowers 1992a), largest amplification occurs along the line of sight to the star. For a stellar radial velocity $V_\star$, the maser profile is then doubly peaked at velocities $V = V_\star \pm V_e$ with the peaks being caps of emission coincident on the plane of the sky at the position of the star. At intermediate velocities the angular radius θ at velocity V is related to the shell radius θ_s by

$$\theta = \theta_s \{1 - [(V - V_\star)/V_e]^2\}^{1/2}. \qquad (1)$$

Equation (1) also is valid for a highly flattened distribution (i.e., a ring), but in this case the maximum difference between V and $V_\star$ is $V_e \cos i$, where i is the inclination of the plane of the ring to the line of sight. The expanding shell model is most applicable at a large distance from the star, but departures from it are seen in all maser shells (e.g., clumpy brightness distributions, asymmetries). In this review I briefly discuss the status of high-resolution studies.

2. SiO

SiO maser emission has been detected from various rotational transitions of the vibrationally excited ($v = 1$–4) states. There is a great deal of literature on the properties of the masers derived from single-dish studies (Bowers 1992a), but interferometric data are sparse. The best studied cases are the Mira R Cas and the supergiant VX Sgr (Lane 1984; McIntosh et al. 1989). For the ($v = 1, J = 1 \to 0$) emission, the size of the maser region is about 50–70 mas for each star ($R \geq 7$ AU for R Cas; ≥ 50 AU for VX Sgr), so the masers are distributed out to a few $R_\star$. The FWHM sizes of the smallest components are about 5–10 mas, giving apparent linear spot sizes of about 1–3 AU for R Cas and 10–20 AU for VX Sgr, $T_B = 0.5$–2×10^{10} K, and probable saturation of the masers. However, the result is model-dependent; Alcock & Ross (1986) suggest that the sizes are about ten times smaller and that the density distribution must be highly clumped. Colomer et al. (1992) subsequently have detected components as small as 0.4 mas (0.1 AU) in circumstellar shells. For R Cas, VLBI data indicate that some regions of the shell have well defined polarization properties, consistent with single-dish measurements (McIntosh et al. 1989). These authors suggest that temporal variations of separated components may place a lower limit of 10^3 km s^{-1} on the speed of the pump mechanism, suggestive of radiative pumping but not conclusive. They also find the angular distribution of components to be considerably different than that obtained six years earlier. For VX Sgr, Lane finds equal sizes for the $(v = 1, J = 1 \to 0)$ and $(v = 2, J = 1 \to 0)$ distributions.

Available data suggest a complex, possibly chaotic velocity field and a random distribution of clumps, but data for more stars with much better sensitivity and time sampling are needed to confirm this; systematic effects may be present. Hall et al. (1990) find comparable position angles of some polarized features in SiO profiles obtained in sequential light cycles of R Aqr, suggesting that some features of the distribution may persist over a time scale of at least 1 yr and that the polarization mechanism may be driven by the stellar magnetic field; spatial polarization maps such as those of McIntosh et al. might conceivably trace the structure of the field. Bowers (1992b) suggests that at least some of the SiO components may have been accelerated nearly to the terminal outflow velocity, in which case such components may be very close to the stellar position. Observations of the $(v = 1, J = 2 \to 1)$ SiO masers from Orion-IRc2 (a presumably stellar but not necessarily evolved source) reveal an impressive symmetry between the angular distributions of the blue- and redshifted components (Plambeck et al. 1990), providing a very accurate estimate of the stellar position relative to the masers as well as possible insight into the velocity field (also see Bowers 1991). The $(v = 1, J = 1 \to 0)$ distribution published by Greenhill et al. (1988) is similar over some velocity intervals. Even if systematic effects generally are not found for evolved stars, the small shell sizes ($\leq 0''.15$) enable accurate determination of stellar positions (Wright et al. 1990).

3. H$_2$O

Interferometric data for the 22 GHz ($6_{16} \to 5_{23}$) ground-state, rotational transition have been obtained from about two dozen stars and reveal spot sizes of about 1 mas distributed over regions $< 0''.5$. Apparent spot sizes are about 0.3–1 AU for Miras and a few AU for supergiants (e.g., Spencer et al. 1979; Diamond et al. 1987), and shell radii range from about 15–50 AU for Miras and 100–400 AU for supergiants (Lane et al. 1987). It is estimated that the masers are highly beamed and not strongly saturated (e.g., Reid & Menten 1990).

Asymmetric, time-varying angular distributions and strong temporal changes of the shapes and velocity ranges of the profiles make interpretation of H$_2$O data very difficult, but systematic effects which are consistent with a component of outflow have been found for several stars. Chapman & Cohen (1986) and Diamond et al. (1987) respectively interpret the H$_2$O masers for the supergiants VX Sgr and S Per to lie in accelerating, thick-shell regions; for both sources there are indications of axial symmetry. Yates (1992) finds evidence of a disk plus polar jet structure for the H_2O maser distribution of the supergiant VY CMa. Ringlike structures near the stellar velocity and roughly centered about the estimated stellar position are apparent for IK Tau (Lane et al. 1987) and W Hya (Reid & Menten 1990). The shell structure has remained relatively stable for IK Tau over a one-year interval but has changed radically for W Hya over a five-year interval (Bowers et al. 1992). Engels et al. (1992) find the H_2O maps for RX Boo to be almost identical over a five-month interval and suggest that the masers are confined to an accelerating part of the outflow at the epoch of their observations. Bowers et al. (1992) suggest that evidence for a component of outflow and comparable velocity ranges of H_2O and OH indicate that some parcels of gas have been accelerated nearly to the terminal outflow velocity in the H_2O shells of RT Vir and IK Tau. They also find the velocity range of H_2O masers from VX Sgr to be about twice that found by Chapman & Cohen (1986)—a result relevant to models accounting for acceleration of the outflow (Netzer 1989; Pijpers 1990). Stellar positions estimated from H_2O masers and from optical measurements agree typically within 0".3 (de Vegt et al. 1987; Bowers et al. 1992), comparable to the sizes of the shells.

4. OH

Interferometric observations of 1612 MHz OH masers have been obtained for several dozen stars, and maps of main line (1665/1667 MHz) emission have been obtained for about a half-dozen stars. The radius of the maser shell depends on both the mass loss rate and UV radiation field (e.g., Netzer & Knapp 1987) and ranges from about 50–500 AU for Miras and 10^3–10^4 AU for AGB OH/IR stars, PPN's, PN's, and supergiants. For Miras, emission distributions are complex and do not conform to a simple model in which the OH is uniformly distributed in a constant, isotropic outflow. Expansion appears to be a significant or dominant kinematical component, but strong asymmetries (or possible indications of axial symmetry) are present (Bowers et al. 1989; Chapman et al. 1991; Collison & Fix 1992).

The most thoroughly studied case is U Ori [$R(OH) \approx 60$ AU]. At 1665 and 1667 MHz, Chapman et al. (1991) find clumpy structures with a characteristic size of about 25 AU and a filamentary structure which appears to project over radial distances of 45 to 140 AU. Proper motions of the masers indicate a transverse expansion velocity comparable to that measured along the line of sight. They and others (e.g., Bowers 1991) conclude that an axially symmetric distribution is indicated, but clumpiness and strong asymmetries complicate comparison with simple models.

A comparison of VLA, MERLIN, and VLBI data at 1612 MHz for two OH/IR stars with large shells (OH127.8-0.0 and OH26.5+0.6) has been made by Bowers & Johnston (1990). The sensitive VLA data indicate relatively smooth regions (1"–2") interspersed with unresolved (< 1") clumps. MERLIN data show the clumps to consist of numerous components with sizes of 100–300 mas (few $\times 10^2$ AU). VLA and MERLIN data combined in the u-v plane for OH127.8 reveal fascinating, fine-scale structure throughout the shell (Migenes et al. 1991). For a few sources, including OH127.8, VLBI data indicate that the most compact component (≤ 40 mas) occurs at the velocity of the blue-shifted peak feature, perhaps indicating amplification of the stellar continuum (Norris et al. 1984; Sivagnanam et al. 1990; Nedoluha & Bowers 1992). Spot sizes for OH/IR stars toward the galactic center are much larger than normal, consistent with interstellar scattering (Van Langevelde et al. 1992).

For many AGB OH/IR stars, a small angular separation of peak features ($< 0.1 - 0.2\theta_s$), an increasing shell radius as V approaches $V_\star$, and partial or complete ringlike structures seen in the brightness distributions indicate that the OH is not closely confined to a plane and that the data can be modeled, to first order, by a roughly spherical shell of gas with constant V_e. (For comparison, a tilted, expanding ring would produce an angular separation of peak features $> 0.2\theta_s$ if $i > 6°$.) Thus the gas is distributed in most or all directions over a timescale of about 10^3 yr. OH127.8 provides one of the best known examples; the shell radius is constant within 10% (rms) as a function of direction from the star. However, the 1612 MHz brightness distribution is asymmetric at most velocities with indications of a large-scale asymmetry between the eastern and western portions of the shell.

Clear evidence for axial symmetry and a strong bipolar outflow is detected in some OH shells, the best examples being OH231.8+4.2 (Morris et al. 1982), OH19.2-1.0 (Chapman 1988), and HD101584 (te Lintel Hekkert et al. 1992). Other cases of interest are W43A (Diamond & Nyman 1988; Likkel et al. 1992) and the PN's NGC 6302 (Payne et al. 1988) and M1-92 (Seaquist et al. 1991). OH also has been mapped from the PN's Vy2-2 and OH0.9+1.3, but there are no clear indications of axial symmetry (Shepherd et al. 1990). Axial symmetry (but not necessarily a strong bipolar outflow) may be present in the OH shells of the supergiants VY CMa (Bowers et al. 1983), VX Sgr (Chapman & Cohen 1986), S Per (Bowers et al. 1989), IRC+10420 (Nedoluha & Bowers 1992), and the OH/IR star OH16.1-0.3 (Welty et al. 1987). A measure of the strength of the anisotropy in the outflow is the ratio of the velocity along the polar axis to that in the equatorial plane, but this is obtained only by modeling. Another measure may be the ratio of the velocity ranges of OH and CO (te Lintel Hekkert et al. 1992). A ratio < 1 could be caused by acceleration of the outflow, as they suggest, or by different latitudinal distributions of the OH and CO. In a strong bipolar outflow CO may be widely distributed, tracing the entire velocity field, while conditions suitable for OH masers (abundance, pumping, organized outflow) may occur preferentially in regions closer to the equatorial plane (see model for OH231.8 by Bowers 1991).

In conclusion, circumstellar masers trace envelope structures which range from nearly spherical, isotropic outflows to strong bipolar outflows. However, even for the simplest cases (AGB OH/IR stars), asymmetries are present not only in the maser distributions but also in the dust distributions (Jones & Gehrz 1990)—the latter result clearly indicating clumpy and/or aspherical mass distributions. In addition to complicating the interpretation of maser data, anisotropies in the density or velocity structure influence estimates of fundamental parameters (stellar position, V_*, V_e). Attempts to develop more realistic models have been published (cf. Bowers 1992a), but complementary high-resolution observations of masers, thermal lines, and dust distributions are needed to constrain the models and achieve a unique description of individual envelopes.

5. References

Alcock, C., & Ross, R.R. 1986, ApJ, 310, 838

Bowers, P.F. 1991, ApJS, 76, 1099

Bowers, P.F. 1992a, in *Astrophysical Masers*, eds.: A.W. Clegg and G. E. Nedoluha, (NY: Springer-Verlag), in press

Bowers, P.F. 1992b, ApJ, 390, L27

Bowers, P.F., Claussen, M.J., & Johnston, K.J. 1992, AJ, in press

Bowers, P.F., & Johnston, K.J. 1990, ApJ, 354, 676

Bowers, P.F., Johnston, K.J., & de Vegt, C. 1989, ApJ, 340, 479

Bowers, P.F., Johnston, K.J., & Spencer, J.H. 1983, ApJ, 274, 733

Chapman, J.M. 1988, MNRAS, 230, 415

Chapman, J.M., & Cohen, R.J. 1985, MNRAS, 212, 375

Chapman, J.M., & Cohen, R.J. 1986, MNRAS, 220, 513

Chapman, J.M., Cohen, R.J., & Saikia, D.J. 1991, MNRAS, 249, 227

Collison, A.J., & Fix, J.D. 1992, ApJ, 390, 191

Colomer, F., Graham, D., Kirchbaum, T.P., et al. 1992, A&A, 254, L17

de Vegt, C., Kleine, T., Johnston, K.J., et al. 1987, A&A, 179, 322

Diamond, P.J., Johnston, K.J., Chapman, J.M., et al. 1987, A&A, 174, 95

Diamond, P.J., & Nyman, L. Å. 1988, in *The Impact of VLBI on Astrophysics and Geophysics*, eds.: M.J. Reid and J.M. Moran (Dordrecht: Kluwer), 249

Engels, D., Winnberg, A., Brand, J., et al. 1992, in *Astrophysical Masers*, eds.: A.W. Clegg and G.E. Nedoluha, (NY: Springer-Verlag), in press

Greenhill, L., Moran, J.M., et al. 1988, in *The Impact of VLBI on Astrophysics and Geophysics*, eds.: M.J. Reid and J.M. Moran (Dordrecht: Kluwer), 253

Hall, P.J., Allen, D.A., Troup, E.R., et al. 1990, MNRAS, 243, 480

Jones, T.J., & Gehrz, R.D. 1990, AJ, 100, 274

Lane, A.P. 1984, in *VLBI and Compact Sources*, ed.: R. Fanti (Dordrecht: Reidel), 329

Lane, A.P., Johnston, K.J., Bowers, P.F., et al. 1987, ApJ, 323, 756

Likkel, L., Morris, M., & Maddalena, R.J. 1992, A&A, 256, 581

te Lintel Hekkert, P., Chapman, J.M., & Zijlstra, A.A. 1992, ApJ, 390, L23

McIntosh, G.C., Predmore, C.R., Moran, J.M., et al. 1989, ApJ, 337, 934

Migenes, V., Cohen, R.J., & Bowers, P.F. 1991, BAAS, 23, 825

Morris, M., Bowers, P.F., & Turner, B.E. 1982, ApJ, 259, 625

Nedoluha, G.E., & Bowers, P.F. 1992, ApJ, 392, 249

Netzer, N. 1989, ApJ, 342, 1068

Netzer, N., & Knapp, G.R. 1987, ApJ, 323, 734

Norris, R.P., Booth, R.S., Diamond, P.J., et al. 1984, MNRAS, 208, 435

Payne, H.E., Phillips, J.A., & Terzian, Y. 1988, ApJ, 326, 368

Pijpers, F.P. 1990, A&A, 238, 256

Plambeck, R.L., Wright, M.C.H., & Carlstrom, J.E. 1990, ApJ, 348, L65

Reid, M.J., & Menten, K.M. 1990, ApJ, 360, L51

Seaquist, E.R., Plume, R., & Davis, L.E. 1991, ApJ, 367, 200

Shepherd, M.C., Cohen, R.J., Gaylard, M.J., et al. 1990, Nat, 344, 522

Sivagnanam, P., Diamond, P.J., Le Squeren, A.M., et al. 1990, A&A, 229, 171

Spencer, J.H., Johnston, K.J., Moran, J.M., et al. 1979, ApJ, 230, 449

Van Langevelde, H.J., Frail, D.A., Cordes, J.M., et al. 1992, ApJ, 396, 686

Welty, A.D., Fix, J.D., & Mutel, R.L. 1987, ApJ, 318, 852

Wright, M.C.H., Carlstrom, J.E., Plambeck, R.L., et al. 1990, AJ, 99, 1299

Yates, J.A. 1992, in *Astrophysical Masers*, eds.: A.W. Clegg and G.E. Nedoluha, (NY: Springer-Verlag), in press

VLBI observations of SiO masers in Europe

Francisco Colomer *†

Abstract

Very long baseline interferometric observations of the SiO maser emission in evolved stars and in the star-forming region Orion A have been performed in Europe since 1990. We find evidence for the existence of complex structures at milliarcsecond scales in the form of multiple maser spots distributed in the region. This is important for the understanding of the physical conditions close to the stellar photosphere and the SiO pumping mechanism. Repeated observations at several epochs should allow the modelling of the circumstellar envelope. We also discuss the suitability of the circumstellar SiO masers as a link between the optical and radio reference frames as they are bright radio sources and their parent stars are sometimes optically bright.

1. Introduction

We performed a spectral line VLBI experiment on the $v = 1, J = 1 \to 0$ line of SiO at 43 GHz towards selected galactic late-type stars and Orion in June 1990. With 15 hours of VLBI observations using three stations (Effelsberg (MPIfR, Germany) - Onsala (OSO, Sweden) - Yebes (CAY, Spain)), we detected 11 out of 19 late-type stars and Orion. Since the baseline range was 330-1740 km, the observations already indicated that the detected maser spots were much more compact than previously expected. The maser emission appeared in distinct regions separated by several tens of milliarcseconds. We found typical (Gaussian) sizes for the maser spots of ~ 1.0 mas. The results of this test are published and further details can be found in [CGK91] and [CGK92].

The experiment of June 1990 was intended as a search for detectable sources. A second experiment (first epoch) was performed on April 9 to 11, 1991, using the same VLBI setup. We observed 10 stars (the best detections in June 1990 and also R Aqr) and Orion using the same VLBI-setup for a longer time (48 hours) in order to obtain visibilities with much better uv-coverage. The data shows that the SiO

---

*Onsala Space Observatory, S - 439 92 Onsala, Sweden
†Now at Harvard-Smithsonian Center for Astrophysics, MS-42, 60 Garden Street, Cambridge, MA 02138, USA.

Table 1. List of collaborators

Institute	Collaborators
Onsala Space Observatory	F.Colomer†, B.O.Rönnäng, R.S.Booth
Max-Planck-Institut für Radioastronomie	D.A.Graham, T.P.Krichbaum, A. Witzel
Centro Astronómico de Yebes	P.de Vicente, A.Barcia, J.E.Garrido
	J.Gómez-González, J.Alcolea†
Observatoire de Bordeaux	A.Baudry, G.Daigne, N.Brouillet
IRAM at Pico Veleta / Grenoble	A.Greve / C.Schalinski, R.Neri
ISTS/SGL of Canada	W.Cannon, P.Leone, R.Wietfeldt

maser sources have complex structures at milliarcsecond scales and that most of the observed objects can be mapped. A preliminary spot map of the supergiant μ Cep can has been published [Col92]. A synthesys map of this source has been produced with *AIPS*, and will be part of a paper in preparation. Absolute positions of the SiO maser spots have not been obtained, as very few extragalactic calibration sources were bright enough to be detected with our system.

A new set of observations (second epoch) has been perfomed during May 27 to 29, 1992. The 30-m telescope of IRAM on Pico Veleta (Spain) also participated. We obtained a bandwidth of 4 MHz by using the MK III-A VLBI system at Onsala and Effelsberg, a VLBA terminal at Pico Veleta, and a S-2 system at Yebes (for the first time in this kind of observations). The correlation of these data will be produced soon at the MK III correlator of the MPI in Bonn. The institutions and scientists involved are listed in Table 1.

2. Future prospects

Our observations show that the SiO maser emission in the circumstellar envelopes of late-type stars has complex compact structures at milliarcsecond scales and can be mapped with VLBI. With better sensitivity, more extragalactic objects can be detected. Their observation is important for calibrating the system in order to reduce the systematic errors that dominate the individual maps obtained from our few, widely separated experiments. This might allow as well the determination of the absolute positions of the SiO maser spots by phase-referencing to these extragalactic objects. We need good u-v coverage to model the relative spatial distribution and proper motions of the maser spots with multi-epoch VLBI monitoring. We have proposed the participation of other telescopes, once front-end receivers are available. The inclusion of Metsähovi (Finland), Cambridge (U.K.), Medicina (Italy), Torun (Poland) or Crimea (Russia) would result in a very good array in Europe for the study of these masers.

Regarding absolute position measurements, it has been recognized for some time that the stellar masers may be suitable as a link between the extragalactic radio reference frame (EGRF) and the galactic optical reference frame because they are bright radio sources and their parent stars are sometimes optically bright (e.g. [BMR84]; [Rön89]). The stellar positions will be measured with the Hipparcos satellite and the radio positions can be linked to those of the quasars by VLBI phase-referencing. The still difficult step in this application is to estimate the position of the central star from the observed distribution and kinematics of the SiO maser emission, as it strongly depends on the actual distribution of the maser spots. We need good maps of the SiO emission in order to study if we can locate the star with reasonable accuracy.

Acknowledgements. We are very grateful to the staffs operating the Effelsberg, Onsala, Yebes, and Pico Veleta observatories.

References

[BMR84] Baudry A.,Mazurier J.M.,Requième Y., 1984, *A new way of tying together the Hipparcos frame to the VLBI frame: Astrometric observations of stellar maser sources*. Proceedings of the IAU Symposium No. 110. VLBI and Compact Radio Sources, Reidel, Dordrecht, pp.355

[CGK91] Colomer F., Graham D.A., Krichbaum T.P., et al., 1991, *Astrometry of SiO masers*. Radio Interferometry. Theory, Techniques and Applications, IAU Colloquium No. 131, ASP Conference Series (Vol. 19), pp. 338

[CGK92] Colomer F., Graham D.A., Krichbaum T.P., et al., 1992, A&A 254, L17

[Col92] Colomer F., 1992, *Recent VLBI observations of SiO masers*. Astronomical masers, Lecture Note Series, Springer Verlag. (in press)

[Rön89] Rönnäng B.O., 1989, *Galactic masers as a tool to relate the Optical and Radio Celestial Reference Frames*. Proceedings of the 7th Working Meeting on European VLBI for Geodesy and Astrometry, Instituto de Astronomía y Geodesia, Madrid, pp.113

Note: While this paper was in press, we found an error in the program used to produce the preliminary fringe-rate map of μ Cep shown in Colomer (1993). The correct coordinate scale is 20 times smaller than the one shown. We have produced a synthesis map of this source. It shows clearly the existence of two SiO spots, separated $\sim$6mas. We cannot distinguish such structure in the corrected fringe-rate map; with the errors obtained, the source appears as a unique spot.

DISTANCE MEASUREMENTS THROUGH OBSERVATIONS OF MASERS

J. M. Moran [*]

Abstract

The two principal methods of directly determining the distances to cosmic masers are described and the results summarized. In the first technique the angular size of the shell of OH maser emission around a late type star determined by interferometry is compared with the linear size of the shell derived from the time delay of emission between the front and back of the shell. The derived distances can be compared with those of the galactic rotation model to estimate R_0. A new analysis of all published data for 10 stars at low galactic latitude gives $R_0=8.8\pm0.9$ kpc. The reduce chi-square in this analysis is unity if the random component of stellar velocity along the line of sight is 17 km s^{-1} (1σ). In the second method, the proper motions of H_2O masers in the envelopes of newly formed stars are measured with VLBI and compared statistically with the linear velocities derived from Doppler shifts. The distances for 7 masers have been measured in this way and the value derived for R_0 is 7.8 ± 0.6 kpc. This result relies heavily on the measurements of masers in Sgr B2 and W49N.

1. Phase Lag Distances

The OH emission at 1612 MHz from late type stars (or OH/IR stars) has a very distinctive double peaked profile. It is well established that these peaks originate in outflowing gas that resides in the front most and rear most parts of a shell that is near the terminal wind speed. The mean velocity of the profile defines the stellar velocity. The radii of these shells are 10^{16}–10^{17} cm or $0\rlap{.}''8$ to $8''$ at a distance of 1 kpc. The light crossing time is 8 to 80 days, and the stellar periods are typically 100–1000 days. The masers are presumably saturated and their temporal intensity variations follow the stellar variations. The ratio of the linear radius derived from the temporal phase lag and the angular size determined from interferometry gives an estimate of the maser's distance. The best estimate of the linear radius comes from an analysis of the time delay or phase lag over the entire spectral profile since

---
[*]Harvard–Smithsonian Center for Astrophysics, 60 Garden Street, Cambridge, MA 02138, USA

the delay is proportional to the line of sight velocity (with respect to the stellar velocity) V_r, that is,

$$\tau = \frac{R}{c}\frac{V_r}{V_e} \, , \qquad (1)$$

where R is the shell radius, c is the speed of light and V_e is the expansion velocity. Similarly, the angular radius of the shell, θ_0, is derived from the radius θ at velocity V_r using all spectral velocities,

$$\theta = \theta_0 [1 - \left(\frac{V_r}{V_e}\right)^2] \, . \qquad (2)$$

This analysis is essential because the emission from the limbs of the shell where $V_r = 0$ is usually very weak. The distance estimate is R/θ_0. As part of his PhD thesis, van Langevelde [Lan 92] made extensive measurements of phase lags and reanalyzed published data of phase lags and angular shell sizes. His results are summarized in Table 1 [LHS 90]. There are several problems with this technique. It assumes spherical symmetry. A possible systematic error arises from non-isotropic opacity effects caused by observing the maser emission in directions parallel and perpendicular to the flow direction. For a finite thickness shell, this effect leads to a slight overestimate of the angular size and an underestimate of the distance [Lan92]. This method is only viable when the effects of interstellar scattering (ISS) are negligible. van Langevelde et al [LFC92] show that the angular sizes of OH/IR stars towards the galactic center region are dominated by ISS.

I have compared the phase lag distances (D_{PL}) with the distances (D_K) from the stellar velocities and a galactic rotation model having a constant velocity of 220 km s^{-1} in order to derive R_0, the Sun-Galactic-Center distance. The sources WX PC (high latitude) and OH44.8-2.3 (deviant velocity) were eliminated from the sample of [LHS92]. A least squares analysis was performed to solve for the parameter b in the relation $D_{PL} = bD_K$ with the appropriate χ^2 minimization when there are errors in both variables [PrT92] given by

$$\chi^2 = \frac{1}{N}\sum_{i=1}^{N}\frac{(D_{PL_i} - bD_{K_i})^2}{\sigma_{PL_i}^2 + b^2\sigma_{K_i}^2} \, , \qquad (3)$$

where σ_{PL_i} are the errors in phase lag distances and σ_{K_i} are the random errors in distance due to the random error in stellar velocity, σ_v. The value of σ_v that makes $\chi^2 = 1$ is 17 km s^{-1}. The minimization gives $R_0 = 8.8 \pm 0.9$ kpc where the errors were calculated from the values of b for $\Delta\chi^2 = 1$. (Note that if the errors in D_K were neglected, the derived value of R_0 would be 7.9 kpc with a χ^2 of 5.8). Our value of

R_0 is close to that derived by Herman *et al.* [HBH85] who obtained 9.2 ± 1.2 kpc based on a smaller and less accurate data set, rather arbitrary statistical weights, and a rotation velocity of 250 km s^{-1}. This method has excellent potential for refining the value R_0 but requires much labor in monitoring the light curves. In the best data phase lags and sizes can be determined to better than 5%. Distant sources on the solar circle are most valuable for determining R_0.

Table 1: Phase Lag Distances to Late Type Variables

Source[1]	V[2] km s^{-1}	$D_{PL}^{[3]}$ kpc	$D_K^{[4]}$ kpc	σ[5] kpc	χ[6]
20.7+0.1	136	5.1±1.4	7.5±0.8	1.6	−1.5
21.5+0.5	116	8.0±1.0	6.6±0.7	1.3	1.2
26.5+0.6	27	1.4±0.2	2.2±1.2	1.2	−0.7
28.7−0.6	46	1.4±0.5	3.3±1.0	1.1	−1.7
32.0−0.5	76	11.9± 2.7	10.0±1.0	2.9	0.7
32.8−0.3	61	5.0±0.3	4.1±1.0	1.0	0.9
35.6−0.3	78	4.2±1.7	5.2±1.1	2.0	−0.5
39.7+1.5	20	1.3±0.3	1.5±1.2	1.2	−0.2
104.9+2.4	−26	2.3±0.4	3.1±1.8	1.8	−0.4
127.9−0.0	−55	2.9±0.6	5.4±2.2	2.3	−1.1

[1]galactic coordinates; [2]stellar velocity; [3]distance from phase-lag analysis ([LHS90], table 3); [4]distance from kinetic model: $v = 220$ km s^{-1}, $R_o = 8.8$ kpc, 17 km s^{-1} random velocity; [5]total error, $(\sigma_{PL}^2 + b^2\sigma_K^2)^{\frac{1}{2}}$; [6]$(D_{PL} - D_K)/(\sigma_{PL}^2 + b^2\sigma_K^2)^{\frac{1}{2}}$.

2. Proper Motion Distances

H$_2$O maser emission arises in the outflows surrounding newly formed stars. The angular velocity of a clump of masing gas moving with transverse velocity V_T and distance D is

$$\theta = 206\left(\frac{V_T}{\text{km s}^{-1}}\right)\left(\frac{D}{\text{kpc}}\right)^{-1} \mu\text{as yr}^{-1} \qquad (4)$$

Hence a velocity of 40 km s^{-1} results in an angular velocity of 1 mas and 10 μas at distance of 8 kpc (∼Galactic Center) and 800 kpc (nearby galaxy) respectively. The synthesized beam of an intercontinental VLBI array has a width θ_r that is limited to about 300 μas for the 22 GHz transition of H$_2$O. However, in the absence of systematic errors, the relative positions of masers can be estimated to an accuracy of about $\delta\theta \sim 0.5\theta_r/\text{SNR}$ where SNR is the signal-to-noise ratio. Only masers with reference features strong enough to be detectable within the interferometer's coherence time can be readily studied. With current instrumentation the coherence time is about 2 minutes and limiting sensitivity is about 1 Jy. There are about 300 known interstellar masers with flux densities exceeding this level. After a reference feature is detected, the SNR improves as the square root of the integration time and features as weak as ~ 20 mJy can be detected. A prime source of systematic error is due to structural changes in the maser as a function of time. That is, two features with velocities separated by less than the linewidth and positions separated by less than the interferometric beamwidth cannot be readily distinguished and any change in their relative amplitudes will give rise to an apparent shift in position. In addition, although phase referencing removes geometric errors to first orders, second order systematic errors remain given by

$$\delta\theta \sim \frac{\Delta B}{B}\Delta\theta + \frac{\Delta\nu}{\nu}\frac{c\Delta\tau}{B} \qquad (5)$$

where $\Delta\theta$ and $\Delta\nu$ are that offsets from the reference feature in angle and frequency, respectively, B is the baseline length, ν is the frequency, and ΔB and $\Delta\tau$ are the errors in the baseline length and instrumental delay. For ΔB and $c\Delta\tau \sim$ 10 cm, $\Delta\theta = 2''$ and $\Delta\nu = 2$ MHz, $\delta\theta$ is below 1 μas. Hence geometric errors are significant primarily for the study of extragalactic masers and for space VLBI (e.g., [Rei84]).

If the motions of the maser cloudlets are random and isotropic then the distance to the source can be estimated by the method of statistical parallax. If σ_μ is the dispersion in transverse angular velocities in one coordinate and σ_z is the the dispersion in line-of-sight velocity then the distance estimate is given by $D = \sigma_z/\sigma_\mu$. Note that the observed dispersion in transverse angular velocity, $\sigma_{\mu_{obs}}$, will be larger than σ_μ because of the effects of error in the position measurements, such that $\sigma_\mu^2 = \sigma_{\mu_{obs}}^2 - \sigma_m^2$, where σ_m is the measurement error in angular velocity. Hence an underestimate of σ_m leads to an underestimate in the distance. This dispersion correction is important primarily for extragalactic work. If σ_m is negligible then the accuracy of the statistical error in the distance estimate is given by the relation $\frac{\sigma_D}{D} \sim \frac{1}{\sqrt{2N}}$ where N is the number of maser spots measured in each coordinate. If the flow is not isotropic and random then this method can be biased. For example, if the maser cloudlets were flowing from a central point with constant flow velocity, V_e, but were confined to the hemisphere facing us, then the spread in angular velocity would be $2V_e/D$ and the spread in radial velocity would be V_e. Hence the distance estimate would be one-half the true distance. Comparison of the ambient velocity of the molecular cloud and the mean velocity of the maser features is an important diagnostic test for assessing the isotropy and randomness of the maser emission.

When the masers show organized motion such as uniform outflow, their velocities can be fit to a parameterized model where a distance parameter relates angular and linear velocities. Such an analysis is not entirely straightforward because it is non-linear and the line-of-sight positions for the features are unknown parameters. Hence an observation of 25 masers features (and a reference feature) yields 75 velocity-component measurements (and 50 coordinate measurements). A model of simple outflow at constant velocity has 7 global parameters and 25 line-of-sight position parameters. The least mean square analysis involves both radial and transverse velocities, which are experimentally quite distinct quantities, and the solution depends on the relative weights assigned to them. The quantity to be minimized can be written [GMR92b]

$$\chi^2 = \frac{1}{3N_m - N_p} \sum \left[\frac{(\dot{\theta}_i - \frac{V_{m_i}}{D})^2}{\sigma_i^2} + \frac{(V_{z_i} - V_{zm_i})^2}{\sigma_{z_i}^2} \right] \quad (6)$$

where $\dot{\theta}_i$ and σ_i are the measured angular velocities and their errors, V_{z_i} and σ_{z_i} are the measured line-of-sight velocities and their errors, V_{m_i} and V_{zm_i} are the model velocities in the transverse and radial directions respectively, N_m is the number of maser spots, and N_p is the number of model parameters. The weights are given by $\sigma_i^2 = \sigma_V^2/D^2 + \sigma_{n_i}^2$ and $\sigma_{z_i}^2 = \sigma_V^2$ where σ_V is the rms turbulent velocity of the flow and σ_{n_i} is the measurement error. We assume that σ_V is greater than measurement errors in radial velocities. Note the unusual situation wherein the weighting depends on the distance parameter. If the σ_{n_i}s can be calculated from the residuals in the position-versus-time analyses for individual features, then σ_V can be found by requiring that $\chi^2 = 1$.

Table 2: Distances of Galactic H$_2$O Masers Derived from Proper Motions

Source	Fit[1]	N[2]	$V_e^{(3)}$	$V^{(4)}$ km s^{-1}	$D_{PM}^{(5)}$ kpc	$D_K^{(6)}$ kpc	$\chi^{(7)}$	Ref.
Orion	M	26	18	8	0.5±0.1	0.7±1.0	−0.2	GRM81
W51M	SP	27	-	63	7.0±1.5	5.1±2.0	0.8	GDS81
W51N	SP	10	-	63	8.3±2.5	5.1±2.0	1.0	SLD81
Sgr B$_2$N	M	24	45	50	7.1±1.5	7.8±0.2	−0.5	RSM88
Sgr B$_2$M	M	27	35	26	6.5±1.5$^{(8)}$	7.8±0.2	−0.9	RGM88
W49N	M	105	18	7	11.4±1.2	10.9±0.7	0.4	GMR92b
W3(OH)	M	42	18	−48	−	−	−	AMM92
Cep A	SP	11	−	−11	0.5± 0.2$^{(8)}$	1.1±0.9	−0.6	CMR92

[1]SP = statistical parallax; M = model fit. [2]Number of maser features. [3]Outflow velocity in km s^{-1}. [4]Systemic velocity. [5]Proper motion distance. [6]Distance from kinetic model: $v = 220$ km s^{-1}, $R_0 = 7.8$ kpc and random velocity = 10 km s^{-1}. [7]Weighted residual. [8]Preliminary result.

The proper motions of eight Galactic masers have been analyzed (see Table 2).

Note that the flow velocities fall in the narrow range of 18–45 km s^{-1}. The flow in W3(OH) is highly collimated and its distance cannot be estimated reliably. The distance estimates for the other masers can be used to derive a value for R_o based on the same analysis used for the phase lag distances (eq. 3). The result is $R_0 = 7.8 \pm 0.6$ kpc. Statistical parallax measurements have also been successfully made for a maser in the nearby galaxy M33 [GMR92a]. I thank M. Reid and L. Greenhill for helpful comments.

References

[AMM92] Alcolea, J., Menten, K.M., Moran, J.M., and Reid, M.J., *Astrophysical Masers*, (ed. A. Clegg and G. Nedoluha) (1992), in press.

[CMR92] Cawthorne, T.V., Moran, J.M., Reid, M.J., *Astrophysical Masers*, (ed. A. Clegg and G. Nedoluha), (1992), in press.

[GDS81] Genzel, R., Downes, D.D., Schneps, M.H. et al *Astrophys. J.*, Vol. 247 (1981), pp. 1039–1051.

[GRM81] Genzel, R., Reid, M.J., Moran, J.M., and Downes, D., *Astrophys. J.*, Vol. 244 (1981), pp. 884–902.

[GMR92a] Greenhill, L.J., Moran, J.M., Reid, M.J., Menten, K.M., and Hirabayashi, H., *Astrophys. J.*, (1992), in press.

[GMR92b] Gwinn, C.R., Moran, J.M., and Reid, M.J., *Astrophys. J.*, Vol. 393 (1992), pp. 149–164.

[HBH85] Herman, J., Baud, B., Habing, H.J. and Winnberg, A., *Astron. Astrophys.*, Vol. 143 (1985), pp. 122–135.

[Lan92] van Langevelde, H.J., (1992), *Ph.D. Thesis*, University of Leiden.

[LFC92] van Langevelde, H.J., Frail, D.A., Cordes, J.M., Diamond, P.J., *Astrophys. J.*, Vol. 396 (1992), pp. 686–695.

[LHS90] van Langevelde, H.J., van der Heiden, R., and van Schooneveld, C., *Astron. Astrophys.*, Vol. 239 (1990), pp. 193–204.

[PrT92] Press, W.H. and Teukolsky, S.A., *Computers in Physics*, Vol. 6 (1992), p. 274–276.

[Rei84] Reid, M.J., in *QUASAT–A VLBI Observatory in Space Proceeding of a Workshop at Gross Enzersdorf*, Austria, European Space Agency, (1984), pp. 181–184.

[RGM88] Reid, M.J., Gwinn, C.R., Moran, J.M., and Matthews, A.H., *Bull. AAS*, Vol. 20 (1988), p. 1017.

[RSM88] Reid, M.J., Schneps, M.H., Moran, J.M., Gwinn, C.R., Genzel, R., Downes, D., and Rönnäng, B., *Astrophys. J.*, Vol. 330 (1988), pp. 809–816.

[SLD81] Schneps, M., Lane, A.P., Downes, D., Moran, J.M., Genzel, R., and Reid, M.J., *Astrophys. J.*, Vol. 249 (1981), pp. 124–133.

Maser Theories

David Field *   *Malcolm Gray* †

Abstract

We review the ingredients of maser models, emphasising the theory of maser propagation and maser beaming. We use maser theories to analyze OH masers in star-forming regions, with an outflowing zone in W3(OH) as a specific example. We show how a geometrical model of maser beaming can explain the relative sizes of 1665 and 1720MHz VLBI features in W3(OH), noting the importance of beaming due to saturation.

1. Introduction

It is more than 25 years since the discovery of masers in space, but there are still important properties such as beaming and polarization which we do not understand, even at a qualitative level. The first species observed as a maser was OH (star-formation, circumstellar shells, megamaser galaxies, comets). Since then ten other masing species have been found, among which are H_2O (star-formation, circumstellar shells, megamaser galaxies), methanol (star-formation), SiO (star-formation, circumstellar shells). Masers offer an opportunity to investigate objects on a uniquely small scale, using VLBI, in a wide variety of environments. Masers may show linear or circular polarization. In particular 100% circular polarization in OH masers in star-forming regions (Barrett & Rogers 1966; Garcia-Barreto et al. 1988) remains unexplained and is a major challenge to theory (Gray et al. 1993). It may also be possible to use the observed distribution of OH maser spot sizes to model the nature of the turbulence close to HII regions (section 4).

General aspects of maser theories are discussed in section 2. In section 3, we use OH masers in star-forming regions to illustrate how maser theories may operate. There are a number of recent VLBI studies of OH maser emission in W3(OH), at least two of which are reported here (Bloemhof et al. 1992; Masheder et al. 1993a). We use data from these observations.

---
*School of Chemistry, University of Bristol, Bristol, BS8 1TS, UK.
†School of Chemistry, University of Bristol, Bristol, BS8 1TS, UK.

2. The Structure of Maser Theories

A complete maser theory would require (a) the calculation of the populations of the molecular energy levels of the masing species (b) a means of treating the propagation of maser rays, including polarized radiation (c) a model of the morphology of the maser zone (d) a treatment of maser beaming (e) inclusion of velocity redistribution of the masing molecules and (f) a treatment of maser temporal variation. No such complete theory exists. Here, we briefly discuss aspects of (a), (b), (c) and (d).

Under (a), populations are calculated by solving the population master equations in the steady state. For this we require (i) rate coefficients for rotationally and rovibrationally inelastic collisions of the masing species with H_2 or H-atoms; (ii) precise energy levels and A-values for all transitions; (iii) the IR and FIR spectrum of dust emission; (iv) a treatment of line overlap; (v) a knowledge of velocity fields in the masing zone; (vi) a method of calculating the mean continuum and line radiation fields . Other inputs into the model are masing species abundance, kinetic and dust temperatures, collision partner number densities, and magnetic field and direction relative to the axis of maser propagation. Of the above, inelastic rate coefficients are are not adequately known for any species but are undergoing continual refinement, for example in the case H_2O (Balasubramanian et al. 1993). The spectrum of dust emission is only crudely characterized, as discussed in Gray et al. 1991. Observations by ISO should improve this situation. The mean line and continuum radiation fields are calculated, almost without exception, using the Sobolev or Large Velocity Gradient (LVG) approximation. The accuracy of this approximation is unclear, and its use poses severe restrictions on maser models, since it introduces strong spatial averaging over distances, for OH, which approach 10^{14}m.

Turning to (b), maser propagation, the first substantial description of the formation of maser radiation was that of Goldreich & Keeley 1972, who treated the radiation field classically and the molecules as classical oscillators. Field & Richardson developed a two-level semi-classical description of maser radiation transport, again treating the maser radiation classically but the response of the molecules quantum mechanically. The use of a semi-classical theory is necessary since the masing molecules do not behave as independent species but are coupled into an ensemble through collisions and radiation. Over a narrow range of frequencies, dictated by the homogeneous linewidth, this ensemble responds collectively to the ambient maser field over a characteristic distance cT_2 where T_2 is the relaxation time of the medium. For OH masers this distance may be of the order of 5×10^9m. A purely classical theory is accurate only in limit of T_2 tends to zero. As the flux of maser radiation grows, so stimulated emission perturbs the relative energy level populations of the masing molecules, leading to saturation. Semiclassical theory shows a more abrupt onset of saturation than does a purely classical theory. In the limiting case of weak saturation, the classical and semiclassical theories agree.

In real many-level systems, masers do not saturate independently. For example, in OH, 1720 and 1667MHz masers share a common upper level, and they compete for the population in this level when saturation becomes important (competitive gain). In SiO, saturation in (say) J=2-1 will tend to increase, or even create, an inversion in J=1-0, by increasing population in J=1. Moreover, as populations change, due to competing saturating effects, so the collisional and radiative coupling to all other levels is modified. In Field & Gray 1988 we treated the general case of the propagation of self-generated partially coherent maser radiation in a many-level system, including all collisional and radiative coupling and involving all masers concurrently present. We use this theory in our investigations of OH masers, section 3.

We turn now to (c) and (d), the morphology of the maser zone and the beaming of maser radiation. Maser beaming takes place for two reasons. In the first place there are purely geometrical effects. Maser propagation in general takes place in a turbulent medium and it would seem appropriate to model the passage of maser rays along a gain length in such a medium (section 4). The second factor affecting beaming is saturation. Masers may be initiated in many physically separated regions. Due to differing gain lengths, rays travelling in particular directions become strong and saturating while, in other directions, rays remain weak. Strong saturating rays suppress the formation of weaker rays, by taking the lion's share of the population inversion, through stimulated emission. Maser radiation therefore beams along the longest velocity-matched lines-of-sight. Since only a small fraction of maser rays, (solid angle subtended by the maser/maser beaming angle), can be detected on earth, an estimate of the maser beam angle is necessary to compare calculated and observed fluxes. The value of the maser beam angle has a direct bearing on whether or not masers are saturated, given an observed flux. For OH masers, we find that saturation becomes significant for flux in a ray exceeding 10^{-15} Wm^{-2}Hz^{-1}. For a maser beam angle of 10^{-4} sr, for a typical VLBI spot in W3(OH)(Garcia-Barreto et al.), then an observation of > 1 Jy implies saturation.

3. Applications to OH Masers in Star-forming Regions

For a detailed description and extensive references to other work, see Gray et al.1991 and Cesaroni & Walmsley 1991. In our model we use the first 48 hyperfine levels of OH, including levels up to 800 cm^{-1} (1150K) above the rotational ground state. Collisions are treated using an approximate theory of Dixon & Field 1979 and Dixon et al.1985. We include a dust radiation field using a modified Planck function generating a far-infrared (FIR) excess up to 80 microns. The mean line and continuum intensities are calculated using the LVG approximation for the escape probability. FIR line overlap, of fundamental importance for OH masers, is included in our analysis, using methods set out in Doel et al.1990. Line overlap arises because certain FIR transitions lie < 1 to a few MHz apart and, in the presence of large velocity gradients, radiation originating in one line may be reabsorbed in a different line, at some distant point in the medium. Often, photons formed in one line may be reabsorbed in a second line at small velocity shifts, or reabsorbed in a third line at larger shifts. Separate parts of an accelerating flow are then influenced in different ways by line overlap, with a succession of inversions being enhanced and removed as we progress along such a flow. Apart from line overlap, the non-Planck nature of the FIR dust radiation field is the major factor in forming population inversions. Collisions and fluorescence alone tend not to favour inversions in the main lines of the ground state, 1665 and 1667MHz. The kinetic scheme is coupled into the theory of Field & Gray in order to calculate factors of amplification as masers rays traverse the maser zone. Our results allow us to associate the appearance of specific maser lines with a defined range of physical conditions. For example 1665MHz is associated with accelerating, cold (< 75K) flows. The well-established association of 4765MHz and 1720MHz also emerges from the model.

In the northern part of W3(OH) there exists a region in which there may be a spatial superposition of 4765MHz (Baudry et al.1991), 1720MHz (Fouquet & Reid 1982; Masheder et al.1993a,b) and a cluster of 1665MHz masers (Bloemhof et al.). 4765MHz peaks at -43.3 kms^{-1}, 1720MHz at -44.85 kms^{-1} and the 1665MHz cluster at -46.3 kms^{-1}, where velocities have been demagnetised as appropriate. Our calculations show that such a superposition, at a succession of velocities close to those observed, arises in an accelerating flow experiencing adiabatic expansion. Initial conditions involve a flow velocity of 0.5 kms^{-1}, a kinetic temperature = 250K, number density of H$_2$= 10^8 cm^{-3}, in the absence of dust local to the zone, but with a small, spatially diluted external component. These conditions generate 4765MHz alone, up to shifts of 0.3 to 0.4 kms^{-1}. Further downstream with kinetic tempertaures < 160K and number densities $< 5 \times 10^7$ cm^{-3}, 1720MHz emerges strongly, as the flow enters a dusty zone, and 4765MHz drops to negligible flux. Still further downstream, with velocity shifts > 2 kms^{-1}, kinetic and dust temperatures < 90K, number density $= 2 \times 10^7$ cm^{-3}, 1665MHz becomes the strongest maser, and 1720MHz decays. The physical conditions mentioned here differ quantitatively from those in Gray et al.1992, because of the new 1720MHz VLBI data reported in this volume. We note that the LVG approximation does not strictly allow us to vary conditions on the scale size of $< 10^{13}$m inferred from the above description. We have now developed Λ-iteration methods (Scharmer & Carlsson 1985) to overcome this difficulty. With regard to the proposed flow model, we note that Bloemhof et al. have measured proper motions of 1665MHz maser clusters and, on quite independent obervational grounds, show that the 1665MHz cluster at -46.3 kms^{-1} is embedded in an outflow, directed towards us.

4. Maser Beaming: geometrical effects

Observations (Masheder et al.1993a,b) show that 1720MHz maser spot sizes are < 1.2 mas in W3(OH). 1665MHz spot sizes range from 3 to 14 mas (Garcia-Barreto et al.). We ask here how this difference may arise. For this purpose, imagine that you are suspended in a turbulent maser zone, at a point where a maser may be initiated. Looking around, you will perceive directions which intercept regions of space which are Doppler-matched to the velocity of your own position. These regions are regions of potential maser gain. If you attempt to identify short Doppler-matched pathlengths, then a large number of such paths may be found. As you require longer paths, so the number of paths will decrease rapidly. Also, you will notice that the average solid angle subtended at your position by short paths is a good deal greater than that subtended by long paths, because of the overlapping requirements. Classifying paths by the number of Doppler-matched regions traversed, and writing N_i as the number of paths traversing i such regions, it is evident that the behaviour of N_i vs i depends on the spectrum of sizes of Doppler-matched regions in the turbulent medium. Equally, the variation of solid angle, Ω_i, with i will also depend on this property of the turbulence. Evidently, the projected area of a maser spot in the plane of the sky will depend both on the gain length required and on the nature of the turbulence.

In order to investigate the variation of N_i and Ω_i with i, we have used a Monte-Carlo technique, for a range of pathlengths over a factor of 10. We have presently performed this analysis only for the case in which the number of Doppler-matched regions is proportional to (region size)$^{-4/5}$ (Scalo 1987) and therefore we do not yet know the sensitivity of our results to the form of the turbulence. Our results may be summarised as follows: calculations, as in section 3, show that 1720MHz requires typically 3 times the gain length of 1665MHz to yield strong saturated amplification. Therefore a 1720MHz ray traverses on average three times as many regions as a 1665MHz ray. The ratio of linear size, in the plane of the sky, of 1665MHz to 1720MHz spots is therefore given by (Ω_i/Ω_{3i}). For i=1,2 and 3, we find that these ratios are 3.4, 7.4 and 13, and therefore the observation of < 1.2 mas for 1720MHz spot size predicts 1665MHz spot sizes of < 4.1 mas, < 8.9 mas and < 15.5 mas, satisfactorily reproducing the observed range of spot sizes. The relative probability of encountering i=1,2 or 3 type paths is also in accord with observations of the prevalence of 1665MHz masers of sizes 3 to 4 mas, 5 to 8 mas and 9 to 14 mas. The calculated maser beam angle is of the order of 10^{-2} sr. This is 2 to 3 orders of magnitude too great to account for observed maser fluxes and saturated beaming must therefore be important in masers.

5. Concluding Remarks

We have presented a very brief and personal overview of some of the areas of current interest in maser theories. We have omitted any consideration of maser polarization, of velocity redistribution, of maser lineshapes or of maser temporal variation. We have also chosen to reference only some of the most recent work. Much of the earlier work, which underpins our account, was developed in the late '70s and early 80's and is reviewed in Cohen 1989 and Elitzur 1982. Future work on maser models will benefit greatly from the new observational capabilities of the VLBA and MERLIN (especially phase referencing) and from new theoretical input, in particular, inelastic scattering rate coefficients, semiclassical descriptions of the transfer of polarized maser radiation, and the use of precise techniques to evaluate the mean line and continuum radiation fields.

References

Barrett A H, Rogers A E E, 1966 Nature **210** 188

Balasubramanian V, Balint-Kurti G G & van Lenthe J H, JCS Faraday Trans. 1993 to appear

Baudry A, Diamond P, Booth R S, Graham D, Walmsley C M, 1989 A & A **201** 105

Bloemhof E E, Reid M J, Moran J M, Ap.J. to appear, and this volume.

Cesaroni R, Walmsley C M, 1991, A & A, **241** 537

Cohen R J, Rep.Prog.Phys. 1989 **52**, 881

Dixon R N, Field D, 1979 Proc. Roy. Soc. Lond. **A368**, 99

Dixon R N, Field D and Zare R N, 1985 Chem.Phys.Lett. **122**, 310

Doel R C, Gray M D, Field D, 1990, MNRAS **244**, 504

Elitzur M, 1982, Rev.Mod.Phys. **54**, 1225

Field D & Richardson I M, 1984, MNRAS **211**, 799

Field D & Gray M D, 1988, MNRAS **234** 353

Fouquet J E & Reid M J, 1982, Astr. J. **87** 691

Garcia-Barreto J A, Burke B F, Reid M J, Moran J M, Haschick A D, Schilizzi R T, 1988, Ap.J. **326**, 954

Goldreich P. & Keeley D A, 1972, Ap.J. **174**, 517

Gray M D, Doel R C, Field D, 1991, MNRAS **252**, 30

Gray M D, Field D, Doel R C, 1992, A & A, **262**, 555

Gray M D, Jones K N, Field D, 1993, J.Chem.Soc.Farad., to appear

Masheder M R W et al.,1993a,this volume

Masheder M R W, Migenes V, Gray M D, Cohen R J, Booth R S, Field D, 1993b, submitted to A & A

Scalo J M, 1987, in Interstellar Processes, Astr.Sp.Sci.Lib., Reidel.

Scharmer G B, Carlsson M, 1985, J.Comp.Phys. **59**, 56

Preliminary Results from VLBI Observations of 1720 MHz Masers in W3(OH)

M.R.W. Masheder [*]   V. Migenes [†]   M.D. Gray [‡]
R.J. Cohen [§]   D. Field [¶]   R.S. Booth [||]

Abstract

VLBI observations of the 1720 MHz transition of OH were made in Nov. 1989. A number of sources were observed, but most of the observing time was devoted to W3(OH) for which interferometer maps were constructed for several velocity channels. The maser zone contains two maser spots, one of which was resolved by the interferometer network and one which appears unresolved on the longest baseline used, corresponding to an angular size of <1.2mas. A comparison is made between the 1720MHz observations reported here and other authors' results.

1. Introduction

VLBI observations of astrophysical masers are important because they yield direct information about the size and structure of the masing objects themselves. This sets limits on the way in which masers grow, how tightly their emission must be beamed and on how they interact with their environment. Of the four ground state masing transitions, the satellite line at 1720 MHz is intrinsically weaker than the main lines at 1665 and 1667 MHz. We might therefore expect, that in order to achieve the same brightness as main line masers, the 1720 MHz masers must propagate, on average, for a greater distance. This expectation is borne out in the set of models discussed in Gray et.al., where 1720 MHz is a common maser but tends to require longer gain lengths to reach its peak intensity than either of the main lines. It seems likely that a longer propagation distance is coupled to a restriction of both the angular size of the maser and the angle into which its radiation is beamed and we therefore expect 1720 MHz masers to have smaller angular size on average than 1665 and 1667 MHz masers from the same object. These observations were designed to test this hypothesis.

[*]Department of Physics, University of Bristol, Bristol BS8 1TL, U.K.
[†]NRAL, Jodrell Bank, Macclesfield, SK11 9DL, U.K.
[‡]School of Chemistry, University of Bristol, Bristol BS8 1TS, U.K.
[§]NRAL, Jodrell Bank, Macclesfield, SK11 9DL, U.K.
[¶]School of Chemistry, University of Bristol, Bristol BS8 1TS, U.K.
[||]Onsala Space Observatory, S-43900 Onsala, Sweden.

2. The Observations

The experiment was carried out for 24 hours on 10th-11th Nov 1989 using 7 receivers worldwide. Various faults reduced the number of receivers which provided useful data to 5 (Effelsberg 100m, Onsala 25m, NRAO Greenbank 43m, Pie Town 25 m, Shanghai 25m). All observations were taken with left-hand circular polarisation and recorded on the VLBI Mark II system. The five operational antennae formed a system with a synthesized interferometer beam some 2.0 mas diameter FWHM. The raw data were correlated on the Mk II correlator at MPI, Bonn in May 1990 and subsequent data reduction used the AIPS software package at Jodrell Bank, Macclesfield, U.K.

Most of the observing time was devoted to W3(OH) yielding the most detailed results, including channel contour maps of source brightness. Two regions of maser emission were found, offset by 0".15 in R.A and 1".15 in declination. These detailed maps will be published elsewhere. Our two regions correspond to peaks at different velocities in a single dish spectrum with the southern region corresponding to the smaller peak at -43.8 kms^{-1} and the northern region to the larger double peak, centered on -45.5 kms^{-1}. The more northerly object was found to be resolved by the interferometer and exhibits structure, in the form of a group of some 5 objects, distributed over several velocity channels clustered within about 15mas. The individual spots in the group, however, appear to be unresolved. The smaller, southern, source was also unresolved, in all channels, on the longest baseline in the system. The unresolved spots have angular sizes of less than 1.2mas, which using a value of 2 kpc for the distance of W3(OH), yields a linear size of 3.6×10^{11}m.

3. Comparison with other Observations

A considerable amount of data is available for comparison at 1665 and 1667 MHz. (See references in Gray et.al. [1]). Fouquet and Reid show data at 1720 MHz. We compare spot positions and sizes from our 1720 MHz data with those at 1720 MHz and 1665 MHz published by other authors. By comparing offsets in position, we make a tentative identification of our two 1720 MHz regions with regions 2 and 9 of the Reid et.al. observations at 1665 MHz. Indeed, the more northerly position is as given by Fouquet and Reid for their spot at 1720 MHz. On the basis of this identification, we compare the spot sizes at 1665 and 1720 MHz. In the southern region, 9 in Reid et.al., we find a single unresolved spot of size <1.5mas. The feature in this region with the better spatial agreement has a 1665 MHz spot size given by Reid et.al. of 8x4 mas. In the northern region we find a cluster of objects each of which appears to be unresolved, with spot sizes <2mas. Reid et.al. found one bright feature and three weak features, in their region 2, with spot sizes of 4,5,4 and (9x6) mas. While it has not been possible to make positive one to one identifications with the 1665 MHz features, the indications are that the spot sizes for 1720 MHz are significantly smaller than for 1665 MHz in similar regions. This is as expected for the reasons outlined above. We are developing a more detailed model for this mechanism.

---

[1] Gray M D, Field D and Doel R C, 1992. *Astron. Astroph*, **262**, 555.

Time Variations in the Flux Density of Spatially Resolved Interstellar OH Masers

E. E. Bloemhof, M. J. Reid, and J. M. Moran [*]

Abstract

We have carried out two multi-station spectral-line VLBI experiments, at epochs 1978 and 1986, to measure the proper motions of the 1665 MHz OH masers surrounding the ultracompact HII region in W3(OH). Visibility coverage was adequate to permit full synthesis mapping with spatial resolution approximately 5 milli-arc seconds; data in both circular polarizations were recorded. As a byproduct of this investigation, we have obtained the flux density at each epoch of individual maser-emitting gas condensations, some of which appear to be spatially resolved. We find significant temporal variations.

Time variation of astrophysical maser emission has been the subject of extensive study (eg. [SuK76]). Such variations might provide clues about underlying physical processes; as one example, they might indicate what pump mechanism is at work in a maser source. Single-dish studies are complicated, however, by the fact that emission in a given spectral channel will generally arise from numerous spatially distinct spots on the sky. This spectral blending makes time variations less apparent and makes their interpretation ambiguous. VLBI mapping of maser emission can overcome this observational problem. We have analyzed our two-epoch VLBI study of the OH masers in W3(OH) [BRM 92] for changes in flux density of individual maser spots, and present for the first time the spatial pattern of time variations in an interstellar OH maser source.

The two epochs of observation were 1978 and 1986, and we have a total of 45 maser spots, either left- or right-circularly polarized, for which flux densities at both epochs were measured. To compare variations among maser spots with widely differing intensities, we convert to a normalized change in flux density:

$$\widetilde{\Delta S} = \frac{(S_{1986} - S_{1978})}{\sqrt{S_{1986}^2 + S_{1978}^2}} \ .$$

---

[*] Harvard-Smithsonian Center for Astrophysics, 60 Garden Street, Cambridge, MA, USA 02138.

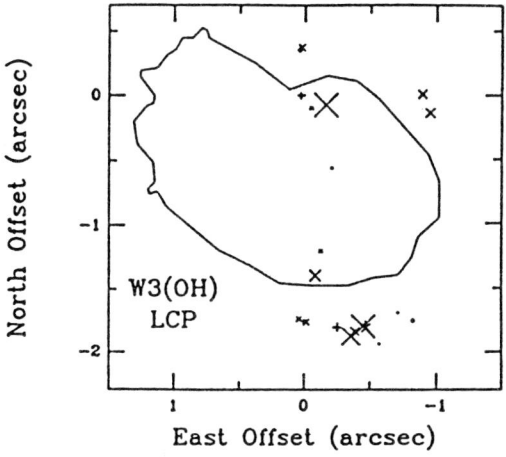 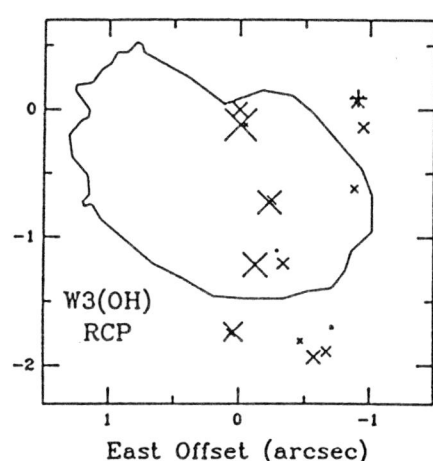

When defined in this way, flux density change is limited to lie between -1 (for a large fractional decrease) and +1 (for a large fractional increase). The detailed spatial pattern of flux variation is shown in the Figure, with maser emission in the two circular polarizations presented separately for clarity. In these maps, (0,0) denotes the position of the reference maser feature of [BRM 92]; the general outline of the ultracompact HII region seen in radio continuum emission is shown. Each maser spot is represented by a + or × symbol according to whether its flux increased or decreased; further, the size of the symbol is proportional to the magnitude of $\widetilde{\Delta S}$.

We searched for a possible pattern of spatial correlation to the flux variations by constructing a variety of correlation and structure functions. The existence of such a pattern, depending on its form, might for example reveal a pump mechanism acting globally over the 60 milliarc second clustering scale of observed maser spots (it is expected on theoretical grounds that the masering region is larger than the observed size of individual spots, and might correspond to this clustering scale). No clear pattern has as yet been identified in the flux changes shown in the Figure. Since masers are highly non-linear phenomena, it may well be necessary to apply a more sophisticated model to compute how a candidate pumping mechanism would affect the spatial pattern of flux variation.

References

[BRM92] Bloemhof, E.E., Reid, M.J., Moran, J.M., *Kinematics of W3(OH): First Proper Motions of OH Masers from VLBI Measurements.* Ap.J., 397 (1992), pp. 500-519.

[SuK76] Sullivan, W.T., Kersholt, J.H., *Time Variations in 18-cm OH Emission Profiles over the Period 1965-1972.* A.A., 51 (1976), pp. 427-450.

VLBI OBSERVATIONS OF METHANOL MASERS AT 6.7 GHz

K. M. Menten, M. J. Reid, P. Pratap, J. M. Moran*, T. L. Wilson[†]

Abstract

We have conducted VLBI observations of the 6.668 GHz maser transition of interstellar methanol toward the ultracompact HII region W3(OH). We have determined accurate maser positions and show that the methanol masers have a distribution similar to the hydroxyl masers in this source. The intrinsic sizes of individual maser spots are $\approx 0\rlap{.}''0014$ (FWHM), or ≈ 3 AU, and are not significantly affected by interstellar scattering. VLBI maps of 6.7 and 12.2 GHz methanol masers can be aligned so that the positions of the strongest 6.7 GHz emission features agree to within a maser spot size with the positions of strong 12.2 GHz features at the same velocities.

1. Introduction

Recently, two transitions of interstellar methanol (CH_3OH), the $2_0 \rightarrow 3_{-1} E$ line near 12.2 GHz [BMM87] and, most prominently, the $5_1 \rightarrow 6_0 A^+$ line near 6.7 GHz [Men91] have been found to show strong maser emission toward numerous star-forming regions. Since high resolution observations of these masers can provide important information on the physical and chemical conditions in their emitting regions, we conducted a VLBI experiment to obtain milli-arcsecond resolution observations of the 6.7 GHz methanol masers toward a number of star-forming regions. In the following we summarize the results of our observations toward the archetypical ultracompact HII region/maser source W3(OH).

2. Observations

The VLBI experiment was conducted between 1992 April 15 and 20 with the NRAO 43-m antenna in Green Bank, WV, the Haystack 37-m antenna in Westford, MA,

---

*Harvard–Smithsonian Center for Astrophysics, 60 Garden Street, Cambridge, MA 02138, USA

[†]Max-Planck-Institut für Radioastronomie, Auf dem Hügel 69, 5300 Bonn 1, Germany

and the MPIfR 100-m telescope in Effelsberg, Germany. Minimum fringe spacings were 0″.011 for the short (NRAO–Haystack) baseline and 0″.0015 for the longest (NRAO–Effelsberg) baseline. The relative positions of individual maser features were determined from aperture synthesis maps of the spectral line data, which were phase referenced to the $v_{\rm LSR}= -45.4$ km s^{-1} channel. We also produced synthesis maps from the 12.2 GHz data of [MRM88]. Analysis of fringe rate residuals yielded for W3(OH) the absolute position of the -45.4 km s^{-1} reference feature as $(\alpha, \delta)_{1950} = 2^{\rm h}23^{\rm m}16^{\rm s}\!.456 \pm 0^{\rm s}\!.008, +61°38'57''\!.76 \pm 0''\!.05$. Errors in the relative positions of maser spots are $\leq 0''\!.001$ for each transition.

3. Results and Discussion

By fitting a Gaussian model to the observed variation of the fringe visibilities with baseline length, we determined a characteristic size of 0″.0014 (FWHM) for our reference feature, corresponding to 3 AU, implying a brightness temperature of 3×10^{12} K. This size is approximately the same as for the 12.2 GHz CH$_3$OH masers [MRM88] and the 1665 MHz OH masers, for which the more compact spots have sizes $\leq$ 0″.003 [RHB80]. Since a variation of the source size proportional to the square of the wavelength would be expected if scattering were important we conclude that the observed spot sizes for both methanol transitions, and probably for the OH masers, are intrinsic sizes and not the result of interstellar scattering.

The results of our synthesis mapping are presented in Fig. 1, where we show the distribution of 6.7 GHz maser spots together with the positions of the 1665 MHz OH masers, the 12.2 GHz methanol masers, and the 15 GHz continuum emission from the ultracompact HII region. We identify about 40 emission features distributed over an 1″.3 × 2″.2 area. Most of the features are clustered in a few regions. The general distribution of the 6.7 GHz CH$_3$OH masers resembles closely that of the OH masers in W3(OH), indicating that CH$_3$OH masers and OH masers arise from the same gas cloud.

Maser emission in the 12.2 GHz line arises from only two regions, which coincide with the regions of most prominent 6.7 GHz emission. The features chosen for phase-referencing the 6.7 GHz and the 12.2 GHz data are at identical velocities and their absolute positions agree within their joint uncertainty. If we assume that the 6.7 and 12.2 GHz reference features arise from *exactly* the same position, we find that another six of the strongest 12.2 GHz maser spots are also aligned (within 1 – 2 milli-arcseconds) with strong 6.7 GHz features *at the same velocities*. The fact that we are observing 6.7 and 12.2 GHz emission arising from the same maser cloudlets should put stringent constraints on any methanol maser excitation model. Similarities in the emission distributions of 6.7 and 12.2 GHz masers in other regions are reported by Norris at this conference.

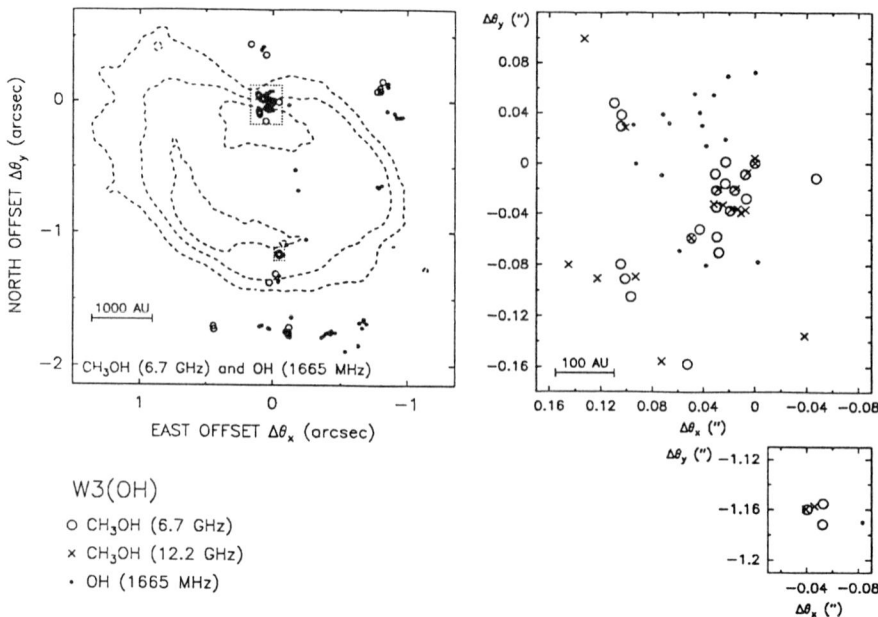

Fig. 1: Angular distribution of OH and CH$_3$OH maser emission toward W3(OH). In the left-hand panel the dashed contour lines represent 15 GHz continuum emission measured with the VLA by C. Masson. All position offsets are relative to the position of the 6.7 GHz CH$_3$OH reference feature at $v_{LSR}= -45.4$ km s^{-1}. The two panels on the right-hand side represent blow-ups of the regions within the dotted rectangles, which contain maser emission from *both* the 6.7 and the 12.2 GHz CH$_3$OH lines. Symbols are explained in the figure.

References

[BMM87] Batrla, W., Matthews, H. E., Menten, K. M., Walmsley, C. M. *Detection of Strong Methanol Masers towards Galactic HII regions*, Nature, Vol. 326 (1987), p. 49.

[Men91] Menten, K. M. *The Discovery of a New, Very Strong and Widespread Interstellar Methanol Maser Transition*, Ap. J., Vol. 380 (1991), p. L75.

[MRM88] Menten, K. M., Reid, M. J., Moran, J. M., Wilson, T. L., Johnston, K. J., Batrla, W. *VLBI Observations of 12 Gigahertz Methanol Masers*, Ap. J., Vol. 333 (1988), p. L83.

[RHB80] Reid, M. J., Haschick, A., Burke, B. F., Moran, J. M., Johnston, K. J., Swenson, G. W., Jr. *The Structure of Interstellar Hydroxyl Masers: VLBI Observations of W3(OH)*, Ap. J., Vol. 239 (1980), p. 89.

Observations of Southern Methanol Masers

R.P.Norris [*]

1. Introduction

Interstellar masers are potentially extremely powerful tools for studying the kinematics of star formation, since these masers sample the velocity and other physical parameters at selected points within the protostellar clouds. However, only modest advances have resulted from the three decades of intensive study of interstellar OH and water masers. The recently discovered strong methanol maser transitions at 6.7 [Men91] and 12.2 GHz [BMM87] promise to add a new impetus to this area of research, since they offer a new tool with which to probe the protostellar cloud.

2. Single-Dish Observations

Single-dish surveys for these masers have started both in Australia and elsewhere, and have resulted not only in dozens of new galactic masers but also in the discovery of two methanol masers in the Large Magellanic Cloud ([SCC92] ; Ellingsen et al., 1992a, in preparation). Perhaps more surprising is that, in a survey by Ellingsen et al. of OH and water megamaser galaxies, none showed any methanol megamaser emission. This implies that the conditions required to produce methanol maser emission differ significantly from those required to produce OH and water emission.

3. Australia Telescope Observations

There has so far been very little high-resolution mapping of the methanol masers because both methanol maser transitions have frequencies which are not generally available on synthesis arrays. The 6-km AT Compact Array [FBW92] is already equipped with receivers which operate in the 6-GHz region, although their nominal specification does not extend to 6.7 GHz. However, because of the enormous maser intensities, even the relatively poor performance of the receivers at this frequency is adequate for studying the masers. Consequently, several sources have been mapped, and relative positions of maser spots measured to an accuracy of about 0.02 arcsec.

---

[*]Australia Telescope National Facility, Radiophysics Laboratory, CSIRO, PO Box 76, Epping, NSW 2121, Australia

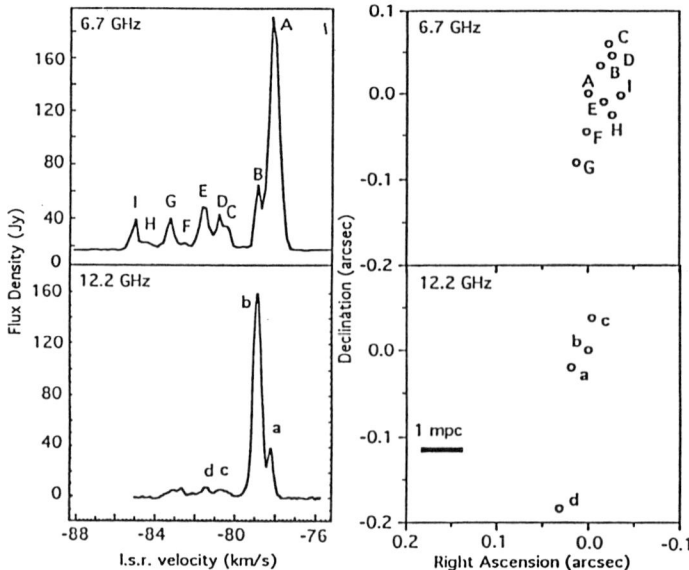

Figure 1. Map and spectra of the 6.7- and 12.2-GHz maser emission from G331.28-0.19. Letters on the maps correspond to the lettered features on the spectra. Relative positional uncertainty is estimated at 0.02 arcsec. The 12.2-GHz data are adapted from [NMC88].

Their absolute positions have also been measured to an accuracy of about 0.5 arcsec. A full account of this work will be published elsewhere [NWC92]. Here I present an example of the results from this program.

The 6.7-GHz maps and spectra of G331.28-0.19 are shown in Fig. 1, along with those obtained with the Parkes-Tidbinbilla Interferometer (PTI) at 12.2 GHz [NMC88]. This source demonstrates two important results which are also common to other sources. First, the spectra at the two transitions are similar, and several maser spots (components a, b, and c) at the same velocity in the two transitions are coincident within tens of milliarcsec. Second, the masers tend to be located along lines or arcs. This appears to be a common effect, occurring in ten of the 16 sources studied. Because it occurs in both 6.7-GHz maps (made with the AT) and the 12.2-GHz maps (made with the PTI) we can rule out instrumental effects as a cause. Instead we ascribe this effect to edge-on protoplanetary disks.

4. VLBI Observations

A number of collaborative VLBI observations of the masers have been made at both 6.7 and 12.2 GHz involving telescopes in Australia, South Africa, China, and Japan. Imaging of G309.92+0.48 (Norris et al. in preparation) indicates a broad agreement with the lower resolution maps. Imaging of other sources continues to look for detailed structure within the masers. Preliminary results (Ellingsen et al. in preparation) on longer baselines indicate that both the 6.7- and 12.2-GHz masers are largely resolved on the longer baselines, indicating maser spot sizes of order 1 milliarcsec

5. Conclusion

The most striking result is that in 10 of the 16 sources studied by [NWC92], the masers are located along lines or arcs. In five of the sources, there is a clear velocity gradient along the line. This effect can be attributed neither to chance nor to instrumental artefacts. We suggest instead that it occurs because the methanol masers are located in edge-on disks surrounding the young stars.

In several cases, the 6.7-GHz and 12.2-GHz masers are coincident to within 20 milliarcsec. This implies that one maser spot is masing in both transitions, which is surprising given that the two transitions are from completely different excitation species. It may indicate that the masers are confined to particular regions such as concentrations of density within a protoplanetary disk.

6. Acknowledgement

I thank my colleagues, collaborators, and co-authors, especially Jim Caswell, Phil Diamond, Simon Ellingsen, and Tasso Tzioumis, for permission to refer to their results prior to publication.

References

[BMM87] Batrla, W., Matthews, H. E., Menten, K. M., & Walmsley, C. M., *Nature* 326 (1987) 49

[FBW92] Frater, R. H., Brooks, J. W., & Whiteoak, J. B. *J. Electr. Electron. Eng. Australia* 12 (1992) 2

[Men91] Menten, K. M., 1991, *Ap. J.* 380(1991) L75

[NMC88] Norris, R. P., McCutcheon, W. H., Caswell, J. L., Wellington, K. J., Reynolds, J. E., Peng, R.-S., & Kesteven, M. J. *Nature* 335 (1988) 149

[NWC92] Norris, R. P., Whiteoak, J. B., Caswell, J. L., Wieringa, M. H., & Gough, R. G. *Ap. J.*(1992),*in press*

[SCC92] Sinclair, M. W., Carrad, G. H., Caswell, J. L., Norris, R. P., & Whiteoak, J. B., *MNRAS* 256 (1992) 33p

Probing the Milliarcsecond-Scale Structure of Molecular Clouds

Alan P. Marscher *   Thomas M. Bania *   Zhong Wang †

Abstract

Monitoring absorption lines in the radio spectra of compact extragalactic sources that lie behind local molecular clouds provides a method for exploring the milliarcsecond-scale structure of the foreground clouds.

1. Introduction

Several bright, compact, extragalactic radio sources happen to lie behind molecular clouds in our Galaxy. In the Earth's reference frame the foreground cloud moves in the plane of the sky as a function of time owing to proper motion, parallax, and the Solar Motion, while the background source remains essentially stationary (See Fig. 1). The background source can therefore be used as a milliarcsecond-scale beam capable of probing the AU-scale structure of the foreground cloud through observations of time variability in absorption-line profiles.

2. A Method for Exploring the Milliarcsecond-Scale Structure of Molecular Clouds

We have begun a program to use the Owens Valley Millimeter Array to monitor the CO absorption and the VLA to monitor the formaldehyde (H_2CO) absorption toward BL Lac, NRAO 140, NRAO 150, and 3C 111, all of which lie behind local molecular clouds. VLBI spectral line observations will be even more powerful, since the probing "beam" will be resolved into the core and compact jet. The background source will therefore serve as a multi-beam array tracing a corkscrew pattern across the cloud.

Structure in H I absorption has been reported on a scale of 25 AU by [Dia89]. [Mar88] has also reported an X-ray absorption event toward the quasar NRAO 140 that

---

*Department of Astronomy, Boston University, 725 Commonwealth Ave., Boston, MA 02215, USA. This work is supported in part by US National Science Foundation grant AST-9116525.

†IPAC, 100-22, California Institutue of Technology, Pasadena, CA 91125, USA.

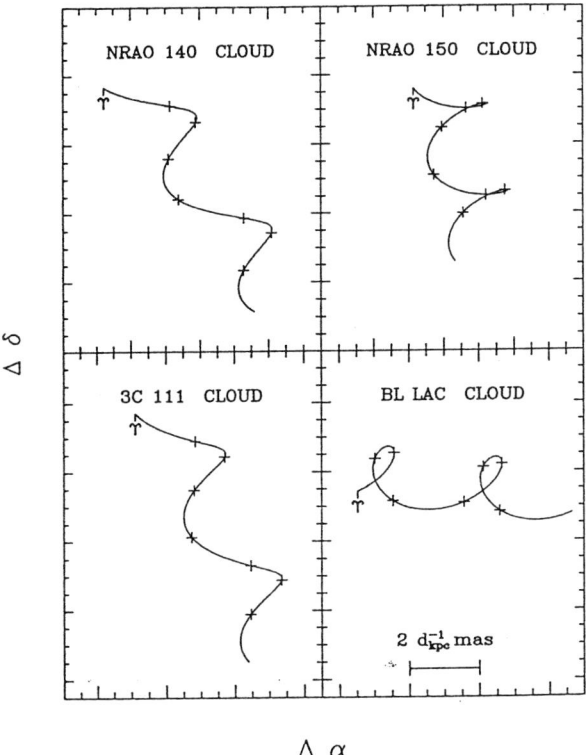

Figure 1. *Apparent path traced by the background compact continuum source relative to the foreground molecular cloud for each source in our observing list. Each path corresponds to the apparent secular motion over a two-year period starting with the vernal equinox (marked as "♈"). Each 3-month period is marked by a cross.*

requires a clump of AU size scale and density exceeding 10^9 cm^{-2} to have passed in front of the quasar in 1980. Theoretically, molecular gas is expected to clump on AU size scales during the star formation process, but such gravitational condensations should be rare for any single chosen line of sight. Turbulence also causes clumping, in which case the size of the smallest clumps would correspond to the inner scale of the turbulence, which is unknown for molecular clouds. Detection of AU-scale structure through these absorption-line observations would therefore be highly significant for our understanding of physical processes in molecular clouds.

References

[Dia89]   Diamond, P., Goss, W.M., Romney, J.D., Booth, R.S., Kalberla, P.M.W., Mebold, U. 1989, ApJ, 347, 302.

[Mar88]   Marscher, A.P. 1988, ApJ, 334, 552.

Interstellar Scattering

Brian Dennison [*]

Abstract

Fluctuations in the density of ionized gas in the interstellar medium (ISM) scatter radio waves from background sources. This results in apparent angular broadening of highly compact sources, thus limiting the achievable resolution. Both diffractive and refractive scintillation occur provided the source angular size does not exceed the appropriate transverse angular scales. For pulsars both types of scintillation are apparent. The angular sizes of compact extragalactic sources are substantially in excess of the diffractive scale, but often comparable with the refractive scale. Therefore, refractive scintillation is expected for these objects and may be manifested as low–frequency variability and flickering at centimeter wavelengths. In addition, certain phenomena, such as extreme scattering events, may indicate the presence of compact, localized structures in the ISM which can refractively produce ray crossings and other strong focusing effects. VLBI observations will be of great importance in detecting refractively induced image distortions.

1. Introduction

The density of the interstellar medium fluctuates spatially over a broad range of scales. This is probably a manifestation of turbulence. The most refractive component of the ISM is ionized gas, with the absolute refractivity increasing as wavelength–squared. Thus radio observations, particularly at low frequencies, are sensitive to the effects of interstellar scattering by density fluctuations.

The spatial power spectrum of the density fluctuations follows a well-known power law dependence of the form

$$P_{\delta n}(q) = C_n^2 q^{-\alpha} \qquad \frac{2\pi}{\ell_o} < q < \frac{2\pi}{\ell_i},$$

where q is the spatial wavenember, and C_n^2 measures the strength of the fluctuations and is in general a function of position. The power law index, α, is in many cases

[*]Center for Stochastic Processes in Science and Engineering, Physics Department, Virginia Tech, Blacksburg, Virginia 24061, USA.

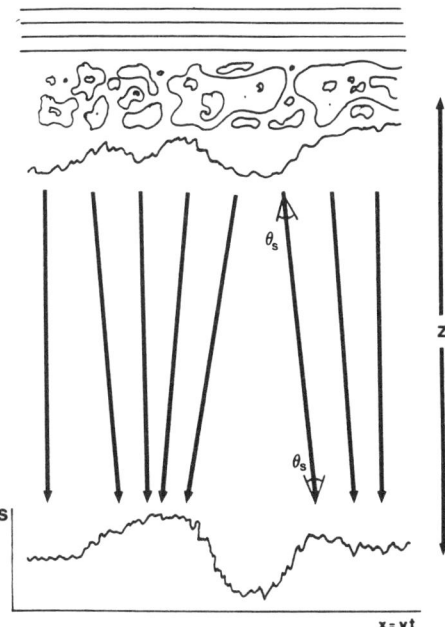

Figure 1. Interstellar scattering. Small scale fluctuations in the emergent wavefront produce angular scattering of width θ_S and diffractive scintillation. Large scale fluctuations produce refractive scintillation.

measured to be quite close to the Kolmogoroff value of 11/3, lending strong support to the notion that the fluctuations are turbulent in origin. Remarkably, this spectrum appears to persist over a vast range of wavenumbers. If the turbulent interpretation is correct, energy is injected at the outer scale, ℓ_o, cascades to smaller scales and is degraded into heat at the inner scale, ℓ_i. The outer scale may be of order parsecs [Sim92], while measurements of the inner scale range from 10^7 cm [SpG90] to 10^{10} cm [GRC92]. The existence of strong refractive phenomena seems to indicate that the simple Kolmogoroff picture is incomplete. (See below.)

2. Diffractive Scattering

A highly simplified picture illustrating propagation through this medium is shown in Figure 1. The ISM is represented as a "screen" of limited thickness at distance, Z.

Initially plane waves from a point source at infinity emerge from the medium corrugated. The corrugations, which are quantified by the fluctuation in phase, naturally cover the full range of scales present in the medium. A correlation scale, ℓ_ϕ, can be defined as the scale over which the rms phase difference is 1 radian. For a Kolmogoroff spectrum

$$\ell_\phi \approx [4\pi^2 r_e^2 \lambda^2 SM]^{-3/5} \quad ,$$

where r_e is the classical electron radius and the scattering measure, SM, is defined as the line of sight integral through the medium: $\int C_n^2(z)dz$. The existence of ℓ_ϕ presupposes that the ensemble averaged phase variance is greater than unity, i.e. $\langle\phi^2\rangle > 1$; this condition is met at all but the very highest of radio frequencies.

In many situations of interest in the ISM, ℓ_ϕ is smaller than the Fresnel scale $\approx \sqrt{\lambda Z}$, the condition of strong scintillation. Phase fluctuations smaller than the Fresnel scale act diffractively to scatter radiation. The medium behaves like a random phase grating on these scales. The scattering angle which characterizes the angular width of the scattered radiation is approximately given by, $\theta_S \approx \lambda/\ell_\phi \propto \lambda^{2.2}$. This is the well-known "λ^2 law" for scattering, or more precisely "$\lambda^{2.2}$ law" appropriate to a Kolmogoroff spectrum. An observer detects radiation arriving from a cone of approximate width θ_S. Hence the source appears scatter broadened to this angular size if its intrinsic size is smaller. For a point source, interference among the received rays results in a random diffraction pattern in the observer's plane with characteristic scale ℓ_ϕ. Motion of the line of sight relative to the screen with velocity v results in a time–changing measured intensity (diffractive scintillation) with characteristic time scale $\tau_d \approx \ell_\phi/v$. The diffraction pattern will be smeared out by sources having angular sizes much larger than the angular size subtended by ℓ_ϕ. Along high–latitude lines of sight we may typically find $\ell_\phi \approx 10^{3.8}$ km $\nu_{\rm GHz}^{1.2}$, which yields a critical angle, $\theta_C \approx \ell_\phi/Z \approx 1$ μas $\nu_{\rm GHz}^{1.2}$. Not suprisingly, pulsars are the only class of object in which diffractive interstellar scintillations have been unambiguously detected.

The relevant angular scales are shown in Figure 2 as a function of frequency. The geometric mean of the scattering angle and the critical angle is the Fresnel angle (the angle subtended by the Fresnel scale). That is, $\theta_f = \sqrt{\theta_S \theta_C} = \sqrt{\lambda/Z}$ For strong scintillation, $\theta_S > \theta_f > \theta_C$. At the frequency of transition to weak scintillation these scales meet, and at higher frequencies the Fresnel scale, $\sqrt{\lambda Z}$, is the dominant scale for weak scintillation. Scattering in this regime is not expected to be a serious problem for VLBI. At lower frequencies however, scattering limits angular resolution. Recovery of intrinsic structure information in the shaded region of the diagram would be extraordinarily difficult. Although scattering preserves information, its extraction becomes a practical impossibilty in this regime since the diffraction pattern is strongly smoothed by the source intrinsic size.

It is important to bear in mind that the angular scales shown in Figure 1 are typical of high galactic latitude lines of sight. At lower latitudes the scattering is usually much stronger resulting in a larger scattering angle, and a depressed critical angle. The effect is to increase the shaded area of the diagram and move the transition point (where $\theta_S \approx \theta_C \approx \theta_f$) to higher frequencies. The available data indicate that turbulent gas is widely distributed in the galaxy, with enhanced turbulence in the

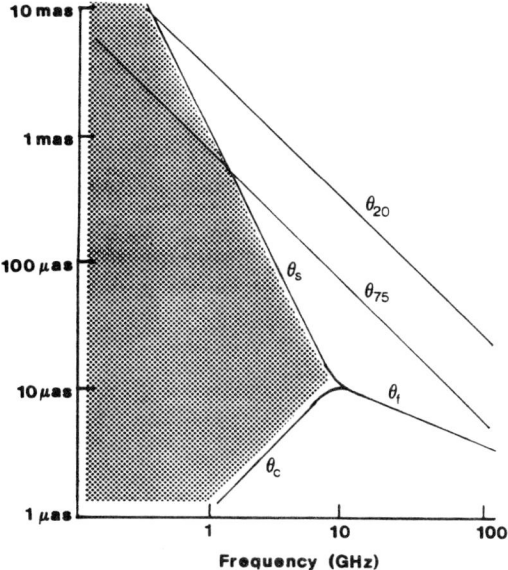

Figure 2. Typical angular scales at high galactic latitude. Also shown are the angular resolutions achieved with baselines of 20×10^3 km (θ_{20}), and 75×10^3 km (θ_{75}).

inner galaxy and clumps of intense turbulence confined to the disk, possibly spiral arms [DTB84] [CWB85] [CWF91] [FSC91].

3. Refractive Scattering

Of course, fluctuations will very probably be present on scales larger than the Fresnel scale, and these can be understood as acting refractively, in which case rays can be constructed perpendicular to the mean wavefront. Slope in the wavefront then signifies refraction, and curvature in the wavefront produces a convergence or divergence of rays, resulting in refractive amplification or deamplification. (See Figure 1.) The importance of refraction in the ISM was recognized in a series of important papers [Sha78] [Hew80] [Sie82] [RCB84].

For a Komogoroff spectrum, the characteristic refraction angle is comparable with the scattering angle [Ric86] and has nearly the same frequency dependence. Compact components with sizes comparable with or smaller than the refraction angle will then exhibit fully formed refractive scintillation. The most compact components ($T_b \approx 10^{12}$ K, $S_\nu \approx 1$ Jy) of extragalactic sources seen at moderate to high galactic latitudes should satisfy this condition at "low frequencies" ($\nu < 1$ GHz). Hence, much apparent "low frequency" variability is refractive scintillation. At GHz frequencies, the most compact components are probably about an order of magnitude larger than the characteristic refraction angle, and the consequent smoothing of the refractive scintillation makes it much weaker, quite plausibly just several percent of the total source flux. At both sub–GHz and GHz frequencies the component sizes largely determine the timescales, about 0.5 year at 300 MHz and 20 days at 3 GHz. Clearly, both of these phenomena can be understood as refractive scintillation. It is interesting to note that at low galactic latitudes ($|b| < 5°$) the GHz variations of compact extragalactic sources appear to be longer (year–timescale) and stronger, as would be expected from the larger scattering angle at these latitudes [DFJ87].

Some refractive phenomena appear, however, to require irregularity spectra with the large scales ($> \sqrt{\lambda Z}$) enhanced, either by steeper spectra ($\alpha > 4$) or additional power on certain scales. These phenomena include quite strong variations at low frequencies in a few sources [ABC84], multipath propagation episodes in pulsars [CoW86], extreme scattering events [FDJ87] and possibly some strong interday variations reported by the Bonn group [QWK89].

The best evidence for localized sites with enhanced refractive scattering comes from extreme scattering events. These are events in which the flux of a compact source is significantly disrupted, evidently by a condensation of interstellar plasma passing across the line of sight which acts as a negative lens (owing to the negative refractivity of ionized gas.) In accord with this simple picture the classic ESE shows a minimum bracketed by maxima on either side. The refracted flux then appears around a refraction "shadow".

4. Conclusions

Interstellar scattering has important implications for sub–arcsecond radioastronomy. Most importantly, diffractive scattering imposes strong limitations upon the achievable resolution. This is typically 1 mas $\nu_{\text{GHz}}^{-2.2}$ at high galactic latitudes, but can be much larger along lines of sight intersecting regions of enhanced turbulence, which are common near the galactic plane. In addition, refractive scattering is expected to produce image distortion, probably on scales similar to the diffractive scattering angle. Finally, refractive scintillation can limit the dynamic range in a map, particularly if the observations span a timescale in excess of that for the scintillation [BNR86].

The other side of the coin is that high resolution radioastronomy provides a sensitive probe of the microstructure of the ISM. Of great importance will be unabiguously detecting time–changing refractive distortions. These distortions will probably be greatest in sources undergoing ESEs. It is also important to identify the regions of enhanced tubulence as well as the sources of turbulence. We are only beginning to reconcile these "microscopic" phenomena (on a.u. scales) with the broad range of interstellar morphologies identified on much larger, typically parsec, scales.

References

[ABC84] Altschuler, D. R., Broderick, J. J., Condon, J. J., Dennison, B., Mitchell, K. J., O'Dell, S. L., Payne, H. E., *Astron. J. Vol. 89 (1984), p. 1784*.

[BNR86] Blandford, R., Narayan, R., Romani, R. W., *Astrophys. J. Lett. Vol. 310 (1986), p. L53*.

[CWB85] Cordes, J. M., Weisberg, J. M., Boriakoff, V., *Astrophys. J. Vol. 288 (1985), p. 221*.

[CWF91] Cordes, J. M., Weisberg, J. M., Frail, D. A., Spangler, S. R., Ryan, M., *Nature Vol. 354 (1991), p. 121*.

[CoW86] Cordes, J. M., Wolszczan, A., *Astrophys. J. Lett. Vol. 307 (1986), p. L27*.

[DTB84] Dennison, B., Thomas, M., Booth, R. S., Brown, R. L., Broderick, J. J., Condon, J. J., *Astron. Astrophys. Vol. 135 (1984), p. 199*.

[DFJ87] Dennison, B., Fiedler, R. L., Johnston, K. J., Spencer, J. H., Waltman, E. B., Angerhofer, P. E., Florkowski, D. R., Josties, F. J., Klepczynski, D. D., McCarthy, D. D., Matsakis, D. N., *Astrophys. J. Vol. 313 (1987), p. 141*.

[FSC91] Fey, A. L., Spangler, S. R., Cordes, J. M. *Astrophys. J. Vol. 372 (1991), p. 132*.

[FDJ87] Fiedler, R. L., Dennison, B., Johnston, K. J., Hewish, A. *Nature Vol. 326 (1987), p. 675*.

[GRC92] Gupta, Y., Rickett, B. J., Coles, W. A., *preprint (1992)*.

[Hew80] Hewish, A., *Mon. Not. Roy. Astr. Soc. Vol. 192 (1980), p. 799*.

[QWK89] Quirrenbach, A., Witzel, A., Krichbaum, T., Hummel, C. A., Alberdi, A., Schalinski, C. *Nature Vol. 337 (1989), p. 442*.

[RCB84] Rickett, B. J., Coles, W. A., Bourgois, G. *Astron. Astrophys. Vol. 134 (1984), p. 390*.

[Ric86] Rickett, B. J. *Astrophys. J. Vol. 307 (1986), p. 564*.

[Sha78] Shapirovskaya, N. Ya. *Sov. Astron. Vol. 26 (1978), p. 151*.

[Sie82] Sieber, W. *Astron. Astrophys. Vol. 113 (1982), p. 311*.

[Sim92] Simonetti, J. H. *Astrophys. J. Vol. 386 (1992), p. 170*.

[SpG90] Spangler, S. R., Gwinn, C. R. *Astrophys. J. Vol. 353 (1990), p. L29*.

A Model for Extreme Scattering Events

Brian Dennison *   *John H. Simonetti* *

Abstract

We propose a model in which extreme scattering events (ESE's) are produced by small scale structures of enhanced density near the periphery of expanding superbubbles. These structures are assumed to be oblate ellipsoids, having been compressed in the radial direction from the superbubble center. Along lines of sight nearly tangent to, but just inside, the superbubble surface refraction effects will be enhanced and timescales minimized. This model explains the apparent angular association of compact extragalactic sources exhibiting ESE's with loops in the galactic continuum brightness [FDJ92].

1. Model

ESEs pose a serious challenge since they appear to require very dense and highly compact clouds of ionized gas in the interstellar medium [FDJ87]. These difficulties may be alleviated somewhat if the causative structures are viewed transversely to their compact dimension(s). In our model ambient clouds are compressed in one dimension by an expanding superbubble. The resulting oblate ellipsoids (with axial ratio, e) lie near the surface of the outward expanding superbubble. (See Figure 1.) For lines of sight near the tangent line, timescales are significantly reduced by a factor, F, and column densities enhanced by F. The refraction angle, which is proportional to the transverse gradient of the column density, is enhanced by F^2 over that of an uncompressed cloud.

We are able to explain ESEs both in terms of the required scattering and frequency of occurence, if the compressed clouds persist behind the front within a shell which is about 3% of the superbubble radius. For a compression factor of 20, F ranges up to 20, depending the geometrical circumstances of an interception. The other parameters of this model follow. All measures of total ionized gas (e.g. emission measure) are well within acceptable bounds.

---

*Center for Stochastic Processes in Science and Engineering, Physics Department, Virginia Tech, Blacksburg, Virginia 24061, USA

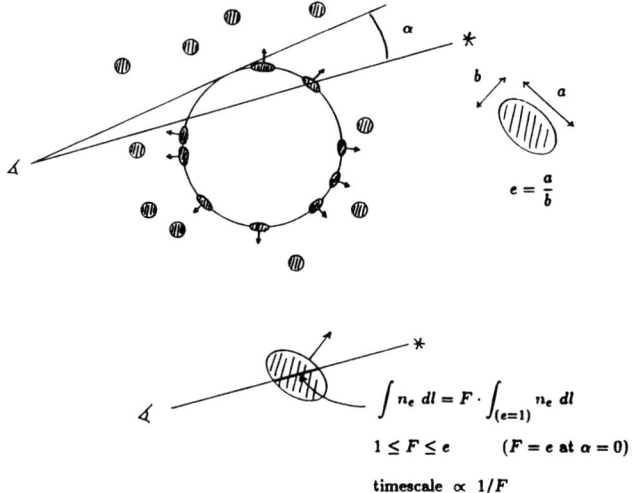

Figure 1. Cloud compression at a superbubble boundary. Note that the cloud dimensions ($\approx$ au) are greatly exaggerated in comparison with the superbubble scale (radius $\approx$ 160 pc).

2. Parameters

Distance to superbubble center = 300 pc
Superbubble radius = 160 pc
Superbubble angular diameter = 65°
Shell Thickness = 5 pc
Pre-compression cloud density = 2.5 cm^{-3}
Post-compression cloud density = 50 cm^{-3}
Pre-compression cloud radius = 5 au
Post-compression cloud semi-minor axis = 0.25 au
Covering Factor ($\alpha = 1°$) = 0.43
Emission Measure ($\alpha = 1°$) = 1.5×10^{-2} pc cm^{-6}
Mass of a cloud = 7×10^{18} gm
Surface density of clouds = 3×10^{-3} au^{-2}
Total mass of clouds = 0.15 solar mass

References

[FDJ87] Fiedler, R. L., Dennison, B., Johnston, K. J., Hewish, A. *Nature Vol. 326* (1987), p. 675.

[FDJ92] Fiedler, R. L., Dennison, B., Johnston, K. J., Waltmann, E. *preprint (1992)*.

Galaxy Centre - Anticentre Asymmetry of Low Frequency Variability Radio Source Distribution: Evidence of Interstellar Matter Influence

N. Shapirovskaya [*]   *O.B. Slee* [†]   *G.S. Tsarevsky* [*]

Low frequency variability of radio sources could be considered as an important clue to the sources compactness. This connection strongly depends on the nature of such variability.

We report an asymmetry in the distribution of low frequency variable (LFV) radio sources towards the centre-anticentre regions of our Galaxy. This study is based on 15 years of monitoring a large sample of sources observed with the Culgoora circular array at 80 and 160 MHz (Slee and Siegman, 1988). This asymmetry is in marked contrast with the symmetrical distribution of non-variable sources from the same sample. Median values of the corresponding modulation indices on both short and long time-scales also show a significant increase towards the Galactic centre compared with the anticentre region (see Table 1). These results suggest that LFV has an interstellar origin and that the most likely mechanism is refractive scintillation (Shapirovskaya, 1978; Rickett, 1986).

---

[*] Astro Space Centre, Lebedev Physical Institute, 117810 Moscow, Russia
[†] Australia Telescope National Facility, CSIRO, PO Box 76, NSW 2121

Table 1

Numbers of Radio Sources, N, and Median Modulation indices, m,
in the Centre (c) and Anticentre (ac) regions
with Galactic Latitude less than 30 Degrees

Sample and sub-sample		N_c	N_{ac}	$\frac{N_c}{N_{ac}}$	χ^2	P	m_c	m_{ac}	χ^2	P
Non-variable radiosources		32	32	1.0	-	-	-	-	-	-
Var. at 80 MHz	m1	12	4	3.0	3.20	0.03	0.19	0.14	1.20	0.130
	m12	7	3	2.3	1.39	0.12	0.30	0.23	0.34	0.280
Var. at 160 MHz	m1	21	11	1.9	2.10	0.07	0.24	0.15	7.50	0.003
	m12	10	6	1.7	0.80	0.18	0.33	0.26	0.74	0.190
Var. at 80 & 160	m1-80	7	2	3.5	3.00	0.06	0.14	0.08	7.37	0.003
	m1-160	7	2	3.5	3.00	0.06	0.24	0.10	1.63	0.100

Notes: 1) Columns χ^2 contain results of ratios N_c/N_{ac} and centre - anticentre sub-samples chi-square test for significance.
2) Columns P contain corresponding chance probabilities (one sided criteria).

We report also an unusual behavior of the modulation index versus Galactic Latitude dependence. The modulation index tend to decrease towards the galactic Plane at the both frequencies. One can see the same behavior in two opposite directions - centre and anticentre.

There is some evidence that maxima of these dependencies are shifted to higher latitude (i) with wave length and (ii) from anticentre to centre.

References

[1] Rickett B.J., 1986, *Astroph. J.*, **307**, 564.

[2] Shapirovskaya N., 1978, *Sov. Astron. J.*, **22**, 544.

[3] Slee O.B., Siegman B.S., 1988, *Mon. Not. R. astr. Soc.*, **235**, 1313.

Speckle VLBI

C. R. Gwinn [*] *K. M. Desai* [†] *J. Reynolds* [‡] *D. Jauncey* [§]
E. A. King [¶] *C. Flanagan* [∥] *G. Nicolson* [**] *R. A. Preston* [††]
D. L. Jones [‡‡]

Abstract

We present first observations of the Vela pulsar in the speckle limit of interstellar scattering. These observations indicate that density fluctuations responsible for scattering are concentrated in the Vela supernova remnant near the pulsar. As we discuss, in the speckle limit the scattering disk can be used as an AU-scale lens, to study the pulsar with nanoarcsecond resolution.

1. Introduction

A pointlike source radiating through an irregular medium will produce a diffraction pattern in the plane of the observer. The observer sees scintillations and, given sufficient angular resolution, a changing speckle pattern, on short timescales. Averages over longer timescales show a smooth scattering disk. These effects are familiar as effects of optical seeing in the earth's atmosphere. Analogous, though somewhat different effects, are predicted for radio-wave scattering by the interstellar plasma [NaG89].

A single-dish dynamic spectrum for a pulsar will show scintillations as variations of flux density with frequency, and with time, as the line of sight moves relative to the scattering material. For the Vela pulsar, we find that these variations have characterisitic timescale $t_{ISS} = 14$ sec and characteristic frequency scale $\Delta\nu = 68$ kHz. The auto-correlation spectrum measured at one antenna is, of course, a real quantity.

---

[*]Physics Department, University of California, Santa Barbara, CA, USA 93106.
[†]Physics Department, University of California, Santa Barbara, CA, USA 93106.
[‡]Australia Telescope National Facility, CSIRO, Epping NSW 2121, Australia.
[§]Australia Telescope National Facility, CSIRO, Epping NSW 2121, Australia.
[¶]Physics Department, University of Tasmania, Hobart, Tasmania 7001, Australia.
[∥]Hartebeesthoek Radio Astronomy Observatory, P.O. Box 443, Krugersdorp 1740, South Africa.
[**]Hartebeesthoek Radio Astronomy Observatory, P.O. Box 443, Krugersdorp 1740, South Africa.
[††]Jet Propulsion Laboratory, California Institute of Technology, Pasadena, California 91109, USA.
[‡‡]Jet Propulsion Laboratory, California Institute of Technology, Pasadena, California 91109, USA.

The cross-power spectrum, measured with an interferometer, is a complex quantity. Interferometric observations with sufficient time and frequency resolution measure phase and amplitude variations due to scintillation; in general these spectra are complex. The amplitude of the correlation function describes source structure, in particular whether the source is resolved. The phase of the correlation function describes source position for a point source, or internal structure for a more complicated source.

2. Observations

We observed the Vela pulsar and several comparison quasars using the SHEVE southern-hemisphere VLBI network on 1 December 1989 and 17 March 1991. Participating antennas included Hobart (26 m diameter), DSS43 (70 m), DSS45 (26 m), Parkes (64 m), and Culgoora (22 m), and Hartebeesthoek (26 m). RCP radiation at sky frequencies between 2.290 and 2.292 GHz was recorded with the Mark II recording system. The data were processed with the JPL/Caltech Block II correlator. Details of the analysis are given by Desai et al. [DGR92]. Time and frequency sampling were denser than characteristic scales of scintillation, so that our observations reach the speckle limit of interstellar scattering.

We observe phase and amplitude variations indicative of structure within the scattering disk, changing on the time and frequency scales of scintillation. Such behavior is expected in the speckle limit. From the variation of phase on short baselines, we calculate the average visibility over many speckles. This in turn yields the angular size of the scattering disk. From the characteristic bandwidth of scintillations, we find the temporal broadening of signals from the pulsar. Angular and temporal broadening both measure the integrated strength of scattering along the line of sight, but with different lever arms; comparision of the two demonstrates that the scattering material is close to the pulsar [DGR92].

3. Discussion

Cosmic-ray acceleration in young supernova remnants may produce strong plasma fluctuations [Bel78]. The variations in electron density that produce interstellar scattering may accompany these fluctuations [SMB86]. We suggest that we are observing such fluctuations directly, by scattering along the line of sight to the Vela pulsar. Further observations of scattering along this line of sight may characterize the spectrum of plasma fluctuations generated by particle acceleration.

In principle, the scattering disk can be used as an AU-scale lens to image the source with nanoarcsecond resolution. In practice, such studies are complicated by the need to correct for the effects introduced by scattering. The diffraction pattern in the plane of the observer is the convolution of such a high-resolution image of the source with the diffraction pattern due to a point soure [CAN89]. Imaging the source is thus a deconvolution problem. Preliminary examination of our data suggests that we may see some effects of resolution of the emission region of the pulsar by the scattering disk.

References

[Bel78] Bell A.R., *The Acceleration of Cosmic Rays in Shock Fronts.* M. N. R. A. S. Vol. 182 (1978), pp. 147-156.

[CAN89] Cornwell T.J., Anantharamiah K.R., Narayan R., *The Propagation of Coherence in Scattering–An Experiment Using Inter-Planetary Scintillation.* J. O. S. A. Vol. A6 (1989), pp. 977-986.

[DGR92] Gwinn C.R., Desai K.M., Reynolds J., King E.A., Jauncey D., Flanagan C., Nicolson G., Preston R.A., Jones D.L., *A Speckle Hologram of the Interstellar Plasma.* Ap. J. Vol. 393 (1992), pp. L75-L78.

[NaG92] Narayan R., Goodman J., *The Shape of a Scatter-Broadened Image.* M. N. R. A. S. Vol. 283 (1989), pp. 963-994.

[SMB86] Spangler S.R., Mutel R.L., Benson J.M., Cordes J.M., *Interstellar Scattering of Compact Radio Sources Near Supernova Remnants.* Ap. J. Vol. 301 (1992), pp. 312-319.

Highly Scattered OH/IR Stars at the Galactic Centre

P.J.Diamond*, D.Frail*, H.J. van Langevelde[†], J.Cordes[‡]

Abstract

VLBA observations of OH/IR star masers towards the Galactic Centre have revealed maser sizes of up to 520 milliarcsec, such sizes are attributed to the effects of interstellar scattering. However some sources observed in our VLBA experiment were resolved, we have used the VLA in order to determine the size of the maser components. We observe scattering sizes of up to 3.7 arcsec, these are the most highly scattered lines of sight in the Galaxy.

1. Introduction

In the past three years we have performed highly successful VLBA observations of a sample of OH/IR stars close to the Galactic Centre (van Langevelde, Frail, Cordes and Diamond, Ap.J., 396, 686, 1992; van Langevelde and Diamond, MNRAS, 249, 7p, 1991) These observations were designed to measure the angular sizes of the compact, blue-shifted OH maser components in the circumstellar shells. Since the sizes measured are all much larger than those found in typical OH/IR stars we have attributed the sizes observed to angular broadening due to fluctuations of the electron density along the line of sight to the stars. Combining our data, with other data taken from the literature, we have been able to determine the distribution of interstellar scattering towards the Galactic centre. In short we find a region of pronounced scattering (Θ_{obs} = 512mas at 1612 MHz) nearly centred on Sgr A*.

2. Observations and Data Reduction

These observations were based upon a sample of 17 OH/IR stars in the close vicinity of Sgr A*. We determined angular sizes for 13 sources, however 4 of our sample

---

*National Radio Astronomy Observatory, Socorro, New Mexico, U.S.A. The National Radio Astronomy Observatory is operated by Associated Universities, Inc. under cooperative agreement with the National Science Foundation.
[†]Sterrewacht Leiden
[‡]Cornell University

Table 1. Maser sizes

Source	Chn	Flux Jy	Maj.axis arcsec	Minor axis arcsec	P.A. degrees
OH0.334-0.181	19	8.7	2.0 ± 0.1	1.7 ± 0.1	30 ± 10
	45	10.3	2.3 ± 0.1	1.8 ± 0.1	10 ± 5
OH359.581-0.240	24	3.3	1.5 ± 0.15	1.2 ± 0.3	63 ± 16
	44	9.9	1.5 ± 0.15	1.0 ± 0.1	51 ± 15
OH0.319-0.040	16	11.8	3.2 ± 0.2	1.5 ± 0.3	49 ± 6
	47	4.8	3.7 ± 0.2	1.5 ± 0.3	23 ± 5
OH359.880-0.007	23	5.9	2.8 ± 0.1	1.1 ± 0.2	40 ± 4
	40	15.6	3.0 ± 0.1	1.0 ± 0.1	36 ± 2

were not detected on our shortest baseline VLA-PT. Acting on a suspicion that these sources were resolved on our shortest baseline we followed up with a small amount of VLA time. We observed all 4 sources (see Table 1), plus two control sources that were detected in the previous VLBA observations. We typically observed each source for 4 × 15 minute scans (the VLA was in its B and C configurations). Our results are tabulated in Table 1, for each source we observe a large angular size ($\Theta_{max} = 3.7 \pm 0.2$ arcsec), several times larger than the maximum size we had determined from the VLBA observations. These are by far the most scattered lines of sight ever seen in our Galaxy. Our control sources were unresolved by the VLA.

Fig. 1 shows the spectrum for the source OH0.334-0.181, Fig. 2a shows the visibility curve for the strongest channel in the maser spectrum. As can easily be seen the resolved nature of the OH maser is immediately evident, in contrast Fig. 2b shows the visibility curve from one of our control sources. We determined the sizes of the OH masers in two ways: (1) by fitting to the phase-calibrated real partion of the complex amplitude; and (2) by making an image, deconvolving it and determining the source size by Gaussian fitting. In general both methods agreed as to the size of the maser features, however it was evident that the maser shapes were not symmetrical Gaussians. As can be seen in Table 1 there is a very clear asymmetry in all of the fitted sizes. This result shows that the scattering discs are highly anisotropic, implying a preferred direction for the electron density fluctuations in the direction of the Galactic Centre. Unfortunately our data had insufficient uv-coverage for us to confidently determine the true shape of the scatter-broadened maser features. Further observations are planned to accurately determine the shape of these features.

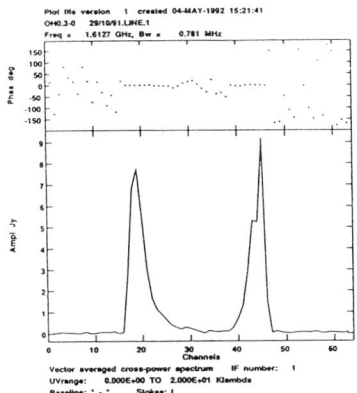

Figure 1. The cross power spectrum of the OH maser towards the star OH0.334-0.181 as observed by the VLA. The top frame shows the phase as a function of velocity, the bottom frame shows the amplitude.

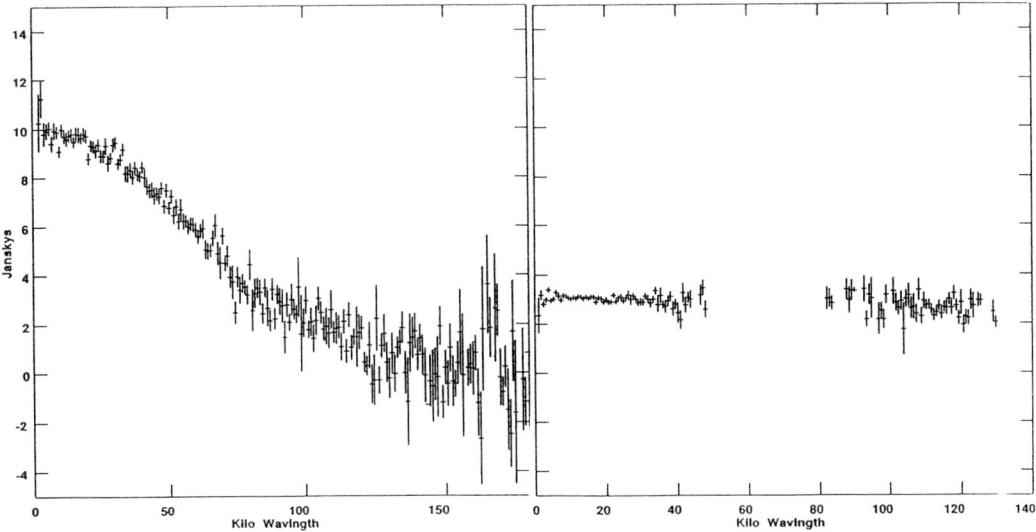

Figure 2. The visibility functions of the strongest spectral channel for 2 sources; the left-hand plot is for OH0.334-0.181 and clearly shows the effects of significant resolution. The right-hand plot is of OH359.762+0.120, one of our control sources; it is obviously unresolved. The vertical scales in each plot are identical.

Spectrum of Interstellar Turbulence: Observations of OH and H₂O Masers in W49N

Ketan M. Desai * *Carl R. Gwinn* [†] *Philip J. Diamond* [‡]

Abstract

We present preliminary angular broadening measurements and limits on detection of substructure within scattering disks for OH masers in W49N. These and previous observations of H_2O masers in W49N probe turbulent fluctuations of electron density in the interstellar medium, on length scales from 100 km to 10 AU.

1. Introduction

Turbulent fluctuations of electron density in the interstellar medium scatter radiowaves. Scattering by fluctuations on scales smaller than ~ 1 AU produces diffractive effects like pulsar scintillation and angular broadening of compact radio sources. Scattering by fluctuations on scales larger than an AU produces refractive phenomena such as slow scintillation, image wander and substructure in the scattering disk. The power spectrum of these fluctuations can be characterized using the phase structure function D_ϕ which measures the relative phase variations induced by the scattering plasma over a spatial scale L.

Fluctuations on small spatial scales angularly broaden compact radio sources; the angularly broadened image is called a scattering disk. Observations of scattering disks measure D_ϕ directly on spatial scales $\sim$ terrestrial baselines.

Fluctuations on large spatial scales focus and defocus the scattering disk; this produces substructure within the disk. Observations of the interferometric visibilities of scattering disks at large spatial frequencies B_λ measure substructure from which D_ϕ is inferred on a scale $L \sim D/B_\lambda$ where D is an assumed distance to the scattering material.

---

*National Radio Astronomy Observatory, Socorro, NM 87801 and Physics Department, University of California at Santa Barbara, CA 93106.
[†]Physics Department, University of California at Santa Barbara, CA 93106.
[‡]National Radio Astronomy Observatory, Socorro, NM 87801.

Fluctuations on large spatial scales act as refractive prisms to deflect images from their true positions; these deflections are called image wander. Observations of the image wander of a moving maser allows D_ϕ on be inferred on scales $L \sim$ linear distance traversed by the maser.

2. Results

We have measured D_ϕ over spatial scales from 10^6 to 10^{15} cm using observations of H_2O masers and OH masers in W49N.

Angular broadening of masers has been used to measure D_ϕ on scales from 10^6 to 10^9 cm. Observations of H_2O masers indicate that D_ϕ has a power-law dependance on baseline length with a best fitting index of $\beta = 1.85$ for baselines from 10^8 to 10^9 cm [GMR90]. Preliminary results of observations of OH masers indicate that this index may steepen below 10^7 cm.

Proper motions of H_2O masers have been used to measure image wander over a period of 7 months [GMR90]. The inferred value of D_ϕ at a scale of 10^{13} cm is roughly in agreement with the predicted value from H_2O angular broadening measurements.

Detection of substructure within the scattering disks of OH masers is expected to yield a value for D_ϕ on a scale of 10^{15} cm. Analysis of our data has not yet placed useful limits of the value of D_ϕ at this scale.

3. Conclusions

In "fractal" theories of turbulence such as the Kolmogorov theory D_ϕ would be expected to have power-law dependance on length scale in the interstellar medium. In such theories energy is injected into the medium at some outer scale, cascades to smaller scales, and is dissipated at some inner scale. Our observations suggest that a power-law spectrum of density fluctuations may extend from $> 10^{13}$ to $< 10^8$ cm.

References

[GN89]   Goodman J., Narayan R. *The shape of a scatter-broadened image - II. Interferometric visibilities.* M.N.R.A.S. 238 (1989) pp. 995-1028.

[GMR90] Gwinn C.R., Moran J.M., Reid J.M. *Interstellar Scattering of Pulsars and Masers.* Radio Astronomical Seeing, ed. J. Baldwin & Wang Shoguan, International Academic Publishers, Beijing. (1990) pp. 225-228

VLBI Structures of Low Frequency Variables: Implication for Refractive Scintillation Theory

Padrielli L.*, Bondi M.*, Gregorini L.*, Mantovani F. *
Eastman W.‡, Shapiroskaya N. †   and Spangler S. ‡

1. Introduction

It is generally accepted that Low Frequency Variability (LFV) ($\nu < 1 GHz$) is mainly due to propagation effects in the interstellar medium. Only in $< 20\%$ of cases it is due to intrinsic mechanisms such as relativistic expansion of synchrotron emitting beams. In these sources the variability is correlated over a very wide frequency range, from millimetric to decimetric wavelengths. In the majority of cases ($> 80\%$) the low frequency activity is completely unrelated to the high frequency one: it is attributed to refractive scintillation from the interstellar medium. In the last years we concentrated our studies to the latter class of variables trying to provide further observational tests for the refractive scintillation hypothesis proposed by Shapirovskaya (1978, Sv.A. 22, 544) and Rickett et al. (1984, A&A, 134, 390).
The analysis of flux density time series of a sample of about 50 extragalactic sources monthly monitored with the "Northern Cross" radiotelescope in Bologna at 408 MHz provided us scintillation indices $m = \sigma_v/S_{mean}$, and variability time scales τ (for several cases two characteristic time scales were measured). The knowledge of the low frequency structure of the varying components is necessary together with m and τ for a rigorous exploration of the refractive scintillation theory.

2. Observations and Results

We obtained 608 MHz VLBI images of eleven low frequency variable sources chosen from the large roster of sources monitored at 408 MHz, the majority of them have large amplitude variability and represent some of the most extreme examples of the LFV phenomenon. MK2 observations were done in June 1988, with the stations of Effelsberg, Westerbork, Jodrell Bank, Green Bank, Iowa, Fort Davis, and Owens

---

*Istituto di Radioastronomia, CNR, Bologna, Italy
†Space Research Institute, Moscow, CSI
‡Dep. of Physics and Astronomy, University of Iowa, USA

Valley. They consisted of a series of snapshots, each of approximately half hour duration. Observing time for each source ranges from 4 to 5 hours. Resolution is about 10 mas in E-W direction and from twice to several times in N-S direction. We estimate that systematic errors in the amplitude calibration are less than or of the order of 5%. The calibrated visibility data were mapped using hybrid mapping techniques adapted from the methods of Cornwell and Wilkinson and the maps were published by Padrielli et al (1991, A&A 249,351).

In all cases (8 sources) in which a comparison can be made with VLBI maps at higher frequency a good correspondence exists in the position angle of the extended structure. Since higher frequencies (18 and 6 cm) are substantially less affected by interstellar scintillation, we conclude that the structures seen in our maps are intrinsic structure rather than scintillation pattern images. These observations are therefore of interest in a wider context than exclusively interstellar medium propagation phenomena, giving a contribution to the study of jet propagation from the cores of these objects to more remote regions, on angular scales from 10 to 100 mas.

We discuss here the use of these maps to investigate the relationship between source structure and LFV. We defined a model for the interstellar medium consistent with that deduced by pulsars characterized by : i) a power law with a Kolmogorof spectrum for the distribution of the plasma turbulence irregularities $C_N^2 = q^{-\alpha-2}$, $\alpha = 5/3$; ii) a scattering measure $C_N^2 \simeq 3*10^{-4} m^{-22/3} kpc$; iii) thin screen at a distance of 250 pc; iv) relative pattern speed, which includes contributions of the observer motion due to solar motion with respect to the local standard of rest and Earth orbital motion, $v \simeq 30 km/s$.

This model and the measured source structure allow us to calculate (theoretically), on a source-by-source basis, the expected scintillation index and variability time scale.

1) In fig. 1 observed m's are plotted against the inverse of the expected ones, the sketched area represents the locus of points predicted by the theory. The horizontal line is the detection limit of the Bologna variability observations. For most of the sources (8/11) the observed variability indices are consistent with those expected from our *model*, for the remaining three sources the values of m are considerably greater than expected. We speculate that these extreme variations are caused by the same type of irregularities which produce occasional refractive phenomenon in pulsars (i.e. steeper spectrum for the irregularities distribution).

2) The comparison between theoretical and observed variability timescales is complicated by the fact that the predicted timescale depends on a further unknown parameter: the direction of the velocity of the medium with respect to the major axis of the source structure. Fig 2 shows the comparison. Observed variability scales are shorter than those predicted by about a factor three. This result, pointed out in earlier examination of the refractive theory (Spangler et al. 1989, A&A 209, 315), persists even with the use of a more sophisticated theory.

3) This analysis gives however a natural explanation for the phenomenon of dual timescale in the variability of some sources. In the cases of the double component

sources 1358+62 and 1117+14, if we assume that the motion of the irregularities is along the source axis, we can predict a double time scale behaviour, very similar to that we found.

4) Finally, if the scale of scintillation pattern is > 1 AU, having assumed that the pattern speed is of the same order as the earth orbital velocity, we expect refractive scintillation to be dependent on the ecliptic latitude and flux density time series to show one year modulation due to earth orbital motion across the diffraction pattern. We have produced the Structure Functions (SF) of 33 homogeneous 408 MHz variable sources and we have performed the weighted average SF shown in fig. 3. The plot shows a flattening around 6 months, that corresponds to the expected modulation of one year. Severe checks for possible instrumental effect have been performed and an upper limit of 0.5% to any spurious annual effect has been estimated, while the effect of fig. 3 is of the order of 4%. Simulated time series confirm the significance of the result.

Qualitatively we interpret the presence of the 6 month time scale in the averaged SF as due to the earth orbital motion around the sun providing a new observational evidence of the refractive scintillation theory.

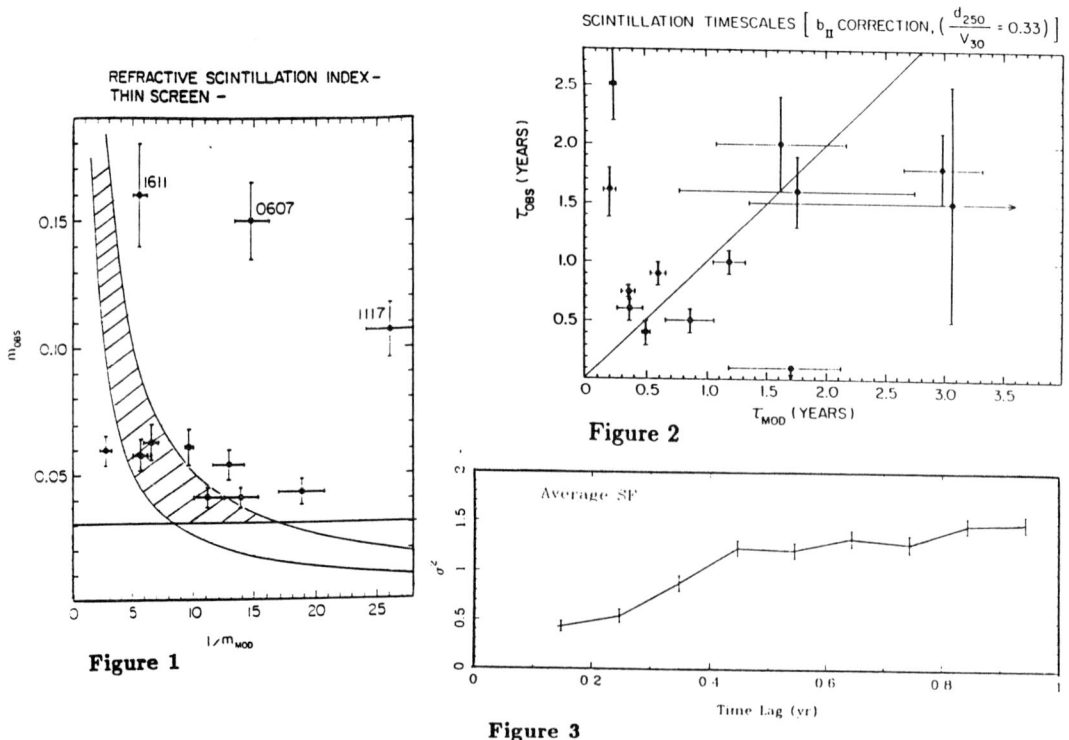

Figure 1

Figure 2

Figure 3

Three Epoch VLBI Observations at 1.6 GHz of a Sample of Low Frequency Variable Sources

Bondi M.*, Padrielli L.*, Fanti R.*, Gregorini L.*, Mantovani F.*
Romney J.D.[†]   Bartel N.[‡]   Weiler K.W.[§]
Nicolson G.D.[¶]

1. VLBI Observations at 18 cm

In 1980 we started a monitoring program at 18 cm on 21 LF variable sources to obtain direct information on their size and surface brightness and to investigate the possible correlations between LF variations, spectral behaviour of variability and structural changes. Table 1 lists the sources of the sample.
First-epoch observations were made in Feb. 1980, second-epoch in Oct. 1981. Further observations involving only 4 sources were made in Oct. 1983 (epoch 2.5) and third-epoch observations of the complete sample in Oct. 1987. The sources were observed with MKII recording system in a snap-shot mode.
Hybrid maps were independently made for each epoch. To evaluate structural changes a careful statistical comparison was directly made on the uv plane: fringe visibility amplitudes and closure phases were smoothed with a "box" of 0.25 Mλ obtaining a set of data points used for a direct comparison. Thus a χ^2 analysis was applied to the subsample of data. The analysis was performed for different baseline lengths and different position angles on the uv plane. The observed changes were described, from the bidimensional distribution of fringe visibility and closure phases, on the basis of simple source models.
1) The results can be summarized as follows: between the first two epochs (1.7 years apart), 9 sources (43%) show statistically significant structural changes and 4 (19%) show probable changes;
2) between the second and the third epoch (6 years apart), 16 (80%) show statistically significant changes and 1 (5%) shows probable changes.

---

[*]Istituto di Radioastronomia, CNR, Bologna, Italy
[†]National Radio Astronomy Observatory, Charlottesville, USA
[‡]Harvard Smithsonian Center for Astrophysics, Cambridge, USA
[§]Naval Research Laboratory, Washington, DC 20375, USA
[¶]National Institute for Telecommunication Research, South Africa

In fig. 1 the χ^2 distribution relative to the I–II and II–III epochs is given.

2. Comparison between VLBI monitoring and spectral variability behaviour

Considering the three epochs, we found variation corresponding to an increase of separation of a double source model with expansion rates from 0.1 to 0.6 mas/year for 3C345, BL Lac, 0224+67, and 0605-08. The rates found for 3C345 and BL Lac are in agreement with high frequency VLBI observations and those of 0224+67 and 0605-08 confirm the first two epoch analysis. All of them show broad band frequency activity. The variability appears to be correlated across the whole radio frequency band consisting of outbursts either quasi simultaneous or regularly drifting in time towards lower frequencies, with somewhat reduced amplitude.

The remaining sources show changes that can be interpreted as flux variations of a source component. All these sources (except 1510-08) have high and low frequency activity unrelated.

The observational scenario is in agreement with the interpretation that in a minority of cases $(10-20\%)$ the LFV is due to relativistic motions close to the line of sight. The relativistic effects increase the apparent brightness temperature and compress the time scale of the events. This model has the advantage of explaining both superluminal motions and LFV behaviour. In the majority of sources, however, the observed LF flux changes are not associated to superluminal motions. They are explained as an effect of focusing by irregularities in the local interstellar medium (refractive scintillation). Relations between angular size, percentage of variability, variability time-scale and interstellar medium properties are under investigation.

Table 1. The sample of low frequency variables.

Name	Id.	z	m_v	Name	Id.	z	m_v
0202 + 149	QSO	–	21.9	0224 + 671	QSO	–	19.5
0316 + 413	Gal	0.02	12.7	0333 + 321	QSO	1.26	17.5
0405 − 123	QSO	0.57	17.1	0422 + 004	BL Lac	–	16.0
0605 − 085	QSO	0.87	18.0	0607 − 157	QSO	0.32	18.0
0723 − 008	QSO	0.13	18.0	0736 + 017	QSO	0.19	16.5
0859 − 140	QSO	1.33	16.6	1055 + 018	QSO	0.89	18.0
1116 + 128	QSO	2.12	19.3	1127 − 145	QSO	1.19	16.9
1504 − 166	QSO	0.88	18.5	1510 − 089	QSO	0.36	16.5
1611 + 343	QSO	1.40	17.5	1641 + 399	QSO	0.59	16.0
1730 − 130	QSO	0.90	18.5	2200 + 420	BL Lac	0.07	14.5
2251 + 158	QSO	0.86	16.1				

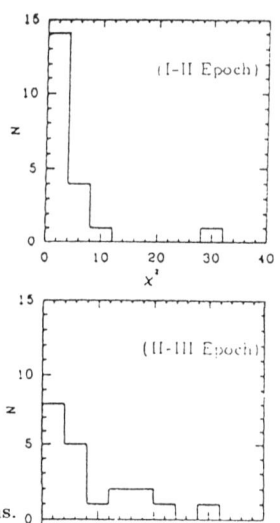

Figure 1. χ^2 distributions.

Interferometric Study of Interstellar Turbulence

Alexander Lazarian *

Abstract

Turbulence in the ISM is different from that in laboratory. Injection of energy from small scales is very important as are other factors such as self-gravity and shock waves. Kolmogorov's simple picture is no longer valid. It is shown here that interferometric data can be used directly to find the spectrum of ISM turbulence, provided that the measuring instruments are sufficiently sensitive.

1. New Method

To study the ISM turbulence, one should use fluctuations of the characteristics, for example of intensity, which are available from observations provided these fluctuations are due to turbulence. Along the line of sight, eddies of different size contribute to parameters measured. This problem was analysed elsewhere ([Laz91]); it was shown that analytical solutions exist expressing the statistical characteristics of turbulence like the correlation function of density $b(r) = \langle \delta n_x \delta n_{x+r} \rangle$ (where δn_x is a fluctuation at the x point) and the spectrum of the turbulence $E(u) = \frac{2}{\pi} \int_0^\infty ur \sin(ur) b(r) dr$ in terms of correlation functions of intensity $K(\sqrt{(x-x')^2 + (y-y')^2}) = \langle \delta I(x,y) \delta I^*(x',y') \rangle$ (δI is, for example, the intensity fluctuation) ([Laz 92]). The turbulence was considered to be isotropic on its characteristic scale. However, it can be shown that anisotropies at large turbulence scales do not influence the solution of the inverse problem at small scales. But to study turbulence one needs high resolution. The characteristic scale of the turbulence, for example, in HI discs is ~ 7 pc ([Bak74]) and is even smaller in molecular clouds. This makes it difficult to study such turbulence with the known statistical method ([Laz91]).

This paper deals with a new method of studying turbulence using radiointerferometric data without preliminary restoring the intensity distribution. It is shown here that a dispersion of the radiointerferometric signal corresponds to the spectrum of the ISM turbulence.

---
*DAMPT, University of Cambridge, Silver Street, Cambridge, UK.

Consider the problem in more detail. Using a radiointerferometer, one can measure the one dimensional Fourier transform of δI:

$$D_i(u) = æ \int \int \delta I(x,y) e^{2i\pi ux} dx dy \qquad (1)$$

where æ is a coefficient. The dispersion of the signal is $\langle S_i(u) \rangle = \langle (\mathcal{R}eD_i(u))^2 \rangle + \langle (\mathcal{I}mD_i(u))^2 \rangle$, where $\mathcal{R}et$ and $\mathcal{I}mt$ are the real and the imaginary part of t. Thus

$$\langle S_i(u) \rangle = æ^2 \int \int \langle \delta I(x,y) \delta I^*(x',y') \rangle \cos(2\pi u(x-x')) dx dx' dy dy' \qquad (2)$$

or

$$\langle S_i(u) \rangle = æ^2 \int \int K(\sqrt{(x-x')^2 + (y-y')^2}) \cos(2\pi u(x-x')) dx dx' dy dy' \qquad (3)$$

where $\langle \rangle$ denotes an ensemble averaging procedure. Denoting $x - x' = p$ and $y - y' = b$ it is easy to see that

$$\langle S_i(u) \rangle = 4\sigma æ^2 \int \int K(\sqrt{p^2 + b^2}) \cos(2\pi up) dp db \qquad (4)$$

where σ is the area over the image of the object over which the signal is collected. Introducing a new change of variables $p^2 + b^2 = \theta^2$ and $p = \eta$ one gets

$$\langle S_i(u) \rangle = 4\sigma æ^2 \int_0^\infty d\eta \cos(2\pi u\eta) \int_\eta^\infty \frac{K(\theta)\theta}{\sqrt{\theta^2 - \eta^2}} d\theta \qquad (5)$$

and changing the order of integration in Eq.(5) it is possible to obtain

$$\langle S_i(u) \rangle = 4\sigma æ^2 \int_0^\infty d\theta K(\theta)\theta \int_0^\theta \frac{\cos(2\pi u\eta) d\eta}{\sqrt{\theta^2 - \eta^2}} \qquad (6)$$

The second integral in Eq.(6) corresponds to the Bessel function of zeroth order $J_0(x)$ and thus the final expression for the $\langle S_i(u) \rangle \rangle$ is

$$\langle S_i(u) \rangle \rangle = 2\pi \sigma æ^2 \int_0^\infty d\theta K(\theta) \theta J_0(2\pi u\theta) \qquad (7)$$

The last expression for the dispersion of the *interferometer signal* $\langle S_i(u) \rangle \rangle$ is proportional to the expression for the *ISM turbulence spectrum* $E(u)$ (e.g. $\frac{E(u)}{u^2} \sim \langle S_i(u) \rangle \rangle$) found previously ([Laz 92]). This result is very important as it shows that one can directly measure the spectrum of the ISM turbulence using an interferometer. Therefore the high resolution power of the ground and space-based interferometers enables one to get important information on the ISM turbulence.

References

[Bak74] Baker, P. L., 1974: Ap.J., **187**, 223.

[Laz91] Lazarian, A., 1991: in *Fragmentation of Molecular Clouds and Star Formation.*, ed. E. Falgarone, F. Boulanger & G. Duvert, Kluwer, p. 65.

[Laz 92] Lazarian A., 1992: *Astron. and Astrophys. Transactions* (in press)

Observations of Gravitational Lenses

Dennis Walsh *

1. Introduction

In the relatively short time (as it seems to me) since the discovery of the first gravitationally lensed system in 1979, the subject has shown almost explosive growth. The numbers of papers published in the five–year periods 1974–78, 79–83 and 84–88 were 36, 191 and 583 respectively [Wa90]. When the figures for 1989–93 are in, I have no doubt this growth will be seen to have continued.

What do we hope to achieve by this effort? The key point is that gravitational deflection of light is sensitive to mass of all kinds, luminous or 'dark', baryonic or not. Thus, some quantities we can hope to determine are:

- Masses of galaxies and clusters by a method independent of traditional, less direct, dynamic ones.
- H_o from the time delay between light–curves of images of a variable source.
- The cosmological parameters Ω and Λ from statistics of lensing as a function of redshift.

This list is not exhaustive, but it indicates the possible uses of lensing. For the first two of these problems, detailed study of individual objects is needed, but of course they have to be found first. For the third, well–defined surveys are needed.

In what follows, I emphasise the rôle of radio studies, which have been particularly valuable for multiple–imaging of QSOs. This is not intended to belittle the importance of optical searches and studies, which have made major contributions, particularly in the study of 'luminous arcs' produced by the lensing action of clusters on background galaxies, but rather to reflect the interests of this conference.

2. Generic Properties

Some properties of a simple lens are helpful in identifying and interpreting the images of lensed systems and will be described without discussing the well–developed underlying theory (see, for example, the review [BN92] and references therein). Most known systems can be understood in terms of a simple elliptical galaxy as lens. For this case the object plane of Fig.1 is divided by two caustics into three regions: a source outside both caustics has a single image, when it is between the caustics three images, and when it is inside the inner 'diamond' caustic five images. The image plane is divided by critical curves; a critical curve is an image of a caustic, the inner critical curve corresponding to the outer caustic. An image of a small source is related to the source by a magnification tensor. Surface brightness is conserved and the flux of an image is determined by its magnification.

It is instructive to move a small test source along a track in the object plane of Fig.1, starting well outside the outer caustic and ending at the centre. Initially the position and flux of the single image A are virtually the same as those of the source, but it

---

*Nuffield Radio Astronomy Laboratory, Jodrell Bank, Macclesfield, Cheshire SK11 9DL, U.K.

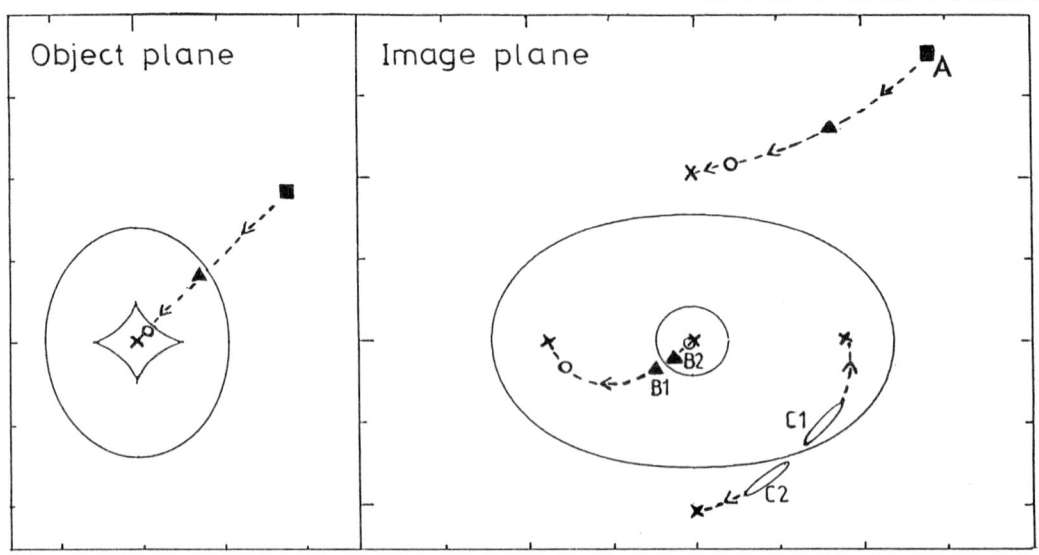

Figure 1. Caustics and critical lines in object and image planes respectively. Corresponding source and image positions are indicated by the same symbol; only for images C1 and C2 is an indication of the magnification given.

stays outside the outer critical curve and ends on one of the principal axes of the system. The magnification varies smoothly along this track. As the source crosses each caustic two new images appear, one each side of the corresponding critical line: first B1 outside and B2 inside the inner critical line as the outer caustic is crossed, giving a three–image system; then C1 inside and C2 outside the outer critical line as the inner caustic is crossed, giving a five–image sytem. The magnification for a point source on a caustic is infinite, but conservation of surface brightness ensures that flux magnification remains finite for an extended source. Each pair of new images have opposite parities (ie mirror–inversion symmetry). A, B2 and C2 have the parity of the source, ie positive parity, B1 and C1 have negative parity. Close to a critical line a pair of images of the source are mirror images of each other with equal flux magnification and images B1 and B2 stretch radially towards each other while C1 and C2 stretch tangentially towards each other (the latter effect is indicated in Fig.1 by using symbols appropriate to images of a circular source). Hence reference to radial and tangential caustics and critical lines. More complicated lenses may give rise to more caustics, and every time a source crosses a caustic two new images appear or disappear. Thus there must always be an odd number of images.

An image track never crosses a critical line and (i) B2 moves to the centre of the image plane; (ii) C2 ends on the same principal axis as A and at a point of symmetry; (iii) B1 and C1 end at symmetrical points on the other principal axis. (This final configuration is the so–called 'Einstein Cross'). The magnifications of new images fall off rapidly as they move away from the critical line where they were born. In practice virtually all known systems have two or four detected images. This is because the B2

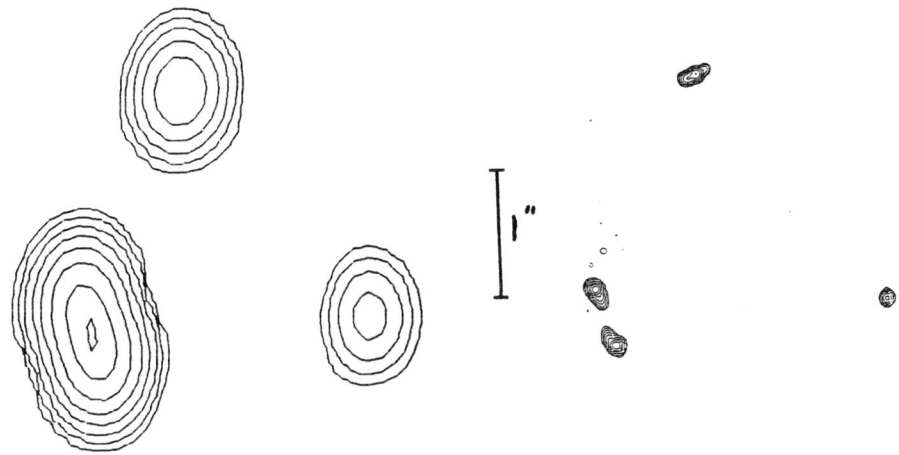

Figure 2. 5–GHz maps of the system 0414+053: left, VLA, 0″.5 beam [HB89]; right, MERLIN, 0″.05 beam [GP93].

image is usually strongly de–magnified. It is very sensitive to the core radius of the lens, the de–magnification being greater for smaller core radii with the flux vanishing in the limit of a point mass or a galaxy with zero core radius.

If a source is sufficiently extended to cover substantially all the inner caustic, the image has the 'Einstein ring' configuration. If somewhat less of the inner caustic is covered, incomplete ring images and extended structures result.

3. Importance of sub–arcsecond radio observations

It is particularly appropriate to address this topic in the context of this conference. The key factor is the range of sizes of known lensed systems: of ~ 14 securely established systems listed later, only two have maximum component separations $\theta_{max} > 2''.3$, and the smallest has $\theta_{max} = 0''.33$. The value of sub–arcsecond observations can be considered under several headings.

(i) **Recognition of multiply–imaged systems.** The resolutions available with the VLA, MERLIN and VLBI enable radio searches to be extended well beyond the limits of available optical resolution. This has the obvious advantage of enabling smaller systems to be found. Less immediately obvious is that radio sources normally have extended structure and the imaging of this can give confidence in the identification, even in the absence of optical counterparts. For example, only two of the five known ring sources have optically identified lens galaxies. Again, the 'quad' configuration arising when a source is just inside the inner (tangential) caustic is quite distinctive, consisting of two close, strong components C1 and C2 of Fig.1 stretching towards each other with mirror symmetry, plus two fainter components A and B1. The second lensed system to be discovered, 1115+080, has this configuration of optical components and 0414+053 is a virtually identical radio system. In Fig.2 the VLA map is strongly suggestive of a lensed system; if any doubt existed , the

MERLIN map, resolving the close components and showing the expected stretching, would remove it (even without the recent optical confirmation). Similarly the system 1938+666 [PB93a] is undoubtedly a lensed system intermediate between a quad and an incomplete ring.

In addition to the geometrical configurations of images, their polarisation characteristics can provide further evidence, particularly when several frequencies are available. This is invaluable in the confident identification from radio data alone of the recently discovered systems 0218+357 and 1422+231 [PB93a, PB93b]. For radio evidence to give identifications in these ways requires many pixels across the source, so resolution of $\leq 0''.1$ is desirable for the range of sizes of known systems.

(ii) Interpretation of multiply–imaged systems. This topic is discussed in the companion paper by Kochanek and little further comment is needed, except to emphasise that the number of constraints on the model of the system depends directly on the number of resolution elements across the image.

4. Searches for lensed systems

Optical searches have been reviewed recently by [Su90, SC92] and will not be discussed here. Two major radio searches are in progress, the MIT survey [Bu90] and the Jodrell Bank survey [PB93a, KB93]. Both use VLA snapshots to select candidates for further study, with more detailed VLA observations for selected candidates. The Jodrell survey has followed up with MERLIN 5–GHz maps with 50 mas resolution, and the MIT group also plan to take advantage of MERLIN.

The major difference between the two surveys is that the Jodrell Bank survey is of flat–spectrum sources only, whereas the MIT survey includes all spectral types. A flat–spectrum source usually has an intrinsic structure dominated by a compact core and multiple imaging of the core is readily recognisable. In the Jodrell Bank survey 90% of the maps of sources are dominated by a single compact component and can readily be rejected. The inclusion of steep–spectrum sources in the MIT search should lead to a larger fraction of lensed systems because the cross–section for an extended source is greater. For example, the discovery of ring images of radio lobes in 1654+134 and 1549+304 would not have been made in the Jodrell Bank survey. However, recognising multiple imaging in an inherently extended structure is more difficult than for a compact source and the Jodrell survey is likely to be more complete, which is important for statistical cosmological studies.

5. Known multiply–imaged systems

Table 1 summarises the properties of securely established lensed systems. I have taken the list of [BN92] (which gives full references) and updated it by the addition of four recently discovered systems. [BN92] adopt the most conservative conditions, excluding candidates that have two optical images but no additional confirmation, such as a plausible lensing galaxy, because distinct objects can have quite similar spectra. They exclude five doubles in the latter category, some of which are strong

Table 1. 'Secure' multiply–imaged systems, July 1992. z_s, z_d – source and deflector (= lens) redshifts; O/R – optical/radio.

Source Name	θ_{max}	No. Images	z_s	z_d	Discovery O/R	Year
0957+561	6".1	2	1.41	0.36	R	1979
2016+112	3".8	3	3.27	1.01	R	1984
1115+080	2".3	4	1.72	?	O	1980
0142–100	2".2	2	2.72	0.49	O	1987
0414+053	2".1	4	2.63	?	R	1988
1131+045	2".1	Ring	?	?	R	1988
1654+134	2".0	Ring	1.75	0.25	R	1989
2237+031	1".8	4	1.69	0.039	O	1985
1549+304	1".7	Ring	?	0.111	R	1991
1413+117	1".4	4	2.55	?	O	1988
1422+231	1".3	4	3.62	?	R	1992
1830–211	0".98	Ring	?	?	R	1988–91
(0952–01	0".95	2	4.5	?	O	1992)
1938+666	0".92	4	?	?	R	1992
(1208+101	0".45	2	3.80	?	O	1992)
0218+356	0".33	Ring	?	0.685	R	1991

candidates, one being 1208+101 [MB92, MS92], and would have excluded the recently announced 0952-01 [MI92], both of which I have listed in Table 1 (in parentheses as a health warning!) to draw attention to these sub–arcsecond candidate systems discovered in the last year.

Since this conference is concerned with resolution, I have listed objects in order of θ_{max}, the maximum separation of components. Some points of interest are: (i) Ten were found as a result of radio observations, two of which were serendipitous; four secure systems were found by optical means, two being serendipitous. (ii) For only five systems are redshifts of both source and lens known. (iii) Four lensed systems were found in six years starting with the discovery of the first in 1979, then six in the next six years prior to the International Conference on Gravitational Lensing held at Hamburg ten months ago. At that conference and subsequently a further four secure systems and two candidates have been reported.

It is striking that the systems reported in the last year are grouped at the low end of the range of θ_{max}, four of the new systems being smaller than any of the previous ones. This suggests that previous searches were resolution limited and that speculation about absence of small–separation systems [Kr89] was premature.

6. Conclusions

How far have we achieved the goals listed in Section 1?
- Probably the best determination of the mass of a galaxy (or at least its inner region) is that of the lensing galaxy of the ring system 1654+134 [LC90]. The

mass/light ratio, M/L, of 15.9 puts on a more secure footing values found by traditional means. Optical studies of luminous arcs and arclets (summarised by [BN92]) have given $M/L \geq 100$ for clusters, again not unsuspected, but putting previous estimates on a sounder basis. They also show that dark matter is reasonably correlated with cluster red light.

- The one case where a time delay between light curves has been determined is for 0957+561. Unfortunately the mass distribution of the lens is quite complicated. However, derived values of H_o of 15–80 kms^{-1}Mpc^{-1} [BN92] overlap those estimated by other means. Kochanek [Ko90] has suggested that the ideal system is a radio ring with a compact variable component. The system 0218+356 [PB93a, PB93b] fits this prescription, and others may be found. I have little doubt that lenses will provide the definitive value for H_o in the next few years.

- The quest for Ω and Λ will undoubtedly be a longer one, requiring more extensive and controlled surveys. However, even with the heterogeneous samples available now, the appetite has been whetted (eg. [FT91]).

Not bad for the first thirteen years!

References

[BN92] Blandford, R.D. & Narayan, R., 1992, *ARA&A*, **30**, 311.
[Bu90] Burke, B.F., 1990, *Gravitational Lensing, Proc. Toulouse Workshop*, p.127, Springer–Verlag, Berlin.
[FT91] Fukugita, M. & Turner, E.L., 1991, *MNRAS*, **253**, 99.
[GP93] Garrett, M.A., Patnaik, A., *et al.*, 1993. This volume, p.128.
[HB89] Hewitt, J.N., Burke, B.F., *et al.*, 1989, *Gravitational Lenses, Proc. M.I.T. Conf.*, Springer–Verlag, Berlin.
[KB93] King, L.J., Browne, I.W.A., *et al.*, 1993. This volume p.144.
[Ko90] Kochanek, C.S., 1990, *Gravitational Lenses, Proc. Toulouse Workshop*, p.244, Springer–Verlag, Berlin.
[Kr89] Krauss, L.M., 1989, *Gravitational Lenses, Proc. M.I.T. Conf.*, Springer–Verlag, Berlin.
[LC90] Langston, G.I., Conner, S.R., *et al.*, 1990, *Nature*, **344**, 43.
[MB92] Maos, D., Bahcall *et al.*, 1992, *Ap.J. Lett.*, **386**, L1.
[MS92] Magain, P., Surdej, J., *et al.*, 1992, *A&A*, **253**, L13.
[MI92] McMahon, R., Irwin, M., & Hazard, C., 1992, *Gemini*, issue 36, 1, RGO.
[PB93a] Patnaik, A., Browne, I.W.A., *et al.*, 1993. This volume, p.137.
[PB93b] Patnaik, A., Browne, I.W.A., *et al.*, 1993. *MNRAS*, **261**, 435.
[Su90] Surdej, J., 1990, *Gravitational Lensing, Proc. Toulouse Workshop*, p.57, Springer–Verlag, Berlin.
[SC92] Surdej, J., Claeskens, J.F., *et al.*, *Gravitational Lenses, Proc. Hamburg Conf.*, p.27, Springer–Verlag, Berlin.
[Wa90] Wambsganss, J., 1990, PhD Thesis, p.21, MPI für Phys. und Astrophys., Garching, FRG.

A (Very) Few Points on the Theory of Gravitational Lenses

Christopher S. Kochanek [*]

1. Introduction

Gravitational lensing is such a rapidly growing field that it is hopeless to provide a complete review. Our goal here is simply to discuss the theory of gravitational lens statistics in outline, with some commentary on the sources of problems in the standard models. A particularly important question is the completeness of surveys for gravitational lenses, and we discuss a simple test that shows the existing radio surveys are incomplete. Finally we give a brief discussion of how to use gravitational lenses to determine the structure of the lens galaxy and possibly to superresolve the underlying radio source. A full review can be found in the book on lenses by Schneider, Ehlers & Falco (1992), and a review of the current observations is given by Walsh in this volume.

2. Cross Sections

The cross section of a gravitational lens is the area on the sky behind the lens that produces multiple images. For a generic lens it looks like the diagram in Figure 1, where the interior of the astroid produces five images, and the region between the ellipse and the astroid produces three images. The lines separating regions producing differing numbers of images are termed caustics. The characteristic physical scale of the cross section and the typical deflection produced by the lens is

$$b = 4\pi \left(\frac{\sigma}{c}\right)^2 \frac{D^A_{LS}}{D^A_{OS}} = 1\rlap{.}''8 \left(\frac{\sigma}{250 \text{ km s}^{-1}}\right)^2 \frac{D^A_{LS}}{D^A_{OS}}$$

where σ is the velocity dispersion of the lens, D^A_{LS} is the angular diameter distance between the lens and the source, and D^A_{OS} is the angular diameter distance to the source (Gott & Gunn 1974). The outer caustic of the lens has a typical radius of b,

---

[*] Harvard-Smithsonian Center for Astrophysics & Department of Astronomy, Harvard College Observatory, 60 Garden Street, Cambridge, MA 02138. The author is partially supported by a fellowship from the Alfred P. Sloan Foundation.

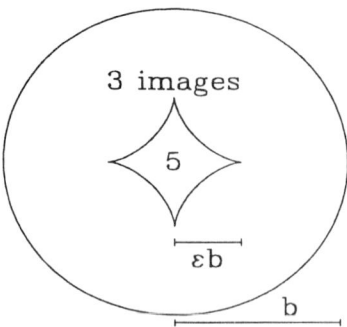

Figure 1. Typical caustic structure of a gravitational lens. The caustics (solid lines) separate regions that produce different numbers of images.

while the inner astroid has size ϵb, where $\epsilon \sim 0.1$ is the dimensionless ellipticity of the lens potential. Naively we expect that only fraction $\epsilon^2 \sim 0.01$ of gravitational lenses are five image lenses.

The major factor that modifies the magnitude of the cross sections is the addition of a core radius s to the lens potential. Core radii become important when the angular size of the core is comparable to the deflection angle b. Since the angular size is $0''.007(s/100 \text{ pc})(r_H/D_{OL}^A)$, where $r_H = c/H_0$ is the Hubble radius and D_{OL}^A is the angular diameter distance to the lens, we find that core radii become important once they are much more than a few hundred parsecs. Evidence from direct observations of galaxy cores (eg Lauer 1985), the number of lenses, and the absence of odd images in the center of lens systems (Wallington & Narayan 1992) all point towards $s \lesssim 300$ pc for a typical L_* galaxy. If we consider an isothermal sphere with a softened core (Blandford & Kochanek 1987) we find that the total cross section, $\sigma_T \simeq \pi b^2 (1-(s/b)^{2/3})^3$, depends on the core radius, while the five image cross section, $\sigma_5 = (3/2)\pi b^2 \epsilon^2$, is largely unaffected if $s \lesssim 0.5b$.

3. Optical Depth

The cross section produced by an individual galaxy must be averaged over the distribution of galaxies between us and the source being lensed. A reasonable model is a Schechter (1976) distribution of isothermal spheres with a Tully-Fisher (1977) relation between luminosities and velocity dispersions. This introduces three parameters: the slope of the Schechter function, $\alpha \simeq -1$, the slope of the Tully-Fisher relation, $\gamma \simeq 4$, and the velocity dispersion scale of an L_* galaxy, $\sigma_* \simeq 250$ km s^{-1}. Under these assumptions, the optical depth can be computed analytically using proper motion distances D_{OS} in all standard cosmologies (Kochanek 1992), and it has the particularly simple form

$$\tau = \frac{1}{30}\tau_* D_{OS}^3 \quad \text{where} \quad \tau_* = 16\pi^3 n_* r_H^3 \left(\frac{\sigma_*}{c}\right)^4 \Gamma[1 + \alpha + 4\gamma^{-1}] \simeq 0.05 - 0.08$$

in flat cosmologies (Turner 1990). The resulting optical depth for a source at a redshift of $z_s = 2$ is about 0.0013 in an $\Omega_M = 1$ Einstein-DeSitter universe, and, at 0.017, it is over ten times higher in a flat universe with $\Omega_\Lambda = 1$. As was first pointed out by Turner (1990), the incidence of gravitational lenses is a powerful means of testing for the presence of a cosmological constant. Now we know that about 1% of bright quasars are lensed, so either we are missing something or $\Omega_\Lambda = 1$!

4. Magnification Bias

Lenses magnify the source, so that an 18th magnitude lensed quasar came from the more numerous fainter quasars. We approximate the differential source distribution in flux S by a simple power law $dN/dS \propto S^{-\alpha}$. If lenses simply magnified by a fixed amount M, then the probability of finding that an object of flux S_0 is a lens is not just the optical depth τ, but the optical depth multiplied by the ratio of the number of sources at the unlensed flux, S_0/M, to the number of sources at the observed flux, S_0, or τM^α. This extra enhancement of the probability of finding lenses in a flux limited sample is called magnification bias (Gott & Gunn 1974, Turner 1990). Real lenses produce a distribution of magnifications, but the differential probability of a lens magnifying a source by M always has the asymptotic form $dP/dM \propto M^{-3}$ (Schneider et al 1992), so the magnification bias has the form $\int dM M^{\alpha-3}$. If the source counts rise more rapidly than the magnification probability drops, then there will be a huge enhancement in the number of lenses. Bright quasars ($m \lesssim 19$) have a differential slope of $\alpha \simeq 3.15$, while faint quasars have a differential slope of $\alpha \simeq 1.7$ so we expect that magnification bias is large for bright quasars and drops off as the magnitude approaches the break at 19. Theoretical calculations of the amount of bias show an enhancement in the probability of lensing bright QSOs by a factor of 20-80. This saves us from $\Omega_\Lambda = 1$ because it increases the probability of finding lenses among bright quasars to about 1% in normal cosmologies.

In real life, you cannot (or at least should not) separate all these effects. For example, five image lenses are more magnified than three image lenses. Therefore the five image systems have more magnification bias, and in a bright sample you find that half of the lenses are five image systems even through the cross section says they should less than a percent of the systems. The presence of five image lenses in the observed samples of lenses is a tell tale sign of magnification bias (Kochanek 1991). Similarly, changing the core radius of the lens changes not only the cross section for producing multiple images, but also the distribution of magnifications. For example, the mean magnification produced by the lens increases as the core radius increases, so the magnification bias increases and partially compensates for the fall in the cross section.

5. Selection Effects

Surveys for gravitational lenses do not find all the lenses present in the sample. Finite resolution and dynamic range limit how many lenses can be found by a survey. The biggest problem, however, is that the costs of confirming that a candidate is a lens are very high. This means that surveys cannot use liberal selection functions when choosing candidates because the survey is then swamped with ambiguous candidates and false positives such as stars, other quasars, secondary radio lobes and so on. Statistical models of surveys should explicitly include a model for the selection effects, and the selection model for the statistics can be chosen to balance survey completeness and contamination by ambiguous candidates. Selection functions lead to significant modifications not only in the cross sections and optical depths but also to the magnification bias.

6. What About Radio Lenses

All these effects are generic to all gravitational lenses even though the discussion was largely phrased in terms of optical quasars. The advantages of quasars for statistical studies is that the redshift and flux distributions of optical quasars are much better characterized than for radio sources. The most important barrier to statistical studies of radio surveys is the paucity of information on the intrinsic properties of the sources.

Radio surveys for lenses also have an additional bias, "size bias", because the sources frequently have extended structure. If the major and minor axes of a source are ℓ_1 and ℓ_2, then the cross section for it being lensed is not just πb^2 but, $\pi(b+\ell_1)(b+\ell_2)$ (Kochanek & Lawrence 1990). This effect doubles the probability of lensing a circular source if $\ell_1 \simeq \ell_2 \simeq 0\rlap{.}''2$. The morphology of the lens varies with the size of the source. A source that is much smaller than the deflection scale b, produces lenses like MG0414 that consist of discrete components. When the source size is comparable to the deflection scale, then the resulting lenses will resemble the ring lenses such as MG1131 and B0218. If the source is much larger than the deflection scale, we can only recognize the lens from the pattern of distortions inside the larger structure – no lens has been found in this limit. Size bias can be as important in radio surveys as magnification bias is in optical surveys.

There is strong evidence that the radio surveys for galactic lenses are incomplete. A simple but very powerful means to demonstrate the incompleteness is to take a known lens, invert it to determine the structure of the lens and the source, and then see how often the observed image morphology is generated when we randomly lens the same source with the same lens. For example we can create a very accurate model of a ring

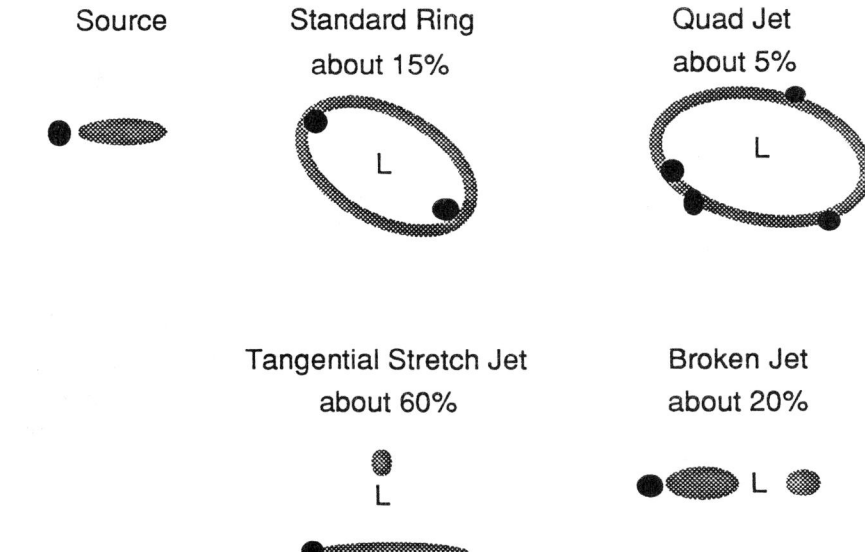

Figure 2. Schematic diagrams of images produced by lensing a core-jet source. The core is solid black, and the jet is shaded. An L marks the center of the lens.

like PKS1830-211 (Kochanek & Narayan 1992) in which the lens image consists of two bright compact cores with a ring, produced by lensing a simple core-jet source (see Figure 2). The readily identifiable normal ring geometry represents only about 15% of the ways in which the fixed lens model can lens the fixed source model. Three times rarer is the "quad-jet" morphology in which we see a low lying ring combined with four images of the point source. The two most common geometries are a "tangentially stretched jet" (60% of the time) in which the core-jet source is stretched out by the lens and a second compact image of the source is formed at right angles to the jet, and a "broken jet" (20% of the time) in which the core jet source is roughly unlensed and a second compact image is formed along the jet axis. It is possible to interpret the quad lenses like MG0414 as being the "quad-jet" sources in disguise, although the surface brightness of the jet would be high enough to see the surrounding ring if the PKS1830 source was lensed into the MG0414 geometry. The dominant geometries are not seen simply because of confusion – there are too many other ways in the sky to have another small blob of radio emission near a core-jet source for the survey groups to pursue these candidates.

7. LensClean

Once you have an extended radio lens, you have a huge number of constraints on the mass distribution in the lens galaxy. The key to making use of these constraints is to have a formal fit of a lens model to the data, including considerations of the noise level and statistical limitations. A "chi by eye" approach to modeling gravitational lenses was justified when we were mainly interested in whether it was possible to produce the observed image morphologies using gravitational lenses, but a more formal approach is required to really make use of the data.

We developed an algorithm (Kochanek & Narayan 1992) based on the Clean algorithm of radio astronomy to invert the extended radio lenses. The presence of multiple images in the lens allows us to simultaneously determine the unlensed structure of the source and the properties of the lens galaxies. This gives us a completely independent way from normal dynamical techniques of estimating the mass distribution in galaxies, and it is probably the only technique that can provide accurate measurements of the asymmetries in the mass distribution. The method automatically performs the calculations required to take into account both the resolution of the radio maps and the noise level. The accuracy with which we can constrain the model is simply a function of the dynamic range in the maps and the number of different maps. Each frequency and polarization emphasizes different parts of the image, which helps to break the degeneracies present in any single map. The advantage of a full statistical method like LensClean is that it makes use of every scrap of information available to it rather than focusing on a few visually obvious (and possibly deceptive) features.

References

[1] Blandford, R.D., & Kochanek, C.S., 1987, *ApJ*, **321**, 658.

[2] Gott, J.R., & Gunn, J.E., 1974, *ApJ (Letters)*, **190**, L105.

[3] Kochanek, C.S., 1991, *ApJ*, **379**, 517.

[4] Kochanek, C.S., 1992, *MNRAS, in press*.

[5] Kochanek, C.S., & Lawrence, C.R., 1990, *AJ*, **99**, *1700*.

[6] Kochanek, C.S., & Narayan, R., 1992, *ApJ*, **401**, *in press*.

[7] Lauer, T.R., 1985, *ApJ*, **202**, 104.

[8] Schechter, P., 1976, *ApJ*, **203**, 297.

[9] Schneider, P., Ehlers, J., & Falco, E.E., 1992, *Gravitational Lensing. Springer-Verlag: Heidelberg.*

[10] Tully, R.B., & Fisher, J.R., 1977, *A&A*, **54**, 661.

[11] Turner, E.L., 1990, *ApJ (Letters)*, **365**, L43.

[12] Turner, E.L., 1980, *ApJ (Letters)*, **242**, L135.

[13] Wallington, S., & Narayan, R., 1992, *ApJ, in press*.

Einstein Rings and Einstein Quads

Bernard F. Burke *Samuel R. Conner* *Jacqueline N. Hewitt* [*]

Joseph Lehar [†]

Abstract

Einstein rings, of which five are currently known, are closely related to another easily-recognized group of lensed images: sources with four images, either symmetrically disposed in a 'cloverleaf' or asymmetric sources with PG1115+080 as the prototype. These are probably composed of five images, but only four are clearly visible, and the term 'quad' is suggested for such objects. Both rings and quads are recognized easily by their radio appearance. The observations of rings and quads also have clear cosmological implications.

1. Observations

The image morphology produced when a foreground galaxy images a distant radio source can be described qualitatively by idealizing the galaxy as an ellipsoidal mass distribution [1, 2, 3, 4]. When the object is sufficiently extended, and is nearly aligned with the foreground galaxy, the resulting image is a ring of the form predicted by Einstein [5]. Hewitt *et al.* [6] presented the first example, MG1131+0456, describing it as an 'Einstein ring'. Kochanek *et al.* [7] demonstrated that the image could be deconvolved, recovering the original radio object. Subsequently, at least four more instances of Einstein rings have been demonstrated: MG1654+1346 (Langston *et al.* [8, 9]), PKS1830-211 (Rao and Subramanyan [10], Jauncey *et al.* [11, 12]), MG1549+3047 (Lehar [13]), and B0218+357 (Patnaik *et al.* [14, 15]). The search for other examples is by no means complete.

Einstein rings are only one aspect of a more general class of lens, typified by a 4-image structure, which can be described in a compact way as 'Einstein quads' in view of their quadrufoil appearance. The 'cloverleaves' G2232+0305 (Huchra et al [16]) and H1413+117 (Magain *et al.* [17]) are the result of a galaxy lensing a well-aligned compact object, while the quads PG1115+080 (Weymann *et al.* [18]) and

---
[*]Department of Physics and Research Laboratory of Electronics, Mass. Inst. of Technology, Cambridge MA 02139 USA

[†]Institute of Astronomy, Cambridge, UK

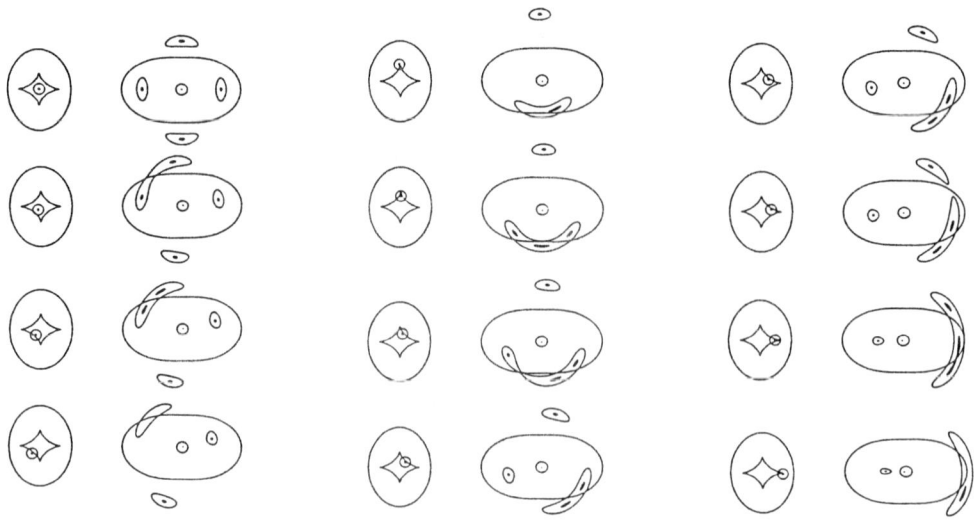

Figure 1. Imaging of a source by a foreground ellipsoidal mass distribution. Columns are designated a,b,c from left to right, and object/image pairs are 1,2,3,4 from top to bottom.

MG0414+0534 (Hewitt *et al.* [19]) are the results of less perfect alignment. In all four of these cases, there are four visible images; there should be a fifth image in theory but in practice it is too faint to be observed.

2. Model Calculations

The various possible morphologies of rings and quads are illustrated in Figure 1, which shows 3 imaging sequences of object-image pairs. The object in each case is a pair of circles, 0.02 and 0.12 arcsec in radius, imaged by a Blandford-Kochanek potential with ring radius b=1 arcsec, core radius $a = 0.05$ arcsec, and eccentricity $e = 0.2$. In the figure, the left-hand sequence starts with a centered source in example 1a, moving progressively SE in example 1b, crossing the diamond caustic between examples 3a and 4a. Example 1a shows how both rings and quads of the symmetrical variety are formed: the extended source is still too compact to form a ring, since all parts are far from the caustic, but if the source were sufficiently extended it is obvious that a ring would gradually fill in. Examples 2a and 3a resemble PG1115+080 and MG0404+0534 strikingly. The second sequence starts at the minor cusp in Example 1b, running just inside the caustic to the midpoint in Example 4b (which is equivalent to 3a). The third sequence continues the motion of the source inside the caustic, reaching the major cusp in Example 4c. Examples 2b and 3c, where the source is close to a cusp, there are still 4 well-marked images when the object is compact, and the appearance is much like that of a new lens, B1422+231, presented by Patnaik *et al.* in this conference. As is well known, the images also resemble 'gravitational arcs'

[20, 21].

3. Observations of Rings and Quads

High-resolution radio surveys, such as those carried out using MERLIN or the VLA, have been a rich source for discovering rings and quads. From the image locations and relative fluxes, the location of the tangential critical line (the critical line corresponding in the image plane to the diamond, or tangential caustic in the object plane. When the redshifts of both the object and the lens can be measured, the system parameters can be remarkably well specified. In addition, if there are extensions of the central source, or compact sources nearby, these give additional information about the lens, and therefore about the distribution of mass in the foreground galaxy. The lensing potential can be expanded in a power series, and the secondary sources, when close to the critical line, give an estimate of the first-order term [22]. Under these conditions, the relative time delay can give a more certain estimate of the Hubble constant, since the lens characteristics can be specified more completely. The derived value is exact to the extent that the lensing galaxy can be represented by an isothermal sphere. The sources B0218+357 and MG1131+0456 show promise in this respect.

The inferred size of the tangential critical line has been estimated for the five rings and four quads listed above [22]. Only three of these exhibit mean radii less than 0.85 arcsec; the remaining six have mean radii up to 1.2 arcsec. The present lens searches have been neither complete nor free of bias, but the work now in progress, both at MIT and elsewhere, should clarify whether the concentration of mean radii in the vicinity of 1 arcsec is real, a result of observational selection, or a statistical variation attributable to the small sample size.

4. Distribution of Matter in the Universe

The mean radius of the tangential caustic gives a measure of all the projected matter that it encloses, and thus is a means of detecting both dark matter and luminous matter. For a lens that is well represented by an ellipsoid of small eccentricity, the mass enclosed within the tangential caustic is

$$M_{enc} = (c^2/4G).(D_d D_s/D_{ds}).r_{tc}^2$$

and if the redshifts of source and deflector are known, the mass can be determined. The total luminosity contained within that locus is also an observable quantity, and thus the mass-to-luminosity ratio is determined (averaged over the mean radius of the tangential caustic). This has been carried out quantitatively for MG1654+1346 by Langston et al. [9] and for G2237+305 by Schneider et al. [23]. The results, quoted for blue absolute magnitude, are $M/L = 15.9\pm2.3h$ and $9.4\pm2.0h$, respectively, with $h = 1$ for $H = 100\ kms^{-1}Mpc^{-1}$, with $<r> = 3.8$ and $0.6h\ Kpc$. These are probably the most accurately specified values of M/L yet measured by any technique. The central velocity dispersions inferred for these two cases are 216 ± 10 and $212 \pm 12\ kms^{-1}$.

The corresponding value of M/L for MG1549+3047 is less well specified because the redshift of the lensed source has not yet been measured accurately. The radio core of the FRII radio source, one of whose lobes is imaged into a ring by the foreground lensing galaxy (redshift 0.111), is coincident with a faint optical object, whose color implies a redshift of about 0.4, but whose radio properties as an FRII radio galaxies imply a redshift greater than 1, and more probably 2. This means that the mass-to-light ratio falls in the range 15-20, with an uncertainty of 20% until the object redshift can be determined more definitively [13]. The corresponding value for the velocity dispersion is $250\ kms^{-1}$. There is also a preliminary value of the velocity dispersion for PKS1115+080 of $234 \pm 25\ kms^{-1}$ (Christian and Schechter, private communication).

As the number of rings and quads with measured redshifts for both source and image increases, the patterns of velocity dispersion and mass-to-light ratio in various foreground galaxies should prove to be a useful tool in examining the arguments for the presence of dark matter in galaxies, and in furthering statistical studies of the occurrence of the lensing phenomenon in the universe.

References

[1] Blandford, R., & Kochanek, C., 1987, *Astrophys. J.*, **246**, 1.

[2] Narayan, R., & Grossman, S., in: *Gravitational Lenses*, Hewitt, J. & Lo, K.-Y., eds., Berlin: Springer Verlag, 1989.

[3] Narayan, R., & Wallington, S., in: *Gravitational Lenses*, p.12, Kayser, R., Schraum, T., & Nieser, L., eds., Berlin: Springer Verlag, 1992.

[4] Bourassa, R.R., & Kantowski, R., 1975, *Astrophys. J.*, **195**, 13.

[5] Einstein, A., 1936, *Science*, **84**, 506.

[6] Hewitt, J.N., et al., 1988, *Nature*, **333**, 537.

[7] Kochanek, C.S., et al., 1989, *MNRAS*, **238**, 43.

[8] Langston, G., et al., 1989, *Astron. J.*, **97**, 1283.

[9] Langston, G., et al., 1990, *Nature*, **344**, 43.

[10] Rao, A.P., & Subramanyan, R., 1988, *MNRAS*, **231**, 229.

[11] Jauncey, D., *et al.*, 1991, *Nature*, **352**, 132.

[12] Jauncey, D., *et al.*, in: *Gravitational Lenses*, p.333, Kayser, R., Schraum, T., & Nieser, L., eds., Berlin: Springer Verlag, 1992.

[13] Lehar, J., *et al.*, 1993, *Astron J.*, in press.

[14] Patnaik, A., *et al.*, in: *Gravitational Lenses*, p.140, Kayser, R., Schraum, T., & Nieser, L., eds., Berlin: Springer Verlag, 1992.

[15] Patnaik, A., *et al.*, 1993, *MNRAS*, in press.

[16] Huchra, J., *et al.*, 1985, *Astron J.*, **90**, 691.

[17] Magain, *et al.*, 1988, *Nature*, **334**, 325.

[18] Weymann, R., *et al.*, 1980, *Nature*, **285**, 641.

[19] Hewitt, J., *et al.*, 1993, *Astron. J.*, in press.

[20] Soucail, G., in: *Advanced Study Institute on Clusters and Superclusters of Galaxies, July 1991*, Dordrecht:Kluwer, 1992.

[21] Fort, B., in: *Gravitational Lenses*, p.267, Kayser, R., Schraum, T., & Nieser, L., eds., Berlin: Springer Verlag, 1992, and references therein.

[22] Burke, B.F., Lehar, J., & Conner, S.R., in: *Gravitational Lenses*, Kayser, R., Schraum, T., & Nieser, L., eds., Berlin: Springer Verlag, 1992.

[23] Schneider, D., *et al.*, 1988, *Astron. J.*, **95**, 1619.

High Resolution Radio Observations of 0957+561

M.A. Garrett,*R.W. Porcas, †
L.J. King,* R. Calder,* P.N. Wilkinson* & D. Walsh*

1. Introduction

0957+561 was the first gravitational lens system to be discovered [WCW79]. Observations by [Sto80] and [YGK80] revealed the lens to be a foreground cluster of galaxies at $z \approx 0.36$, the brightest G1 lying just 1″ from the B image. Further observations suggest a second cluster at $z \approx 0.50$ [GWC92]. It is unfortunate that the lens system for which we have the most comprehensive data (including the redshifts of the lens/source and the time delay [RLH91]) should also turn out to be the most complicated. Nevertheless observations of this system at various wavelengths (optical, X-ray & radio) and on widely different scales continue to provide new additional constraints for lens modelling. In this paper we present new constraints from radio observations of the small scale VLBI core-jet structure.

2. Global VLBI Hybrid Maps of the A and B Images

In November 1989 we observed 0957+561A,B at λ18cm with a global VLBI array. The hybrid maps are shown in figure 1. Each component identified in the A image has a counterpart in the B image and vice versa. The radio structure in both images changes smoothly from the core to the far jet. The kink observed in the A image in the far jet is also seen in the B image but with opposite parity. [GCS88] first presented observational evidence that the images had opposite parities but their data were not conclusive. It is obvious from the hybrid maps presented here that the images have opposite parities. The kink in the B image is not as pronounced as the kink in the A image since the B image is strongly de-magnified in the direction perpendicular to the core-jet structure. It is possible to map the kink from one image to the other using [GCS88]'s original magnification matrix.

[WaP92] have recently simulated the effects of $10^6 M_\odot$ black holes in the halo of G1 on the VLBI jets of 0957+561. We do not observe any of the rings or holes predicted

---

*University of Manchester, NRAL, Jodrell Bank, Macclesfield, Cheshire, SK11 9DL, UK.
†Max-Planck-Institut für Radioastronomie, Auf dem Hügel 69, DW 5300, Bonn 1

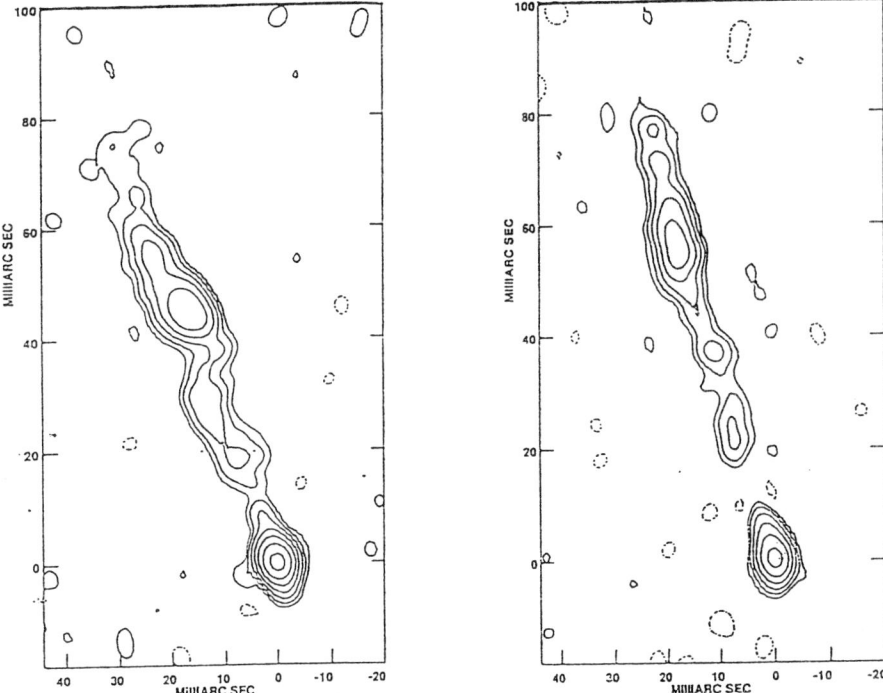

Figure 1. Global VLBI observations at $\lambda 18$ cm of 0957+561A,B. Contours at -0.12, 0.12, 0.24, 0.46, 0.92, 1.84, 3.68, 7.36 mJy/beam. The restoring beam is circular (4 mas).

by [WaP92] or detect any features that are present in one jet but not the other. We believe that the kink in the jet and the other bends are intrinsic to the original source and are *not* due to milli-lensing. [WaP92]'s simulations assumed radio maps with a dynamic range of 100:1 and a resolution of 3.5 mas. Our maps have a dynamic range of 85:1 for the B image and 170:1 for the A image. The restoring beam used in both maps is 4.0 mas for which the corresponding Einstein deflector mass is $\sim 4 \times 10^6 M_\odot$. Following [WaP92]'s analysis and simulations we conclude that black holes of this mass or greater do not form a large fraction ($> 10\%$) of the dark matter in galactic halos.

3. Magnification Matrices and Core/Jet Flux Ratios

It was [GCS88] who first showed that the brightness distribution in one image could be mapped to the other by means of a magnification matrix. They assumed that this matrix would not vary over the length of the jets (~ 80 mas). As in previous VLBI observations we measure the core and jet flux ratios to be significantly different (B/A$_{core} \approx 0.8$ & B/A$_{jet} \approx 0.66$). We note that the total (core + jet) VLBI flux ratio is 0.70 which is consistent with VLA observations [RLH91]. Our results support the hypothesis of [CLB92] that the difference in the core and jet flux ratios is due to a variation in the magnification matrix over the core-jet VLBI structure.

Let us assume that our core flux ratio (0.8) corresponds to the determinant of the matrix at the core and similarly that the far jet flux ratio (0.66) is the determinant of the matrix which applies at the far jet. Since our new maps show features along the length of the entire core-jet structure it is possible to estimate the variation in the matrix in the direction parallel to the core-jet by inspection of the relative positions of the components in the A and B images. We have fitted six-component gaussian models to the A and B image data to allow us to specify accurately the brightness distributions. We find that the magnification in the direction parallel to the core-jet increases by $\approx$ 7% as we move from the core to the outer jet. If we accept that the flux ratio at the core and at the jet is just the determinant of the magnification matrix then the magnification perpendicular to the jet varies between -0.70 in the core and -0.54 in the far jet *i.e.* it decreases by 23%.

[GCS88]'s matrix accurately describes the image transformation at the far jet but cannot be applied at the core. The brightness distribution must be modelled by a magnification matrix and the derivatives of its components. This will supply an additional six constraints to lens models. More sensitive observations may be required to measure all six constraints accurately. Higher resolution observations are required to rule out the existence of black holes $< 4 \times 10^6$ M$_\odot$ in the halo of G1. It is clear that 0957+561 will continue to be a fruitful topic of study for some time to come.

References

[CLB92]   Conner, S.R., Lehar, J. and Burke, B.F. 1992, *ApJ*, **387**, L61.

[GWC92]   Garrett, M.A, Walsh, D. and Carswell, R.F. 1992, *MNRAS*, **254**, 27.

[GCS88]   Gorenstein *et al.* 1988 *Ap.J.*, **334**, 42.

[PBB81]   Porcas, R.W., Booth, R.S., Browne, I.W.A., Walsh, D., Wilkinson, P.N. 1981, *Nature*, **289**, 758.

[RLH91]   Roberts, D.H., Lehar, J., Hewitt, J.N. and Burke, B.F. 1991, *Nature*, **352**, 43.

[Sto80]   Stockton, A. 1980, *Ap.J.*, **242**, L141.

[WaP92]   Wambsganns, J and Paczynski, B. 1992, *Ap.J*, **397**, L1

[WCW79]   Walsh,D., Carswell, R.F. and Weymann. 1979, *Nature,*, **279**, 381

[YGK80]   Young, P.J., *Ap.J.*, **241**, 507.

Global 6 cm VLBI Investigations of 0957+561

R. M. Campbell [*]   B. E. Corey [†]   E. E. Falco [‡]   I. I. Shapiro [‡]

M. V. Gorenstein [§]   P. Elósegui [‡]   J. M. Marcaide [¶]   K. Alvi [*]

Abstract

We have completed 6 cm VLBI observations of the images of the gravitationally lensed quasar 0957+561 at two epochs in a search for correlated structural evolution from which to estimate the time delay, $\Delta\tau_{BA}$. However, we detected only marginal increases in the separation between the core and the innermost jet component in each image over the two-year interval. Because monitoring brightness variations has led to apparently useful estimates of $\Delta\tau_{BA}$, we are refocussing our efforts towards estimating the relative magnification matrices for the inner and far-jet regions of the images to provide additional constraints on the mass distribution of the lens, which remains the principal contributor to the uncertainty in estimating H_0 from $\Delta\tau_{BA}$.

The Hubble Constant is inversely proportional to the difference in propagation times, $\Delta\tau_{BA}$, for the A and B images of a gravitationally lensed source; the lens' mass distribution sets the constant of proportionality (see *e.g.*, [FGS91]). Previous determinations of $\Delta\tau_{BA}$ for 0957+561 have been based on cross-correlations of the two images' optical or radio light curves, but were beset, until recently, with complications in interpreting the results (see [PRH92] and references therein). Detection of correlated evolution within the milliarcsecond (mas) scale brightness distribution of the images through VLBI might avoid some of the difficulties that light-curve analysis has encountered.

Primarily for this purpose, we made simultaneous VLBI observations of the A and B images on 1987 September 26 and on 1989 September 26–27 using Mark III mode A recording with a central frequency of 4983.99 MHz and a bandwidth of 56 MHz. The 1987 VLBI array was composed of individual antennas at Medicina, Effelsberg, Haystack, Green Bank, Owens Valley, and phased arrays at Westerbork and the VLA. The 1989 array was identical except that Medicina was omitted as it was not then configured to observe at 6 cm.

For each epoch, we modelled the image structure with two elliptical gaussian components in the inner region (within 2 mas of the core) and one or two elliptical gaussian components in the outer, far-jet, region ($\sim$ 50 mas from the core). We investigated two classes of models: one in which we estimated all components in both images independently (Model I), and

---

[*]Department of Astronomy, Harvard University, Cambridge, MA 02138 USA
[†]Haystack Observatory, Westford, MA 01886 USA
[‡]Harvard-Smithsonian Center for Astrophysics, Cambridge, MA 02138 USA
[§]Millipore Corporation, Waters Chromatography Division, Milford, MA 01757 USA
[¶]Universitat de València, Valencia, Spain

one in which we introduced a 2 × 2 relative magnification matrix (M_{BA}) which relates the parameters of specified components in one image to the corresponding components in the other (Model II — see [GCS88]). The relative magnification matrix may be used, along with the position of the images with respect to the lens, to set model-dependent constraints on the lens' mass distribution. Figure 1 shows the brightness distributions for the 1987 model I solution using two far-jet components; the inverse parity of the two images predicted by the gravitational lensing hypothesis appears clearly.

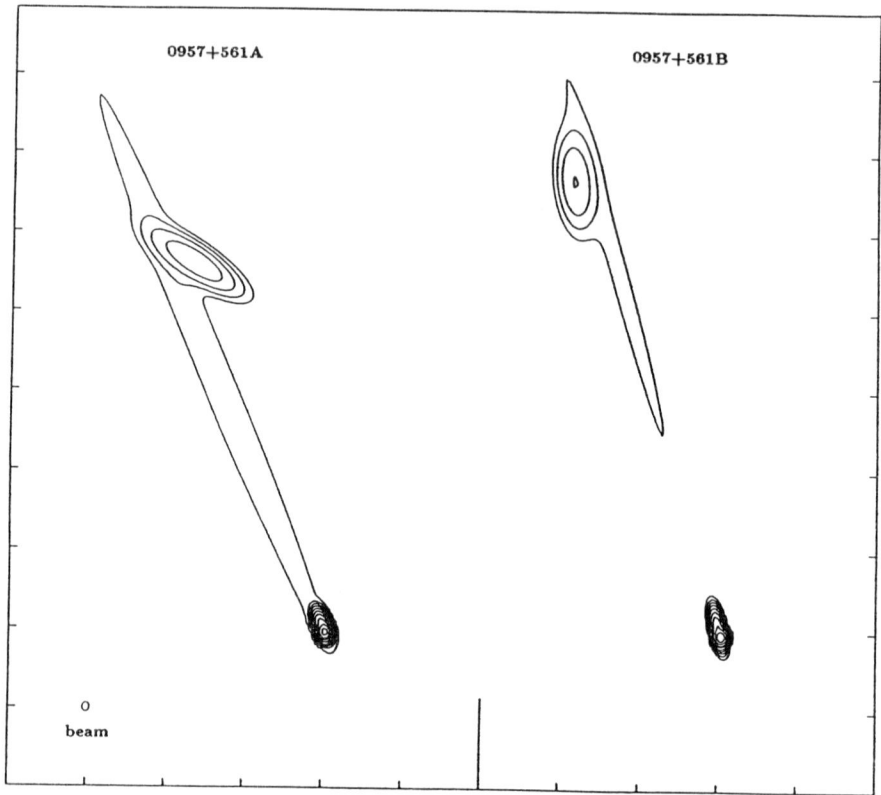

Figure 1. Elliptical gaussian component model for the 1987 6 cm observations of 0957+561 A and B. Contours are drawn at 1/8, 1/4, 1/2, 1, 2, 4, 8, 16, 32, and 64 times 0.1 mJy per beam. The beam (1.14 × 0.94 mas) is shown in the lower left corner. The scale is 10 mas per tick. North is up, east is to the left.

Principal results from these two 6 cm VLBI observations include:

- The detection of only marginal ($< 0.7\sigma$) increases in the separation between the core and the innermost jet component over the two-year interval, which bodes ill for the estimation of $\Delta\tau_{BA}$ from this component's motion within a reasonable timeframe.

- The detection of a significantly asymmetric brightening in the inner regions of the two images: $31 \pm 2\%$ in the A image as compared to $12 \pm 2\%$ in the B image. In other

- extragalactic compact radio sources, episodes of core brightening generally precede the emission of a new jet component, thus providing some, albeit slim, promise for the successful application of the structural evolution method in the future.

- The estimation of a relative magnification matrix for the far-jet region from the 1987 6 cm observations which agrees with that estimated from the 1981 13 cm observations [GCS88] to within the standard errors, as would be expected in the case of static and achromatic gravitational lensing.

The lack of significant observed proper motion between the core and the inner jet component in each image leaves constraint of the lens' mass distribution as the principal way at present VLBI may contribute to the estimation of H_0. Indeed, systematic errors in the estimate of H_0 resulting from the model used to describe the lens' mass distribution may equal or exceed those stemming from the effects of the standard errors of the observables which parameterize H_0 in these models: $\Delta\tau_{BA}$ and the velocity dispersion of the principal lensing galaxy [FGS91], [Koc91]. Two epochs of observations separated by $\Delta\tau_{BA}$ (e.g., the [PRH92] value) would allow estimation of the relative magnification matrix for both the inner and far-jet regions. The asymmetric placement of the two images with respect to the principal lensing galaxy causes the elements of M_{BA} to vary appreciably across the extent of the image structure (see [CLB92]), for example as much as 20% in moving from the core to the center of the far-jet region using the [FGS91] lens model. We have refocussed our efforts towards the measurement of these variations within the relative magnification field, with the goal of establishing better constraints for the model of the lens' mass distribution. We conducted the first of two such "phased" observations on 1992 March 21–22. We gratefully acknowledge support from NASA grant NGT-50663 (RMC), NSF grant AST89-02087 (RMC), and a NATO fellowship (PE).

References

[CLB92] Conner S.R., Lehár J., Burke B.F., *Reconciling the Image Brightness Ratios in the Gravitational Lens System 0957+561*. Ap. J., Vol. 387 (1992), pp. L61–L64.

[FGS91] Falco E.E., Gorenstein M.V., Shapiro I.I., *The Time Delay of Gravitational Lens System 0957+561: Bounds on Masses of a Possible Black Hole and Dark Matter and Prospects for Estimation of H_0*. Ap. J., Vol. 372 (1991), pp. 364–379.

[GCS88] Gorenstein M.V., Cohen N.L., Shapiro I.I., Rogers A.E.E., Bonometti R.J., Falco E.E., Bartel N., Marcaide J.M., *VLBI Observations of the Gravitational Lens System 0957+561: Structure and Relative Magnification of the A and B images*. Ap. J., Vol. 344 (1988), pp. 42–58.

[Koc91] Kochanek C.S., *Systematic Effects in Lens Inversions: $\aleph_1$ Exact Models for 0957+561*. Ap. J., Vol. 382 (1991), pp. 58–70.

[PRH92] Press W.H., Rybicki G.B., Hewitt J.N., *The Time Delay of Gravitational Lens 0957+561 I. Methodology and Analysis of Optical Photometric Data II. Analysis of Radio Data and Combined Optical-Radio Analysis*. Ap. J., Vol. 385 (1992), pp. 404–420.

Southern Hemisphere Observations of PKS1830-211.

David L. Jauncey *   and the SHEVE team †

Abstract

We present radio, optical and infrared observations from continued monitoring of the strong Einstein ring/gravitational lens radio source PKS1830-211. The radio source has decreased in intensity from 10 to 9 Jy at 2.3 GHz over the past two years, but has increased at 8.4 GHz from a minimum of 5 Jy in Feb 1991 to a maximum of 11 Jy in March 1992. Optical and infrared observations reveal no obvious identification of either the lensed or lensing object.

VLBI, MERLIN and VLA observations of the strong, flat-spectrum radio source PKS1830-211 have established it is an Einstein ring/gravitational lens [J91], following the earlier suggestion that it may be a gravitational lens [R88]. The radio source consists of two compact VLBI components on opposite sides of a one arcsecond diameter ring of radio emission. The MERLIN and VLA images have been successfully modelled with the "LensClean" algorithm and indicate a magnification factor of about 10 for a background core-jet radio source [K92].

Observations have been made with the SHEVE array [P93] at 0.843, 2.3, 8.4 GHz several times over the past four years, as well as at 5 GHz from the north [J93]. Observations at 0.843 GHz were made in order to investigate the effects of interstellar scattering, which is apparently causing angular broadening of the two compact components at 2.3 GHz [J92]. While the angular size of the two compact components is indeed greater at 0.843 GHz, the increase appears to be less than expected from a simple λ^2 proportionality.

The source flux density has been monitored with the 26 m antenna of the Mt Pleasant observatory, Hobart, at 2.3 and 8.4 GHz since July 1990, and the results are presented in Figure 1. At 8.4 GHz the initial decrease was followed by a subsequent increase to

---

*Australia Telescope National Facility, P.O. Box 76, Epping, NSW 2121, Australia. The Australia Telescope National Facility is operated in association with the Division of Radiophysics by CSIRO.
†Members of the Southern Hemisphere VLBI Experiment (SHEVE) team are listed as co-authors on the paper by R. A. Preston *et al.* in these proceedings.

a peak of 11 Jy in early 1992 with a decline since then, and the change in intensity is quite striking. There is no evidence for a secondary peak during 1992, suggesting that either the delay between the components is small or that the second image has not yet peaked.

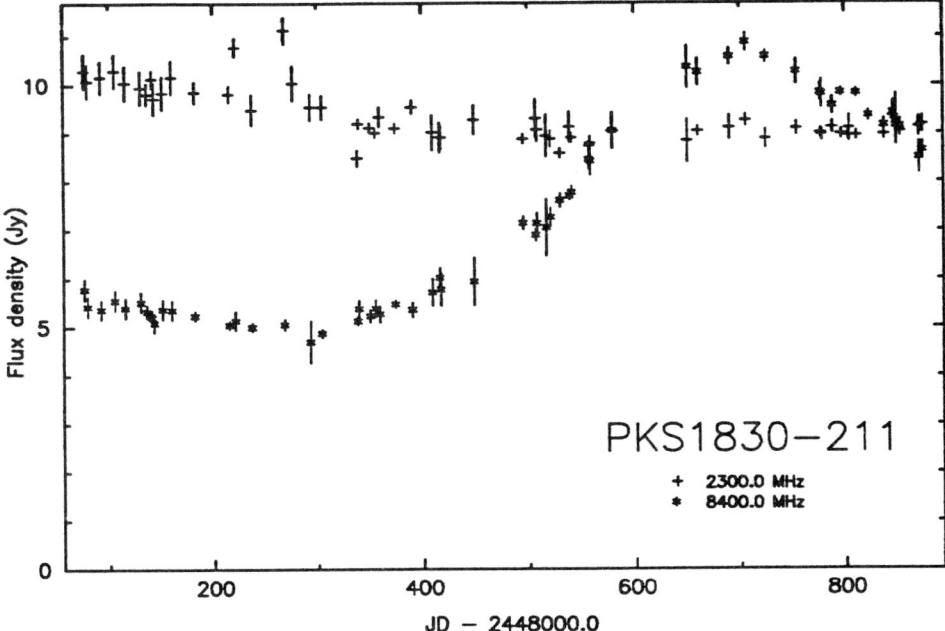

Figure 1. Plot of the flux density variations of PKS1830-211 at 2.3 and 8.4 GHz

The clear signature in the variability at 8.4 GHz indicates that a direct measurement of the time delay between the two compact VLBI components may be feasible in the future, provided the observations have sufficient resolution to separate the two components. The results from the two-epoch Northern Hemisphere VLBI images [J93], show that this variability was accompanied by a similar dramatic change in the structure of the northeast component.

The outburst was clearly present at 8.4 GHz before appearing at 2.3 GHz. We therefore expect that monitoring of the source flux density at mm wavelengths in the future will allow any such outbursts to be recognized ahead of time so that VLBI observations can be scheduled at cm wavelengths to track the flux density changes in the two compact components individually.

Optical and infrared observations have been undertaken in an attempt to identify the optical counterparts of the radio image as well as the lensing galaxy. A faint red object was found within one arcsecond but proved to have the spectrum of a Galactic M3 dwarf, and is clearly not the identification [J92]. A deep K' infrared image taken at the Anglo-Australian Telescope at a wavelength of 2.1 μm indicates that the M star image appears to be slightly extended to the south west and close to the radio position of the south-western radio component, confirming the possible detection of this component reported recently [D92]. This object is very red and may be either the lensing galaxy or the lensed object. It is clear that the optical identification of PKS1830-211 is very faint and very red. Spectroscopy of this object will prove difficult due to the close proximity to the M3 dwarf.

References

[D92] Djorgovski, S., Meylan, G., Klemola, A., Thompson, D.J., Weir, W.N., Swarup, G., Rao, A.P., Subrahmanyan, R., and Smette, A., *A search for the optical/IR counterpart of the probable Einstein ring source PKS1830-211.* Mon. Not. R. astr. Soc., Vol. 257 (1992) pp 240-244.

[K92] Kochanek, C.S. and Narayan, R., (1992) *Submitted to Astrophys. J.*

[J91] Jauncey, D.L., Reynolds, J.E., Tzioumis, A.K., Muxlow, T.W.B., Perley, R.A., Murphy, D.W., Preston, R.A., King, E.A., Patnaik, A.R., Jones, D.L., Meier, D.L., Bird, D.J., Blair, D.G., Bunton, J.D., Clay, R.W., Costa, M.E., Duncan, R.A., Ferris, R.H., Gough, R.G., Hamilton, P.A., Hoard, D.W., Kemball, A., Kesteven, M.J., Lobdell, E.T., Luiten, A.N., McCulloch, P.M., Murray, J.D., Nicolson, G.D., Rao, A.P., Savage, A., Sinclair, M.W., Skjerve, L., Taaffe, L., Wark, R.M., White, G.L., *An unusually strong Einstein ring in the radio source PKS1830-211.* Nature, Vol. 352 (1991), pp. 132-134.

[J92] Jauncey, D.L., Reynolds, J.E., Tzioumis, A.K., Sadler, E.M., Jones, D.L., Klein, M.J., Meier, D.L., Preston, R.A., Stelzried, C.T., Campbell-Wilson, D., Meadows, V., Smith, M.G., Hughes, D.H., Muxlow, T.W.B., and Patnaik, A.R., *Recent observations of PKS1830-211.* In: Gravitational Lenses, eds Kayser, R., Schramm, T. and Nieser, L., Springer Verlag. Berlin, (1992) pp 333-337.

[J93] Jones, D.L., and the SHEVE Team, *Northern hemisphere VLBI observations of PKS1830-211,* This volume, (1993).

[P93] Preston, R.A., and the SHEVE Team, *The Southern Hemisphere VLBI Experiment (SHEVE),* This volume, (1993).

[R88] Rao, A.P., and Subrahmanyan, R., *1830-211 - a flat-spectrum radio source with double structure* Mon. Not. R. astr. Soc., Vol 231 (1988), pp 229-236.

Search for Small Separation Gravitational Lens Systems

A. Patnaik, I. Browne, L. King, T. Muxlow, D. Walsh, P. Wilkinson *

We have been carrying out a VLA and MERLIN survey of flat spectrum radio sources with two main objectives: (i) to select a grid of phase calibrators for MERLIN, (ii) to find a well–defined sample of gravitational lenses. So far we have observed ∼1750 sources with the VLA at 8.4 GHz with a resolution of 200 milliarcsec (mas) [PBW92a, PBW93]. Flat spectrum sources generally have simple radio structure, a dominant core sometimes with a weak jet. Unlike the situation for steep spectrum sources, any lensed flat spectrum source is easy to recognise since multiple–cores are rare.

The lens candidates selected to have multiple components are reobserved with the VLA to find the component spectral and polarisation characteristics, and with MERLIN to map their subarcsec structure. MERLIN observations at 5 GHz with 50 mas resolution are crucial in finding subarcsec lenses. Real lensed systems have some or all of the following: (i) unusual shaped images such as rings or arcs, (ii) similar radio (and optical) spectra, (iii) similar fractional polarisation and (iv) similar position angle of polarisation after taking differential Faraday rotation into account.

Below we describe the sources B0218+35.7, B1422+23.1 and B1938+66.6, and show that the observations can be understood in terms of lensing hypothesis.

- **B0218+35.7** This source has two flat spectrum compact components A, B, separated by 335 mas (Fig 1)[PBK92]. The weaker one, B, is surrounded by a steep spectrum ring of emission of diameter 330 mas. Both A and B are variable. The flux density ratio B/A is similar, 0.3, at 1.6, 5, 8.4, 15 and 22 GHz indicating that both have similar spectral index. Note that these observations have rather different resolutions, from 5 mas to 200 mas. The fractional polarisation of A and B are similar at 8.4, 15 and 22 GHz, each polarised by 10 per cent. The position angle difference follows a λ^2–law with the difference RM of 900 rad/m^2. EVN observations at 5 GHz (5 mas resolution) indicate that A is elongated in pa –30° as expected from lens models (D. Narashima private comm.). The source is identified with a 20 mag red object with unknown redshift.

The above arguments strongly suggest that B0218+35.7 has been lensed by a gas–rich (high RM) nucleus of a galaxy. This is the smallest separation lensed system yet found. The relative flux densities of A and B have been seen to vary and this means

---

*Nuffield Radio Astronomy Laboratories, Jodrell Bank, Macclesfield, Cheshire SK11 9DL, U.K.

it should be possible to measure the time delay which should be ∼5 to 10 days. This system may be suitable for the determination of the Hubble constant, H_0.

- **B1422+23.1** This is a peaked spectrum source with 4 compact components within 1.3 arcsec (Fig 2) and is identified with a 15.5 mag (V) quasar at a redshift of 3.62. The three brighter components, A, B and C have similar spectral index –0.71 between 5 and 8.4 GHz, similar fractional polarisation (2.7 per cent), and similar position angle of polarisation (–35°) [PBW92b]. The fourth component D is weak and hence its polarisation properties are not known. However, 2.2μm infrared observations [LNW92] detect emission from D and optical observations using the NOT at V and R bands [PAA92] detect all the four components and show that the flux density ratios from radio to optical bands is essentially the same. Thus the evidence presented above confirms that images in B1422+23.1 are due to gravitational lensing.

- **B1938+66.6** This source has peculiar arc–like structure, four compact components, and a jet–like feature connecting one of the compact sources and the arc (Fig 3). The entire structure lies within a diameter of 0.95 arcsec. The arc appears unresolved across its width. The arc appears to be similar to optical 'giant' arcs. The source is associated with a 23 mag (r) object of unknown redshift. The observed structure can be reproduced with two component source, one component being in the 5–image system, the other is the 3–image system.

In summary, we have presented three cases of gravitational lensing. The maximum image separations in all these cases is close to or less than 1 arcsec. Our observations show that small separation lenses exist and the study of these lensed systems would provide useful information on low mass galaxies ($\sim 10^9$ $M_\odot$).

References

[LNW92] Lawrence C.R., Neugebauer G., Weir N., Matthews K., Patnaik A.R., 1992, MNRAS, 259, 5p

[PBW92a] Patnaik A.R., Browne I.W.A., Wilkinson P.N., Wrobel J.M., 1992, MNRAS, 254, 655

[PBW93] Patnaik A.R., Browne I.W.A., Wilkinson P.N., Wrobel J.M., 1993, in preparation

[PBK92] Patnaik A.R. et al. 1992, MNRAS, in press

[PBW92b] Patnaik A.R. et al. 1992, MNRAS, 259, 1p

[PAA92] Patnaik A.R., Akujor C.E., Ardeberg A., 1992, in preparation.

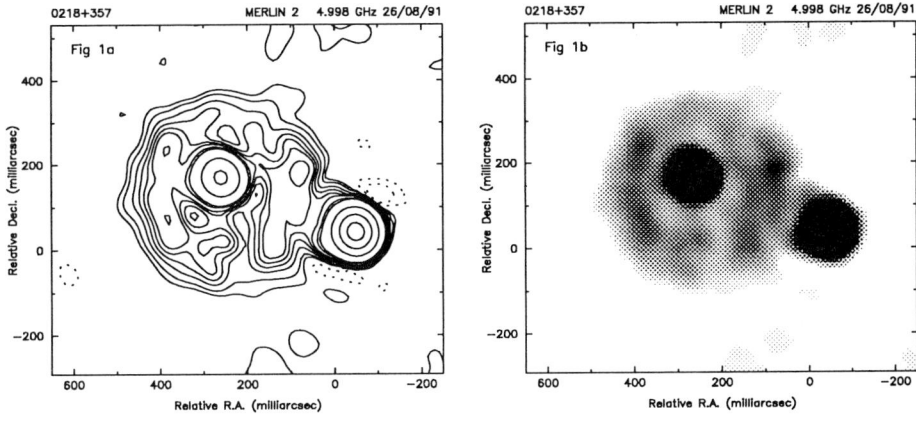

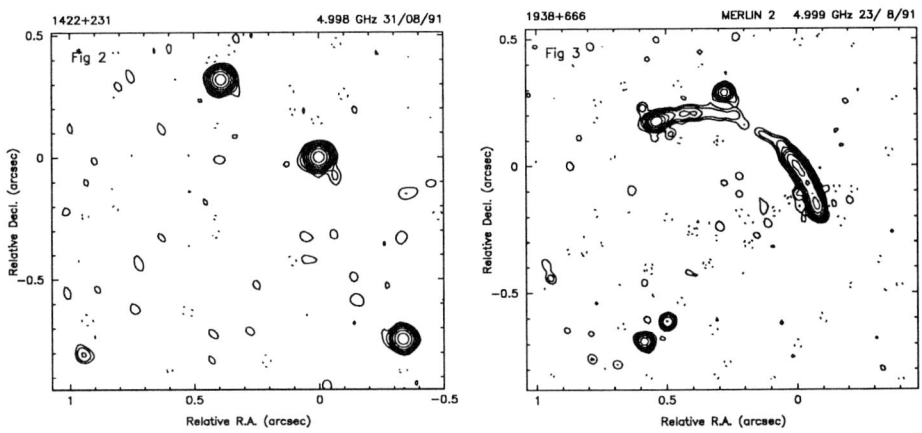

Fig 1a Map of B0218+35.7. Peak flux density = 727 mJy/bm. Contours at 0.5 mJy/bm × -2,-1,1,2,4,6,8,10,15,20,100,400,1000. **Fig 1b** Grey scale map of B0218+35.7. Grey levels between 0.5 mJy/bm and 10 mJy/bm. **Fig 2** Map of B1422+23.1. Peak flux density = 219 mJy/bm. Contours at 1 mJy/bm × -2,-1,1,2,4,8,16,32,64,128,256. **Fig 3** Map of B1938+66.6. Peak flux density = 36 mJy/bm. Contours at 0.2 mJy/bm × -2,-1,1,2,4,8,16,32,64,128,256.

MULTIFREQUENCY RADIO IMAGES OF MG1131+0456

Grace H. Chen and Jacqueline N. Hewitt [*]

Abstract

We present multifrequency VLA images of the Einstein ring gravitational lens system, MG1131+0456. Continuum data at 5, 8, and 15 GHz are presented. A new unresolved radio source, near the center of the ring, is evident in the 8 GHz image. Linearly polarized emission in the 5 and 8 GHz images yields Faraday rotation measurements across the ring and reveals complex structure that may provide strong constraints in lens modelling.

1. INTRODUCTION

The radio structure of MG1131+0456, the first Einstein ring gravitational lens detected (Hewitt *et al.*, 1988), consists of at least four components. One extended radio component is lensed into the ring, one compact radio component is lensed into components A and B, and a fainter extended component is lensed into component C (fig. 1). Optical (Hammer *et al*, 1991) and infrared (Annis 1992) emission has also been detected. Radio observations with the Very Large Array (VLA) [1] can provide detailed information on the morphology and polarization of MG1131+0456 at many wavelengths, which can give strong constraints on lens models.

2. OBSERVATIONS and RESULTS

We observed MG1131+0456 with the VLA during 12 observing sessions over a time period from September 1987 to May 1992. The observations were carried out in the A, B, and C configurations at 5, 8, and 15 GHz. The main purpose of these observations was to investigate MG1131+0456's variability. For presentation at this

---

[*] Department of Physics and Research Laboratory of Electronics, Massachusetts Institute of Technology, Cambridge, Massachusetts 02139

[1] The VLA is part of the National Radio Astronomy Observatory, which is operated by Associated Universities, Inc., under cooperative agreement with the National Science Foundation.

conference, we select three observations (see Table). The selection criteria were based on the total integration time, the u-v coverage and the overall quality of the data.

The contours of the total continuum intensity in Fig. 1 display the features recognized by Hewitt et al. (1988). The source consists of two compact components (A and B), a ring, and one extended source (C). In addition to components A, B, C and the ring, a new component, D, at the center of the map is detected at 8 GHz. At 5 GHz, although there is a hint of the existence of D, it is not cleanly resolved from the ring; at 15 GHz, component D is too faint to be detected. Polarized emission of MG1131+0456 is detected at 5 and 8 GHz. At 5 GHz, components A and B are 5% polarized, and component C is 27% polarized. At 8 GHz, component A and B are 10% polarized, and component C is 30% polarized. No polarization in D is detected at either frequency. From the thermal noise of the 8 GHz map (25 μJy), we estimate an upper limit to the percentage polarization in D of 10%.

What is component D? We consider three possible explanations for the existence of D: (1) the second image of C, (2) radio emission from the lensing galaxy, and (3) the third image of some combination of A, B and the ring.

If components C and D are indeed images of the same source, both must possess the same polarization properties. However, as seen in the 8 GHz map, component C is 30% polarized whereas the upper limit to the percentage polarization in D is only 10%. To make this hypothesis valid, an external Faraday screen at the position of D is needed to depolarize D from 30% to 10%. This implies a large magnetic field and electron density in the Faraday screen.

Since D is at the expected position of the lensing galaxy, it is possible that D represents radio emission from that galaxy. From the flux density of D at 8 GHz and considering a range of reasonable redshift for the lensing galaxy ($0.4 \leq z_l \leq 0.7$), the estimated luminosity of the lensing galaxy at a rest frame frequency of 8 GHz is

$$L_\nu \sim \frac{10^{22}}{h^2} \sim \frac{10^{23}}{h^2} \frac{W}{Hz},$$

where $H_o = 100 h \frac{Km}{Mpc-s}$. Therefore, if D is the lensing galaxy, the lensing galaxy must radiate strongly at radio wavelengths. Since most normal galaxies don't show strong radio emission, component D is not likely to be the radio image of the lensing galaxy, though the possibility can not be excluded without further data.

The position of component D and the upper limit to the percentage polarization support the remaining hypothesis. It is important to investigate the nature of component D further. For example, detection of any polarized emission in D would test this hypothesis.

The complex polarized emission of the ring at 8 GHz provides new, strong constraints in modelling. Fig. 2 shows the Q, U, and the total polarized intensity maps of MG1131+0456 at 8 GHz. In the Q and U maps, the ring breaks up into several arcs. This indicates that the extended radio source, lensed into the ring, contains at least two or three polarized sub-components. In addition to these polarized sub-components, we also see an evidence for one unpolarized sub-component in this extended radio source. In the total polarized intensity map, a gap, which separates the ring into an inner ring and an outer ring, represents the unpolarized emission and gives evidence for the unpolarized sub-component.

From a pair of observations at 5 and 8 GHz, we compute the Faraday rotation angle pixel by pixel at positions where the signal to noise ratio is higher than 3:1. For the purpose of resolving the $n\pi$ ambiguities, we also computed two polarization maps for the 5 GHz observation, one at each IF frequency. Fig 3(a) and 3(b) show the rotation angles across the ring. Large rotation angles appear in parts of the ring. Using these rotation angles, we compute the rotation measure across the ring, pixel by pixel, and obtain a polarization map, corrected for Faraday rotation, of MG1131+0456 (see fig 3(c)).

3. CONCLUSION

We have detected a new radio source at the center of the ring. A plausible explanation is that D is the third image of some combination of A, B and the ring. If this is the case, the size of radial caustics of the lensing potential must be tightly bound in a narrow range. Since an understanding of of component D can provide interesting information on the lensing galaxy, future study of D is important. The complex polarized emission also offers constraints in modelling. After correction for the Faraday rotation, parts of the ring with the same polarized emission must come from the same region of the source. This information allows us to identify images of the same source, which is usually difficult in extended gravitational lenses.

4. TABLE: Summary of data presented

Date	Freq(GHz)	Array Configuration	Resolution
Sep, 87	5	A	0.39"
Sep, 87	15	A	0.12"
Jul, 91	8	A	0.19"

References

[1] J. Annis, *Ap. J.*, **391**, L17, 1992.

[2] F. Hammer, O. Le Fèvre, M. C. Angonin, G. Meylan, A. Smette and J. Surdej, *Astron. and Astroph.*, **250**, L5, 1991.

[3] J. N. Hewitt, E. L. Turner, D. P. Schneider, B. F. Burke, G. I. Langston and C. R. Lawrence, *Nature,* **333**, 537, 1988.

MULTIFREQUENCY RADIO IMAGES OF MG1131+0456: CHEN & HEWITT

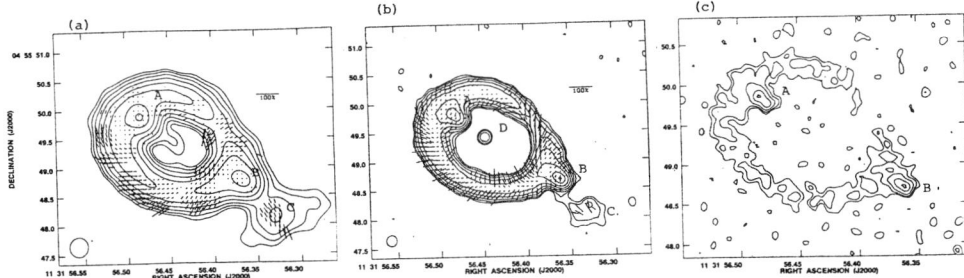

Figure 1: the total continuum intensity (the contour lines) and the polarized intensity (the vectors) at (a) 5 GHz, (b) 8 GHz, and (c) 15 GHz. The contour levels represent 1, 2, 4, 8, 16, 32, 64, and 95% of the peaks in (a) and (b), and 8, 16, 32, 64, and 95% in (c). The scale of 100% polarization is shown in the upper right hand corner.

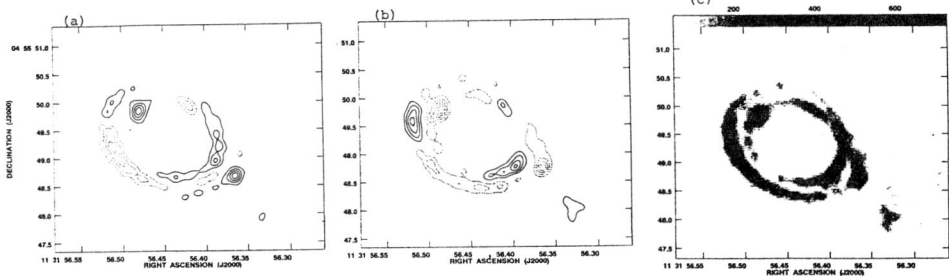

Figure 2: the (a) Q, (b) U, and (c) the total polarized intensity maps at 8 GHz. The grey scale in (c) is in units of μJy.

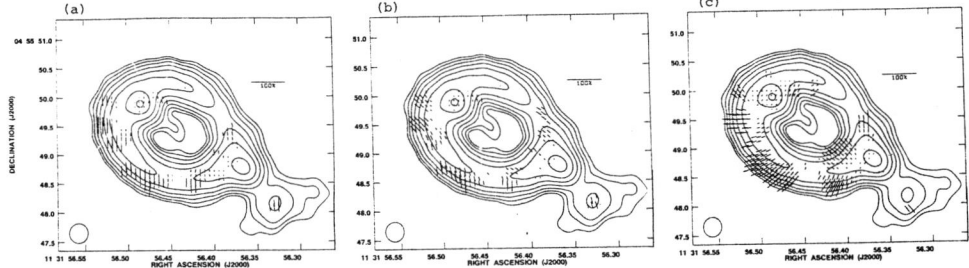

Figure 3: The Faraday rotation angle (a) between 4885 MHz and 4835 MHz, and (b) between 8440 MHz and 4835 MHz are superimposed on the 5 GHz total intensity contour lines. The 5 GHz polarization map corrected for Faraday rotation is shown in (c).

The Jodrell Bank Lens Survey

Lindsay Jane King [*] *Ian W. A. Browne* *Alok Patnaik*
Tom W. B. Muxlow *Dennis Walsh* *Peter N. Wilkinson*

Abstract

We are surveying ≈ 2500 flat spectrum radio sources with the VLA in A configuration at 8.4GHz with 200 mas resolution for gravitational lens systems. So far, ≈ 1500 have been observed. We have followed up 14 candidates selected from the first 840 sources with the VLA and MERLIN.

1. Motivation

Flat spectrum source surveys facilitate the identification of multiple imaging events as strong secondary components or "extended structure" are not expected. The completeness of such a survey should therefore be greater than that of a survey with no spectral index selection criterion. We are incorporating the number of lensing events in our survey into an analysis which sets confidence limits on various values of Λ and Ω for a flat (k=0) cosmology as in Fukugita and Turner [F91].

Optical surveys have a typical angular resolution of 1", depending on the seeing. Radio surveys, however, have lower limits. The number of lens systems (N) vs. the image angular separation ($\Delta\theta$) is determined by the distribution of masses and core radii of the lens population [T84]. The low $\Delta\theta$ end of the observed N vs. $\Delta\Theta$ distribution suffers from angular resolution bias; obviously this is decreased as the angular resolution is improved.

2. Selection of candidates

There are several positive indicators of gravitationally lensed sources in a flat spectrum survey. We look for multiple compact components and/or extended structure. Lensed components should have very similar spectral indices, percentage polarisations and position angles of polarisation (when corrected for Faraday rotation).

The candidate selection was primarily based upon the presence of multiple compact components or "unusual morphology" in the survey maps. The polarisation data were largely discounted at this stage, since the observations were generally of a signal to noise too low to detect the polarisation of subsidiary components. 14 candidates were selected and reobserved with the VLA at one or more of the frequencies 1.5, 5, 8.4, 15, 22 GHz. A subset was observed with MERLIN at 5 GHz.

---

[*]N.R.A.L., Jodrell Bank, Macclesfield, Cheshire SK11 9DL, England

3. Results

The current status of our follow-up is that we have identified 2 lens systems (0218+357, 1938+666) (see [P92]), 10 assorted core–jet, compact double and triple sources, and 2 "live" candidates. Three maps of a candidate which has been ruled out (1750+509) are shown below; the contours are logarithmic and the scale bars represent 1 arcsecond.

		A:B (total)	A (peak:total)	B (peak:total)	$3\sigma(Jy/beam)$
fig.1	VLA 15GHz	1.45	0.73	0.97	1.35e-03
fig.2	VLA 8.4GHz	1.52	0.79	0.82	9.1e-04
fig.3	MERLIN 5GHz	1.55	0.48	0.86	3.9e-04

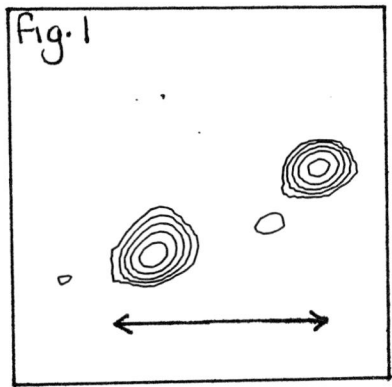
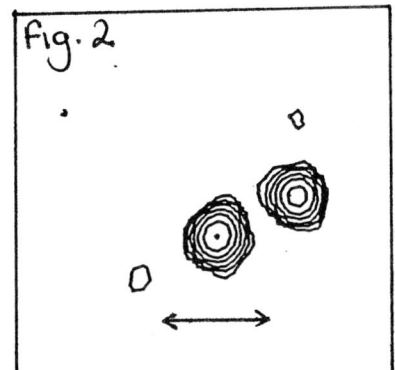
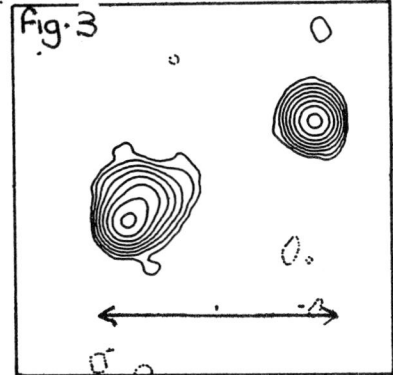

References

[F91] Fukugita M., Turner E.L., *MNRAS 1991, 253, 99*

[P92] Patnaik A. et al., *This proceedings*

[T84] Turner E.L., Ostriker J.P., Gott J.R., *Ap.J. 1984, 284, 1*

MERLIN Observations of 3 Gravitational Lens Candidates

M.A. Garrett,*A.R. Patnaik,
T.W.B. Muxlow, P.N. Wilkinson & D. Walsh

We present λ6cm MERLIN images of three gravitational lens candidates which first emerged from the MIT/VLA lens survey. The maps shown in Figure 1 reveal new details of the radio structure in MG0414+056 [HTL92], 2016+112 [LSS84] and 1042+178 [Hew86]. MG0414+056 exhibits extended emission in the A and B images which appears to form an unusual "S-shape" structure. Our new map of 2016+112 shows the southern component to be extended (east-west) on a scale of $\approx$ 150 mas as predicted by recent lens models [LFA91]. Observations of 1042+178 [Hew86] detect the northern components of the source but appear to resolve out the weaker southern components. This result suggests that 1042+178 is not a gravitational lens system.

The observations described here were made during the new MERLIN's commissioning phase. Further observations are planned with the 'full blown' MERLIN system. We expect that these will be twice as sensitive as the observations described here and will include polarization data. We hope that this will alow us to map the "S-shape" structure in MG0414+056 in more detail. Observations at λ18cm are also planned.

References

[Hew86]   Hewitt, J.N., *Ph.D. Thesis Massachussetts Institute of Technology, 1986.*

[HTL92]   Hewitt, J.N., Turner, E.L., Lawrence, C.R., Schneider, D.P. and Brody, J.P. 1992, *A.J.*, **104**, 968.

[LFA91]   Langston, G.I., Fischer, J. and Aspin,C., 1991, *Astr. J.*, **102**, 1253.

[LSS84]   Lawrence, C.R., Schneider, D.P., Schmidt, M., Bennet, C.L., Hewitt, J.N., Burke, B.F., Turner, E.L. and Gunn, J.E. 1984, *Science*, **223**, 46.

---

*University of Manchester, NRAL, Jodrell Bank, Macclesfield, Cheshire, SK11 9DL, UK.

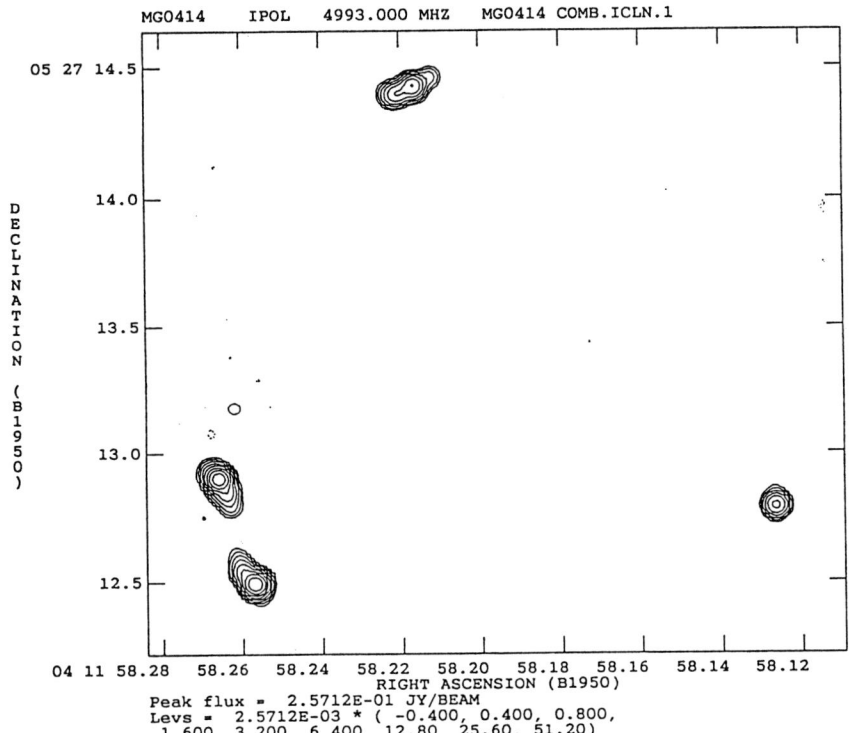
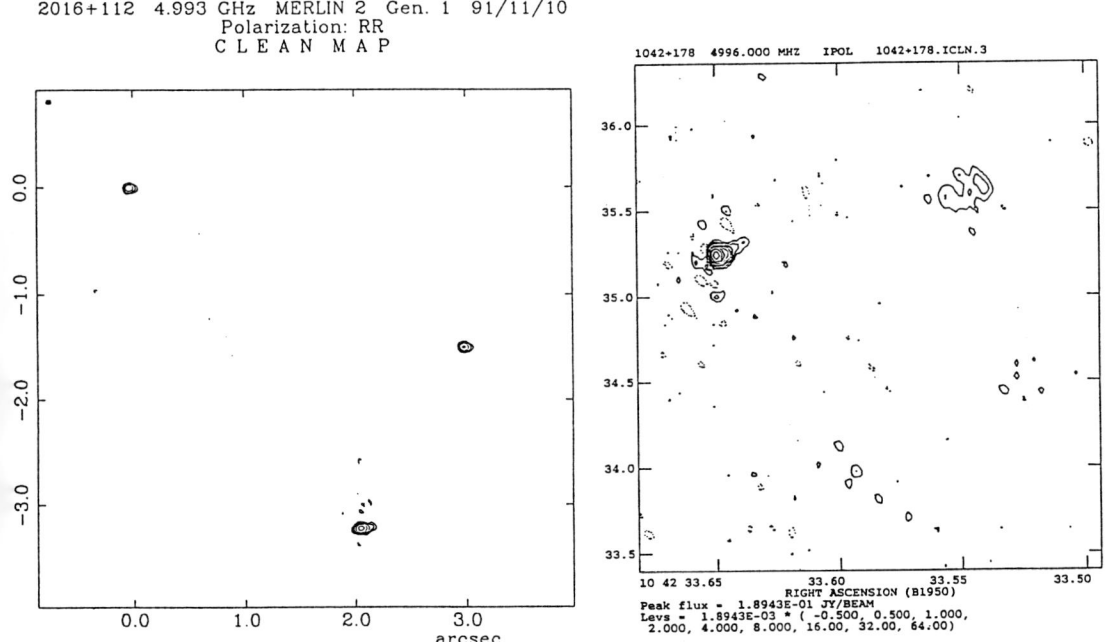

Figure 1. Top: MERLIN λ6 cm images of MG0414+056; bottom left: 2016+112 and bottom right: 1042+178.

Is the BL Lacertae Object 1308+326 a Microlensed Quasar?

Denise C. Gabuzda [*] *Ronald I. Kollgaard* [†]

Abstract

VLBI polarization maps of the BL Lacertae object 1308+326 have revealed both VLBI polarization structure and tentative superluminal speeds that are characteristic of *quasars* rather than BL Lacertae objects. This behaviour can be understood if this source is actually a gravitationally microlensed image of a distant quasar. Our observations suggest a speed for one knot of $\sim 21h^{-1}c$, which is among the highest speeds ever observed; it is possible that this may also be understood as a consequence of gravitational microlensing.

BL Lacertae objects are compact AGN whose primary distinguishing characteristics are a lack of strong line emission, high polarization, extreme variability, and a lack of a UV excess. Some authors have suggested that many sources that have been classified as BL Lacertae objects are in reality images of distant quasars which are being microlensed by stars in intervening galaxies (e.g., OV91). In this scenario, the optical continuum is amplified by the lensing, so that the quasar's intrinsically strong emission lines are buried by the amplified continuum. Systematic differences in VLBI polarization structure (Gab92), arcsecond-scale radio structure (Kol92), and X-ray spectra (WW91) rule out the possibility that the majority of BL Lacertae objects are lensed quasars. However, there may be a few that are microlensed: such sources should have many features characteristic of BL Lacertae objects while exhibiting certain "quasar-like" properties. 1308+326, which has been proposed as a microlensing candidate by Stickel et al. (Sti91), appears to be such a source.

Mark III VLBI observations of a number of BL Lacertae objects were made at 5.0 GHz in March 1989, using an eight-antenna global array. The resulting VLBI total intensity and polarization maps of 1308+326 show polarized flux in the inner portion of the jet which is nearly *perpendicular* to the jet (Gabuzda et al., in preparation). This is the usual orientation for *quasars*, not for BL Lacertae objects (Gab92). A comparison of these maps and those made from earlier observations (Gab92) suggests speeds

---

[*]Department of Physics and Astronomy, University of Calgary
[†]Department of Astronomy and Astrophysics, Pennsylvania State University

of order $8h^{-1}c$, $21h^{-1}c$, and $4h^{-1}c$ for three knots in the VLBI jet. The available evidence indicates that the speeds of the two fastest knots are considerably greater than those typically seen in BL Lacertae objects. Indeed, the tentative speed for the middle knot is among the largest ever measured: it is intriguing that under certain circumstances, lensing can amplify intrinsic component motions, increasing the observed apparent speeds (e.g., GS91).

Stickel et al. (Sti91) identified 1308+326 as a microlensing candidate based in part on discovery of faint extended optical emission, suggestive of an intervening system. In addition, 1308+326 is one of a small number of powerful BL Lacertae objects displaying FR II luminosity and arcsecond-scale morphology, which is typical of core dominated, variable quasars (Kol92). Another piece of circumstantial evidence comes from the work of Brown et al. (Bro89), who report that 1308+326 was the only one of five BL Lacertae objects monitored at a large number of wavelengths which exhibited evidence for a slight optical/UV excess, or "blue bump", which had never before been observed in a BL Lacertae object.

The VLBI polarization structure and the other properties discussed above strongly indicate that 1308+326 is intrinsically a quasar in which the optical continuum is much stronger than usual. The optical evidence presented by Stickel et al. (Sti91) suggests that the origin for the unusually strong continuum emission is amplification by microlensing. Whether or not 1308+326 is indeed a microlensed quasar, it is certainly an unusual object worthy of further study.

References

[Bro92]   Brown, L, et al., *Ap. J. Vol. 340, p. 129.*

[Gab92]   Gabuzda, D., Cawthorne, T., Roberts, D. and Wardle, J., *Ap. J. Vol. 388 (1992), p. 40.*

[GS91]   Gopal-Krishna and Subramanian, K., *Nature, Vol. 349 (1991), p. 766.*

[Kol92]   Kollgaard, R., Roberts, D., Wardle, J., and Gabuzda, D., *A. J. in press.*

[OV91]   Ostriker, J. and Vietri, M., *Nature, Vol. 344 (1991), p. 45.*

[Sti91]   Stickel, M., Padovani, P. Urry, C., Fried, J., and Kühr, H., *Ap. J. Vol. 374 (1991), p. 431.*

[WW91]   Worral, D. and Wilkes, B., *Ap. J. Vol. 360 (1991), p. 396.*

Northern Hemisphere VLBI Observations of PKS 1830-211

Dayton L. Jones * and the SHEVE team [†]

Abstract

We present 5-GHz VLBI images of the two compact components in the Einstein ring PKS1830-211. Comparing images from two epochs separated by ten months shows that the morphology of the northeast component changes significantly while the southwest component changes very little (if at all). A possible explanation is that a differential time delay between the two components is responsible for the different morphological evolution.

The recently discovered Einstein ring gravitational lens PKS1830-211 [J91] is one of the strongest compact radio sources in the sky ($\sim$ 10 Jy). It consists of two VLBI components on opposite sides of a one arcsecond diameter ring of radio emission.

We have imaged PKS1830-211 at 5 GHz with a global array of telescopes in North America and Europe during two epochs (15 November 1990 and 21 September 1991) separated by ten months. Images of the two compact components from each epoch are presented in Figure 1. The morphology of the northeast component can be seen to have changed significantly, while the southwest component remained essentially unchanged.

The four VLBI images in Figure 1 show the northeast and southwest components of PKS1830-211 in 1990 (left side) and 1991 (right side). In all images the contour levels are -5, 5, 10, 15, 25, 35, 50, 70, and 90 percent of the peak, and the restoring beam is 10×5 milliarcseconds with the major axis along position angle 26°. The tick marks along the axes are 12 milliarcseconds apart.

Note the dramatic change in the structure of the northeast component. There is a short extension to the northwest in 1990 (this is the same direction as extensions seen

---

*Jet Propulsion Laboratory, California Institute of Technology, Pasadena, CA. 91109, U.S.A. This research was carried out at the Jet Propulsion Laboratory, California Institute of Technology, under contract with the National Aeronautics and Space Administration.

[†]Members of the Southern Hemisphere VLBI Experiment (SHEVE) team are listed as co-authors on the paper by R. A. Preston *et al.* in these proceedings.

in earlier 1.6 and 2.3 GHz VLBI maps), but in 1991 the brightest feature is close to the center of the component, and the total extent of emission is nearly twice as large. This may be due to motion of knots along a jet, but with only two epochs we can not determine proper motions unambiguously.

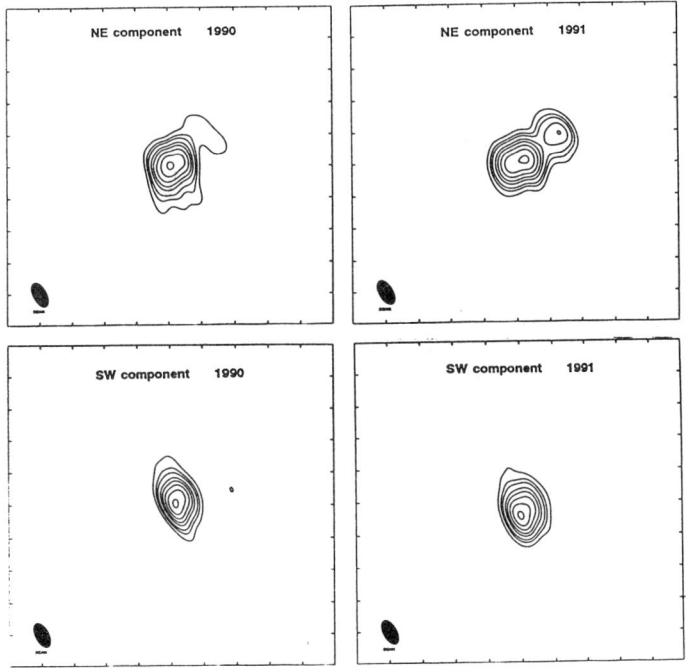

Figure 1. 5-GHz VLBI maps of PKS1830-211

The southwest component, which has always appeared more compact than the northeast component in VLBI images, is very similar in our 1990 and 1991 images. In both cases, a short extension to the northwest can be seen.

The separation between the northeast and southwest components was 975 ± 2 milliarcseconds in 1990 and 973 ± 2 milliarcseconds in 1991, an insignificant difference.

References

[J91] Jauncey, D.L., Reynolds, J.E., Tzioumis, A.K., Muxlow, T.W.B., Perley, R.A., Murphy, D.W., Preston, R.A., King, E.A., Patnaik, A.R., Jones, D.L., Meier, D.L., Bird, D.J., Blair, D.G., Bunton, J.D., Clay, R.W., Costa, M.E., Duncan, R.A., Ferris, R.H., Gough, R.G., Hamilton, P.A., Hoard, D.W., Kemball, A., Kesteven, M.J., Lobdell, E.T., Luiten, A.N., McCulloch, P.M., Murray, J.D., Nicolson, G.D., Rao, A.P., Savage, A., Sinclair, M.W., Skjerve, L., Taaffe, L., Wark, R.M., White, G.L., *An unusually strong Einstein ring in the radio source PKS1830-211*. Nature, Vol. 352 (1991), pp. 132-134.

A VLBI Survey of Southern Hemisphere Peaked Spectrum Sources

E. A. King [*]   and the SHEVE team [†]

Abstract

The SHEVE array has been used to study a sample of 29 Parkes radio sources which have spectral peaks in the range 0.1 to 2.0 GHz. All the objects have been observed on the Tidbinbilla to Hobart baseline at 2.3 GHz. Imaging observations with the full array have been undertaken for 10 of these. Many are doubles with two high brightness temperature components separated often by much more than 150 mas. The discovery of a strong Einstein ring in this sample, together with the wide separation and high brightness temperatures, raises the question as to how many of these sources may be gravitational lenses.

The sources in this sample have been selected on the basis of a low frequency turnover in the range 0.1 to 2.0 GHz. The wide range of turnover frequencies allows an examination of the variation of source properties as a function of turnover frequency. To ensure suitability for investigation with the SHEVE array the sources have declinations $< 0°$ and a 2.7 GHz flux density > 1.5 Jy.

Ten of the objects in the sample have been imaged with the full SHEVE array at 2.3 GHz and/or 8.4 GHz. The ratio of the separation of the compact components to their angular size is frequently greater than 10:1. All the objects imaged have peak brightness temperatures $\sim 10^9$ K. Four images, for PKS0023–263, PKS1151–348, PKS1245–197 and PKS1830–211, are presented in Figure 1. The observations were made using a four antenna array operating at 2.3 GHz in April 1991. PKS0023–263 and PKS1830–211 in particular, exhibit a large value of the ratio of component separation to component size.

In parallel with the investigation of the VLBI structure, a flux monitoring program at 2.3 and 8.4 GHz has been conducted using the 26 metre antenna at Hobart. The

---

[*] Department of Physics, University of Tasmania, Hobart, 7001, Australia
[†] Members of the Southern Hemisphere VLBI Experiment (SHEVE) team are listed as co-authors on the paper by R. A. Preston et al. in these proceedings. This research was carried out in part at the Jet Propulsion Laboratory under contract to NASA.

majority of the sample exhibits the low flux variability typical of GPS sources studied in the Northern Hemisphere. The most dramatic variable is the strong Einstein ring PKS1830–211 [J91] which has fluctuated by more than a factor of two in little more than a year at 8.4 GHz. PKS1830–211 is a particularly good example of a wide compact VLBI double and we are investigating the possibility that similar sources in our sample are also gravitational lenses. Also variable, PKS0023–263 is one such candidate.

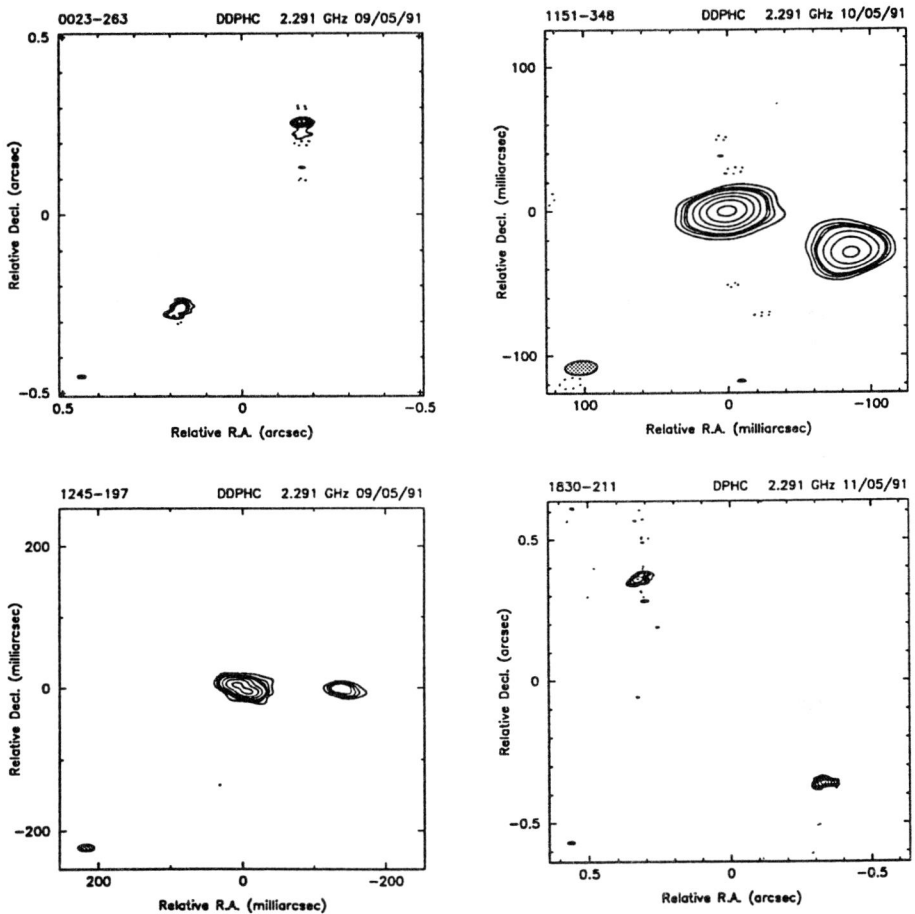

Figure 1. 2.3 GHz images of PKS0023–263, PKS1151–348, PKS1245–197 and PKS1830–211.

References

[J91] Jauncey, D.L., et al *An unusually strong Einstein ring in the radio source PKS1830–211. Nature, Vol. 352 (1991), pp. 132-134.*

Undetected Lens Systems in the MIT-Green Bank-VLA 5 GHz Lens Search Sample

*Samuel R. Conner   André Fletcher   Lori Herold
and Bernard F. Burke* [*]

Abstract

We discuss the question of undetected lens systems present in the MIT-Green Bank-VLA Lens Search Sample by extrapolating from the properties of detected systems.

The MIT-Green Bank-VLA (MGV) gravitational lens search program [1,2] has led to the discovery of 4 lens systems in the 5 GHz sample of 3500 objects [3,4,5,6] and one lens system in an 8 GHz sample of 360 objects [7]. All of these lens systems are on the order of 2 arc-seconds or larger in image separation, which contrasts with theoretical predictions that most lens systems should be substantially smaller than this [8,9].

Most of the lens systems detected thus far in the MGV survey are highly magnified ring or quad systems which appear in our single-dish finding list only because they have been substantially magnified. Under the approximation that the image magnification is roughly linear in the scale of the image deflection, we can empirically extrapolate the observed properties of our detected lens systems to determine whether we should expect to find less highly magnified systems — either smaller image separations or lower image multiplicities — in our sample. The sampling function of the MGV 5 GHz sample falls rapidly below $S_{5\mathrm{GHz}} = 100$ mJy. The two ring systems whose flux is dominated by the highly magnified ring, MG1131+0456 and MG1654+1346, have fluxes of 150 and 200 mJy, respectively, and so would not be in our sample if their sizes had been much less than 1.5" The ring system MG1549+3047 (detected in the 8 GHz survey) is dominated by large-angular-extent emission which is not multiply imaged and so would be present in our sample for arbitrarily small image separation. The quadruple compact image system MG0414+0456 has a flux

---

[*]Department of Physics and Research Laboratory of Electronics, Massachusetts Institute of Technology, Cambridge, Massachusetts 02139

of 740 mJy and would be present in our sample were its image separation as little as .3", below the resolution limit of the 5 GHz search, though not of the 8 GHz search.

From this we speculate that there should be several yet-to-be-recognized ring systems with angular diameters greater than 1" in our 5 GHz sample. Systems smaller than this may have too little flux to appear in our single-dish finding list, unless, like MG1549+3047, they are dominated by singly-imaged emission.

There may also be a number of small multiple-compact-image lens systems in our sample. The small size of the background source in such systems makes large magnifications possible (the background source in MG0414 is only about 15 mJy), with the result that image configurations substantially smaller than the limit for our ring systems will still have enough flux to appear in our sample.

Two-image systems with sources like those in our ring-dominated systems may have fluxes below the 5 GHz sample limit, regardless of their angular size.

Experiments we have performed on simulated 5 GHz data suggest that with optimal reduction we should be able to detect ring systems down to the $\approx 1''$ angular size limit below which they are likely to be too faint to appear in our sample. Compact multiple image systems should be detectible down to image separations of as little as .5" We are recalibrating and remapping the entire 5 GHz survey in order to find these systems [10]. The 8 GHz extension of the lens search is being pushed to a lower flux limit than the 5 GHz search and will contain and be able to detect smaller lens systems as well as two-image configurations.

References

[1]   Lawrence, C.R., et al. *Ap J. Suppl. Ser.*, **61**:105-157, 1986 May

[2]   Hewitt, J.N., PhD. Thesis, MIT, June 1986

[3]   Lawrence, C.R. et al. *ApJ Lett.* **278**, L95, 1984

[4]   Hewitt, J.N. et al. *Nature* **333**:357 1988

[5]   Langston, G.I. et al. *AJ* **97**: 1283-1290, 1989 May

[6]   Hewitt, J.N. et al., submitted, 1992

[7]   Lehár, J., PhD. Thesis, MIT, September 1991

[8]   Turner, E.L., Ostriker, J.P. and Gott, J.R *ApJ* **284**:1-22, 1984 September

[9]   Kochanek, C.S. and Lawrence, C.R., *AJ* **99**:1700-1708, 1990 June

[10] Conner, S.R., PhD. Thesis, MIT, 1993, in prepration

The Unusual Radiosource MSH 04 − 71

John E. Reynolds* *and the SHEVE team* †

Abstract

VLBI and Australia Telescope Compact Array (ATCA) observations of the strong steep-spectrum radiosource MSH 04−71 show two compact components with brightness temperatures in excess of $\sim 10^9 K$, separated by 6 arcsec. Deep CCD images taken with the 3.9m Anglo Australian Telescope (AAT) reveal a faint galaxy or galaxies near the radio centroid and two fainter objects coincident within 1 arcsec of the radio components. MSH 04 − 71 may be another 10 Jy gravitational lens, although other explanations cannot yet be ruled out.

The radio source MSH 04 − 71 is listed in the early southern radiosource catalogue of Mills, Slee & Hill [M61] with a flux density of 87 Jy at 85 MHz, and an angular size <20 arcsec determined with a radio-linked interferometer [G60]. It is thus one of the brightest compact steep-spectrum sources in the southern sky and has a radio spectrum $\alpha \sim -0.85$ $(S \sim \nu^\alpha)$.

Observations with the Parkes-Tidbinbilla Interferometer (PTI) at 2.3 GHz in 1988 showed a substantial fraction of the total flux to be contained in components of angular size less than ~ 0.1 arcsec. Subsequent observations with the SHEVE VLBI array [P93] in 1991 indicated a 6 arcsec double with an unusually high ratio of separation to component size ($\sim 300:1$). The components have brightness temperatures in excess of $\sim 10^9 K$ and angular sizes ~ 20 milliarcsec. This high ratio of separation to size is unusual in extragalactic radiosources, which typically host a single core of high brightness temperature, and is similar to the ratio observed in the Einstein ring / gravitational lens PKS1830-211 [J91].

Independently, MSH 04−71 was imaged at 4.8 GHz in a programme to find calibration sources for the ATCA [D92], which also led to the suggestion of a lensing origin for the source. A more recent ATCA image at 8.6 GHz in Figure 1. clearly shows the two components with the SE component the brighter by a factor of ~ 1.5. Additional

---

*Australia Telescope National Facility, Epping, N.S.W., Australia.
†Members of the Southern Hemisphere VLBI Experiment (SHEVE) team are listed as co-authors on the paper by R. A. Preston *et al.* in these proceedings.

ATCA images at 1.4 GHz and 4.8 GHz show that the components have similar spectral indices ($\Delta\alpha < 0.1$) below 5 GHz, but that the index of the NW component steepens above 5 GHz. There is no central core visible at a level of $\sim 0.1\%$ of the total flux density at 8.6 GHz.

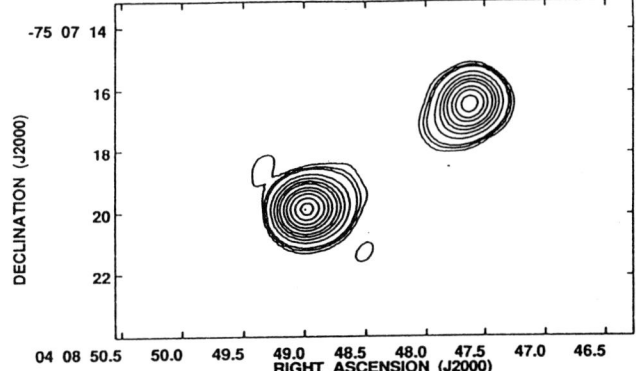

Figure 1. Australia Telescope Compact Array (ATCA) image of MSH 04 − 71 at 8.6 GHz Peak flux is 1.2 Jy/Beam; contours are ±.3 .6 1 3 6 10 15 25 35 50 70 and 90 % of peak.

Deep CCD images taken with the AAT in B and R are shown in Figure 2. The optical morphology is complex but reveals an elongated red object G (R$\sim$21, B-R$\sim$1) close to the radio centroid which appears to be a giant galaxy or galaxies in a distant cluster. The B-band image shows fainter peaks of B$\sim$23 to the NW and SE of this galaxy, each within 1 arcsec of the radio components. The angular vector between peaks A and B agrees closely ($\sim$0.2 arcsec in each coordinate) with that between the two radio components.

The object(s) B/C are neutral or slightly blue in colour, and are visible on the Schmidt J survey image at 21.5^m $\sim 22^m$, significantly brighter than in the more recent CCD frames. This implies optical variability, which would be consistent with a (lensed) QSO or BLLac. On the other hand the R image shows object A to be quite red (B-R$\sim$1). This large colour difference between A and B/C may be due to the presence of a galaxy close to the line of sight to object A, a possibility supported by an apparent shift in the position of A between the R and B images.

While the evidence for gravitational lensing within MSH 04 − 71 is not conclusive, neither are alternative interpretations entirely satisfactory. The optical morphology is most unlike any classical double radiosource, such as Cygnus A, and the linear size is apparently an order of magnitude larger than for CSS sources. Optically, perhaps the closest resemblance lies with powerful distant radiogalaxies of Chambers & Miley, such as 3C 368 [C88], which often have extended optical morphologies aligned with the radio axis. However, the strong spatial coincidence of the radio and optical emission in MSH 04 − 71, as well as its unusual radio structure, appear to set it apart from this class as well. Sensitive optical spectroscopy and more detailed VLBI imaging are required to determine the nature of this intriguing object.

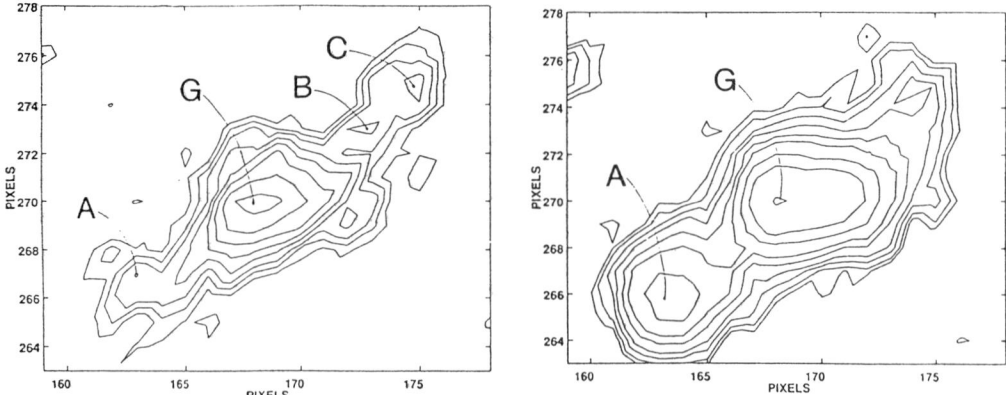

Figure 2. CCD images of MSH 04 − 71 in B (left) and R (right). Scale is 0.5 arcsec/pixel. Lettering is explained in text

References

[C88] Chambers, K.C., Miley, G.K., Joyce, R.R., *2.2 micron image of 3C 368 at $z = 1.13$, a galaxy with aligned radio and stellar axes.* Astrophys. J. Lett. Vol. 329, pp. L75-79.

[D92] Duncan, R.A., Sproates, L., *ATCA images of southern radio sources.* Proc. Astron. Soc. Aust. (in press).

[G60] Goddard, B.R., Watkinson, A., Mills, B.Y., *An interferometer for the measurement of radio source sizes.* Aust. J. Phys. Vol. 13 (1960), pp. 665-675.

[J91] Jauncey, D.L., Reynolds, J.E., Tzioumis, A.K., Muxlow, T.W.B., Perley, R.A., Murphy, D.W., Preston, R.A., King, E.A., Patnaik, A.R., Jones, D.L., Meier, D.L., Bird, D.J., Blair, D.G., Bunton, J.D., Clay, R.W., Costa, M.E., Duncan, R.A., Ferris, R.H., Gough, R.G., Hamilton, P.A., Hoard, D.W., Kemball, A., Kesteven, M.J., Lobdell, E.T., Luiten, A.N., McCulloch, P.M., Murray, J.D., Nicolson, G.D., Rao, A.P., Savage, A., Sinclair, M.W., Skjerve, L., Taaffe, L., Wark, R.M., White, G.L., *An unusually strong Einstein ring in the radio source PKS1830-211.* Nature, Vol. 352 (1991), pp. 132-134.

[M61] Mills, B.Y., Slee, O.B., Hill, E.R., *A catalogue of radio sources between declinations $-50°$ and $-80°$.* Aust. J. Phys. Vol. 14 (1961), pp. 497-507.

[P93] Preston, R.A., & the SHEVE Team, *The Southern Hemisphere VLBI Experiment (SHEVE).* Proceedings of the Sub-Arcsecond Radio Astrononomy symposium, Manchester 1992.

Intraday Variability and High Brightness Temperatures

A. Witzel[†], S. Wagner[*], R. Wegner[†], W. Steffen[†], T. Krichbaum[†]

Introduction

Intraday variability (IDV) has been detected at radio wavelengths in 1985 (Witzel et al., 1986; Heeschen et al., 1987) and has been shown to be a common phenomenon at radio and optical bands for flat spectrum radio sources since then (see e.g. Wagner & Witzel, 1992a, Krichbaum et al., 1992, Witzel 1992). The initial measurements were concentrated on the brighter sources from the S5–sample of radio sources, which contains extragalactic objects north of declination of +70 degrees (Kühr et al., 1981). Subsequently, sources at lower declinations were added so that a statistically complete sample of flatspectrum radio sources with $\delta \geq +35$ degrees is formed, which is investigated for IDV at least for one epoch. Later, additional "promising" candidates were added to this source compilation. These are blazars with very compact structures. In general, about one third of the brighter compact sources seem to exhibit 'lightcurve'-type variability (type II as defined by Heeschen et al., 1987) with amplitudes exceeding 2-3 % at any given epoch (Quirrenbach et al., 1992). Due to the short timescales of this type of variability it was obvious from the beginning that extremely high brightness temperatures – vastly in excess of the inverse Compton limit of 10^{12} K – will be deduced if a source–intrinsic interpretation of the variability is adopted.

Multifrequency observing campaigns have been carried out with the 100 m radio-telescope, the VLA, and various optical telescopes (e.g. the 1.2 and 2.2 m at Calar Alto, the 0.7 m at Heidelberg). In addition, data were taken in the UV (IUE), X-ray (ROSAT) and γ-ray (GRO) regimes. These observations revealed numerous characteristic properties, which will be discussed in the following.

Radio observations

Repeated observations aimed to monitor short timescale variability were done with the 100 m radio-telescope at Effelsberg and the VLA at Socorro. The wavelengths ranged from 2 to 90 cm. The observations with the 100 m telescope covered timespans of a few days to up to one week. Twice our VLA observations covered longer intervals: about 4 weeks in 1991 and about 3 weeks in 1992. In addition to the variability of the total flux density, the changes in the polarized flux density and in the polarization angle were measured since 1988. The exceptionally good agreement of the measured flux densities and of the - considerably fainter - polarized flux densities obtained with the two conceptually different instruments is demonstrated in Fig. 1 in the article of R. Wegner (this volume).

Furthermore, a lot of additional data is available for many of the sources which have

---

[*]Landessternwarte Königsstuhl, Heidelberg, Germany
[†]Max-Planck-Institut für Radioastronomie, Bonn, Germany

shown IDV, e.g. radio spectra up to the mm-regime, brightness distributions from arcsecond to milliarcsecond scales (see e.g. Witzel et al., 1988; Pearson & Readhead, 1988) and the multi-frequency variability studies on longer timescales (e.g. Aller et al., 1992; Teräsranta et al., 1992).

Optical/UV observations

Stimulated by the detection of IDV in the radio regime, we started simultaneous radio/optical observing campaigns in 1989. Parallel optical monitoring over considerably longer timespans – than at present available from radio observations – were carried out. In Fig. 1 we show the optical variability of the BL Lac 0716+71 during about half a year in 1990/91. The object belongs to the most strongly variable sources known. As indicated in Fig. 1, simultaneous UV and X-ray observations were done just at a time when the source was in a high state of optical activity.

Additional simultaneous observations at optical and UV frequencies were carried out on 0716+714 in January 1992.

X-ray and γ-ray observations

Most of the objects showing IDV have been observed by ROSAT. We observed 0716+714 for 21000 sec (on source) in March 1991 and found it to vary by a factor of 2 within 24 hours in intensity (see Fig. 3). Significant variations on timescales $\leq$ 1000 sec were detected in only two of 14 continuous observing intervals. The GRO satellite detected 0716+714 in January 1992. The γ-ray luminosity exceeds by far the optical and X-ray luminosity. Such a behaviour seems to be a common phenomenon of IDV sources. Most of the AGN detected by EGRET up to now are rapid optical variables, and for about half of the sources, IDV has been found in the radio bands, too.

Correlations

In Fig. 2 the variations of the total flux densities in 0716+714, 0917+624 and 0954+658 are shown as measured with the VLA in January 1990. The power spectrum analysis of these data revealed typical discrete timescales in the range from 0.8 days to up to 2 weeks (indicated on the right hand side of Fig. 2). Generally, there is no unique dependence of the amplitude of variability on the wavelength, except that at longer wavelengths (20 cm) the variations seem to be washed out (see also e.g. Fig. 2 in Wegner and Witzel, this volume). In 0716+714 the variability amplitude increases with frequency, whereas in 0917+624 the amplitudes are largest at 11cm (Witzel, 1992). The time sequence for the individual peaks seems to be canonical in 0954+658, i.e. the maxima appear first at the higher frequencies, but in 0716+714 there is some indication that the peaks at 6 cm precede the maxima at 3.6 cm!

In the quasar 0917+624 we observed an anticorrelation between the total and the polarized flux densities prior to 1988 (Quirrenbach et al., 1989) and a transition to correlated variability of the two quantities in May 1989. For 0716+714 our observations revealed correlated variability between S_{tot} and S_{pol} except for one measurement in 1989 when the quantities were anticorrelated. Changes in the polarized flux densities are usually accompanied by changes in the polarization angle.

At optical wavelengths all of the IDV sources show detectable amounts of linearly polarized emission. Their variability is similar to that in the radio range (Wagner & Witzel, 1992a). The fractional variations are usually larger in the optical than in the radio (20 % and 10 % respectively). The degree of variability in the total and polarized flux density is similar in the optical but almost an order of magnitude different in the radio regime. A conclusive evidence for an intrinsic origin of the variability of the emission would be derived from correlated variability at radio and optical wavelengths. 0716+714 showed a transition between two modes of variability in 1990 both in the optical and in the radio regime. This was interpreted as a proof for correlated variability (Quirrenbach et al., 1991). A detailed statistical analysis strengthened this conclusin and showed that any time lag between the variations in the radio and the optical is smaller than two days (Wagner & Witzel, 1992a). In the source 0954+658 observations at two epochs give some hints of correlated radio/optical variability, too: In 1989 such a correlation is suggested by – unfortunately – heavily undersampled data. In 1990 similarly an undersampling of the radio data due to maintenance periods at the VLA prevents a convincing correlation analysis of the otherwise highly suggestive data (see Wegner et al., 1992 for details).

The variability of the total intensities in the optical and the UV was directly correlated in the source 0716+714 in February 1991 and in January 1992.

Fig. 3 displays the X-ray lightcurve of 0716+714 and the simultaneous optical observations. It is obvious that both lightcurves are sampled too sparsely and do not cover a sufficient time-span to obtain meaningful cross-correlations. The variability in both bands has similar power-spectra. Rapid variations (on time-scales of a few minutes to several hours) have been found only in the X-ray band. The dominant time-scales are of the order of a day in either band. This suggests that the emission in the X-ray may be closely linked to the optical emission, which owing to its polarization properties could be interpreted as synchrotron radiation.

Interpretation

The central question in view of the topics of this conference is that of the origin of the IDV emission. At least in the case of 0716+714 the correlated variability in the radio and optical domains – and perhaps also in the X-ray regime – clearly favors an intrinsic explanation and thus leads to apparent brightness temperatures in the objects far in excess of the inverse Compton limit of 10^{12} K. Nevertheless, it has to be remembered that for sources of such high compactness, effects of the interstellar medium must exist, too (e.g. Witzel & Quirrenbach, 1993). Microlensing as the dominant cause for the IDV has been ruled out by Wagner & Witzel 1992b on the basis of the short timescales.

Models invoking incoherent synchrotron emission (SSC models) (e.g. Qian et al., 1991; Marscher 1992) have to explain the brightness temperatures of up to 10^{19} K. Relativistic beaming with Doppler factors of ≈ 100 could do so, but these values are uncomfortably high (one order of magnitude higher than observed). Qian et al., 1991 use a cylindrical jet geometry and propagating thin shocks to reconcile the observed high brightness temperatures with the Compton–limit. This leads to Doppler-boosting factors of the order of 10. The quasi–periodic nature of the variability (Wagner & Witzel, 1992b) favors geometric scenarios as discussed by e.g. Steffen, 1992, & Steffen et al. this volume and Camenzind & Krockenberger, 1992. A totally different attempt to explain IDV has been proposed by Benford 1992 and Lesch 1992: coherent radiation as produced by an electron maser mechanism has no difficulties concerning the derived high brightness temperatures.

Acknowledgements We would like to thank Dr. A. Quirrenbach and A. Zensus for their help in observing and calibrating the VLA data and J. Heidt for his help in reducing the optical data. We also thank S. Britzen and C. Naundorf for useful discussions.

References

Aller, H. et al., 1992, in *Variability of Blazars*, p. 126
Benford, G., 1992, *ApJ*, **391**, L59
Camenzind, M. & Krockenberger, M. (1992), *A&A*, **255**, 59
Heeschen, D.S. et al., 1987, *AJ*, **94**, 1493.
Krichbaum, T.P. et al., 1992, in *Variability of Blazars*, p. 331
Kühr, H. et al., 1981, *A&A Suppl*, **45**, 367
Lesch, H., 1992, in *Physics of Active Galactic Nuclei*, Springer, p. 579
Marscher, A.P. et al., 1992, in *Variability of Blazars*, p. 85
Pearson, T.J. and Readhead, A.C.S, 1988, *ApJ*, **328**, 114
Qian, et al., 1991, *A&A*, **241**, 15
Quirrenbach, A., et al. , 1989, *A&A*, **226**, L1
Quirrenbach, A., et al. , 1991, *ApJ*, **372**, L71
Quirrenbach, A., et al. , 1992, *A&A*, **258**, 279
Steffen, W., 1992, Diplomarbeit, Univ. Bonn
Teräsranta, H. et al., 1992, in *Variability of Blazars*, p. 126
Wagner, S. & Witzel, A., 1992a, in *From Beams to Jets*, in press
Wagner, S. & Witzel, A., 1992b, in *The Nature of Compact Objects in AGN*, in press
Wegner, R., et al., 1992, in *Physics of Active Galactic Nuclei*, Springer, p. 600
Witzel, A., et al., 1988, **206**, 245
Witzel, A., 1986, *Mitt. d. Astron. Gesellsch.*, **65**, 239.
Witzel, A., 1992, in *Physics of Active Galactic Nuclei*, Springer, p. 484
Witzel, A. & Quirrenbach, A. , 1993, in *Propagation Effects in Space VLBI*, in press

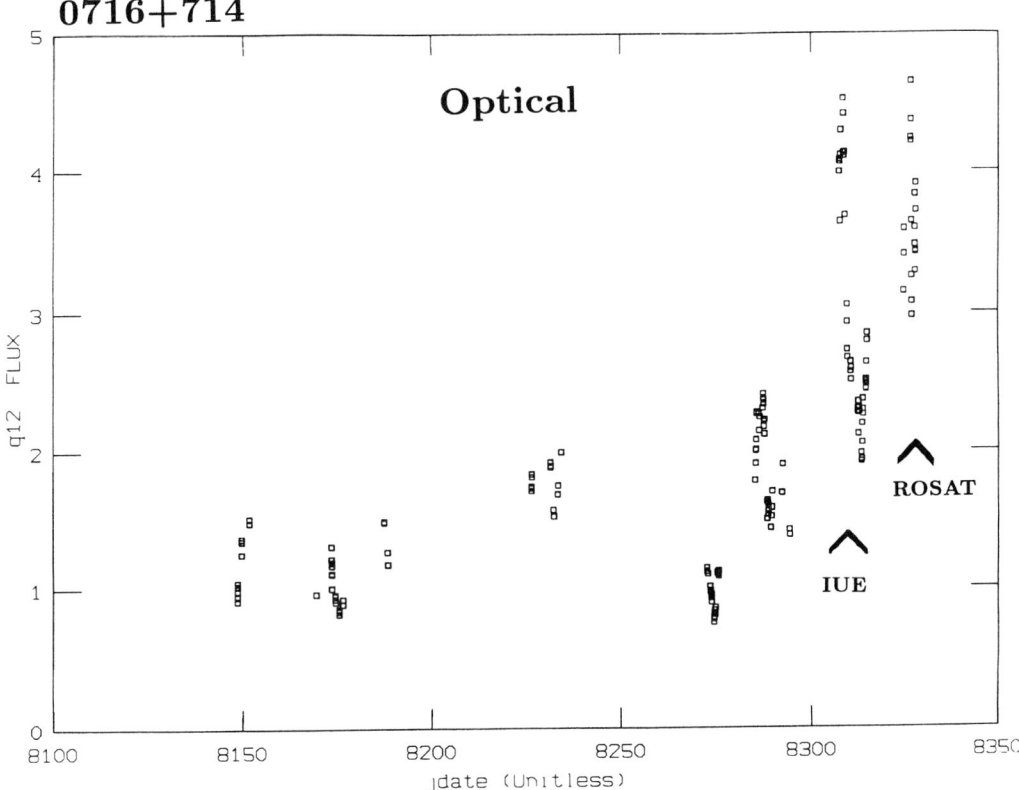

Figure 1: The graph shows the optical lightcurve of the BL Lac 0716+714. Relative flux density is plotted versus Julian date - 2 440 000 and dates of parallel observations with IUE and ROSAT are marked.

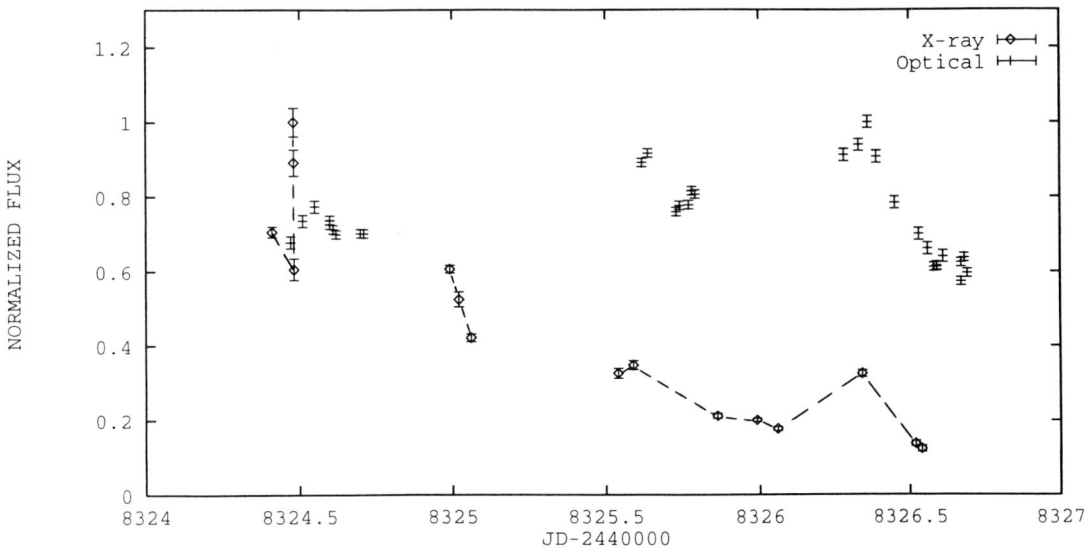

Figure 3: X-ray and optical lightcurve of 0716+714. Both are normalized to the maximum value. Note the rapid variation of the X-ray flux at JD=8324.5.

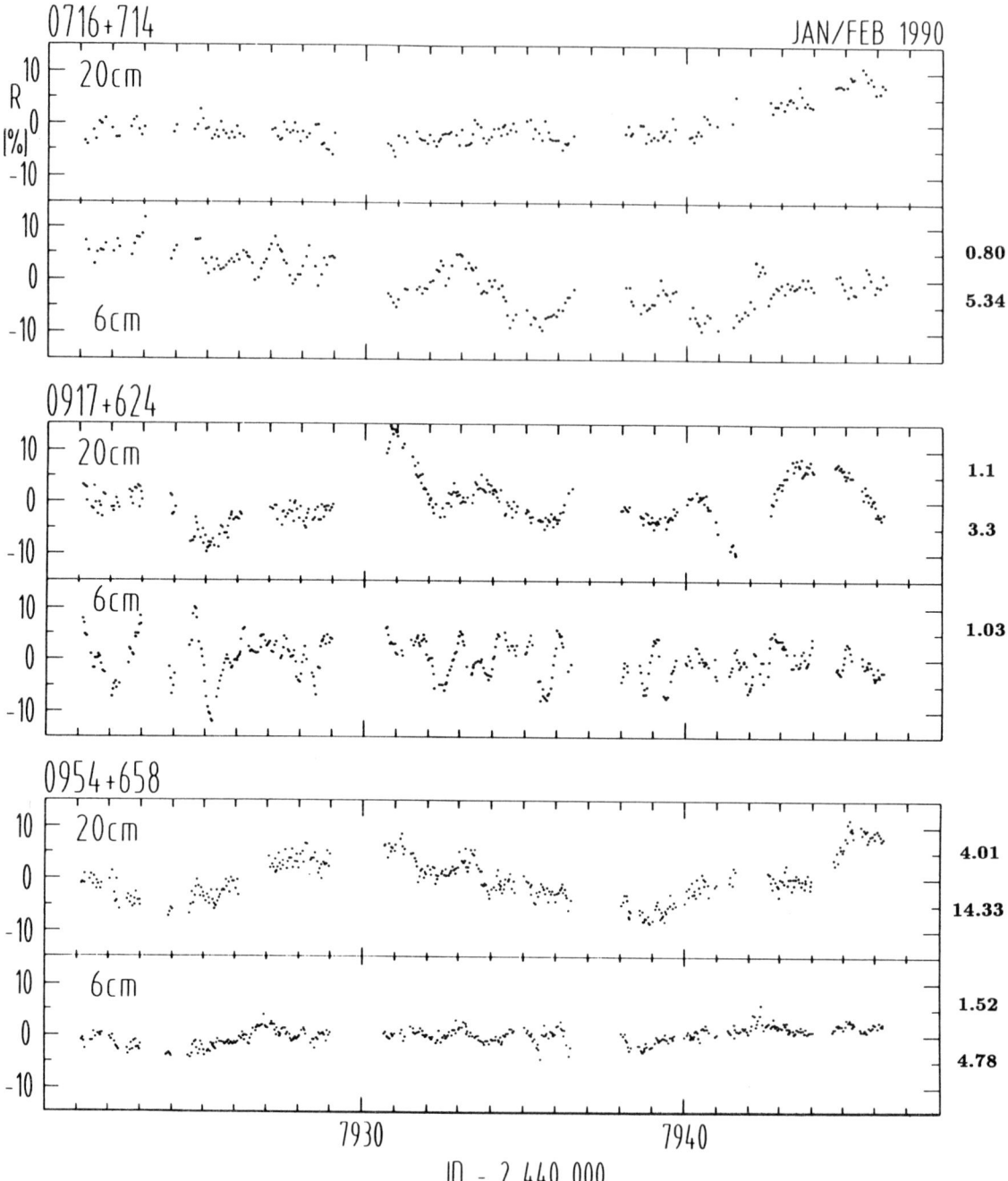

Figure 2: Variabilities of the total intensities of 0716+714, 0917+624 and 0954+658. The residual flux densities (R) given in percent are plotted at 6 and 20 cm wavelength. The contributing time periods determined by power-spectrum analyses are shown on the right.

IDV: Polarization and Spectral Indices

R. Wegner* and A. Witzel*

New radio polarimetry and 'spectral-index-curves' are presented for the best studied intraday variable (IDV) sources 0716+714 and 0917+624.
The quasar 0917+624 has shown an anticorrelation between total and polarized flux density [W92a]. In contrast to 0917+624, the BL Lac-object 0716+714 is directly correlated in total flux density, polarized flux density, and polarization angle [W92b]. Fig. 2 a) presents the lightcurves of 0917+624 at 20, 11, 6 and 3.6 cm in May 1989 [Q91]. Fig. 2 b) shows the corresponding spectral index curves which resulted by calculating the spectral index α ($S \propto \nu^\alpha$) of all points of two light-curves which have been measured at the same time.

Fig. 1 presents the polarimetry of 0917+624 for the same observing run at 6 cm. The variations of the polarized flux density and the total flux density are anticorrelated, as already observed in 1988. Furthermore, the polarization angle swings up to 40 degrees (drastic changes during minima of polarized emission) and is correlated to the variation of the polarized flux density. As a test of reliability the VLA and Effelsberg observations are plotted together for all three quantities (S,P,χ). (S plotted as R%).

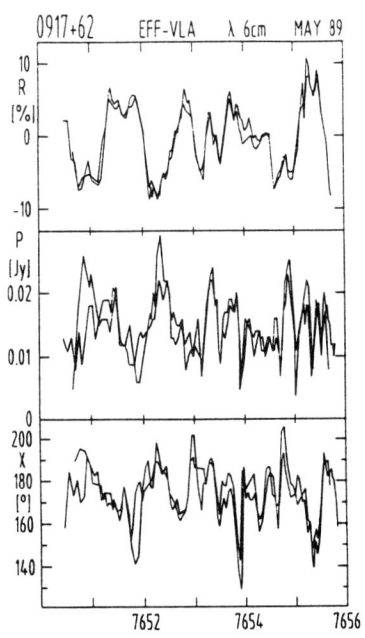

Fig. 1: Polarimetry of 0917+624 (May 1989)

Fig. 2 c),d) show the lightcurves and corresponding spectral index curves for 0716+714. Turning from higher to lower frequencies the spectral index becomes negative (concave spectrum) and the maximum of spectral index curves seems to be shifted on the time axis. Because of the flat curve at 20 cm, which may be due to optical depth effects, the shape of α_{11}^{20} curve is predominantly given by the 11 cm curve and should therefore be neglected when discussing shock in jet models. Thus, there would remain an anticorrelation between the α_6^{11} and $\alpha_{3.5}^6$ curves.
In 0917+624 the spectral index becomes more positive with decreasing frequencies (convex spectrum). This anticorrelation can also be seen in 0917+624 for α_{11}^{20} and α_6^{11}, whereas $\alpha_{3.5}^6$ seems to be correlated to α_6^{11}. An anticorrelation of S and α is predicted by shock-in-jet models [K. Mannheim, priv. comm.].

---

*Max-Planck-Institut für Radioastronomie, Bonn, Germany.

References

[Q91] Quirrenbach, A. (1991). in: *Variability of Active Galaxies*, eds. W.J. Duschl and S.J. Wagner, Springer Heidelberg, 131

[W92a] Witzel, A. (1992) in: *Physics of Active Galactic Nuclei*, eds. S.J. Wagner and W.J. Duschl, Springer Heidelberg, 484

[W92b] Wegner, R. et al. (1992) in: *Physics of Active Galactic Nuclei*, eds. S.J. Wagner and W.J. Duschl, Springer Heidelberg, 600

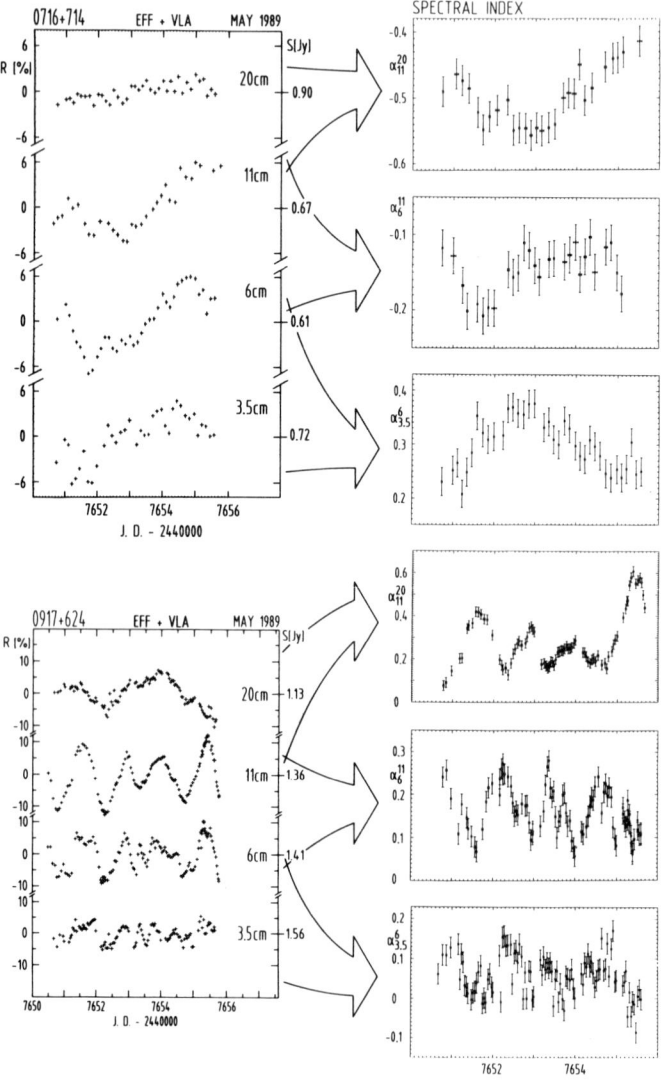

Fig. 2 a)–d): Lightcurves and corresponding spectral index lightcurves of 0716+714 and 0917+624

Continuum vs. VLBI: a comparison between data derived from flux monitoring and VLBI observations

Esko Valtaoja and Harri Teräsranta [*]

Continuum flux variations and structural changes in VLBI maps both result from the same mechanism, shocks in relativistic jets. While interferometry is necessary for detailed structural information, much of other data such as component sizes, fluxes, brightness temperatures and zero epochs can at least in principle also be derived from multifrequency flux monitoring. Continuum observations are much easier and cheaper, and information is available in real time, in stark contrast to VLBI.

Most VLBI observations are obtained at too low frequencies and have too little resolution to be very useful in studying the physics of AGN cores. A typical example is given by the multifrequency behavior of 3C 273 (Lainela et al. 1992). The outbursts are strongest (> 20 Jy) at mm-wavelengths; at the highest standard VLBI frequency, 5 GHz, they become observable (resolved from the core) first a year or two later as new emerging ~ 2 Jy components. Only a minuscule fraction of the total (radiated) energy remains; in VLBI one is studying the interaction of the last senile shock remnants with the environment, not the physics of the shocks themselves.

In advance of space and millimeter VLBI, flux variability thus remains the only way to study young shocks and other very small ($\ll 1$ mas) or bright ($\gg 10^{12}$K) structures in AGN. At high frequencies the outbursts typically last less than a year, leading to a simple source structure. In the first approximation, we see either one component (the core or mas jet) or two components (the core and the most recent shock), making possible to estimate the source parameters from continuum data. The core flux is given by the quiescent flux (no outbursts) and the shock flux ΔS by the variable flux. The variability timescale is given by $\tau_{obs} =\mid dt/d(lnS) \mid$, and the brightness temperature by $T_b \propto \lambda^2 \Delta S/\tau_{obs}$. The Doppler boosting factor $D = (T_b/10^{12}K)^{1/3}$, the intrinsic variability timescale $\tau_{int} = \tau_{obs}[D/(1 + z)]$ and the characteristic linear size $L = 2c\tau_{int}$.

How reliable are such continuum estimates? The paucity of mm-VLBI data prohibits extensive comparisons. A comparison of our continuum data with the 22 GHz VLBI

---

[*]Metsähovi Radio Research Station, Helsinki University of Technology, SF-02540 Kylmälä, Finland

sample of Lawrence et al. (1985) shows that simultaneous ΔS_{cont} is a better predictor of correlated VLBI flux than previous VLBI observations (Valtaoja et al. 1992). In the few existing individual cases the agreement between VLBI- and continuum-derived values is also good. For example, the VLBI zero epoch, shock size, flux and brightness temperature estimates for the outburst in 3C 273 in 1988-1989 (Bååth et al. 1991, Krichbaum et al. 1990) agree well with continuum-derived values (Valtaoja 1991), except for the 43 GHz VLBI flux which is most likely due to a misidentification of the shock and the core in the Krichbaum et al. VLBI map (cf. an alternate map in Bååth 1992). Such misidentifications are not unusual in VLBI maps with inadequate time coverage; another example is the C7 component of 3C273, for which the anomalously low expansion speed (Zensus et al. 1990) is really due to a misidentification, as revealed by comparison with continuum data (see also Abraham et al., these proceedings).

While continuum monitoring cannot replace VLBI, it can be a most useful tool for optimizing, planning, reducing and interpreting high frequency ground and space VLBI observations, as well as for predicting what we will eventually be able to see with tomorrow's techniques. For example,using a complete monitoring sample (Teräsranta and Valtaoja, these proceedings), one can construct a 22 GHz "continuum map" of a typical bright source at the time of shock maximum, 0.5-1 years after the onset of the flare. The core has a flux ≤ 1 Jy and $T_b < 10^{12}$ K. The shock is separated from the core by ~ 2 l.y. (assuming average superluminal speeds), its size is ~ 1 l.y. ($\sim 40\mu as$ at z = 0.5), flux ≤ 1 Jy and $T_b \sim 10^{12} - 10^{13}$K; these give the minimum requirements future VLBI should aim at.

References

Bååth, L.B., 1992, *Variability of Blazars*, eds. E. Valtaoja and M. Valtonen, Cambridge University Press, p. 229

Bååth, L.B., et al., 1991, A&A 241, L1

Krichbaum, T.P., et al., 1990, A&A 237, 3

Lainela, M. et al., 1992, *Variability of Blazars*, eds. E. Valtaoja and M. Valtonen, Cambridge University Press, p. 102

Lawrence, C.R., et al., 1985, ApJ 296, 458

Valtaoja, E., 1991, *Frontiers of VLBI*, eds. H. Hirabayashi et al., Universal Academy Press, p. 209

Valtaoja, E., Lähteenmäki, A., Teräsranta, H., 1992, A&AS (in press)

Zensus, J.A., et al., 1990, AJ 100, 1777

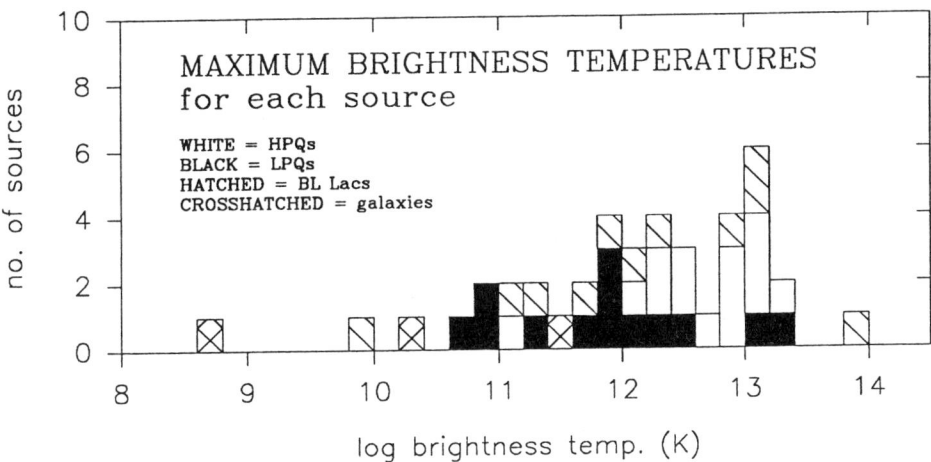
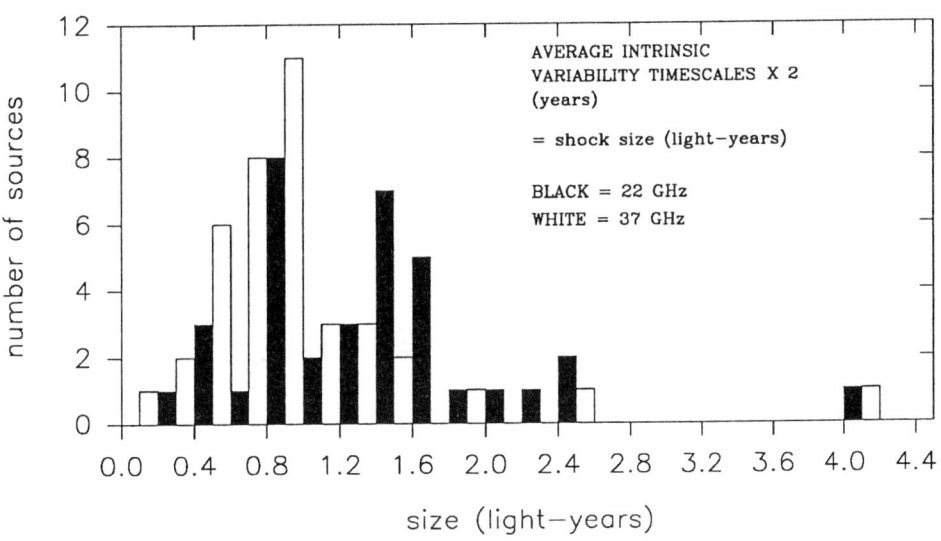

Brightness temperatures and sizes of the flaring components in AGN from Metsähovi continuum monitoring

Harri Teräsranta and Esko Valtaoja [*]

Abstract

Using continuum monitoring data, we derive shock component linear sizes and brightness temperatures for a complete sample of bright AGN. The size distribution is sharply peaked around 1 light-years. The brightness temperature distributions for various types of AGN support unified models.

1. The data and the calculations

In the Metsahovi monitoring program over 15000 flux measurements of about 60 AGN at 22 and 37 GHz have been obtained since 1980 (Teräsranta et al. 1992). The basic monitoring sample is a complete flux-limited 2 Jy sample of northern hemisphere flat-spectrum AGN (Valtaoja et al. 1992). In addition, a few bright equatorial sources and fainter BL Lacs are included. We have used this monitoring data to derive variability timescales, component sizes and brightness temperatures for compact extragalactic radio sources (see Valtaoja and Teräranta, these proceedings). We included only well-defined single radio outbursts related to the creation of new shocks, later visible as newly ejected VLBI components.

2. The results

A typical flare in our 2 Jy sample has a maximum flux of ~ 1 Jy at 22/37 GHz (= the flux in the new VLBI component), comparable to the constant flux in the core (= the VLBI or mas jet). We find that the typical observed variability timescales are rather sharply peaked around 0.5 years, with only a few outbursts lasting longer than a couple of years. The fluxes and timescales are similar for radio galaxies, BL Lacs, ordinary quasars (LPQ) and highly polarized blazar-type quasars (HPQ).

---

[*]Metsähovi Radio Research Station, Helsinki University of Technology, SF-02540 Kylmälä, Finland

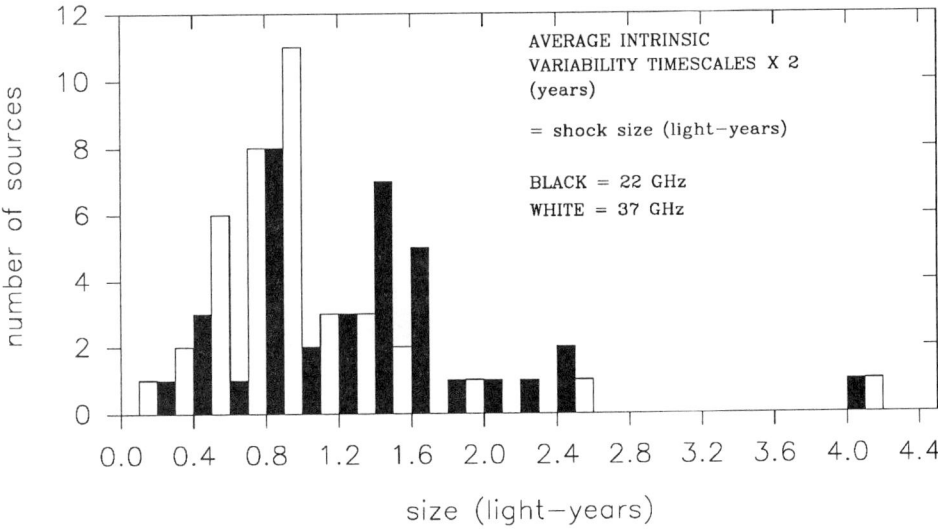

Figure 1. Intrinsic variability timescales for the 2 Jy sample

The intrinsic variability timescales also cluster around 0.5 years (Figure 1). Since our monitoring program can equally well detect variations from timescales of weeks to several years, we conclude that this is a real effect and the shock mechanism produces a surprisingly narrow range of characteristic shock sizes. The average shock size ($2c\tau_{int}$) increases from ~ 0.8 l.y. at 37 GHz maximum phase to ~ 1.2 l.y. at the later occurring 22 GHz maximum, reflecting the expansion of the shock with time. The sizes are similar for different classes of AGN and show no dependence on redshift (a new distance indicator?).

The highest calculated $T_{b,obs}$ are often $> 10^{12}$K, indicating a need for Doppler boosting in many sources. The most interesting thing, however, is that the $T_{b,obs}$-distributions are significantly different for different classes of sources (Figure 2). Radio galaxies have lowest brightness temperatures, $< 10^{12}$K, followed by ordinary quasars (LPQ) and finally HPQs with $T_{b,obs} > 10^{12}$K in all 14 sources except one. The most straightforward interpretation is that different classes of sources have different average viewing angles, as suggested by unified models. To illustrate this, we assume that the strongest outbursts in all sources have $T_{b,int} = 10^{12}$K, and that the intrinsic Lorentz factor $\Gamma \sim 10$. The Doppler boosting (or de-boosting) required to produce the observed brightness temperatures then gives the viewing angles. For these values the average viewing angle for HPQs is $17°$, for LPQs $26°$, and for radio galaxies $50°$. A radio galaxy appears as a quasar if $\theta \leq 43°$ and as a blazar(HPQ) if $\theta \leq 25°$. These values are similar to the ones derived by Barthel (1989) using different arguments.

For sources with measured v/c (from VLBI) and D (from continuum), θ and Γ can be directly calculated. Such data is available for 16 sources. We find that HPQs, LPQs,

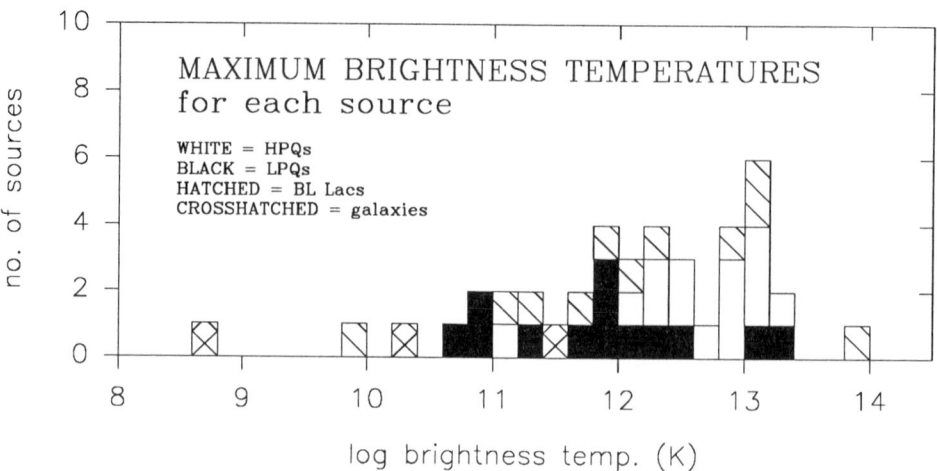

Figure 2. Largest observed brightness temperatures for the sample sources

and radio galaxies occupy different regions in the (v/c,D)-space, in accordance with orientation-dependent scenarios. For quasars, $\Gamma \sim 10 \to 40$ (with $H_o = 100$), and the viewing angle decreases as one moves from radio galaxies to LPQs and HPQs. These values are in good agreement with those derived from other methods (e.g., Urry, these proceedings). The BL Lacs occupy a different regime, with $\Gamma \sim 3 \to 15$ and $\theta \leq 40°$, indicating that they have different Lorentz factors and a different parent population. Thus, the small superluminal speeds of BL Lacs are not due to especially small viewing angles, but to smaller Lorentz factors (cf. Mutel 1990). The v/c and D -values for 1308+326 (Gabuzda, these proceedings) indicate that it is not a microlensed BL Lac but a quasar with $\theta < 10°$, in accordance with suggestions that many distant 'BL Lacs' in reality are quasars with small viewing angles (cf. Valtaoja et al. 1992).

References

Barthel, P.D., 1989, ApJ 336, 606

Mutel, R.L., 1990, *Parsec-scale radio jets*, eds. J.A. Zensus and T.J. Pearson, Cambridge university press, p. 98

Teräsranta, H., et al., 1992, A&AS 94, 121

Valtaoja, E., Lähteenmäki, A., Teräsranta, H., 1992, A&AS (in press)

Valtaoja E., et al., 1992, *Variability of blazars*, eds. E. Valtaoja and M. Valtonen, Cambridge University Press, p. 70

Parsec-Scale Jets

Anthony C.S. Readhead *

Abstract

The parsec-scale radio structures now being revealed by Very Long Baseline Interferometry are as complex as those seen on scales 1000× larger by conventional interferometry. They cry out for more realistic physical models, and computer modelling, and provide the first opportunities for watching the detailed evolution of astrophysical jets. In addition, new structures are being found in the most compact regions which cannot simply be explained on the usual jet models. In some nearby objects we are probing regions strongly influenced by the potential of the central engine.

1. Introduction

It is a commonplace that radio loud quasars and galaxies are powered by a central engine of dimensions much less than a parsec, and that energy is transported out to distances of kiloparsecs to megaparsecs (in some cases), along well collimated narrow channels, or "jets". On both kiloparsec and parsec scales these jets exhibit a wide variety of morphologies, and the stability and collimation of these jets pose interesting problems in relativistic magnetohydrodynamics. One finds one-sided and two-sided jets, both relativistic and non-relativistic, jets which propagate for only a few tens of parsecs before breaking up, and jets which maintain their stability over megaparsecs. There are straight jets, bent jets— some of which apparently bend through 180°; helical jets, jets with and without hotspots, born again jets, and so on.

What can we hope to learn from all this variety? Are there interesting physical conclusions that can be drawn from these morphologies — for that is often all we have to go on in the radio emission regions — or are these merely a reflection of fairly minor differences in jet strength and the density and temperature of the ambient environment? The Fanaroff & Riley classification of large-scale jets [FR74] is clearly telling us something fundamental about these objects, since it correlates so strongly with luminosity. This is one morphological result on large-scales which must be tied to some important property of the central engine itself.

---

*California Institute of Technology, Pasadena, California 91125, USA.

When we turn to smaller scales and study the *parsec-scale* structures things look promising, especially among the steep spectrum compact objects where there seem to be *fundamentally* different types of morphologies *viz:* core-jet objects [WRPA77], compact double objects [PM82], compact triple objects [PR88, CPR et al.92] and complex objects [NSV et al.91, VRMB93].

There are four major areas in which the last few years have seen important developments and in which we can anticipate major advances:

1) The study of magnetohydrodynamics of relativistic jets is leading to a much better understanding of the physics of the emission regions and the stability and collimation of jets. Here the parsec-scale jets afford one clear advantage over the studies of large-scale jets — they change on timescales of months, rather than 10^5 years, so that the time dimension is added and dynamical evolution can be studied directly in individual cases.

2) Nuclear jets — particularly in the Steep Spectrum Compact Objects — should be invaluable probes of the surrounding medium, especially when coupled with high resolution optical spectral line observations. Radio polarization observations will be important here.

3) The classification of different types of radio-loud active galaxies, which is absolutely vital in the study of so-called "Unified Theories", should enable us to determine the evolutionary histories of the different types as well as the cosmological evolution of different classes of object.

4) The nature of the central engine has always been a prime motivation for studies of nuclear radio emission regions — very high resolution observations combined with observations of large samples plus some recent theoretical developments promise to shed new light on this important problem.

I have picked out a few examples to illustrate the ways in which recent observations of parsec-scale jets are impacting each of these four areas in §2–5 below.

2. Helical Filaments and Limb-Brightened Jets

My first example is the superb image of the nuclear jet in M87 [RBJ et al.89] shown in Figure 1, which has a dynamic range of 2200:1. Analysis of this image shows that the jet oscillates from side to side and that it is limb-brightened at distances beyond 50 milliarcseconds (4 pc). Furthermore, greyscale images show clearly that between 40 and 90 milliarcseconds there are sinuous, sometimes multiple, filaments. In another fine example, 3C273 [DUM91], we see evidence of a similar type — the oscillating ridgeline is clear, but in this case the resolution perpendicular to the jet is insufficient to show limb brightening if it is present. Similar oscillations are seen in 3C120 [WBU87, Walker, private communication] and 3C345 [Unwin, private communication]. In an interesting development it has now been demonstrated that in 3C 345 successive ejecta do not follow the same path [ZCU93].

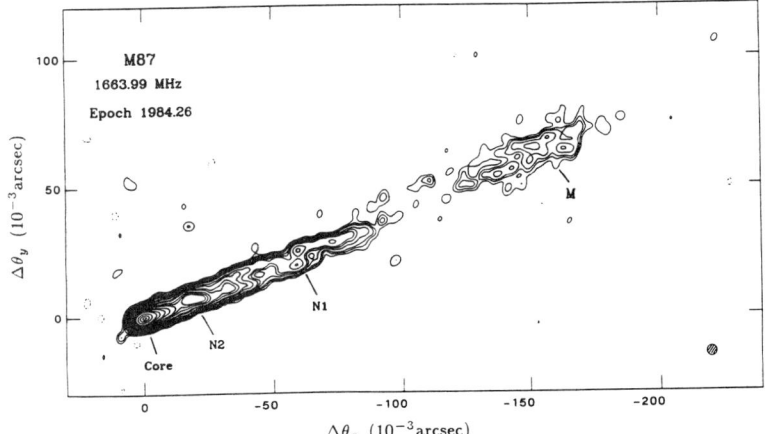

Figure 1 — High dynamic range map of the small-scale jet in M87 [RBJ et al.89] showing both limb-brightening and an oscillating ridge line (see text).

Thus far we have been looking at rather small oscillations in fairly straight jets. In the case of 3C309.1 [KWPR90] the jet bends through 90° and there is evidence from the low X-ray emission demonstrating that the bulk flow along the jet is relativistic. The large bends in this case may very well be due to projection [CM93]. Another good example of a highly bent jet is the case of 1347+539 [XPR et al. 93]. This could well be a quasi-helical jet which bends through an apparent angle of 180°.

The difference in position angle between the parsec-scale and the kiloparsec-scale jets appears to be bimodal [PR88, WCU et al.92, ILT91], which could be due to a combination of relativistic and projection effects [CM93].

One of the most important recent developments in high resolution radio observations has been observations of the polarization in radio galaxies, quasars and BL Lac objects [RKB et al.90, GCRW92, CWR et al. 93a, CWR et al.93b]. In radio loud galaxies and quasars the magnetic field direction is along the jet. The determination of the magnetic field direction in the core, jet and knots should be most helpful in understanding the physics of these jets.

Computer simulations of extragalactic jets in three dimensions, taking into account both the magnetic field and relativistic effects are now possible. A recent study [HC92] shows that excitation of helical modes of the Kelvin-Helholtz instability is likely to be common, and the simulations show, in addition, a wealth of structures such as bifurcation and trifurcation of filaments which twist circumferentially about the surface of the jet, reminiscent of those seen in M87 [RBJ et al.89].

3. Interaction with the Interstellar Medium

Perhaps the finest example of a parsec-scale jet which is disrupted by interaction with the environment is 3C 48 [WTB et al.91]. Here, after 150 pc the initially well-collimated jet flares out and it appears to become turbulent. Possible interpretations are that this jet has crossed a standing shock, in the manner first discussed in [NBS88], or that it has ploughed into a dense molecular cloud and entrained a lot of material. The object is extraordinarily luminous at infrared wavelengths [NSM85], and it is likely that this emission comes from the turbulent region imaged by VLBI.

In general the steep spectrum compact objects are the most likely candidates for strong interaction with the surrounding medium, and Hubble Space Telescope observations of these objects, which will have resolution comparable to many features in these objects which have been mapped with low frequency VLBI, should make it possible to study the dynamics and energetics of these regions in some detail.

4. The Classification of Compact Structures

On large physical scales, apart from the interesting division of objects into the two Fanaroff & Riley classes, there are no clear physical differences between objects although there is a wide variety of morphologies. In the case of the compact structure the situation looks more promising — the majority of objects belong to the "asymmetric core-jet" class which has had such an impact on our thinking about radio loud quasars and galaxies (see §5), but the steep spectrum compact objects have a number of interesting subclasses, some of which appear to be physically distinct. If there are physically distinct types of active galaxies it is clearly important to classify objects correctly before attempting (1) to understand the evolution of these objects as a class, (2) to understand their cosmological evolution or (3) to "unify" different objects through aspect-dependant effects. Often it has been assumed that there are no significantly different classes, but this assumption is unjustified.

A number of groups have been actively carrying out surveys which are useful in this regard. In particular the steep spectrum compact objects have been intensively studied [FFP et al.85, FFS et al.90b, FFS et al. 91, NSF et al.91], as have flat spectrum objects [SBE et al.87, Wit87]; complete flux-limited samples [PR88]; and orientation unbiased samples of quasars [ZP87, HR89]. The cumulated results have been illuminating. It is clear that the class of steep spectrum compact objects [PW81] is particularly interesting. Bright jets are common in compact steep spectrum quasars ($\sim 60\%$) and less common in compact steep spectrum galaxies ($\sim 20\%$) [FFS et al. 89, FFS et al.90, NSF et al.91]. Complex structures are also more common in quasars. The compact double class of compact steep spectrum objects [PM82] appears to be quite distinct, although many objects have been misplaced in this class. An interesting recent development has been the recognition of "Compact Triple" objects as an important class [PR88, CPR et al.92], which may well be related to the compact double class. It is now clear that there is a class of *symmetric* compact objects. Perhaps the best case is that shown in Figure 2 [WPR et al.93].

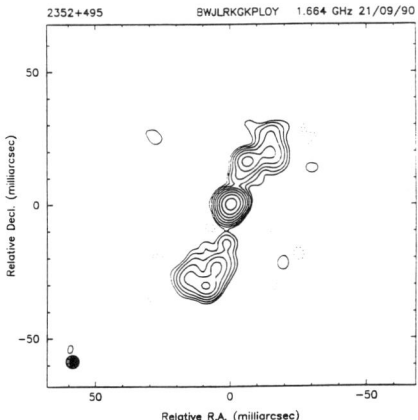

Figure 2 — Map of 2352+495 [WPR et al.93] showing unusual symmetric structure for such a compact object. It belongs to the compact triple class of objects which are likely a new class of short-lived powerful active galaxies.

This is a remarkable structure on a scale of tens of parsecs. It is very similar to structures which are common on scales 100 – 1000 times larger, but very rare on these scales. It shows clearly that not all compact radio structures in active galactic nuclei are one-sided jets. The object bears a remarkable resemblance to the quasar 3C 186 [FFS et al.90a] except that it is forty times smaller. Why should this structure be so remarkable? Because the symmetry clearly shows that the radiation from the oppositely directed jets is largely isotropic, so that we are not here seeing a central compact structure which has been relativistically boosted, as is the case for the core-jet sources and even for steep spectrum compact objects which have jets, like 3C 309.1. There is not space here to discuss the possible explanations for this structure, but these are more likely a new class of short-lived powerful radio sources than precursors of Fanaroff & Riley Type II objects or "frustrated jets", as has been suggested [PM82, HM87, BOMD90].

5. Clues to the Nature of the Central Engine

It is sometimes stated that high resolution radio observations are unlikely to provide interesting clues to the nature of the central engine in active galactic nuclei [Phi84]. It should be remembered that superluminal motions [CCP et al.71, WSR et al.71, MGRL72] when combined with the one-sided core-jet morphology first observed in active galactic nuclei when the first hybrid maps were made [WRPA77, RPC et al.79, PUC et al.81] led immediately to the acceptance of relativistic outflow in jets; and these images showed, moreover, that the basic collimation occurs on subparsec scales — two of the strongest pieces of circumstantial evidence for the existence of supermassive black holes.

I conclude that simple morphology and careful classification, combined with the fact that we can observe motion in these objects on timescales of months, has already played a very major role in our acceptance of the standard model for the powerhouses in these objects.

Before the morphology of the central radio emission regions was known, it was very hard to discriminate between possible models of superluminal motion, but the discovery of the asymmetric core-jet morphology, which is such a natural consequence of relativistic beaming, led immediately to the first suggestions that the steep-spectrum lobe-dominated objects could be unified with the flat spectrum core-dominated objects [RCPW78, BK79, SR79, Rea80] bringing about a unification between radio loud quasars and galaxies [RCPW78, Rea80] and that radio quiet and radio loud quasars could be similarly unified [SR79]. It is becoming more and more obvious that all of these unifications are true at some level. Detailed studies of the implications of unification have been carried out, and considerable progress has been made [Ant93]. Historically, it was the radio work which launched this important branch of enquiry, which has now embraced the whole electromagnetic spectrum [HBF et al.92].

It is clear that the phenomenology is complex, and much progress has been made in exploring the wide range of effects that impact unification theories, as we saw in Urry's excellent review [these proceedings].

There is a point to emphasizing the fact that in the early days of milliarcsecond radio astronomy zeroth order problems were being solved at a rapid pace — we are about to see a quantum leap in both the quality and the quantity of information on the scales of milliarcseconds to arcseconds with the advent of the VLBA and the upgraded MERLIN. These advances are making possible studies which were simply inconceivable in the past, and we are entering an era in which we will be able to attack new zeroth order questions in this field.

In addition to progress through the study of samples and classes of objects, detailed studies of individual objects are important. The structures recently detected in NGC 1275 at 22 GHz [MBW et al.89, VRMB93] and at 43 GHz [BDK et al. 88, KWG et al. 92] look very like helical filaments, although better images are needed to confirm this; but they also show a complex structure within 10^{18} cm of the central engine, which is almost certainly within the dynamical realm of the central engine [VRMB93]. Detailed studies of this region, and of the conical jet region could well give important information about the central engine itself.

Two recent theoretical studies hold out the hope of seeing some radio structures associated with *thermal* emission regions in active galactic nuclei. The effects of induced Compton scattering in these objects have been studied [CBR93], and it has been shown that these can be important if there is thermal plasma illuminated by an object with a brightness temperature greater than 5×10^9K, in which case the electron recoil must be taken into account. This introduces a Compton shift in frequency $\Delta \nu \sim h\nu^2/m_e c^2$ which can lead to backscattering of the synchrotron radiation from thermal emission regions which might be visible at the level of $\sim 1\%$ of the direct synchrotron emission. The effect of stimulated Raman emission, in which longitudinal plasma waves are generated by radio waves, has also been calculated [TBR93]. It happens that stimulated Raman emission would only be important for thermal plasma illuminated by objects with brightness temperatures greater than 10^{12}K — somewhat brighter than is seen in NGC 1275, for example, so that in this object Raman backscattering is probably less important than induced Compton backscattering. It therefore appears possible that thermal structures, such as an accretion disk could be seen through backscattered radiation, and this could be very important in determining the disposition of the thermal matter relative to the non-thermal jets.

6. Conclusion

The improvements in imaging techniques, and the development of the polarization, snapshot and other techniques, coupled with the explosion of information through studies of large samples of objects, has brought us to the brink of a new era in the study of active galaxies. In this respect the advent of the upgraded MERLIN and the VLBA are opportune. Definitive tests of many important aspects of unification theories are now possible, and different classes of radio loud active galaxies and quasars can be studied in depth. The rapid advances we are seeing now are likely to reveal new phenomena and focus attention on new areas over the next decade.

References

[Ant93] Antonucci, *ARAA (1993) in press*
[BDK et al.88] Bartel, Dhawan, Krichbaum, et al., *Nature (1988) 334, 131*
[BK79] Blandford, & Königl, *APJ (1979) 232, 34*
[BOMD90] Baum, O'Dea, Murphy & de Bruyn, *A&A (1990) 232, 19*
[CBR93] Coppi, Blandford, & Rees, *MNRAS (1993) in press*
[CCP et al.71] Cohen, Cannon, Purcell, et al., *APJ (1971) 170, 207*
[CM93] Conway & Murphy, *These Proceedings*
[CPR et al.92]CPR Conway, Pearson, Readhead, et al., *APJ (1992) 396, 62*
[CWR et al.93a] Cawthorne, Wardle, Roberts, et al., *APJ (1993) in press*
[CWR et al.93b] Cawthorne, Wardle, Roberts, et al., *APJ (1993) in press*
[DUM91] Davis, Unwin & Muxlow, *Nature (1991) 354, 374*
[FFP et al.85] Fanti, Fanti, Parma, et al., *A&A (1985) 143, 292*
[FFS et al.90a] Fanti, Fanti, Schilizzi, et al., *Parsec Scale Radio Jets (1990) 146*

[FFS et al.90b] Fanti, Fanti, Schilizzi, et al., *A&A (1990) 170, 10*
[FFS et al.91] Fanti, Fanti, Schilizzi, et al., *A&A (1991) 231, 333*
[FR74] Fanaroff & Riley, *MNRAS (1974) 167, 31P*
[GCRW92] Gabuzda, Cawthorne, Roberts & Wardle, *APJ (1992) 388, 40*
[HBF et al.92] Hartmann, Bertsch, Fichtel, et al., *APJL (1992) 385, L1*
[HC92] Hardee & Clarke, *APJL (1992) 400, L9*
[HM87] Hodges & Mutel, *Superluminal Radio Sources, Zensus & Pearson (1987) 168*
[HR89] Hough & Readhead, *AJ (1989) 98, 1208.*
[ILT91] Impey, Lawrence & Tapia, *APJ (1991) 375, 46*
[KWG et al 92] Krichbaum, Witzel, Graham, et al., *A&A, (1992), in press*
[KWPR90] Kus, Wilkinson, Pearson & Readhead, *Pc-Scale Radio Jets (1990) 161*
[MBW et al.89] Marr, Backer, Wright et al., *APJ (1989) 337, 671*
[MGRL72] Moffet, Gubbay, Robertson & Legg, *IAU Symp 44, (1972) 228*
[NBS88] Norman, Burns & Sulkanen, *Nature (1988) 335, 146*
[NSF et al.91] Nan Rendong, Schilizzi, Fanti, et al., *A&A (1991) 252, 513*
[NSM85] Neugebauer, Soifer & Miley, *APJL (1985) 295, L27*
[NSV et al.91] Nan Rendong, Schilizzi, van Breugel, et al., *A&A (1991) 245, 449*
[Phi84] Phinney, *Astroph. of Active Galaxies & QSOs, Miller (1985) 453*
[PM82] Phillips & Mutel, *APJL (1982) 257, L19*
[PR88] Pearson & Readhead, *APJ (1988) 328, 114*
[PUC et al.81] Pearson, Unwin, Cohen et al., *Nature (1981) 290, 365*
[PW81] Peacock & Wall, *MNRAS (1981) 194, 331*
[RBJ et al.89] Reid, Biretta, Junor et al., *APJ (1989) 336, 112*
[Rea80] Readhead, *IAU 92 (1980) 165*
[RCPW78] Readhead, Cohen, Pearson & Wilkinson, *Nature (1978) 276, 768*
[RPC et al.79] Readhead, Pearson, Cohen, et al., *APJ (1979) 231, 299*
[RKB et al.90] Roberts, Kollgaard, Brown et al., *APJ (1990) 360, 408*
[SBE et al.87] Schalinski, Biermann, Eckart, et al., *IAU (1987) 287*
[SR79] Scheuer & Readhead, *Nature (1979) 277, 182*
[TBR93] Thompson, Blandford & Rees, *APJ (1993) in press*
[VRMB93] Venturi, Readhead, Marr & Backer, *APJ (1993) in press*
[WBU87] Walker, Benson & Unwin, *APJ (1987) 316, 546*
[WCU et al.92] Wehrle, Cohen, Unwin, et al., *APJ (1992) 391, 589*
[Wit87] Witzel, *Superluminal Radio Sources, Zensus & Pearson (1987) 83*
[WPR et al.93] Wilkinson, Polatidis, Readhead et al., *These Proceedings*
[WRPA77] Wilkinson, Readhead, Purcell & Anderson, *Nature (1977) 269, 764*
[WSR et al.71] Witney, Shapiro, Rogers et al., *Science (1971) 173, 225*
[WTB et al.91] Wilkinson, Tzioumis, Benson, et al., *Nature (1991) 352, 313*
[XPR et al.93] Xu, Polatidis, Readhead, et al., *These Proceedings*
[ZCU93] Zensus, Cohen & Unwin, *in preparation*
[ZP87] Zensus & Porcas, *Superluminal Radio Sources, Zensus & Pearson (1987) 126*

New Results from VLBI at 43 GHz

T.P. Krichbaum, A. Witzel, D.A. Graham[*], C.J. Schalinski[†], and J.A. Zensus[‡]

The joint collaboration for global 43 GHz VLBI observing (for a list of participants and more details see e.g. Krichbaum and Witzel, 1991&1992) was continued in order to monitor structural changes on sub-parsec scales in AGN with high angular resolution (~ 0.1 mas). Limiting factor of observations prior to 1991 was the small number of participating stations and the lack of sensitivity. The situation now has improved owing to the participation of sensitive antennas: in April 1991 the VLBA antenna at Pie Town participated for the first time successfully in a global 43 GHz VLBI observation, in May 1992 (experiment not yet completely analyzed) the participation of the VLBA antennas at Pie Town, Los Alamos, Kitt Peak and North Liberty increased considerably the uv-coverage and the detection sensitivity. First fringes were detected at 43 GHz with the IRAM 30-m antenna at Pico Veleta in September 1991. These test-observations demonstrated the superior sensitivity of the 30-m telescopes for mm-VLBI (see also C. Schalinski, this volume) and led to a first 43 GHz map of the inner jet of Cygnus A (Krichbaum et al., 1993). In the following we summarizes some of our recent results:

4C 39.25: The first 43 GHz VLBI map of this source reveals the presence of a new component $\underline{d}$, located ~ 2.6 mas west of the eastern stationary component $\underline{a}$. $\underline{d}$ is also seen at similar positions in new 22 GHz VLBI data (Alberdi et al., 1993). The 'inverted' spectrum of $\underline{d}$ ($\alpha_{22/43\,GHz} = 1.3$, $S_\nu \propto \nu^\alpha$) suggests that $\underline{d}$ is either the self-absorbed VLBI-core or a most recently ejected new jet component.

3C 84: In Fig. 1 new maps of the nucleus of 3C 84 are shown. Comparison with data obtained in 1986-1988 and extrapolation of the motion seen during this period (Krichbaum et al., 1992) allows to identify components (see labels in Fig. 1). If this identification scheme is correct, components N, B, and C have further separated from A along the inner 1 mas-long jet ($P.A. \simeq 210°$). Component D, however, seems to have changed its direction of motion considerably: between 1988.5 and 1991.3 D has moved from $r = 1.0 \pm 0.1$ mas, $P.A. = 214 \pm 10°$ to $r = 1.9 \pm 0.2$ mas, $P.A. = 167 \pm 10°$, clearly suggesting subluminal motion along a curved path ($\Delta P.A._{1991-1988} \simeq 50°$). Jet curvature between $1 \leq r \leq 2$ mas is also indicated by the 0.5 mas-resolution map obtained in 1991.7 (Fig. 1, right). In Fig. 2 we superimposed all component positions obtained during 1985-1991 in a single plot, in order to visualize the shape of the mean jet axis. Despite the still somewhat large error bars, the figure suggests considerable jet-curvature on mas-scales. Up to five position measurements for the distinct jet components of 3C 84 obtained during 1985-1991 now allow to measure their relative velocities with confidence. In Fig. 3 we plot the angular separation rate μ in $[mas/yr]$ of each component versus their separation r in $[mas]$ from A (at epoch 1991.3). In

---

[*]Max-Planck-Institut für Radioastronomie, Bonn, Germany.
[†]Institut de Radioastronomie Millimetrique, Grenoble, France.
[‡]National Radio Astronomy Observatory, Socorro, USA.

the region $0.5 \leq r \leq 6$ mas a linear increase of μ from 0.085 mas/yr to 0.7 mas/yr $(d\mu/dr = (9.0 \pm 1.2) \cdot 10^{-2}\,\mathrm{yr}^{-1})$, corresponding to an apparent acceleration from $0.07\,c$ to 0.5-$0.6\,c$ (for $H_0 = 100$ km s^{-1} Mpc^{-1}, $q_0 = 0.5$) is evident. Adopting a purely geometrical interpretation, bending towards the observer's line of sight could explain this effect.

1803+78: 3.6 cm VLBI observations since 1979 revealed a stationary component (labeled 'C1') at $r = 1.3$ mas located west of the core A of this BL Lac object. An upper limit for the apparent velocity of C1 is $\beta_{app} \leq 0.7$ (Schalinski et al., 1988). In Fig. 4 new maps obtained from observations at 22 GHz and 43 GHz are shown. The quality of this maps increases from top to bottom reflecting the better quality of the data obtained in 1990&'91. A tentative component identification is indicated by straight lines. Between 1988 and 1991 the separation of A and C1 still remained constant, however new components C2 and C3 seem to move superluminally towards C1 with $\beta_{app} = 3.0 \pm 0.9$, respectively $\beta_{app} = 1.5 \pm 1.1$. The relative position of C2 with respect to A changed between 1989 and 1991 from $P.A. = 250 \pm 10°$ to $P.A. = 291 \pm 5°$. Thus C2 moves along a line inclined by about $\sim 30°$ versus the mean jet axis ($P.A. = 270°$). This indicates motion along a bent trajectory. The coexistence of moving and stationary components and motion along curved paths provides further support for motion along three-dimensionally bent paths, as e.g. suggested for 4C 39.25 (A. Alberdi, this volume) and 3C 345 (Krichbaum and Witzel, 1992).

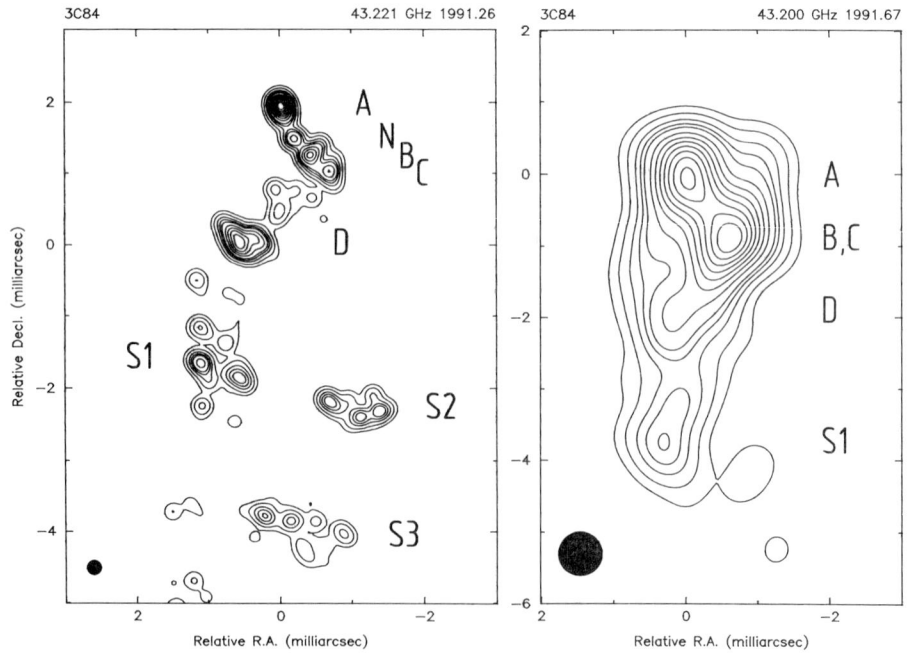

Fig. 1: 43 GHz-maps of 3C 84 obtained from a global VLBI experiment (left) and from a European VLBI experiment including Bonn, Onsala and Pico Veleta (Krichbaum et al., 1993).

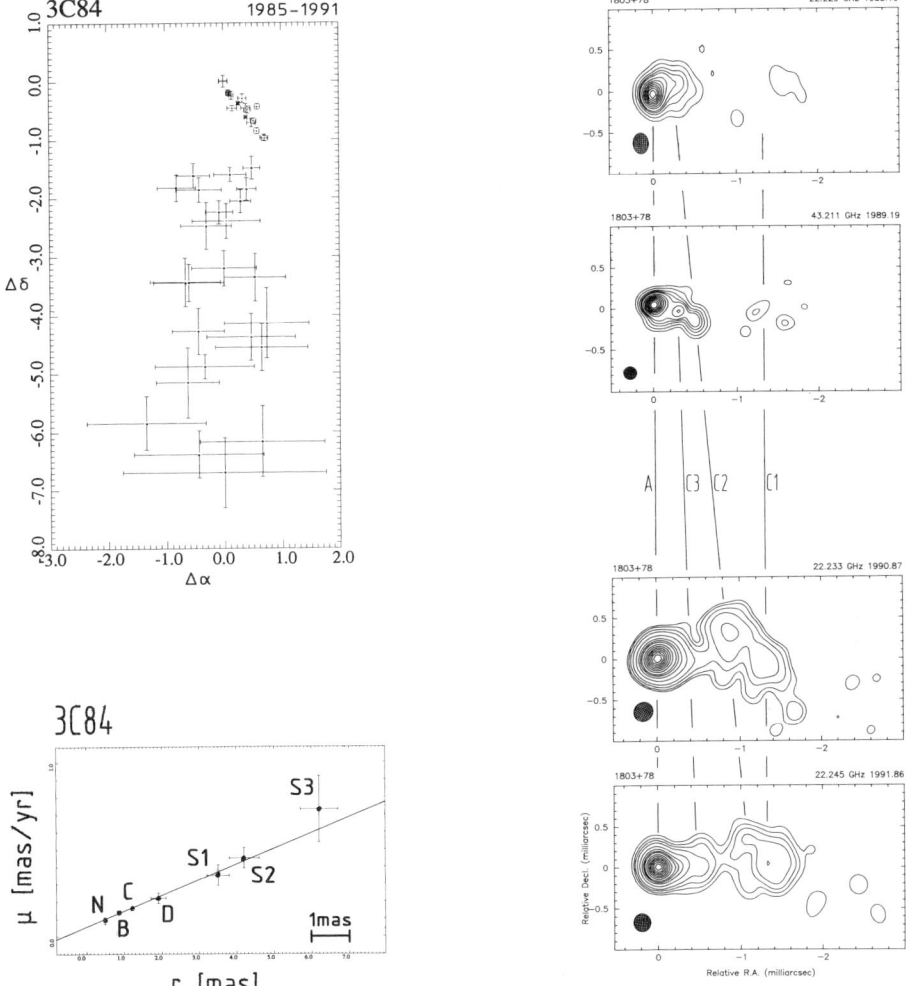

Fig. 2: Left, top: Mean ridge line of the jet of 3C 84. Fig. 3: Left, bottom: Angular separation rate plotted versus core-separation. Fig. 4: Right: Structural variations in 1803+78 at 43 and 22 GHz.

References

Alberdi, A., Krichbaum, T.P., Marcaide, J.M., *et al.*, 1992, *A&A*, in press.
Krichbaum, T.P., and Witzel, A., 1991, in: *Frontiers of VLBI*, Frontiers Science Series No. 1, ed. H. Hirabayashi, M. Inoue and K. Kobayashi (Universal Academy Press, Tokyo), p. 297.
Krichbaum, T.P., and Witzel, A., 1992, in: *Variability of Blazars*, ed. E. Valtaoja and M. Valtonen (Cambridge University Press), p. 205.
Krichbaum, T.P., Witzel, A., Graham, D.A., *et al.*, 1992, *A&A*, **260**, 33.
Krichbaum, T.P., Witzel, A., Graham, D.A., *et al.*, 1993, submitted to *A&A*.
Schalinski, C.J., et al., 1988, in: *IAU Symposium 129, The Impact of VLBI on Astrophysics and Geophysics*, ed. M. J. Reid and J. M. Moran (Dordrecht: Kluwer), p. 359.

The 86 GHz VLBI Test with Pico Veleta: First Detection of the Quasar 3C454.3 at 3 mm

C.J. Schalinski, A. Greve, M. Grewing, H. Steppe [*],
D.A. Graham, T.P. Krichbaum, K. Standke, A. Witzel, [†] *A. Alberdi* [‡]
R.S. Booth, L. Bååth, F. Colomer [§]

The investigation of the *sub*-parsec-scale radio jets of *active galactic nuclei* requires high angular resolution observations at mm- and submm-wavelengths, since the compact central cores are still self-absorbed at short cm-wavelengths. Following the successful mm-VLBI test of the IRAM 30m-RT at Pico Veleta (Spain: "X") with the 100m-RT at Effelsberg (Germany: "B") and the 20m-RT at Onsala (Sweden: "T") in September 1991 at λ7 mm (Krichbaum et al., 1993), we used the same interferometer at λ3.6 mm on three consecutive nights July 7-10, 1992 (July 7-8: B-X, July 8-10: B-T-X, 18-8 UT). Baseline lengths are ranging from 832 km (B-S) to 2511 km (S-X), corresponding to a maximum resolution of $\sim$ 0.14 milli-arcseconds (1 mas= 4.3 pc at z= 1, with $H_0 = 100h$ km s^{-1} Mpc^{-1}, $q_0 = 0.5$), the resolution of a *global* VLBI-array at λ1.3 cm. The telescope at Pico Veleta was equipped with a 14 Video-Converter VLBA-recording terminal[1] controlled by a PC-field system[2], and a cooled Schottky-mixer receiver with a single-sideband temperature of $\sim$220 K was used. The observations were performed in a snap-shot mode, with an interferometric measurement of 6.5 min duration every 15 min, the remaining 8.5 min being sufficient for pointing and system calibration measurements.

10 extragalactic sources with flux densities ranging from 2-30 Jy in the 86-115 GHz band were selected on the basis of maps from high frequency VLBI campaigns (see Krichbaum et al., this vol. and refs. therein) and previously determined flux density measurements with the 30m-RT and the IRAM Interferometer on Plateau de Bure (Guilloteau et al., 1992). Tab. 1 shows in Col. 1 the source name, in Col. 2 the flux density at 86 GHz at the observing epoch, and Cols. 3-5 list the resulting SNRs for the three baselines.

These *raw-data*[3] prior to the application of sophisticated fringe-fitting procedures already clearly demonstrate the advantage of using *large, sensitive* antennas in mm-VLBI: all well-known (or candidate) superluminal sources with $S_{tot}^{86GHz} \geq 4$ Jy have been detected, among them for the first time at 3mm wavelength 3C454.3 (Pauliny-Toth et al., 1987) and 0528+134, which have recently been identified as strong γ-ray

---
[*]Institut de Radio Astronomie Millimétrique, Grenoble, France.
[†]Max-Planck-Institut für Radioastronomie, Bonn, Germany.
[‡]Instituto de Astrofisica de Andalucia, Granada, Spain.
[§]Onsala Space Observatory, Sweden.
[1]B and T with the MkIII-system in track-density upgrade, $\nu_{obs} = 86212.99\ldots 86316.99$ MHz
[2]developed by P. Burgess at Jodrell Bank
[3]a preliminary calibration indicates a detection limit of order $\sim 0.5 - 1$ Jy, i.e. $1\sigma \sim 0.1$ Jy may be used as estimate for the SNRs in Tab.1

SOURCE	$S^{86GHz}_{[Jy]}$	SNR_{X-S}	SNR_{X-B}	SNR_{B-S}
3C273	27.2	15-54	39-73	17-27
3C279	13.6	25-33	43-72	13-19
3C454.3	11.4	≤ 6.9	12-22(70)	17-27
3C345	8.9	≤ 7.2	12-32	8-15
3C84	8.2	≤ 5.9	≤ 5.8	≤ 6
0528+134	6.5	12-15	13	≤ 6.4
NRAO530	4.3	≤ 7.6	7.5-14	≤ 6
3C446	3.3	≤ 6.4	≤ 5.8	≤ 6
DA193	3.8	≤ 5.8	≤ 6	≤ 6.5
CTA102	2.1	≤ 5.9	≤ 6.7	≤ 6.0

Tab.1: Signal-to-noise ratios integrated over 6.5 min on the three baselines at 86 GHz for July 9/10. X: 30m-RT at Pico Veleta, S: 20m-RT at Onsala, B: 100m-RT at Effelsberg (the inner 60m can be used for mm-observations). The detection threshold for the Mk-III system is SNR= 7. The value in brackets for 3C454.3 was obtained on July7/8. X used right hand circular polarization (RHC-P), T: linear polarization, in order to match the RHC-P. at B.

emitters by the Compton-satellite (Kanbach, Lichti: priv. com.). Although the GST-range is generally too short (≤ 2 hrs except for 3C345, 3C454.3 and 3C273) to display strong variations of the visibilities, the baseline dependent SNRs in Tab. 1 are indicative of source structure effects, which can also account for the non-detection of the complex jet of 3C84. Source dependent lower limits of $\sim 1\ldots 7$ Jy may be estimated for compact core components, indicating that *direct* imaging of even "weak" sources will be possible in the future.

References

Guilloteau, S., Delannoy, J., Downes, D., Greve, A., Guélin, M., Lucas, R., Morris, D., Radford, S.J.E., Wink, J., Cernicharo, J., Forveille, T., Garcia-Burillo, S., Neri, R., Blondel, J., Perrigouard, A., Plathner, D., Torres, M.,1992: *A&A*, **262**, 624
Krichbaum, T., Witzel, A., Graham, D.A., Standke, K., Schwartz, R., Lochner, O., Schalinski, C.J., Greve, A., Steppe, H., Brunswig, W., Butin, G., Hein, H., Navarro, S., Peñalver, J., Grewing, M., Booth, R.S., Colomer, F., Rönnäng, B.O., 1993: *A&A*, in press.
Pauliny-Toth, I.I.K., Porcas, R.W., Zensus, J.A., Kellermann, K.I., Wu, S.Y., Nicolson, G.D., and Mantovani, F., 1987: *Nature* **328**, 778

Compact Structure in 3 C273

Zulema Abraham[*]   *Everi A. Carrara* [*]   *J. Anton Zensus* [†]
Stephen C. Unwin [‡]

Abstract

VLBI maps of the quasar 3C273 are presented for the period 1988-1991. Several components present in previous maps are still visible at the positions expected from constant expansion velocities. New components can be seen, its appearance related to the flares detected at higher frequencies in single dish observations.

1. Observations and Results

Since the first 10.7 GHz maps of 3C273 were obtained in 1977, at least eight different superluminal components were detected in addition to the compact core, they were labeled C1-C8 [UC85], [BC85],[CZ87]. A high dynamic range 5 GHz map [ZB88] showed simultaneously C2-C5, extending to a projected distance of 46 h^{-1} parsecs from the core. At higher frequencies C7 and C8 were detected at 22 GHz [ZU90] and at 43 GHz, Krichbaum et al. [KB90] found a new component (C9) in 1988.43 and other two weaker sources near the compact core in 1989.12. Structure at the microarcsecond level was obtained at 100 GHz in 1988.21, two months after a strong multifrequency outburst, suggesting the birth of component C9 very close to the core [BP91].

The observations that we present here were made at 10.7 GHz with the Global VLBI Network, including Itapetinga in Brazil, for the epochs 1988.17, 1989.26, 1990.17 and 1991.15. The data were recorded with the Mark II system, correlated at the "Block II" correlator in Pasadena and processed at the Astronomy Department of São Paulo University in Brazil, using the NRAO-AIPS and Caltech VLBI software.

---

[*]Instituto Astronômico e Geofísico, Universidade de São Paulo, Av. Miguel Stefano, 2400, São Paulo, SP 04301, Brazil. These authors were partially supported by brazilian agencies CNPq and FAPESP.
[†]National Radio Astronomy Observatory, P.O.Box 0, Socorro, NM 87801, USA.
[‡]Owens Valley Radio Astronomy, California Institute of Technology, Pasadena, CA 91125, USA.

Fig. 1 shows the four maps; for the last two epochs no fringes were obtained with Itapetinga. The best map, for epoch 1989.26 included ten stations. For epochs 1988.17 and 1989.26, model fitting of the visibilities and phase closures gave good results with five components, an extra component near the core was required for the last two epochs. From the extrapolated positions in previous maps we identify the first five components as the core D, C9, C8, C7 and C5. The extra component at the last two epochs would be C10, it corresponds to the component seen at 43 GHz near the core in 1989.12 [KB90]. The expansion velocity seems to be similar for all components, about 0.9 mas yr^{-1} or 6c (for H_0=100 km s^{-1} Mpc^{-1} and q_0=0.5).

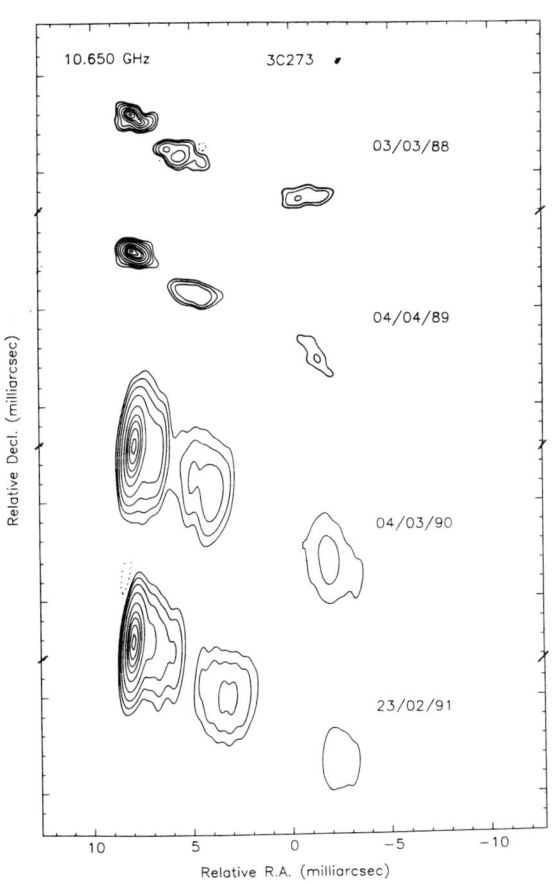

Fig. 1 10.7 GHz maps of 3C 273

At this point, we should mention a discrepancy with the velocity previously reported by Zensus et al. [ZU90] for C7 (around 0.6 mas yr^{-1}). This is probably due to a confusion between this component at 22 GHz and a short lived component called C7b by Cohen et al. [CZ87]. Extrapolation of the positions of C9 and C10 to the origin gives the ejection time for these components as the beginning of 1988 and 1990 respectively. This shows that there is a very good correlation between the strong multifrequency outbursts and the ejection of superluminal components from the nucleus. The maximum flux density measured for C9 was 8 Jy, it decreased rapidly to 2 Jy in 1991.15. The flux density for C10 was around 7 Jy at the two epochs in which it was detected. Both C9 and C10 were optically thin between 22 and 43 GHz.

References

[BC85] Biretta J.A., Cohen M.H., Hardebeck H.E. Kaufmann P., Abraham Z., Perfetto A.A., Scalise Jr. E., Schaal R.E., *Observations of 3C 273 with high north-south resolution.* ApJ Lett., Vol. 292 (1985) pp. L5-L8.

[BP91] Baath L.B., Padin S., Woody D., Rogers A.E.E., Wright M.C.H., Zensus J.A., Kus A.J., Backer D.C., Booth R.S., Carlstrom J.E., Dickman R.L., Emerson D.T., Hirabayashi H., Hodges M.W., Inoue M., Moran J.M., Morimoto M., Plambeck R.L., Predmore C.R., Ronnang B., *The microarcsecond structure of 3C 273 at 3 mm.* A&A Lett., Vol. 241 (1991) pp. L1-L4.

[CZ87] Cohen M.H., Zensus J.A., Biretta, J.A., Comoretto G., Kaufmann P., Abraham Z., *Evolution of 3C 273 at 10.7 GHz.* ApJ Lett., Vol. 315 (1987) pp. L89-L92.

[KB90] Krichbaum T.B., Booth R.S., Kus A.J., Ronnang B.O., Witzel A., Graham D.A., Pauliny-Toth, I.I.K., Quirrenbach A., Hummel C.A., Alberdi A., Zensus J.A., Johnston K.J., Spencer J.H., Rogers A.E.E., Lawrence C.R., Readhead A.C.S., Hirabayashi H., Inoue M., Morimoto M., Dhawan V., Bartel N., Shapiro I.I., Burke B.F., Marcaide J.M., *43 GHz-VLBI observations of 3C 273 after a flux density outburst in 1988.* A&A, Vol. 237 (1990) pp. 3-11.

[UC85] Unwin S.C., Cohen M.H., Biretta J.A., Pearson T.J., Seielstad G.A., Walker R.C., Simon R.S., Linfield R.P., *VLBI monitoring of the superluminal quasar 3C 273, 1977-1982.* ApJ, Vol. 289 (1985) pp. 109-119.

[ZB88] Zensus J.A., Baath L.B., Cohen M.H., Nicolson G.D., *The inner radio jet of 3C 273* Nature, Vol. 334 (1988) pp. 410-412.

[ZU90] Zensus J.A, Unwin S.C., Cohen M.H., Biretta J.A., *The milliarcsecond structure of 3C 273 at 22 GHz.* AJ, Vol. 100, pp. 1777-1784.

Kinematics of the Parsec-scale Jet in 3C 345

Stephen C. Unwin *

The quasar 3C 345 ($z = 0.595$) is one of the brightest high-declination core-jet sources available to the 'global' VLBI array, and structure variations in the jet on timescales of months to years make it an ideal candidate for the study of jet dynamics on scales of a few parsecs. A high dynamic-range VLBI image of the nucleus has been made from a 5-GHz observation in 1991.70 with 1-mas (4 pc) resolution. This image is the latest in a series of monitoring observations designed to follow the evolving jet in some detail. Recent 5-GHz images in this series (from 1989.28 and 1990.18) have been published by Unwin and Wehrle [UW92].

Figure 1 shows the 1990.18 and 1991.70 images. Measured relative to the noise level, the dynamic range is $> 1000:1$; the lowest contour of 1.6 mJy/beam is about 3-σ and should be reliable except near the brightest peaks. The 1991.70 image confirms some tentative conclusions drawn from earlier images, and shows some new features. There are a number of important points to notice:

(1) Both maps show the well-known core plus one-sided jet [BMC86]; sampling in time is sufficient that features are clearly identifiable between epochs.

(2) Component C5 has decayed and moved out in P.A. $\approx -100°$, at apparent speed $\beta_{\mathrm{obs}} \approx 6$ ($H_0 = 100\ h$ km s^{-1}Mpc^{-1} and $q_0 = 0.5$) consistent with the motion away from the nucleus (D) of earlier features C3 and C4 [ZCU92].

(3) The C4 - C2 region is now clearly confirmed to be *edge-brightened*, and is suggestive of emission in a boundary layer between the relativistic jet material and the confining interstellar medium of the host galaxy [UW92].

(4) Component C1 is very extended, in contrast with components closer to the nucleus, which define a much narrower jet opening angle.

(5) Between C2 and C1 the jet brightness drops abruptly below our detection threshold, though the brightness of C1 (integrated perpendicular to the jet axis) is consistent with extrapolation of a power-law fitted to the region within 12 mas of the nucleus.

(6) We confirm that C1 has a flattened edge facing back to the nucleus. Since previous observations suggest that the 'edge' is moving superluminally, its appearance is

---

*Caltech 105-24, Pasadena, CA 91125 U. S. A. This work was supported in part by the NSF under grant AST-9117100.

affected by relativistic aberration [UW92].

(7) Using geometric constraints from superluminal motion, relativistic aberration, jet / counter-jet ratio, and X-ray emission (which yields a lower limit to the Doppler factor), we derive $\gamma = 7.2 \pm 1.1$; $\theta = 6.8° \pm 1.5°$ for the C4-C2 region.

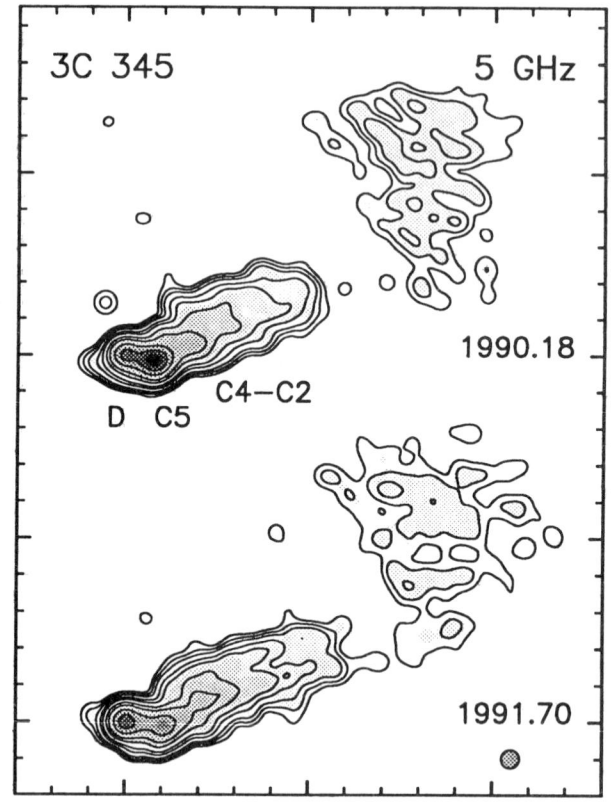

Fig. 1 — VLBI images of 3C 345 at 5.0 GHz made using global VLBI arrays at epochs 1990.18 and 1991.70. Tickmarks are drawn every 2 mas (7.6 pc); contours are shown at 0.1, 0.2, 0.4, 1, 2, 5, 10, 25, 50, and 85 % of the peak brightness (1.77 Jy/beam). Restoring beam is 1.0 mas circular. Noise level on both maps, away from bright features, is ≈ 0.6 mJy/beam, close to the expected thermal noise. Persistent features are labelled using the convention of [ZCU92].

References

[BMC86] Biretta, J. A., Moore, R. L., & Cohen, M. H. 1986, ApJ, 308, 93.

[UW92] Unwin, S. C., & Wehrle, A. E. 1992, ApJ, 398, 000.

[ZCU92] Zensus, J. A., Cohen, M. H., & Unwin, S. C. 1992, ApJ, in preparation.

Superluminal Jet In 3C279

Everi A. Carrara* *Zulema Abraham* * *Stephen C. Unwin* [†]
J. Anton Zensus [‡]

Abstract

Hybrid maps of the Quasar 3C279 at frequencies 10.7 and 22.2 GHz are presented at five epochs between 1987.42 and 1990.17. A new structure at 10.7 GHz is identified near the nucleus at epoch 1989.26 and probably is correlated with the increase in the activity detected at various frequencies. Using X-rays observations we were able to determine a lower limit to the Doppler boosting factor δ. The jet opening angle and the observer's line of sight angle were also derived.

1. Results

We report the results of five epoch observations at frequencies 10.7 GHz and 22.2 GHz and discuss the structural changes in the jet of 3C279 after the multifrequency outburst that ocurred at epoch 1988.0. The maps are presented in the figures 1 and 2. We fitted optically thin sphere models to the data and the best fit was obtained only after the introduction of a new component (labeled C5) at the last two epochs. The superluminal expansion of components C3 and C4 were measured giving 0.16 ± 0.01 mas yr^{-1} and 0.15 ± 0.01 mas yr^{-1} respectively, corresponding to $V_{\text{app}} \approx 2.8 h^{-1} c$ ($H_0 = 100 h$ km s^{-1} Mpc^{-1} and $q_0 = 0.5$). The geometry of beaming models gives an upper limit to the line of sight angle $\phi_{\max} = 39°$. To obtain a lower limit to ϕ, we used the model of DM88, which relates the opening jet angle ψ and ϕ by the expression $\tan\phi = \tan(\psi/2)/\tan(\psi_{\text{app}}/2)$. Our data give $27° \leq \psi_{\text{app}} \leq 38°$ and the calculations following DM88 give $23° \leq \psi \leq 36°$. Applying these values to expression for ϕ we obtain $30° \leq \phi \leq 39°$.

Using the X-rays measurements performed with the GINGA satellite at the energy range 2-20 kev (MKH89) we found $\delta \geq 1.4$, indicating a moderate relativistic bulk motion.

---

*Instituto Astronômico e Geofísico, Universidade de São Paulo, Av. Miguel Stefano, 2400, São Paulo, SP 04301, Brazil. These authors were partially supported by brazilian agencies CNPq and FAPESP.
[†]Owens Valley Radio Astronomy, California Institute of Technology, Pasadena, CA 91125, USA.
[‡]National Radio Astronomy Observatory, P.O.Box 0, Socorro, NM 87801, USA.

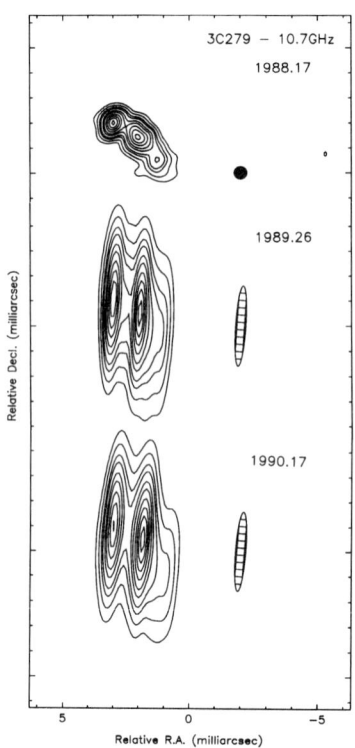
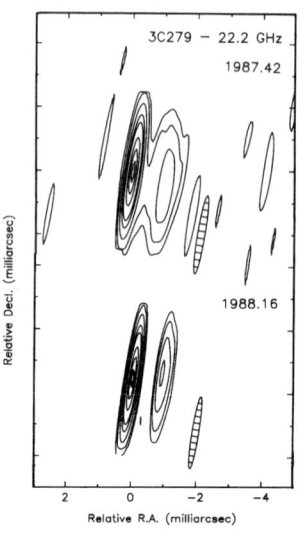

Figure 1 - Hybrid maps of 3C279 at frequency 10.7 GHz at epochs from 1988.17 to 1990.17. The contour levels are 2,5,10,15,25,35,50,65,80 and 95% of the peak brightness in the maps, which are 2.1,3.7 and 3.6 Jy per beam, respectively. There is no emission below -1.5% on any of these images, and the restoring beams used are elliptical Gaussians with FWHM of 0.5 x 0.5 mas for epoch 1988.17 (because of inclusion of Itapetinga Station) and 0.4 x 3.2 mas, in P.A. $-4°$ for subsequent epochs.

Figure 2 - Hybrid maps of 3C279 at frequency 22.2 GHz at epochs 1987.42 and 1988.16. Contours are 4,7,14,25,35,60,80 and 95% of the peak brightness in the maps (1.8 and 4.3 Jy per beam respectively). There is no emission below -3.0% on any of these images, and the convolving beam is the same for both epochs, an elliptical Gaussian with FWHM of 0.2 x 2.5 mas, in P.A. $-9°$.

References

[DM88] Daly R.A., Marscher A.P., *The Gasdynamics of Compact Relativistic Jets*. Ap. J., Vol. 334 (1988), pp. 539-551.

[MKH89] Makino F., Kii T., Hayashida K., Inoue H., Tanaka Y., Ohashi T., Makishima K., Awaki H., Koyama K., Turner M.J.L., Williams O.R., *X-Ray Outburst of the Quasar 3C279*. Ap. J., Vol. 347 (1989), pp. L9-L12.

[UCBHZ89] Unwin S.C., Cohen M.H., Biretta J.A., Hodges M.W., Zensus J.A., *Superluminal Motion in the Quasar 3C279*. Ap. J., Vol. 340 (1989), pp. 117-128.

The Search for Superluminal Motion in a Complete Sample of Lobe-dominated Quasars

David H. Hough [*]   Rene C. Vermeulen [†]
Anthony C. S. Readhead [‡]

Abstract

We are engaged in a global VLBI survey with wideband recording at 8/10 GHz to measure the distribution of parsec-scale jet velocities in a complete sample of 25 3CR lobe-dominated quasars. We discuss pilot observations of 20 nuclei, which yield detections in every case, and first-epoch maps of 11 sources, which reveal the presence of core-jet structures. Multiple-epoch maps of five objects are discussed. These indicate comparatively low superluminal speeds of $\sim$1.5-5c ($H_0 = 100$ km/s/Mpc, $q_0 = 0.5$) in three cases and upper limits of $\sim$1-3c in the other two. These results are consistent with the relativistic beaming hypothesis.

Superluminal motion at speeds of $\sim$5-10c is most often observed in extragalactic radio sources with bright, compact nuclei. Simple relativistic beaming models assuming a narrow emission cone can account for the occurence of large apparent speeds in these core-dominated sources if the jet axes are oriented near our line of sight (e.g., Scheuer and Readhead 1979). One observational test of the beaming hypothesis is to search for superluminal motion in objects free of orientation bias; in fact, *large superluminal motion (>2-3c) should seldom occur in a complete sample of randomly oriented sources.*

To this end, we have chosen a complete sample of 25 3CR double-lobed quasars on the basis of their dominant, presumably unbeamed extended emission at low frequency (Hough and Readhead 1989). *All* the nuclei of these quasars are sufficiently bright to be mapped with wideband VLBI recording and, for the faintest objects, phase-referencing techniques. An additional virtue of this sample is that it is also under

---

[*]Department of Physics, Trinity University, San Antonio, Texas 78212, U. S. A. This author was supported by Trinity University, Research Corporation, Max-Planck-Institut für Radioastronomie, a NASA-ASEE fellowship and NRC associateship at the Jet Propulsion Laboratory, and a NSF grant at the California Institute of Technology.
[†]California Institute of Technology, 105-24, Pasadena, California 91125, U. S. A.
[‡]California Institute of Technology, 105-24, Pasadena, California 91125, U. S. A.

intensive study with the VLA to obtain kiloparsec-scale images (Bridle et al. 1992) and with the Palomar 5-meter telescope to obtain optical spectra.

The observations began in 1981 and have been conducted at $\nu = 8.4$ or 10.7 GHz using Mark III (wideband) recording at 28 or 56 MHz bandwidth. Telescopes of the DSN, EVN, U. S. VLBN, and VLBA have been used, and correlation has been performed on the Bonn and Haystack Mark III/IIIA processors.

The results of pilot observations of 20 nuclei yield detections in every case and evidence of resolved structure in virtually every object. First-epoch maps for 11 nuclei exhibit core-jet structures. Super-resolved maps often show two distinct components that make these objects amenable to searches for superluminal motion. In fact, multiple-epoch maps of five such nuclei have revealed superluminal motion in three objects and set upper limits on the motion in two others: 3C245 ($\sim$3-5c), 3C263 (1.3c), 3C334 (1.6c), 3C204 (<1c), and 3C205 (<3c).

We conclude that the nuclear structures in these lobe-dominated quasars are very compact core-jets. The nuclear jets point toward, and lie on the same side of the VLBI core as, the large-scale jets, suggesting that the asymmetries on both scales are related. A trend continues to emerge in which the parsec-scale jet velocities in lobe-dominated objects are substantially lower than those in core-dominated objects, consistent with simple beaming models. The pace of the survey will now rapidly accelerate with the completion of the VLBA. We note that if our selection of quasars over radio galaxies introduces any orientation bias, then this may be explicitly accounted for in particular scenarios of quasar-radio galaxy unification (e.g., Barthel 1989). Thus our survey – and its extension to radio galaxy nuclei using phase-referencing – will not only serve to test unified beaming models in quasars (e.g., Orr and Browne 1982), but it will also permit a direct test of proposals for quasar-radio galaxy unification.

References

[Bar89] Barthel, P. D. 1989, ApJ, 336, 606

[Bri92] Bridle, A. H., Hough, D. H., Lonsdale, C. J., Burns, J. O., & Laing, R. A. 1992, in preparation

[HR89] Hough, D. H., & Readhead, A. C. S. 1989, AJ, 98, 1208

[OB82] Orr, M. J. L., & Browne, I. W. A. 1982, MNRAS, 200, 1067

[SR79] Scheuer, P. A. G., & Readhead, A. C. S. 1979, Nature, 277, 182

Two-epoch VLBI Maps of Three Weak Nuclei in Lobe-dominated Quasars

D. H. Hough [*]   J. A. Zensus [†]   R. C. Vermeulen [‡]

A. C. S. Readhead [§]   R. W. Porcas [¶]   A. Rius [∥]

Abstract

We are undertaking two largely independent VLBI surveys to measure parsec- scale jet velocities in complete samples of lobe-dominated quasars. We report on a joint effort to obtain sensitive 8 GHz maps of the weaker nuclei in these objects, especially those common to the 3CR and Jodrell Bank samples. Pilot survey detections of three objects are discussed. We present analysis of two-epoch maps of the core-jet structure in three sources – 3C204, 3C205, and 0839+616 – that yield upper limits to the jet speeds of ~1-3c ($H_0 = 100$ km/s/Mpc, $q_0 = 0.5$). These comparatively low speeds are in accordance with simple relativistic beaming models.

Superluminal motion at speeds of ~5-10c is most often observed in extragalactic radio sources with bright, compact nuclei. Simple relativistic beaming models can account for large apparent speeds in these core-dominated sources if the jet axes are oriented near our line of sight (e.g., Scheuer and Readhead 1979). One observational test of the beaming hypothesis is to search for superluminal motion in objects free of orientation bias; in fact, *large superluminal motion (>2-3c) should seldom occur in a complete sample of randomly oriented sources.*

Zensus and Porcas (1987) have initiated a study of a complete sample of double-lobed quasars from the Jodrell Bank 966 MHz survey and Hough and Readhead (1989) have started a similar study using the 3CR 178 MHz survey. Both studies

---

[*]Department of Physics, Trinity University, San Antonio, Texas 78212, U. S. A. This author was supported by Trinity University, Research Corporation, Max-Planck-Institut für Radioastronomie, and a NASA-ASEE fellowship and NRC associateship at the Jet Propulsion Laboratory.
[†]NRAO, P. O. Box O, Socorro, New Mexico 87801, U. S. A.
[‡]California Institute of Technology, 105-24, Pasadena, California 91125, U. S. A.
[§]California Institute of Technology, 105-24, Pasadena, California 91125, U. S. A.
[¶]Max-Planck-Institut für Radioastronomie, Auf dem Hügel 69, D-5300 Bonn 1, Germany
[∥]NASA-INTA, Orense 4, 9 Planta, E-28020 Madrid, Spain

seek to minimize orientation bias by the selection of objects on the basis of their dominant, presumably unbeamed extended emission at low frequency. The faint nuclei (<30 mJy) afford the most stringent test of beaming models but require very sensitive observations. Since a number of them are common to both samples, we have made a joint effort to observe the weak nuclei in six lobe-dominated quasars.

The observations were made in 1988 December and 1991 March at 8.4 GHz using Mark III recording at 56 MHz bandwidth. DSN telescopes at Madrid and Goldstone were used with Effelsberg, Haystack, Green Bank, and the VLA. Correlation was performed on the Bonn and Haystack Mark III/IIIA processors.

Pilot observations in 1988 of three nuclei – in 3C175, 3C215, and 0821+447 – reveal resolved structure and yield position angles for 3C215 and 0821+447. Two-epoch maps at conventional resolution of three more nuclei – in 3C204, 3C205, and 0839+616 – show core-jet structures. Super-resolved maps suggest the presence of two distinct components at each epoch, but there is no discernible internal proper motion (-0.02 ± 0.04, -0.01 ± 0.09, and -0.00 ± 0.03 mas/yr, respectively).

We conclude that the weak nuclei in these lobe-dominated quasars all have resolved structure. The nuclear structure is aligned with the large-scale jet or hot spot; where observed, the nuclear jet lies on the same side of the VLBI core as the large-scale jet, indicating the same cause of the asymmetries on both scales. Upper limits to the apparent tranverse velocity are ~c in 3C204 and 0839+616 and ~3c in 3C205. These results bolster an emerging trend in which the parsec-scale jet velocities in lobe-dominated objects are substantially lower than those in core-dominated objects. This trend is in accordance with simple beaming models. We note two selection criteria that may introduce orientation bias to these surveys, yet augment their value for tests of source models: (1) the choice of quasars over radio galaxies, in which case the predicted bias may be used to test scenarios of quasar- radio galaxy unification (e.g., Barthel 1989); and (2) their different selection frequencies (966 vs. 178 MHz), in which case any systematic difference in jet velocities between the two samples may shed light on the degree of beaming in the large-scale structure.

References

[Bar89] Barthel, P. D. 1989, ApJ, 336, 606

[HR89] Hough, D. H., & Readhead, A. C. S. 1989, AJ, 98, 1208

[SR79] Scheuer, P. A. G., & Readhead, A. C. S. 1979, Nature, 277, 182

[ZP87] Zensus, J. A., & Porcas, R. W. 1987, in Superluminal Radio Sources, ed. J. A. Zensus and T. J. Pearson (Cambridge: Cambridge Univ. Press), 126

Extreme Superluminal motion in CTA102 ?

Fredrik. T. Rantakyrö *   *Lars. B. Bååth* †

October 6, 1992

Abstract

We present observations of proper motion in five epoches of VLBI observations of CTA102. The highest observed proper motion was μ=0.64 mas/year, corresponding to a tranverse velocity of $v = 18.0 \pm 4\,c$.

Superluminal motion

The proper motion of the components relative to the core (figure 2) was obtained by fitting Gaussian point sources to three epoches at 49 cm (Bååth 1987) and to two epoches at 18 cm (figure 1). In the gaussion fits to the 18 cm maps the initial positions for the gaussians were taken from 6 cm observations by Wehrle *et al.* 1989). At each epoch the measured distances are relative the brightest Gaussian component in the fit, labelled D. The highest observed proper motion was for C1, μ_{C1}=0.64 mas/year (figure 2). The tranverse velocity for C1 are then v $\approx$ 18.0 $\pm$ 4 c. Separate LSQ–fits to the two frequencies gives $\mu_{C1}^{49\ cm} = 1.25 \pm 0.3$ mas/year and $\mu_{C1}^{18\ cm} = 1.04 \pm 0.4$ mas/year. Suggesting that spectral index effects do not significantly affect the high observed proper motion. The component C2 closer to the core has a much smaller observed proper motion, $\mu \sim 0.09$ mas/year. This indicates that the proper motion closer to the core is less than what we observe further out (Wehrle *et al.* 1989). We suggest that this may be caused by a change in the viewing angle of the jet. The large proper motion makes CTA102 a unique object, conspicuous on plots of proper motion vs. redshift (Cohen *et al.* 1987). However, this is not a very strong objection though. CTA102 could very well be at the extreme end of a distribution of γ's.

References

Bååth, L.B., 1987, in *Superluminal Radio Sources*, eds. Zensus, J.A., Pearson, T.J, Cambridge University Press, p306

Cohen, M.H. 1987, in *Superluminal Radio Sources*, eds. Zensus, J.A., Pearson, T.J, Cambridge University Press, p306

Wehrle A.E., Cohen, M., 1989, *ApJ*, **346**, L69

---

*Department of Astronomy/Astrophysics, Chalmers, S-412 96 Göteborg, Sweden.
†Onsala Space Observatory, S-439 00 Onsala, Sweden.

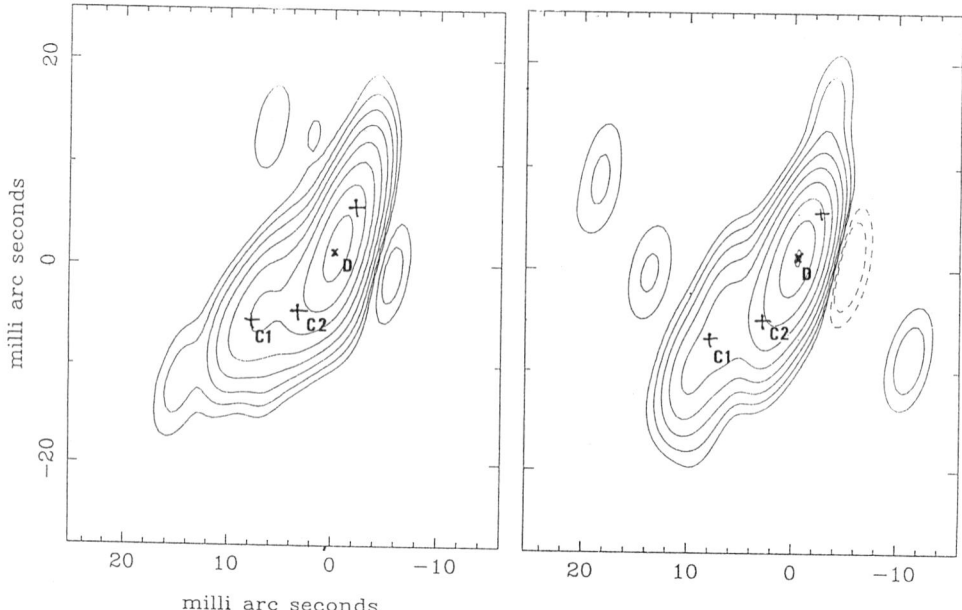

Figure 1: 18 cm maps: May 1988 (left) and June 1989 (right). Contour intervals are (-1.0, -0.5, 0.5, 1.0, 2.0, 4.0, 8.0, 16.0, 32.0, 64.0, 99.0)×32 mJy per beam. The restoring beam was a 8×3 mas (FWHM) elliptical Gaussian with a position angle of -11°. The coordinates refer to the phase centre of the map.

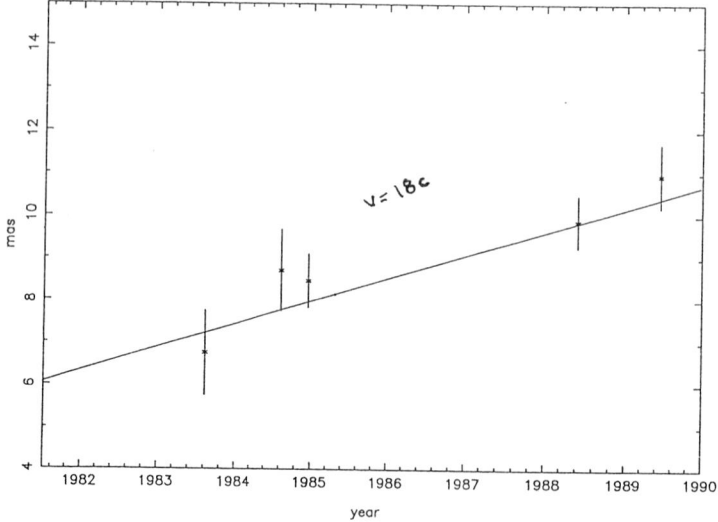

Figure 2: Distance from the core plotted as a function of epoch for component C1. The line represents the solution from a least-squares fit to the data. The errorbars drawn in the figure are three times the standard error obtained from the Gaussian fit.

Geodetic VLBI Monitoring : Parsec-Scale Structure of Extragalactic Radio Sources

C. Schalinski [*], *S. Britzen* [†], *A. Witzel* [†], *W. Alef* [†], *J. Campbell* [‡]

Geodetic VLBI observations of about 20 extragalactic radio sources are performed regularly - every four days for IRIS-A - since more than 10 years now. The total number of sources observed in the IRIS campaigns is 36, since the source list is occasionally updated. The enormous data base allows to study the milliarcsecond structures and their variations at $\lambda 3.6/13$cm simultaneously on short time scales. Furthermore, since the accuracy of baseline determination and measurement of earth rotation parameters reached a limit where source structure contributions become significant, there is substantial interest to monitor and eventually correct for the source structure "effects". Therefore we perform a regular analysis of sources selected from IRIS-A, IRIS-S and EUROPE campaigns covering a maximum time base of almost 10 years (1983-1992). A description of observational details, calibration and mapping along with a summary of the data up to 1988 can be found in Schalinski et al. (1988a,b), and Schalinski (1990). In the following we present a status report of our work. A more detailed summary will be given in the *Annual Reports of the International Earth Rotation Service*. The table lists the source names in IAU-convention (col. 1), other names (col. 2), Identification (col. 3) and redshift (col. 4). Col. 5 gives the number of maps or modelfits (at "X" $\lambda 3.6$cm and/or "S" $\lambda 13$cm), and col. 6 the corresponding maximum time base in years. The resolution is 0.5mas at X-band, and 2mas at S-band respectively. The dynamic range at present is up to 50 : 1, improvements are expected for a combination of IRIS-A (high resolution), IRIS-S (high north south resolution and almost circular beam for low declination sources) and EUROPE campaigns (high sensitivity due to large telescopes).

At present apparent superluminal motion can be confirmed for the quasar 4C39.25 (z=0.699; $0.18 \pm 0.01 mas/yr = 4.0 \pm 0.2c$, $H_0 = 100$km/s/Mpc, $q_0 = 0.5$) (Schalinski et al., 1988), and the OVV 0212+735 (z=2.37; $0.05 \pm 0.04 mas/yr = 2.2 \pm 1.7c$) (Schalinski et al., 1992). The blazar 1803+784 (z=0.68) displays on maps at 13/18cm a jet of at least 30mas length consisting of mulitple components with intensities decreasing with core distance. At X-band (e.g. Schalinski et al., 1988a) the source is dominated by two components of 1.4mas separation. The angular separation at 6cm is 1.2mas indicating opacity effects, and has been found to be constant within the

---
[*]Institut de Radio Astronomie Millimétrique, Grenoble, France.
[†]Max-Planck-Institut für Radioastronomie, Bonn, Germany.
[‡]Geodätisches Institut der Universität Bonn, Germany.

errors, so that an upper limit of 0.004 ± 0.028mas/yr ($0.09 \pm 0.61c$) at $\lambda 3.6cm$, and 0.006 ± 0.028mas/yr ($0.13 \pm 1.0c$) at $\lambda 6cm$ could be derived thus ruling out superluminal motion for this component. A component at ~ 0.5mas, detected in the 1.3cm map of 1985.8, if identified with a major outburst in 1983.2, may display apparent superluminal motion with $\leq 2.9c$. New maps at $\lambda 7mm$ (Krichbaum et al., this vol.) from 1988 to 1991 clearly show apparent superluminal components with speeds of $3.0 \pm 0.9c$ and $1.5 \pm 1.1c$ respectively, in agreement with bulk relativistic motion estimated from deficiency of observed inverse Compton X-rays ($\gamma \sim 4$). Because of the prominent stationary components 1803+784 can be used for structural calibration of the epochs 1983-1992 (and afterwards, unless the Geodetic monitoring results reveal significant changes). It is planned to incorporate flux density monitoring at S/X band – currently only done for 0212+735 and 1803+784 – in order to improve calibration and to derive physical parameters of components.

IAU Name	Name	ID	z	Images	Time base
0106+013		BL?		1(X)	1988.33
0212+735		OVV	2.37	5(S),11(X)	1983.4-1992.15
0229+131		QSO	2.065	3(X)	1988.33-1992.15
0300+470		BL Lac		1(X)	1991.12
0420−014		QSO	0.915	1(X)	1991.12
0454−234		BL?		1(X)	1992.15
0528+134		BL?		3(X)	1988.33-1992.15
0552+398	DA193	QSO	2.365	4(X)	1988.33-1992.15
0727−115		BL?		1(X)	1991.12
0851+202	OJ287	BL Lac	0.31	4(X)	1988.33-1992.15
0923+392	4C39.25	QSO	0.699	5(X)	1990.94-1992.15
1226+023	3C273	QSO	0.158	1(X)	1988.33
1334−127		QSO	0.541	1(X)	1988.33
1404+286	OQ208	QSO	0.077	4(X)	1988.33-1992.15
1611+343	DA406	QSO	1.401	1(X)	1991.12
1641+399		QSO	0.704	1(X)	1988.33
1741−038		QSO	1.057	9(S),4(X)	1988.95-1992.52
1803+784		BL Lac	0.68	13(S),15(X)	1983.4-1992.27
2121+053		QSO	1.878	3(X)	1988.95-1992.15
2134+004		QSO	1.936	1(X)	1988.33
2145+067		QSO	0.99	1(S)	1992.27
2200+420	BL Lac	BL Lac	0.07	1(X)	1988.33
2234+282		QSO	0.795	3(X)	1988.95-1992.52
2251+158	3C454.3	QSO	0.859	1(X)	1988.33

References

Schalinski, C.J., et al., 1988a, in *IAU Symposium 129*, p. 359
Schalinski, C.J., et al., 1988b, in *IAU Symposium 129*, p. 39
Schalinski, C.J., 1990: Dissertation, Universität zu Bonn
Schalinski, C.J., et al., 1992, in *Variability of Blazars*, ed. Valtaoja & Valtonen, p. 225

Multi-Epoch 8.4 GHz VLBI Observations of the Nucleus of Centaurus A

David L. Meier [*] and the *SHEVE team* [†]

Abstract

We present the results of several 8.4 GHz VLBI observations of the nucleus of Centaurus A. We find that the source possesses a classical core-jet structure with the inner portion of the jet expanding at a proper motion of 4.0 mas yr^{-1} or an apparent velocity of 0.26c along the jet.

1. Introduction

At a distance of only 4 Mpc [H84] the radio source Centaurus A in the peculiar elliptical NGC 5128 is the closest active galaxy to the Milky Way. It therefore affords potentially the highest linear imaging resolution (1 mas = 0.019 pc) for such an object and, hence, the best prospects for studying an active nucleus near the central source. On very large scales Centaurus A is an FR I double-lobed source with structure out to 250 kpc at a position angle of 0° [Ju92]. On intermediate (10 kpc) scales, a smaller, apparently younger, double-lobed source lies at a position angle of 51°, roughly perpendicular to the dark dust lane which bisects the elliptical galaxy [B83]. The jet feeding these lobes appears one-sided, in the northeasterly direction, and has been detected at X-ray and infrared wavelengths as well as radio [B83,Jo91]. The compact nucleus has been studied by us with VLBI at 2.3 GHz on 0.2 - 2.0 pc scales and found to possess a jet at the same position angle and with one-sidedness in the same sense [M89]. No significant changes in structure have been detected at 2.3 GHz over an eight year period from 1980 to 1988, although during that time the shortest baseline (with 100 mas fringe spacing) was the only one to be sampled consistently and with sufficient signal-to-noise to detect motion in the jet.

---

[*] Jet Propulsion Laboratory, California Institute of Technology, Pasadena, CA. 91109, U.S.A. This research was carried out in part at the Jet Propulsion Laboratory, California Institute of Technology, under contract with the National Aeronautics and Space Administration.

[†] Members of the Southern Hemisphere VLBI Experiment (SHEVE) team are listed as co-authors on the paper by R. A. Preston *et al.* in these proceedings.

2. The 8.4 GHz VLBI Observations

Our 8.4 GHz observations of Centaurus A span nearly a decade and include two imaging experiments at epochs 1991.17 and 1991.90, each with a 3 × 7 mas beam. For comparison we also have several single-baseline experiments at epochs 1982.30 with 30 mas resolution, 1990.05 with 8 mas resolution, and 1991.78 with 7 mas resolution. The 1982 observations, employing the Deep Space Station (then 64m) antenna at Tidbinbilla and the 64m at Parkes, showed an elongated jet structure similar to that seen at 2.3 GHz and the 3 Jy flux at 8.4 GHz implied a spectral index of $\alpha = -0.8$, typical of optically thin non-thermal synchrotron sources [M89]. There was also some evidence in 1982 for a weak (0.5 Jy) component 100 mas away from but in line with the jet which was visible at 8.4 GHz but not at 2.3 GHz.

The recent 8.4 GHz observations employed the 70m or 34m at Tidbinbilla, the Parkes antenna, the 26m in Hobart, Tasmania, the 22m at Mopra, and one of the 22m antennas of the Australia Telescope at Narrabrai. Significant changes in both source strength and structure are readily seen in the multi-epoch visibility data. The flux of the Centaurus A nucleus has increased from 3 to ~ 11 Jy over the ten year period. Also, in the early 1990 Tidbinbila-Hobart data, there appeared to be a compact, unresolved component close to the jet which had not been present in 1982, and which became rather resolved by late 1991.

3. Analysis and Discussion

We have calibrated and imaged the two largest datasets (1991.17 and 1991.90) using standard hybrid mapping procedures. The two maps are shown in Figure 1 with contours at ±1, ±2, ±4, 8, 16, 32, 50, 60, 70, 80, and 90 percent of the peak flux, which is 2.91 and 2.98 Jy per beam, respectively. The beam size in both is 3 × 7 mas at a position angle of $\sim 90°$. Note that the linear extent of these images (~ 1 pc) is comparable to the inferred size of many broad line regions of active galaxies or quasars, and our resolution (0.06 pc) is substantially smaller than that. The change in elongation of the core over the nine month period, which is responsible for the visibility changes discussed above, is a full beam width (3 mas) in extent, yielding a speed of 4.0 ± 8 mas yr^{-1}, or v = 0.26c. The direction of the motion appears to be along the jet. There is also some indication that outer components in the jet are moving at a similar speed, but this result is less certain due to the lower signal-to-noise there.

Finally, we wish to point out that there is no evidence in our recent 8.4 GHz data for a 0.5 Jy flat-spectrum or self-absorbed component 100 mas away from the brightest portion of the jet, as was indicated in the 1982 data. Also there appears to be no evidence for an accretion disk perpendicular to the jet with a brightness temperature greater than $3 \times 10^7 K$ at distances 0.1 pc or more from the core.

In summary, the Centaurus A nucleus appears to be a subluminal version of the classic core-jet VLBI source – a property consistent with the entire object being a low power FR I source.

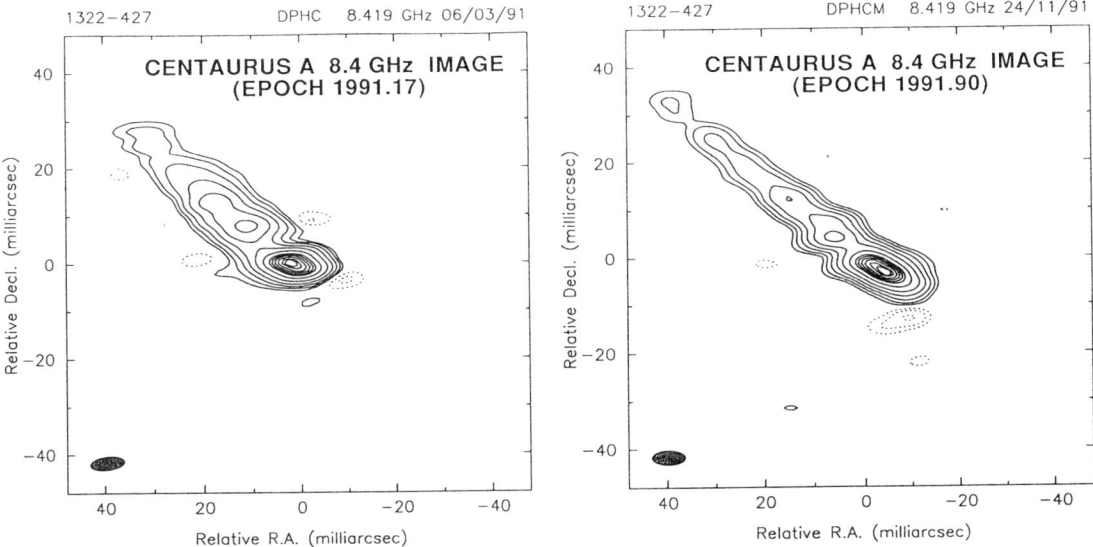

Figure 1.   8.4 GHz VLBI images of Centaurus A

References

[B83] Burns J.O., Feigelson E.D., and Schreier E.J., *The inner radio structure of Centaurus A: clues to the origin of the jet X-ray emission.* Astrophys. J., Vol. 273 (1983), pp. 128-153.

[H84] Harris G.L.H., Hesser J.E., Harris H.C., and Curry P.J., *The NGC 5128 globular cluster system: star counts in U, V, and R.* Astrophys. J., Vol. 287 (1984), pp. 175-184.

[Jo91] Joy M., et al., *An infrared jet in Centaurus A: a link to the extranuclear activity in distant radio galaxies?* Astrophys. J., Vol. 366 (1991), pp. 82-87.

[Ju92] Junkes N., Haynes R.F., Harnett J.I., and Jauncey D.L., *Radio Polarization Surveys of Centaurus A (NGC 5128) 1: The complete radio source at 6.3 cm wavelength.* Astron. Astrophys., submitted.

[M89] Meier D.L., et al., *The high resolution structure of the Centaurus A nucleus at 2.3 and 8.4 GHz.* Astron. J., Vol. 98 (1989), pp. 27-35.

Parsec-Scale Properties of FR-1 Radio Galaxies

Tiziana Venturi,[*] *Luigina Feretti,*[*] *Gabriele Giovannini,* [*]
Ann E. Wehrle [†]

Abstract

We present a preliminary discussion on the parsec-scale morphology and properties of FR-1 radio galaxies. Nine radio galaxies are presented, using our data and data from the literature. The occurrence of one-sidedness on the parsec-scale is remarkable. Six out of 9 radio galaxies have a one-sided parsec-scale jet pointing towards the strongest kiloparsec-scale jet. The possibility that Doppler boosting is responsible for the observed asymmetries is discussed. Two radio galaxies, i.e. NGC315 and 3C338 are presented in detail.

1. The Sources

VLBI observations of low luminosity (FR-1) radio galaxies have been gathering, making it possible to gain some information on the parsec-scale morphology of such objects. We have been studying in detail, at various radio frequencies and at VLBI resolution a complete sample of radio galaxies, selected and defined in [GFC90]. The sample includes 14 FR-1 radio galaxies. Information on the milliarcsec morphology is now available for 9 of these 14 FR-1s.

The information available on the kiloparsec and parsec-scale morphology for these 9 FR-1 radio galaxies is summarised in Table 1, where the last column gives the reference for the parsec-scale maps.

2. Considerations on the Parsec-Scale Morphologies

In two cases, i.e. **0755+37** and **1144+35**, the parsec-scale morphology is unclear from our maps at 6 cm. Observations at other frequencies have already been planned, for an unambiguous classification of these two sources.

Most of the sources in Table 1 (6 out of 9) are classified as one-sided on the basis of multifrequency observations. In **4C29.30**, **3C264** and **3C465** the parsec-scale jet

---

[*]Istituto di Radioastronomia-CNR, Bologna, Italy
[†]IPAC-JPL, Pasadena, CA, USA

Table 1 - Source morphologies

Source	kpc scale	pc scale	Reference
NGC315	2-sided asymmetric	1-sided	[VGF93]
3C66B	2-sided asymmetric	1-sided	Bartel, priv. comm.
0755+37	2-sided asymmetric	ambiguous	[GCF91]
4C29.30	2-sided asymmetric	1-sided	[VFG93]
3C264	1-sided	1-sided	[VFG93]
1144+35	2-sided asymmetric	ambiguous	[GCF91]
3C274	1-sided	1-sided	[BOC89]
3C338	symmetric	2-sided	[FCG92]
3C465	2-sided asymmetric	1-sided	[VFG93]

points towards the strongest kiloparsec-scale jet. Continuity between the small scale and large scale asymmetry is shared also by **3C274** and **3C66B**.

The high percentage of one-sided sources suggests that the parsec-scale emission in FR-1 radio galaxies is Doppler boosted. However this explanation is not entirely satisfactory. It is well known that the intrinsic flow velocity in the kiloparsec jets of FR-1 radio galaxies is of the order of few thousands km/sec, so if we think that the parsec-scale asymmetry is due to Doppler boosting, we must conclude that two different mechanisms are responsible for asymmetries observed on the two scales. In this case, since the asymmetries on the two scales are unrelated, then we should expect to see some parsec-scale jets pointing towards the weakest kiloparsec-scale jet, but no such case is known among FR-1s. The hypothesis that parsec-scale jets in FR-1s are relativistic should be supported by observational evidence. Among the sources listed in Table 1 there is at least one case, i.e. **NGC315**, which contradicts this hypotesis.

NGC315 was observed at 18, 6 and 3.6 cm with the global VLBI array. The source is one-sided at all frequencies ([VGF93]). The jet to counter-jet brightness ratio at 18 cm (≥ 50) could be due to Doppler boosting assuming an angle between the source axis and the line-of-sight $\theta \sim 58°$ and a factor $\beta \rightarrow 1$. The highly intrinsic relativistic speed required by the Doppler boosting is in contrast with the very stationary structure of the source, as discussed in [VGF93], who place an upper limit of $v \sim 0.5c$ on the intrinsic flow velocity. It seems more likely that the flow speed along the parsec-scale jet is only mildly relativistic and that the cause of the parsec-scale asymmetry should be searched for elsewhere.

3C338 is the only source in the sample which is two-sided on the parsec-scale: the jet to counter-jet flux ratio in the 6 cm map is 1.1 (see [FCG92]). Such value suggests that Doppler boosting is not important, and this means either that the source is in the plane of the sky or that the parsec-scale jets in the source are not relativistic, or both. Assuming that the jets are relativistic, then we would expect to detect proper motion in the brightness knots along the jets ([FCG92]). A second epoch observation will allow us either to derive the intrinsic velocity flow in the parsec-scale structure, or to put upper limits on such velocity.

To summarize, one-sidedness seems to be the most common feature in the parsec-scale morphology of FR-1 radio galaxies, which is therefore very similar to the morphology observed in the high power radio loud quasars and galaxies. Doppler boosting could be invoked to explain the observed morphologies. Confirmation of highly relativistic plasma speeds in such objects is still missing, and the subject is presently under investigation. We point out that continuity between small and large scale asymmetry is a matter of concern, since we expect the same mechanism to account for the asymmetry on both scales. Alternatively, the jets are intrinsically asymmetric, or kiloparsec scale jets have much higher intrinsic velocities than predicted by current models.

References

[BOC89] Biretta J.A., Owen F.N., Cornwell T.J., *A search of motion and flux variations in the M87 jet.* ApJ 342 (1989), 128.

[FCG92] Feretti L., Comoretto G., Giovannini G., Venturi T., Wehrle, A.E., *VLBI observations of a complete sample of radiogalaxies. II. The two-sided milliarcsec structure of 3C338.* submitted to ApJ (1992).

[GFC90] Giovannini G., Feretti L., Comoretto G., *VLBI observations of a complete sample of radio galaxies. I. Snapshot observations.* ApJ 358 (1990), 159-163.

[GCF91] Giovannini G., Comoretto G., Feretti L., Marcaide J., Venturi T., Wehrle A.E., *Observations of 5 radio galaxies.* in Active Galactic Nuclei, Heidelberg, 1991.

[VGF93] Venturi T., Giovannini G., Feretti L., Comoretto G., Wehrle A.E., *VLBI observations of a complete sample of radiogalaxies. III. The parsec-scale structure of NGC315.* ApJ (1993), in press.

[VFG93] Venturi T., Feretti L., Giovannini G., Lara L., Marcaide J.M., Rioja M., Trigilio C., Umana G., *Observations of radio galaxies in the S/X band with the triangle Medicina-Noto-Madrid.* these proceedings.

VLBI Polarization Observations of 3C138: Preliminary Results

W. Cotton, [*]   D. Dallacasa, C. Fanti, R. Fanti, [†]
R. Spencer, [‡]   R. Schilizzi, T. Foley [§]

1. Introduction

The small size of the compact steep spectrum sources is frequently interpreted as being due to interactions with a dense interstellar medium. This would suggest substantial Faraday rotation effects in the ionized material surrounding the jet. We present preliminary polarization sensitive VLBI measurements of the compact steep spectrum quasar 3C138 at a wavelength of 6 cm.

2. Results

The observations were made in September 1989 using a 9 antenna network including Bonn, Westerbork, Onsala, Jodrell Bank, Haystack, Greenbank, the phased VLA, Pietown, and OVRO. All antennas except Onsala recorded dual polarization. The data was recorded on the MkIII system and processed on the Bonn MkIII correlator. All post correlation analysis including fringe fitting was carried out in the NRAO AIPS software package.

The results are shown in Figure 1. The core is very weakly polarized and the jet becomes increasingly polarized especially towards the end and the northern edge. The orientation of the E-vectors is relatively constant over most of the length of the jet but exhibits complex structure near the end of the jet.

Our observations resolve the jet in the transverse direction so it is possible to examine the polarization structure. The total intensity is center brightened whereas the fractional polarization is strongly edge brightened. In addition, there is a tendency for the polarization vectors to be across the jet in the center and along it at the edges.

---

[*]NRAO, 520 Edgemont Road, Charlottesville, VA 22903, USA.
[†]CNR, Via Irnerio, 46, 40126 Bologna (Italy)
[‡]NRAL, Jodrell Bank, Macclesfield, Cheshire, SK11 9DL, UK
[§]NFRA, Postbus 2, NL-7990 AA Dwingeloo, Netherlands

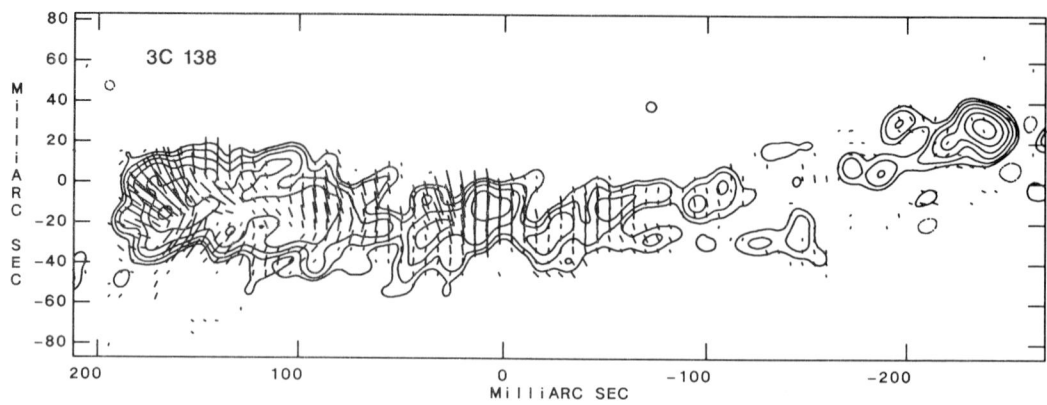

Figure 1. This figure shows the total intensity image of 3C138 in contours with "E-vectors" superimposed. The length of the vectors is proportional to the polarized intensity and the direction shows the orientation of the electric field. The observations have been tapered to a resolution of 10 mas to improve the surface brightness sensitivity. 100 mas corresponds to a projected linear size of $\approx$ 370 pc.

The E-vectors shown in Figure 1 are orthogonal to the magnetic field in the absence of Faraday rotation and optical depth effects. The relative constancy of the observed E-vectors can be taken as evidence that Faraday rotation and/or optical depth effects are minimal over most of the jet and that the dominant component of the magnetic field is along the jet. Towards the end of the jet the observed polarization orientations cannot be unambiguously interpreted without further observations at other wavelengths to separate the effects of Faraday rotation, optical depth and magnetic field structure.

The optical spectrum of this object (R. Gelderman, private communication) shows very broad [OIII] forbidden lines which are approximately 1100 km/sec wide. These features are being interpreted as indicating strong interaction of the jet with the interstellar medium. This suggests that there is a considerable amount of ionized material associated with the jet. However, the relatively constant nature of the polarization structure of the bulk of the jet seems to suggest that it is not in front of this portion of the jet. If the width of the [OIII] lines is due to interactions with the jet then the region near the end of the jet should show substantial Faraday rotation. Further observations are needed to confirm or deny this prediction.

The Results of a Study of the Milliarcsecond Scale Linear Polarization Properties of Extragalactic Radio Sources.

T.V. Cawthorne [*]   J.F.C. Wardle [†]   D. H. Roberts [‡]
D.C. Gabuzda [§]

Abstract

The main results of a survey of the milliarcsecond scale linear polarization properties of extragalactic radio sources are reported. The analysis is based upon observations of 24 sources from the sample defined by Pearson and Readhead (1988). Of the sources observed there are 12 quasars, 8 BL Lac objects and 4 galaxies. The polarization properties on milliarcsecond scales are found to depend strongly upon the optical classification to which they belong. This suggests that the polarization properties observed may be related to the parsec-scale environments in which the radio sources are found. Full details of this work may be found in two papers recently submitted to *Ap. J.*, Cawthorne *et al.* 1992a, and Cawthorne *et al.* 1992b.

1. The Differences between Quasars and BL Lac objects.

Both quasars and BL Lacertae objects have jets that are appreciably linearly polarized. However on milliarcsecond scales, the cores of quasars are consistently less strongly polarized than those of BL Lacertae objects. No quasar was found to have a degree of core polarization in excess of 2 percent, while 5 out of 8 BL Lacertae objects showed degrees of core polarization in excess of 2 percent. The probability that this difference is due to chance is less than 0.04.

---

[*] Harvard Smithsonian Center for Astrophysics, 60 Garden Street, Cambridge MA 02138, U.S.A. (Present address, Department of Physics and Astronomy, University of Central Lancashire, Preston, Lancashire, Great Britain.)
[†] Department of Physics, Brandeis University, Waltham, MA 02254, U.S.A.
[‡] Department of Physics, Brandeis University, Waltham, MA 02254, U.S.A.
[§] Department of Physics and Astronomy, University of Calgary, Calgary, Alberta T2N 1N4, Canada.

The most obvious explanation for the difference is that the line-emitting medium in the quasars depolarizes their cores.

In the cores, the distribution of $|\chi - \theta|$, the offset between the electric field of linear polarization and the direction of the jet, is fairly uniform for both classes of object. However in the jets the distributions of $|\chi - \theta|$ are strikingly different.

In quasars the electric fields are nearly transverse for most sources, indicating a magnetic field that is close to being longitudinal. Significant departures from this trend have been found in a few quasars. It seems likely that these offsets are due Faraday Rotation in the nuclear region of the source. Rotation on these scales is probably not well represented by the standard VLA measurements which are strongly influenced by structure on larger scales. However we must await the results of VLBI measurements at several wavelengths to confirm this hypothesis.

In the BL Lacertae objects, the inferred magnetic fields in the jets are nearly perpendicular to the jet axes. This is consistent with the model developed by Hughes, Aller & Aller (1985) in which the polarization arises from regions in which a tangled field is compressed by a plane perpendicular shock.

The degree of polarization in quasars is often greatest in the weakest components. The origin of the knots has often been attributed to transverse squeezing of the jet, for example by pinch-mode instabilities. However if the field in the jet is predominantly longitudinal with a small tangled component superimposed, then we note that this model predicts that the brightest knots should be the most strongly polarized, a trend in the opposite sense to that observed.

2. Radio Galaxies.

In all cases we have studied, radio galaxies turn out to be unpolarized on milliarcsecond scales. Further work is required to determine whether this is due to Faraday depolarization or internal tangling of the field. However we note that the few compact radio sources associated with radio galaxies show the 'core-jet' structures that are common in quasars and BL Lacertae objects.

References

[1] Hughes P.A., Aller H., Aller M. *Ap.J. vol. 298 (1985)*, p. 301.

[2] Cawthorne T.V., Wardle J.F.C., Roberts D.H., Gabuzda D.C., Brown L.F. *Ap.J. (1992)* submtitted.

[3] Cawthorne T.V., Wardle J.F.C., Roberts D.H., Gabuzda D.C. *Ap.J. (1992)* submtitted.

[4] Pearson T.J., Readhead A.C.S. *Ap.J. (1988) vol. 328 p. 114.*

3.6 cm VLBI Polarization Observations of BL Lacertae Objects

Denise C. Gabuzda [*]   *Timothy V. Cawthorne* [†]

Abstract

6 cm VLBP observations of BL Lacertae objects have revealed surprisingly characteristic and distinctive polarization structure, suggesting that relativistic shocks are common in the VLBI jets of these sources. We are currently making the first 3.6 cm VLBP images of BL Lacertae objects. These images also point toward the presence of relativistic shocks, but indicate that the jets may be more curved than previously realized. There is also evidence that the high core polarization of BL Lacertae objects (compared to that of quasars) may be due to highly polarized knots which are blended with the core at 6 cm.

There are now some 18 BL Lacertae objects for which VLBP images have been made at 6 cm (Gab92 and references therein). In every BL Lacertae object in which polarization structure has been detected, the polarization position angles in knots in the jets are nearly parallel to the VLBI structural axis (the one exception is 1308+326; see Gabuzda and Kollgaard, these proceedings). The magnetic fields inferred by this orientation are nearly perpendicular to the direction of the jet, suggesting that the knots are associated with relativistic shocks that compress an initially tangled magnetic field as they propogate down the VLBI jet (Laing 1980).

3.6 cm Mark III VLBI polarization observations of a number of BL Lacertae objects were made in June 1990, using a ten-antenna global array. The first images from this experiment are currently being made, and promise to shed considerable light on the trends seen at 6 cm. For example, we have included here our 3.6 cm VLBI total intensity and polarization images of 1803+784 (see Figure). These images show the well-known dominant structure in position angle $-90°$ (Wit88, Gab92), but also reveal the core to be extended in position angle $-60°$. This extention is seen more noticeably on the polarized flux map, and is polarized with electric vector closely aligned to the direction of the extention. This suggests that we are seeing a new shocked component emerging along a direction that is quite different from that seen at

---

[*]Department of Physics and Astronomy, University of Calgary
[†]Department of Physics, University of Central Lancashire

larger distances from the core. The emergence of this component has been confirmed by mm VLBI observations of Krichbaum et al. (these proceedings).

The degrees of polarization observed in BL Lacertae object VLBI cores at 6 cm are typically larger than those observed in quasar cores. One possibility is that what is observed as the VLBI "core" at 6 cm is actually the superposition of a weakly polarized "quasar-like" core and highly polarized knots that are not resolved from it. A number of our 3.6 cm images (e.g., that of 1803+784), show evidence for newly emerging polarized knots, which would have been blended with the core at 6 cm. Our 3.6 cm observations suggest that the new knots are strongly polarized ($\sim 15 - 20\%$) while the cores are somewhat more weakly polarized than at 6 cm; the "average" polarization of the cores blended with these new knots would be $\sim 3 - 6\%$, which is consistent with the "core" polarizations observed in these sources at 6 cm (Gab89).

The 6 cm cores of BL Lacertae objects are appreciably polarized, but do not show any preferential orientation for their polarization. Some of our 3.6 cm images (e.g., that of 1803+784) are suggestive that in at least some cases there may be bending of the jets on very small scales. If in fact the "core" polarizations observed at 6 cm are actually dominated by polarized emission from knots which are not yet resolved from the core, these polarization position angles may be aligned with the *local* jet direction for the emerging knots.

References

[Gab92] Gabuzda, D., Cawthorne, T., Roberts, D. and Wardle, J., *Ap. J. Vol. 388 (1992), p. 40.*

[Gab89] Gabuzda, D., Cawthorne, T., Roberts, D. and Wardle, J., *Ap. J. Vol. 347 (1989), p. 701.*

[Lai80] Laing, R., *M. N. R. A. S. Vol. 193, p. 439.*

[Wit88] Witzel, A., Schalinski, C., Johnston, K., Biermann, P., Krichbaum, T., Hummel, C., and Eckart, A., *A & A, Vol. 206, p. 245.*

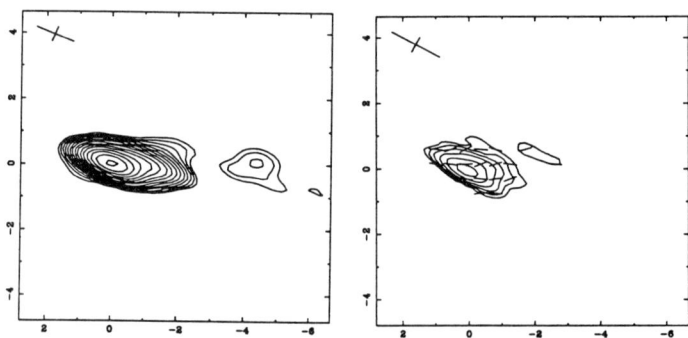

Figure 1. Total intensity and polarization images of 1803+784.

The Caltech–Jodrell Bank VLBI survey

Peter N. Wilkinson,[*] *Antonis G. Polatidis,*[*]
Anthony C. S. Readhead,[†] *Wenge Xu,*[†] *Timothy J. Pearson.*[†]

The Pearson & Readhead (1988) [1] (PR) survey sample contains 65 objects of which 47 were observed with VLBI. This sample provided the first well-defined classification scheme for the milli-arcsecond scale morphologies of strong radio sources. However the size of the sample is too small for some of the most interesting statistical studies.

The Caltech–Jodrell Bank VLBI survey (CJ survey) is an extension of the PR survey whose major aim is to treble the number of objects studied. It contains 135 objects selected from the NRAO-MPI **S4** and **S5** surveys with flux density 1.3 Jy$> S_{5GHz} >$0.7 Jy, declination $\delta > 35°$ and galactic latitude $|b^{II}| > 10°$. Examining spectral and arc-second scale structural information we identified 94 sources that can be imaged with Mark 2 VLBI. The remaining objects are large, lobe-dominated sources that do not have strong enough (i.e. S>0.2 Jy) compact central components to be mapped using the Mark 2 system.

In combination with the PR sample, which contains 65 objects, this survey forms a complete flux–limited sample of 200 objects of which 131 objects can be mapped with Mark 2 VLBI. We are also carrying out a program to obtain optical identifications and redshifts for all of the objects. We expect that at the end of this program we will have complete optical identifications and 95% of the redshift information for the sample.

The main object of this survey is to obtain images of sufficient quality to enable us to :

1. Classify the objects via their radio structure and to add to the present classification scheme any new types of objects we might observe.

2. Provide a sample large enough to make interesting statistical tests of objects by class. For example the cosmological evolution of different classes of objects can be studied using the luminosity–volume test. Testing the applicability of "Unified Schemes" is another obvious goal for this study.

3. Provide first epoch observations of candidate sources for superluminal motion to enable statistical studies of superluminal motion with redshift to be made.

We decided to make observations at both 6 cm and 18 cm in order to clarify ambiguous cases and hence to increase the reliability of the classification. Observations at 6 cm and 18 cm were made for all the CJ sources for which no VLBI map was available in the literature. In addition we observed 21 PR sources for which no previous 18 cm

---

[*]University of Manchester, Nuffield Radio Astronomy Laboratories, Jodrell Bank, Macclesfield, Cheshire, SK11 9DL, U.K.
[†]Owens Valley Radio Observatory, California Institute of Technology, Pasadena, California 91125, U.S.A.

observations were made. In total we have observed 78 sources at 6 cm and 113 sources at 18 cm. The observations were made using the Mark 2 recording system and the data were cross–correlated in the 16–station Caltech/JPL–Mark 2 correlator.

Given the limited time available in VLBI networks a large project like this would be impossible to complete with 'full track' observations. In order to make efficient use of the network time we made 'snapshot' observations carefully scheduled to minimize the time spent on each source without compromising the quality of the images. Three 30 minutes scans were scheduled for each source, separated in hour angle to maximize the uv coverage. Depending on the observing session 13 – 16 antennas were used. To test the reliability of the maps we observed two objects with complicated structure (3C147 and 3C309.1) for which full track observations at the same frequency were available. Mapping following the standard procedures established for the rest of the survey we obtained images of quality similar to those from the long track observations.

First epoch observations at both frequencies are now completed. Virtually all data have been processed and so far 77% of the sources have been mapped. The typical the rms noise level in uniform weighted maps is $\simeq$ 1 mJy/beam and in naturally weighted maps is $\leq$ 0.5 mJy/beam yielding images with dynamic ranges $\geq$ 100:1. The noise levels achieved are close to the theoretical thermal noise.

A preliminary analysis based on the 77 % of the sample for which maps are available, indicates that the basic classification scheme introduced by PR is still valid. The structures can be summarised as follows: 70 % of the sources have a basically "core–jet" structure; 9 % are compact; 10% have a compact ($\leq$5 Kpc) double or triple morphology and about 11 % are not classified yet (sources with complex structure, resolved sources etc.). Figure 1 shows some examples of the structures found in the survey. Fig 1a shows the 6 cm naturally weighted map of the 'core-jet' source 2255+415. Figure 1b shows the 18 cm map of the compact steep spectrum source 0404+768 which shows a 'core-jet' structure with a lobe on the side opposite to the jet. Figure 1c is the 18 cm tapered map of the BL Lac object 1652+398 showing a bright jet that bends through 180° embedded in low level emission. Figure 1d shows the 6 cm map of the compact double source 0646+600.

An important result of the survey is the confirmation of the existence of symmetric, two-sided structure as suggested by Conway *et al.* (1992) [2]. Our 18 cm map of the source 2352+495 from the PR sample (Figure 1e) clearly shows symmetric, steep-spectrum emission on both sides of a strong core. We have also detected another source, 1943+54, with two sided structure (Figure 1f).

References

[1] T. J. Pearson and A. C. S. Readhead, 1988, *Ap.J.*, **328**,114.

[2] J.E. Conway, T.J. Pearson, A.C.S. Readhead, S.C. Unwin, W. Xu and R.L. Mutel, 1992, *Ap.J.*,**396**, 62.

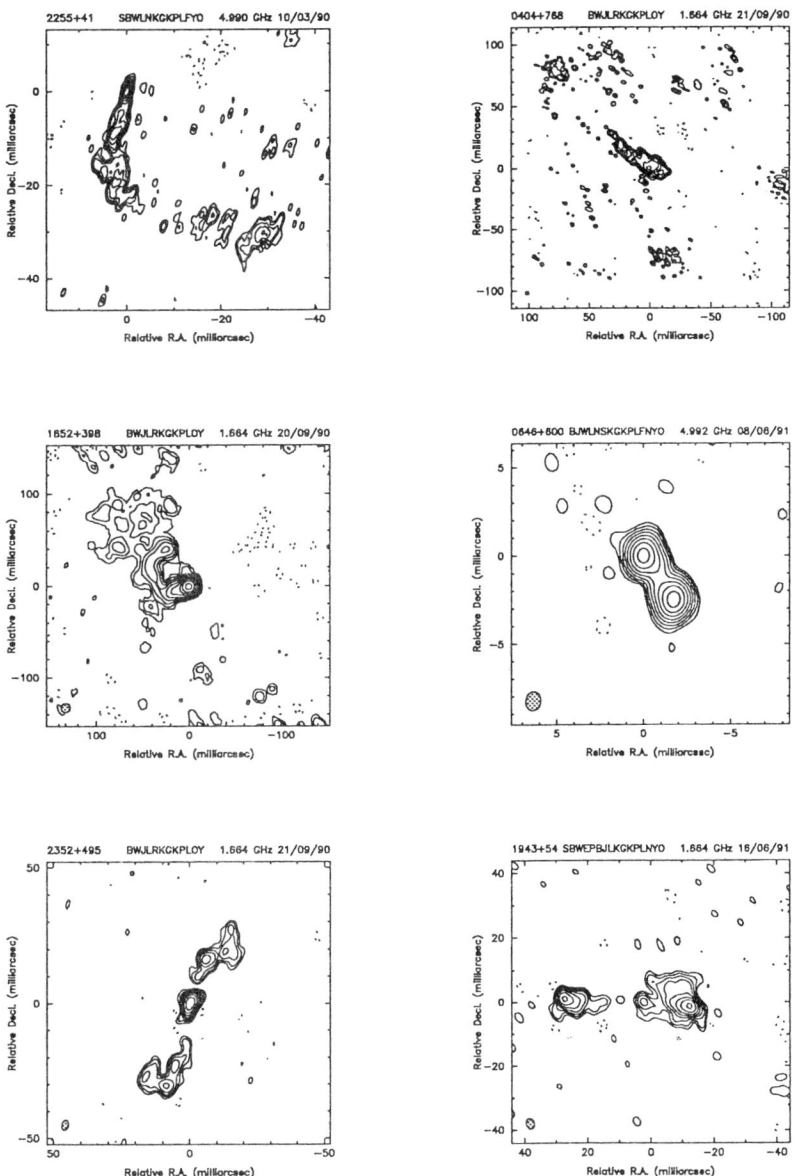

Figure 1. Maps from the CJ survey (All contours are in 1, 2, 4, 8, 16, 32, 64, 128, 256 × 3σ): (a) 6 cm 'naturally' weighted map of the core–jet source 2255+415, peak=206 mJy/bm, beam=1.87×0.82 mas, 3σ=0.98 mJy/bm, (b) 18 cm map of the core-jet source 0404+768, peak=240.2 mJy/bm, beam=2.8×2.4 mas, 3σ=2.7 mJy/bm, (c) 18 cm tapered map of the core-jet source 1652+398 peak=592.3 mJy/bm, beam=10×10 mas, 3σ= 1.7 mJy/bm ,(d) 6 cm map of the compact double source 0646+600, peak=400.7 mJy/bm, beam=1.1×0.89 mas, 3σ=2.1 mJy/bm (e) 18 cm map of the compact triple source 2352+495, peak=672.9 mJy/bm, beam=3.76×2.48 mas, 3σ=5.3 mjy/bm, (f) 18 cm map of the compact triple source 1943+54, peak =537.9 mJy/bm, beam =3.42 ×2.6 mas, 3σ=3.6 mJy/bm.

Two Core–Jet Sources with Large Misalignment

Wenge Xu[*], *Antonis G. Polatidis*[†]
Anthony C. S. Readhead[*] *Peter N. Wilkinson*[†] *Timothy J. Pearson*[*],

In the majority of objects mapped on both milli-arcsecond and arcsecond scales, the milli-arcsecond scale nuclear jet connects with the large scale jet. This has been taken as strong evidence of relativistic beaming in both small and large scale jets, because a number of lines of evidence indicate that the bulk motion in small scale jets is relativistic. Any counter example in which the small and large scale jets are on opposite sides of the center of activity would cast doubts on the relativistic beaming model. For this reason any examples of oppositely directed small and large scale jets merit careful study.

In the Caltech–Jodrell Bank VLBI (CJ) survey (cf. Wilkinson *et al.* in this proceedings), we have found two such objects, 1347+539 and 1418+546. To complement the 6 cm and 18 cm VLBI observations from the CJ survey, additional VLBI observations at 3.6 cm and MERLIN observations at 6 cm have been made. Due to the limited space of this paper, selected images are presented.

1347+539 is identified as a 17.3 mag. quasar with a redshift of 0.978. The VLA observations at 6 cm at the A configuration [2] show that a kilo-parsec scale jet extends to the northwest from a compact component (Fig. 1a). The VLBI observations show that a parsec scale jet extends towards the southeast from a flat spectrum core (Fig. 1b). The cusp-like feature in the jet, shown in the 3.8 cm VLBI image (Figure 1c), strongly suggests that the jet is quasi-helical. MERLIN observations at 6 cm (Figure 1b) show that the parsec scale jet turns through $\sim$ 180° and apparently connects with the large scale structure on the opposite side of the core. This can be explained as a strong projection effect, i.e. a gently curving jet orientated close to the line of sight can appear to bend through a large angle.

1418+546 is identified as a BL Lac object with a redshift of 0.152. 20 cm observations with the VLA at A configuration [2] show an extended component to the west (Fig. 2a). The VLBI observations show that a small scale jet extends toward to the southeast of a flat spectrum core (Fig. 2c). The 50 mas resolution, 6 cm MERLIN image (Figure 2b) does not show any evidence of a turn-around of the small scale jet. 1418+546 remains an interesting candidate in which the small scale jet does not appear to connect with the large scale structure.

---

[*]Owens Valley Radio Observatory, California Institute of Technology, Pasadena, California 91125, U.S.A.

[†]University of Manchester, Nuffield Radio Astronomy Laboratories, Jodrell Bank, Macclesfield, Cheshire, SK11 9DL, U.K.

High dynamic range observations are needed to detect any further extension and possible bending of the small scale jet.

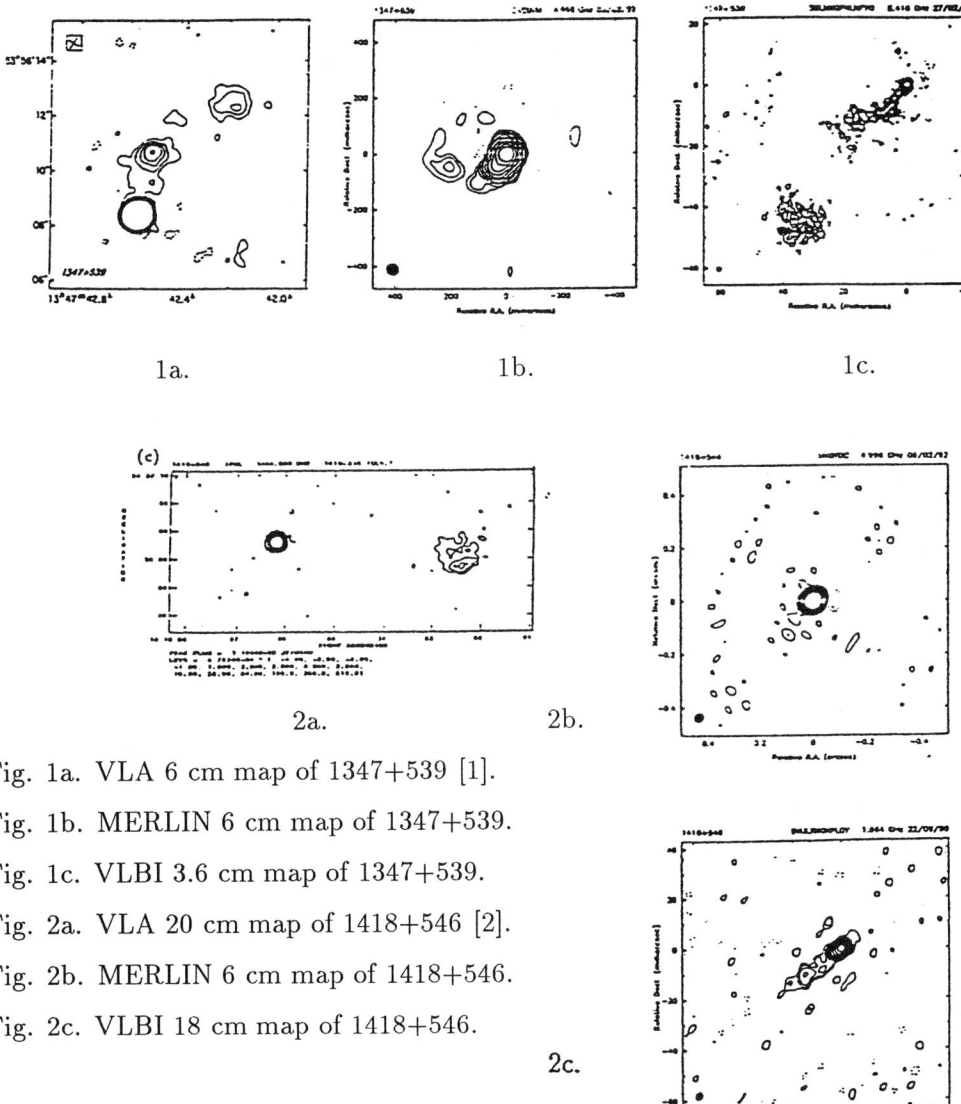

Fig. 1a. VLA 6 cm map of 1347+539 [1].

Fig. 1b. MERLIN 6 cm map of 1347+539.

Fig. 1c. VLBI 3.6 cm map of 1347+539.

Fig. 2a. VLA 20 cm map of 1418+546 [2].

Fig. 2b. MERLIN 6 cm map of 1418+546.

Fig. 2c. VLBI 18 cm map of 1418+546.

References

[1] Murphy, D. 1988, Ph. D. thesis, Univ. of Manchester.

[2] Owen, F. N., and Puschell, J. J., 1984, *A.J.* **89**, 932.

The spectral indices of the parsec-scale jet components of 3C273

P. Charlot [*]

Abstract

Simultaneous 2.3 and 8.4 GHz hybrid VLBI maps of 3C273 have been produced from the observations of two Crustal Dynamics Program geodetic VLBI experiments conducted on North-Pacific baselines in April 1986 and March 1987. For each epoch, the 2.3 and 8.4 GHz maps are compared to determine the spectral indices of the jet components of 3C273. The spectral index ($S \propto \nu^\alpha$) ranges from 0.6 ± 0.1 close to the core to -1.7 ± 0.5 at ~ 12 mas from the core. Variation of the spectral index in the region of the jet close to the core is detected between the two epochs.

The dataset analysed in this report consists of the observations acquired on 3C273 during two dual-frequency (2.3 and 8.4 GHz) Crustal Dynamics Program (CDP) experiments carried out on 1986 April 9 (1986.27) and 1987 March 25 (1987.23), with a 6-station VLBI array (Mojave, Vandenberg, Hatcreek, Gilcreek, Kauai, Kashima). Fig. 1 shows the 2.3 and 8.4 GHz hybrid VLBI maps produced with these data. All maps in Fig. 1 are convolved with a circular restoring beam of 1.5 mas to compare positions and flux densities of the various jet components of 3C273. The 2.3 GHz maps are superresolved by a factor of 1.8 along the jet, whereas the 8.4 GHz maps are smoothed by a factor 2. For details on the identification of the components see [C92].

The relative spatial location of the 2.3 and 8.4 GHz maps at epoch 1986.27 have been determined by registering the gaps of emission located between regions C8/C7 and X, and between regions C6/C5 and C4/C3 (see Fig. 1). We note that the peaks of brightness at 2.3 and 8.4 GHz of region C6/C5 are aligned, whereas those of region C8/C7 are displaced by about 0.7 mas with such a registration. This displacement is the same as that measured by [MS84] for quasars 1038 + 528 A, B. At epoch 1987.23, no similar displacement is found when registering the peaks of brightness at 2.3 and 8.4 GHz of region C6/C5 (Fig. 1). It is likely that this change is related to the decrease of the spectral index in region C8/C7 between April 1986 and March 1987 (see below).

---

[*] Observatoire de Paris - CNRS/URA 1125, 61 Avenue de l'Observatoire, 75014 Paris, France.

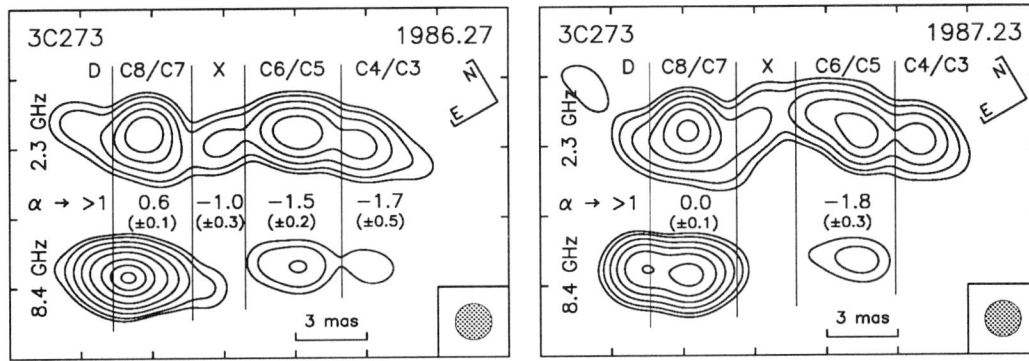

Figure 1. 2.3 and 8.4 GHz hybrid VLBI maps of 3C273 at epochs 1986.27 and 1987.23, rotated counterclockwise by 30°. Spectral indices ($S \propto \nu^\alpha$) of regions corresponding to components D, C8/C7, X, C6/C5 and C4/C3 are printed in between the 2.3 and 8.4 GHz maps. Contour levels are 0.15, 0.3, 0.75, 1.5, 3, 6, 9 and 12 Jy/beam.

The mean spectral indices of the five regions shown in Fig. 1 have been calculated from the integrated fluxes at 2.3 and 8.4 GHz. Uncertainties are evaluated by comparing the peaks of brightness of each region to the dynamic range of the maps with a minimum value of 0.1 accounting for calibration errors. Fig. 1 shows that the mean spectral index ranges from 0.6 ± 0.1 for region C8/C7 (epoch 1986.27) close to the core, to -1.7 ± 0.5 for region C4/C3 at ~ 12 mas from the core. Only a lower limit of the spectral index of the core region D can be determined (> 1), because the core is weak and not well defined at 2.3 GHz. The spectral index of region C6/C5 appears not to have changed significantly between April 1986 and March 1987 (-1.5 ± 0.2 and -1.8 ± 0.3), but that of region C8/C7 decreased from 0.6 ± 0.1 to 0.0 ± 0.1. This change is expected since components C8 and C7 are moving away from the core [C92]. In the future, we plan to extend such a study to additional epochs with data of other CDP experiments, to determine variations of the spectral indices on time scales of a few years.

We are indebted to the Crustal Dynamics Program for providing the high quality data analysed in this report, and to the Caltech VLBI group, especially T.J. Pearson, for providing the Caltech Package, used to produce the hybrid maps shown in Fig. 1.

References

[C92] Charlot P., *Monitoring of extragalactic radio source structures with astrometric/geodetic VLBI experiments. The case of 3C273. Extragalactic radio sources, From beams to jets* (1992), pp. 112-118.

[MS84] Marcaide J.M., Shapiro I.I., *VLBI study of 1038 + 528 A and B: discovery of wavelength dependence of peak brightness location.* Astrophys. J. 276 (1984) 56-59.

Three-frequency VLBI spectral-index maps of the quasars 1038+528A,B

J.C. Guirado [*]   *J.M. Marcaide* [†]   *P. Elósegui* [‡]
J.L. Gómez [*]

Abstract

Rigorous VLBI spectral-index maps in quasar pairs can be obtained combining precise differential astrometry and hybrid mapping. The relative position of the pair of quasars 1038+528A,B, as determined from several epochs and frequencies, shows a wavelength dependence. We present spectral-index maps obtained from three VLBI observations of the quasars 1038+528A,B over a two-year span and we interpret the results by means of a numerical relativistic jet model.

The spectral-index maps (α-maps) consist of a pixel by pixel estimation of α ($S \propto \nu^\alpha$) from flux density maps made at different frequencies which can be registered. The different resolution of the flux density maps and the need to register them correctly are the main difficulties to obtain good VLBI α-maps. Three phase reference VLBI observations of the quasars 1038+528A,B were carried out at the following wavelengths: $\lambda = 3.6/13$cm (X/S bands) in 1981 and 1983 and $\lambda = 6$cm (C band) in 1982. These observations constitute almost a perfect scenario to obtain VLBI α-maps given the accurate determinations of the separation of the quasars and the high dynamic range of the maps [1],[4].

We have applied the method described by Marcaide & Shapiro [3]; Figure 1 shows some of the α-maps obtained for both quasars. The morphology of all the α-maps is rather similar: the contour levels become lines perpendicular to the brightness distribution axis. Since transversal inhomogeneities in the jet are not expected, the gradient of the α-distribution should mark the ridge line of the jet. In fact, we have used the structure of the α-map contour levels as a further criterion to estimate the error in finding the correct registration of the maps: 0.1 mas for X/C α-maps and 0.3 mas for X/S and C/S α-maps.

---

[*]Instituto de Astrofísica de Andalucía, Apdo. 3004, 18080 Granada, Spain.
[†]Dpto. de Matemática Aplicada y Astronomía, Universitat de València, 46100 Burjassot, Spain.
[‡]Harvard-Smithsonian Center for Astrophysics, 60 Garden St., 02138 MA, USA.

The spectral-index distribution constitutes a strong constraint in modelling the physical parameters of a radiosource. The quasar 1038+528A has been modelled using a computer code [2] which calculates the emission of bent shocked relativistic jets. The northeastern component of 1038+528A has been modelled as a plane shock wave travelling along the jet, being detected in 1981 but not in 1983 due to its adiabatic expansion. The parameters estimated by the computer code account simultaneously for the wavelength- and time-dependent shifts of the peak of brightness, found astrometrically, as well as for the slopes of the α-map profiles along the jet ridge line. Those parameters are the following: particle density at jet position 0.5 pc from jet apex ($e^- cm^{-3} erg^{-1}$), 0.0035 ± 0.0004 (1981 and 1982 observations) and 0.0050 ± 0.0004 (1983); magnetic field (gauss), 0.020 ± 0.002 (1981,1982) and 0.022 ± 0.002 (1983); opening angle, $5°\pm0.2°$; angle to the observer, $20°\pm3°$ and jet Lorentz factor, 9 ± 1.

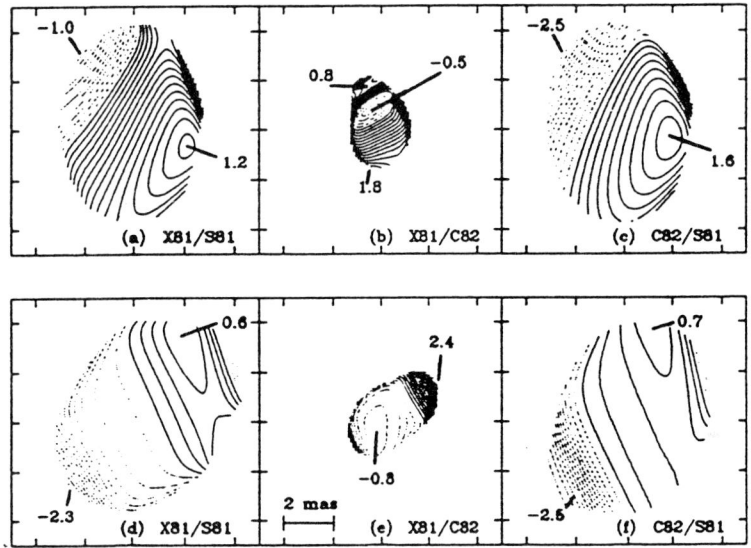

Figure 1. Spectral-index maps of 1038+528A (a,b,c) and 1038+528B (d,e,f).

References

[1] Elósegui P., 1991 *Ph. D. thesis*. Universidad de Granada.

[2] Gómez, J.L., Alberdi, A., Marcaide, J.M. 1993, *Astron. Astrophys.*, submitted.

[3] Marcaide, J.M. and Shapiro, I.I. 1984, *Astrophys. J.*, **276**, 56.

[4] Marcaide, J.M., Shapiro, I.I., Corey, B., Cotton, W.D., Gorenstein, M.V., Rogers, A.E.E., Romney, J.D., Schild, R.E., Bååth, L., Bartel, N., Cohen, N.L., Clark, T.A., Preston, R.A., Ratner, M.I., and Whitney, A.R. 1985, *Astron. Astrophys.*, **142**, 71.

Compact Structure of 3C286, 3C309.1 and 3C380 CSS Quasars

A.J. Kus - TRAO, Nicolaus Copernicus University, Torun, Poland. [*]
R.S. Booth - Onsala Space Observatory, Sweden.
A. Marecki, R.Maszkowski - TRAO Torun, Poland.
R.W. Porcas - MPIfR, Bonn, FRG.
T.J. Pearson, A.C.S. Readhead - OVRO Caltech, USA.
P.N. Wilkinson - NRAL Jodrell Bank, U.K.

Abstract

We present new results from a high frequency (8.4 GHz) VLBI study of 3 luminous CSS quasars. Two of the sources (3C309.1, 3C380) have been observed at two epochs spaced one and two years apart; for one (3C286) we have a single epoch map only. In 3C309.1 and 3C380 we find strong and compact cores whereas the 3C286 core region has relatively large extension and a very weak unresolved component. We confirm SLM (Super Luminal Motion) in 3C380, as reported earlier, and find evidence for an unusually large SLM in 3C309.1. These two quasars show core-jet morphology with core flux domination above 5 GHz and are two of three sources in the class which show well documented SLM. Although the mas structures are similar, there are quite different kinematic models to explain both the observed velocities and the brightness distribution.

The group of Compact Steep Spectrum (CSS) sources is now well established as a separate class of objects (Peacock and Wall 1981) and comprises up to 30% of all extragalactic radio sources. Although Fanti et al. (1990) have argued convincingly, on statistical ground, that CSS objects are not simple double lobed sources viewed end-on, there still remain open questions. It is important to determine finally whether CSS objects are (a) an entirely different type of sources, (b) basically similar to the other classes (ie powered by two sided jets) but smaller or (c) are similar to double-lobed sources viewed at small inclination angles. There is observational evidence suggesting enormously large flows of cold gas into a parent galaxy (Fabian et al. 1984) and if such a phenomenon is common, then CSS sources might be direct indicators of cooling flows.

It is highly probable that CSS radio sources are embodied in dense, gaseous disks, which are capable of breaking and bending jets. This may account for the observed morphologies, but jet precession can remain as a viable explanation of the arc sec structural complexity. For the innermost part of the source (sub-kpc scales) the interaction with the ISM could be responsible for the observed jet flow instabilities

---

[*]This author kindly acknowledges Onsala Space Observatory and the Swedish Institute for a Visiting Scientist Fellowship.

and jet fragmentation. The optically thick medium makes it impossible to observe the core region directly in the dm-wavelength range. High frequency studies allow us to observe fine details in parsec scale jets and to go deeper into the central emitting region. The purpose of this project is to use these sources as a powerful diagnostic to contribute to more fundamental issues of CSS astrophysics.

We have selected the three strongest CSS quasars which are most suitable for doing high frequency VLBI with the MkII recording system. All of them were observed in 1989.89 at 8.4 GHz session, 3C309.1 had also been observed at 1988.76. An additional run, for 3C380 only, was performed in 1992.25 but the map is not yet available. These are the highest frequency and highest angular resolution observations ever made for 3C286 and 3C309.1. We had 8-10 telescopes in the global network. Correlation was done at Caltech and standard reduction and mapping using the AIPS package on Convex computers. Our new maps are presented in Figs 1c-3c.

Although the sources are members of the same class they seem to represent three very different cases. 3C286 has a very weak core (S=90mJy) and a diffuse extended region around it containing most of the observed flux. The source can be explained as having a small intrinsic size (subgalactic) seen at a large angle to the line of sight (see also A.Marecki, poster no: 57). 3C309.1 has a medium strength unresolved core (S=200mJy, $\theta < 0.2$mas) and like 3C380 shows fast structural changes in the mas jet. Earlier indications (Kus et al. 1990) of fast structural changes near the core were fully confirmed by the new observations. Based on 8.4 GHz maps we present evidence that the SLM could be as high as 11c. Such large values imply high Lorentz factors (of the order of 11) which is much higher than usually assumed for CSS objects. Helical flow on scales up to 500pc can plausibly explain the positions and structure of the stationary components as well as the jet PA and SLM of the mas jet close to the core. We find no indication for a change in PA of the core over the one year interval. Such changes could have explained the presence of an initial perturbation in the jet flow causing the appearance of the helical pattern further down the flow.

3C380 is the strongest of all three sources. The compact core is very bright (S=800mJy, $\theta < 0.2$mas) with a peak above 10 GHz and currently it might be undergoing an outburst. The bright mas jet suggests minor outbursts in about 1983-86. We have not mapped the second epoch data but a comparison with earlier 5 GHz maps confirms the direction of motion and velocities (see Fig.3) reported earlier by Wilkinson et al. (1991). The observed motion seems to be a combination of ballistic and helical flows. Presumably an initial ballistic ejection from a precessing central engine changes to quasihelical flow due to interaction with a dense medium. Differences between the strengths of the cores in the above sources could be ascribed to differences in inclination angle. In general the relatively weak cores observed in CSS sources are probably seen at large angle to the LOS ($> 15°$) and as reported recently by Wills et al. (1992) the low polarisation (as compared with core dominated sources) is observed in visible light. This could also suggest that Doppler beaming is less important.

References

[1] Fabian et al., *Cooling flows in clusters of galaxies*, Nature, **310**, 30, 1984.
[2] Fanti et al., Astron. Astrophys., **231**, 333, 1990.
[3] Kus et al., *Parsec-Scale Radio Jets*, Cambridge University Press, 1990.
[4] Peacock and Wall, Mon.Not.R.astr.Soc., **194**, 331, 1981.
[5] Wilkinson et al., Mon.Not.R.astr.Soc., **248**, 86, 1991.
[6] Wills et al., Astrophys.J., **398**, 454, 1992.

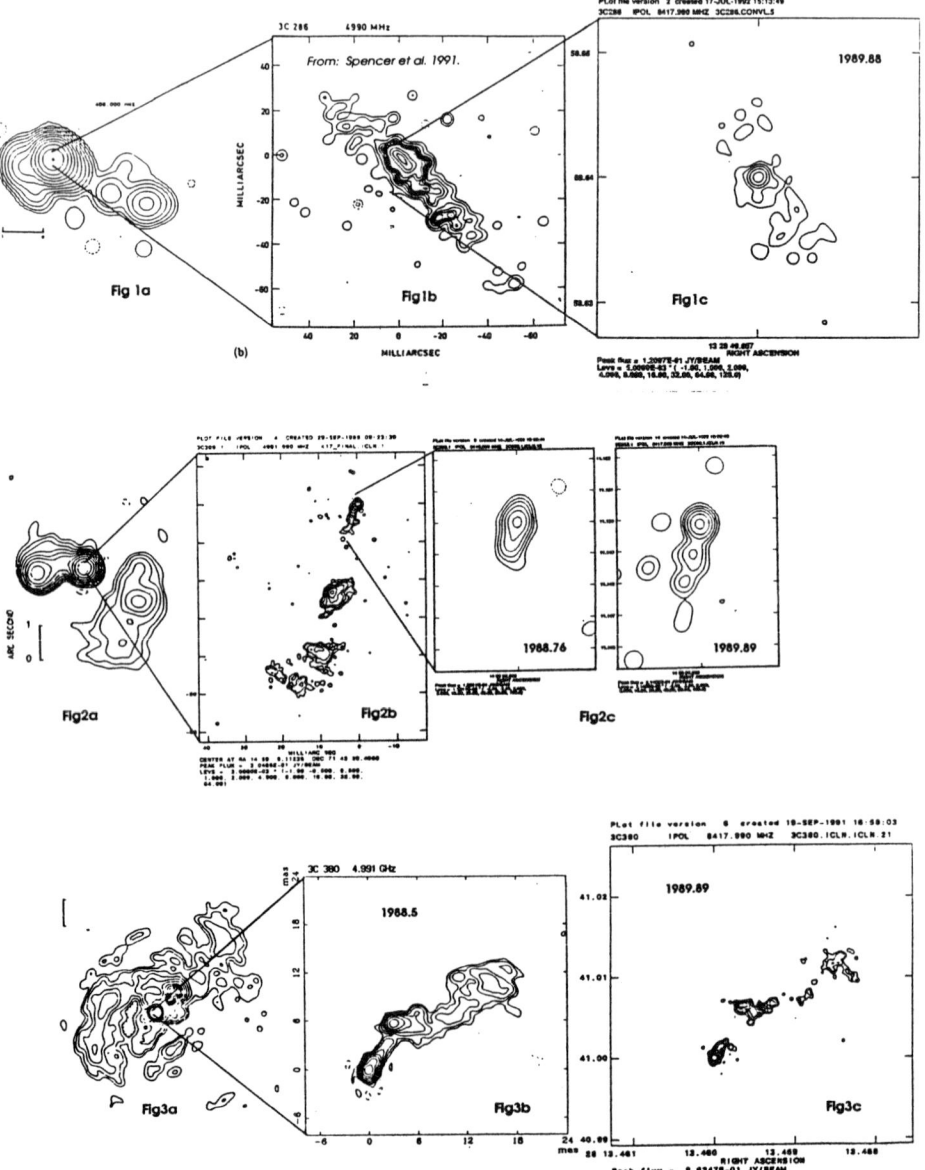

3C380: Motion down a twisted channel; not a second nucleus

Antonis G. Polatidis * *Peter N. Wilkinson** *Chidi E. Akujor* [†]

The quasar 3C380 (B1828+48.5) (z=0.692) is a Compact Steep Spectrum Source with a convoluted kiloparsec–scale structure (Wilkinson et al. 1991) [1]. The milliarcsec–scale structure is a one–sided core–jet structure which extends over 25 mas ($100h^{-1}$ pc). Wilkinson et al. (1990) [2] present VLBI maps at 6 cm from 2 epochs (1982.9 and 1988.4). They note significant changes, the most dramatic being the brightening of component 'A' in the jet by a factor of 2. In 1988 'A' also appeared to become detached from the underlying jet. [2] suggested that 'A' might be a second nucleus that became active between 1982.9 and 1988.4. If both 'C' and 'A' were active nuclei they should have flat spectra and appear unresolved at high frequencies.

Here we present recent VLBI images made at 18 cm, 1.3 cm and a third epoch image at 6 cm. All of the observations were recorded using the Mark 2 system and were cross–correlated at the Caltech/JPL–Mark2 correlator.

The third epoch 6 cm observations were made in November 1990 (epoch 1990.8) with 9 telescopes. Figures 1a and 1b show the 1982.9 and 1988.4 images from [2] and figure 1c shows the 1990.8 image. For consistency all three epoch maps are convolved with the same circular beam of 1 mas. In 1990.8 the core has brightened by a factor of 2 while the flux of A has remained the same. Component 'A' has moved away from the core by about 0.9 ± 0.06 mas between 1988.4 and 1990.8, implying a velocity $u = 8.8 \pm 1.4h^{-1}c$ (for $H_0 = 100h$ Km sec^{-1} Mpc^{-1} and $q_0 = 0$). Further comparison of the three 6 cm images shows that the region 'B' has been moving outwards with a velocity $u \simeq 6 \pm 1h^{-1}c$.

The 18 cm observations were made in Nov 1989 (epoch 1989.8) with 9 telescopes. Figure 2 shows the 18 cm image convolved with a 2.5 mas circular beam. The image shows a structure consistent to that observed at 6 cm. Component 'A' is clearly detected at a position angle of $\simeq 30°$ although it is not detected in an earlier 18 cm map of Readhead and Wilkinson (1980) [3].

Figure 3 shows the 1.3 cm image from the November 1990 (epoch 1990.8) data taken with 9 telescopes. The image (beam = 0.36 × 0.26 mas) shows a bright unresolved core and a jet that emerges at a position angle $-15°$ and bends to $-32°$. The extent of the jet is 2.2 mas. Component 'A' is clearly resolved and is edge–brightened to the East. Component 'C', unresolved at 1.3 cm, has a flat spectrum between 18 cm and 1.3 cm (spectral index $\alpha_{1.3}^{18}=$ -0.15) and can be unambiguously identified with the core.

---

*University of Manchester, Nuffield Radio Astronomy Laboratories, Jodrell Bank, Macclesfield, Cheshire, SK11 9DL, U.K. AGP was partially supported by a British Counsil Research Fellowship
[†]Onsala Space Observatory, Onsala S–439 92, Sweden

In contrast 'A' is a resolved, steep spectrum ($\alpha_{1.3}^{18}=0.76$) component and is certainly not a second active nucleus. The kinematics and the flux evolution of 'A' suggest instead that it is a component in the jet moving down a fixed 'twisted' channel (see also Kus *et al.*, this proceedings).

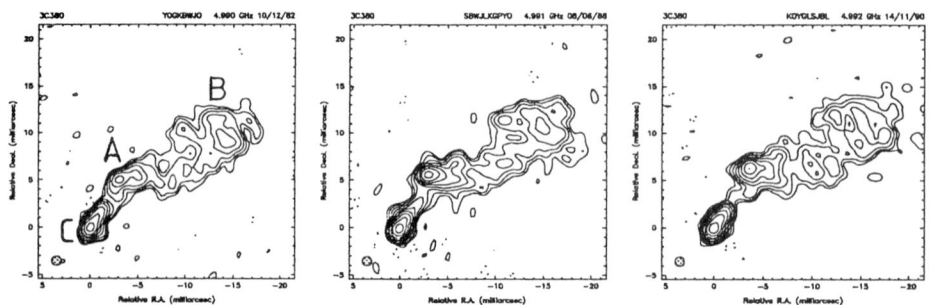

Figure 1. 6 cm maps from epoch 1982.9 (a), 1988.4 (b) and 1990.8 (c). Contours are at 1, 2, 4, 8, 16, 32, 64, 128, 256 × 3σ (3σ = 2.7, 1.8 and 2.7 mJy/beam respectively)

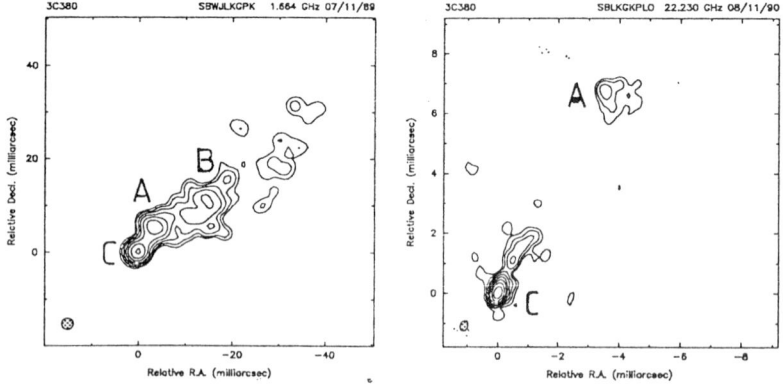

Figure 2. (a) The 18 cm and (b) the 1.3 cm maps. Contours are at 1, 2, 4, 8, 16, 32, 64, 128, 256 × 3σ (3σ = 3.0 and 1.7 mJy/beam respectively)

References

[1] Wilkinson P.N., Akujor C.E., Cornwell T.J., Saikia D.J., 1991, *MNRAS*, **248**, 86.

[2] Wilkinson P.N., Tzioumis A.K., Akujor C.E., Benson J.M., Walker R.C. & Simon R.S, 1990. In *Parsec-Scale Radio Jets*, 152, eds. Zensus J.A. & Pearson T.J., Cambridge University Press.

[3] Readhead A.C.S. & Wilkinson P.N., 1980, *Ap.J.*, **235**, 11.

Core Activities of Compact Steep-Spectrum Radio Sources

Seiji Kameno [*]   *Makoto Inoue* [†]   *Hiroshi Takaba* [‡]
Rendong Nan [§]   *Rechard T. Schilizzi* [¶]

Abstract

We report the observational results of Compact Steep-Spectrum sources (CSS's) by VLBI at 22 and 43 GHz and single dish at 22, 43 and 92 GHz. We mention to the core activities of CSS's.

1. Introduction

As summarized by Saikia (1989) and Fanti et al. (1990), Compact Steep-Spectrum radio sources (CSS's) are characterized by both compact linear sizes ($\leq$10 kpc) and steep spectra ($\alpha \leq -0.5; S \propto \nu^{\alpha}$). Although recent high resolution cm-wave maps have revealed complex jets and lobes of CSS's (e.g. Fanti et al. 1985), but there is little information about their core activities. Thus it is important to obtain mm-wave spectra and to search flat-spectrum cores of CSS's.

2. Observations and Results

We observed 18 of the sample of 46 CSS sources defined by Fanti et al. (1990) except 3C295. We used the Nobeyama 45-m and the Kashima 34-m Radio telescopes for 22.2- and 42.8-GHz VLBI. This 200-km east-west baseline provides minimum fringe spacing of 14 mas at 22 GHz and 7 mas at 43 GHz. We used K-4 recording terminals with bandwidth 2 MHz × 16 channels and made correlations by National

---

[*]Department of Astronomy, Tokyo University, Bunkyo-ku, Tokyo 113, Japan.
[†]Nobeyama Radio Observatory, National Astronomical Observatory, Minamimaki-mura, Minamisaku-gun, Nagano 384-13, Japan.
[‡]Kashima Space Research Center, Communications Research Laboratory, 893-1 Hirai, Kashima, Ibaraki 314, Japan.
[§]Beijing Astronomical Observatory, Academia Sinca, Beijing, Peoples Republic of China.
[¶]Netherlands Foundation for Research in Astronomy, Radiosterrewacht Dwingeloo, Postbus 2, 7990 AA Dwingeloo, The Netherlands.

Astronomical Observatory's COrrelator (NAOCO). Fringes were detected in 9 of 18 sources at 22 GHz and 4 at 43 GHz. At Nobeyama we also measured the integrated flux densities at 22, 43 and 92 GHz. Calibrations of the flux densities were carried out by observations of Mars and 3C273.

3. Discussion

Fig. 1 is a typical example of CSSs' spectra (3C380). The integrated spectra of most CSS's remain steep even in mm-waveband. This is consistent with Steppe et al. (1990). Assuming that VLBI correlated flux densities correspond to core emissions, their spectra are flat or inverted. This supports high core activities on CSS's. However, they are still lobe-dominant so that their integrated spectra are straight.

We find a good correlation between the spectral index in mm waves and the VLBI flux density to integrated one (Fig. 2). The correlation coefficient is about 0.80. This supports that a CSS has a flat spectrum core and steep spectrum lobes like extended radio galaxies, and the intensity ratio of these components results in the integrated spectral property. Thus compactness of CSS's is interpreted by compact radio lobes.

References

[1] Fanti, C. et al: 1985, *Astron. Astrophys.*, **143**, 292.

[2] Fanti, R. et al: 1990, *Astron. Astrophys.*, **231**, 333.

[3] Saikia, D.J.: 1989, in H.R. Miller, P.J. Witta (eds.) *Active Galactic Nuclei*, Springer-Verlag, P. 317

[4] Steppe, H. et al: 1990, in C. Fanti, R. Fanti, C.P. O'Dea, R.T. Schilizzi (eds.) *Compact Steep-Spectrum and GHz-Peaked Spectrum Radio Sources*. Consiglio Nazionale delle Ricerche Istituto di Radioastronomia, Bologna

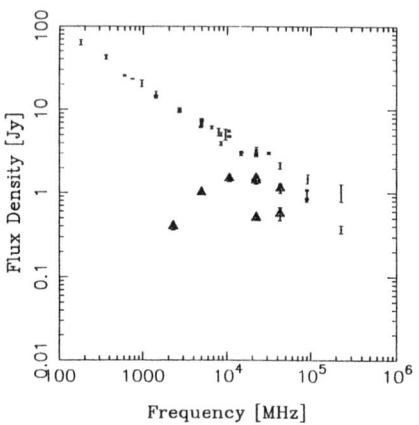

Figure 1: The radio spectrum of the CSS 3C380. Integrated flux density (*dot*) and VLBI correlated flux density (*triangle*) are plotted. VLBI spectrum is due to a flat-spectrum active core.

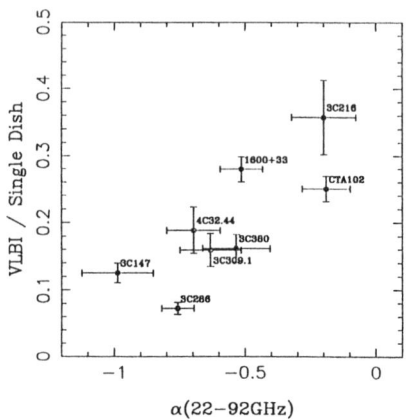

Figure 2: The diagram between spectral index in mm waves (*holizontal*) and VLBI-to-integrated flux density ratio(*vertical*). Spectral indices are derived from the flux densities at 22, 43, and 92 GHz.

Compact Steep-Spectrum Radio Sources from the Peacock & Wall Catalog

D. Dallacasa, C. Fanti, R. Fanti, [*]   R. T. Schilizzi [†]
R. E. Spencer [‡]

Abstract

Here we comment on the results of VLBI observations at 18 cm of a sample of Compact Steep-Spectrum (CSS) radio sources selected from the Peacock and Wall catalog [PW81]. Jets and core candidates are present in most structures. There is morphological evidence for strong interaction of the radio emitting plasma with the surrounding medium.

1. Introduction

The Compact Steep-Spectrum radio sources are small (<15 kpc), extremely powerful and distant objects (z> 0.2), with intrinsic subgalactic dimensions. They are supposed to be strongly confined by a dense environment which prevents expansion and produces distortions in the structures and a great amount of depolarization. Statistical arguments are against the possibility that CSS radio sources are indeed large size objects seen at small angles to the line of sight [Fa90]. Therefore their characteristics are related to the physical conditions of the environment and to the jet interaction with the ambient gas. Unfortunately most of the optical counterparts are faint objects and the spectroscopic work is therefore hard to do. Hence, most of the observational information comes from the radio band, and in particular from VLBI owing to the compactness of the CSS's. High sensitivity VLA observations have shown the presence of haloes or extended components for a few CSS's. Their contribution to the total flux is marginal, but has to be taken into account when considering the mechanism that govers the evolution of these radio sources. Here we summarize some results derived from a study of 16 sources extracted from the "Peacock & Wall extension" of the 3CR sample of CSS's, listed in Table 1. Mark-2 VLBI observations at 18 cm, mapped by means of AIPS, have revealed new aspects such as the presence of prominent jets in both quasars and galaxies.

---

[*]Istituto di Radioastronomia-CNR, Bologna, Italy.
[†]NFRA-Dwingeloo, The Netherlands.
[‡]NRAL-Jodrell Bank, Macclesfield, Cheshire, U.K.

TABLE 1. CSS's from the PW Catalog			
0223+341	0316+161	0319+121	0345+337
0404+768	0428+205	1225+268	1323+321
1358+624	1413+349	1442+821	1600+335
1819+39	1829+29	2230+114	2342+821

2. Results and Discussion

The sources belonging to the sample discussed here (the "PW extension") are in general smaller and have on average a higher turnover frequency (the two things are correlated) than the 3CR CSS's. They were included in our study in order to tackle the bias against strongly self-absorbed objects in the 3CR catalog, due to the selection at low frequency (178 MHz). The 3CR and the PW extension therefore form a complete sample of about 50 CSS's.

The PW morphologies reveal some differences from the scheme drawn up for the 3CR CSS's. There is a wide range of structures (core-jet, double, triple, core-lobe, complex), and in particular there are bright jets and core candidates in the PW radio galaxies, while the relevance of the lobe contribution to the total flux density is much lower than in 3CR CSS's. This may be a selection effect due to the compactness of the PW CSS's.

Most sources show features which could be attributed to the interaction with the ambient medium, and this occurs at the very beginning of the radio jet, and even very close to the candidate nucleus, on scales of few tens of pc or even smaller. At least 2/3 of the sources discussed here show misalignment between components of more than $10°$.

One example is the galaxy 1819+39 (fig. 1) in which a 400 pc-long jet dominates the VLBI emission and is resolved transverse to the jet. Its structure is characterized by several bends, is limb brightened, and the northern end points towards a relatively faint lobe (about half arcsecond away) which is completely resolved out.

On the other hand, there are sources that do not show such behaviour. For example 1413+349 (no optical identification is available) in fig. 2 which has a linear core–jet structure without any significant perturbation, though there is some low surface brightness emission on the west side of the core candidate. The morphological segregation between galaxies and quasars in the PW sample does not seem to be as clear as for the 3CR CSS's. Although the optical work is still incomplete and some identifications are uncertain or changing (see [DP90]), it is clear that there are galaxies with relatively strong cores and bright jets, accounting for most of the total flux density. This is in contrast to 3CR CSS galaxies which have faint cores

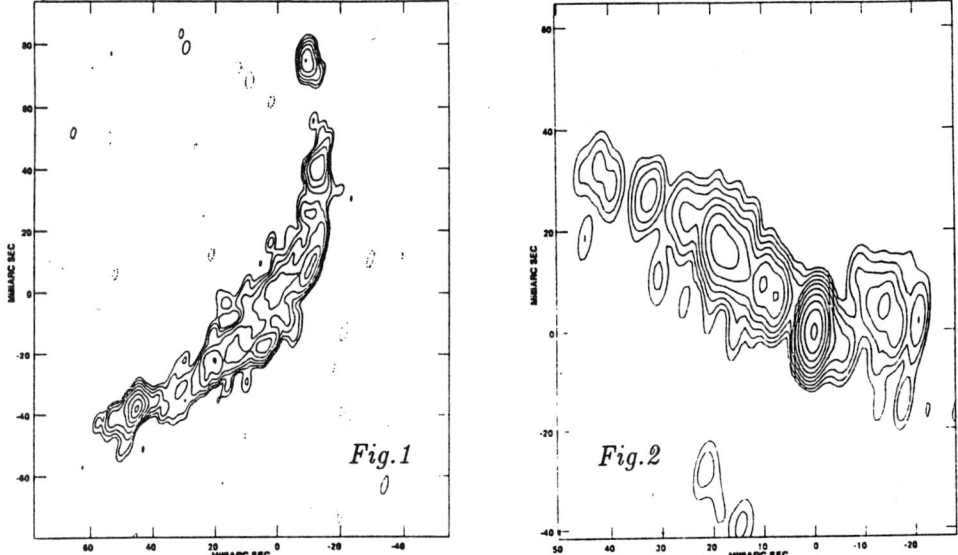

Fig.1: VLBI map of 1819+39 at 18 cm. HPBW= 5.2x2.2 mas, p.a.= -14 deg.; contour levels are: -3, -1.5, 1.5, 3, 6, 12, 25, 50, 100, 180 mJy/beam. Peak= 196.3 mJy/beam. *Fig.2:* VLBI map of 1413+349 at 18 cm. HPBW= 8.0x2.3 mas, p.a.= - 9 deg.; contour levels are: -1, 1, 2, 4, 8, 15, 30, 60, 125, 250, 500 mJy/beam. Peak= 574.6 mJy/beam.

and jets which are not particularly prominent. The occurrence of powerful pc-scale jets is somehow related to the turnover frequency, being more prominent for Giga-Hz Peaked Spectrum (GPS) radio sources (a subclass of CSS), rather than in 3CR CSS's.

References

[Fa90] Fanti R., Fanti C., Schilizzi R.T., Spencer R.E., Nan Rendong, Parma P., van Breugel W.J.M., Venturi T.: (1990), *Astron. & Astrophys.*, **231**, 333.

[PW81] Peacock J.A. & Wall J.V.: (1981), *Mon. Not. R. Astr. Soc.*, **194**, 331.

[DP90] Dunlop J.S. & Peacock J.A.: (1990), *Mon. Not. R. Astr. Soc.*, **247**, 19.

VLBI Observations of Compact Sources with a Limited Number of Antennas

D. Dallacasa, M. Bondi, C. Stanghellini, P. L. Cerchiara *
G. Umana, C. Trigilio, [†]   F. Rantakiro [‡]

1. INTRODUCTION

A hundred sources selected from different samples studied in Bologna, containing Low Frequency Variable radio sources, Compact Steep Spectrum radio sources, GHz Peaked Spectrum radio sources and radio Quasars have been observed at 5 GHz with the MkII recording system, using the two Italian VLBI antennas of Medicina and Noto. The 25 meter Onsala antenna joined the experiment for about 2 of the 7 days of observation.

2. 2. RESULTS and DISCUSSION

In the following we present a brief summary of the results obtained. We obtained the maps of the objects observed with 3 antennas and for a minimum of 3 scans. Comparison with other maps or relevant parameters from literature for the sources previously observed, tests the reliability of the structures seen in the images.

0108+388. The map (Fig. 1a) shows two components separated by 5.4 mas in p.a. 60°. The peak of northeastern component is about twice that of the central component. The same separation and a similar ratio between the peaks of the two components are inferred by a VLBI global map at the same frequency [C90].

1358+624 (4C62.22). The map (Fig. 1b), clockwise rotated by 45°, shows a compact component and a straight jet nearly 50 mas long. Maps at 18 [D93] and 50 cm [P91] show a very similar morphology. We identify the core with the compact emission at the right edge of the source. Model fitting for that component gives a size of 2.0x1.0 mas in p.a. 127° (unrotated) and an integrated flux density of 0.59 Jy. The total flux density of the sources is accounted both in model and map.

---

*Istituto di Radioastronomia-CNR, Bologna, Italy
[†]Istituto di Radioastronomia-CNR, Noto, Italy
[‡]Onsala Space Observatory, Onsala, Sweden

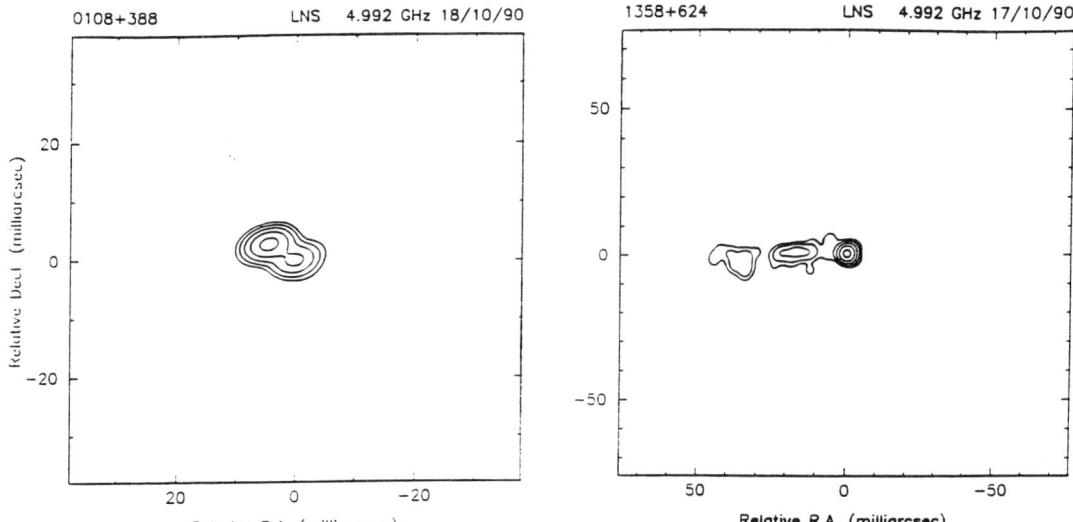

The contour levels of the maps are -5,5,10,20,40,80 % of the peak for each source. Fig. 1a: 0108+388; Beam 5.4 × 3.3 mas p.a. −74°. Peak = 0.75 Jy/beam. Fig. 1b: 1358+624; The map is rotated CW by 45°; beam 4.4 × 4.0 mas p.a. 30°. Peak = 0.53 Jy/beam.

The lack of short spacings and the relatively low sensitivity do not allow mapping of faint and relatively extended components. The quality of the map is correlated to the source complexity: in fact, the best results are obtained for sources like 0108+388 or 0333+321 in which all the emission comes from very well defined and compact regions. Other sources with complex or wide structure, are mapped with a lower fidelity. However all this may not be required by some projects like this or the monitoring of strong and compact sources.

References

[C90] Conway, J.E., Unwin, S.C., Pearson, T.J., Readhead, A.C.S., Xu, W., 1990 *Proc. of Dwingeloo Workshop on CSS and GPS Radio Sources*, eds. Fanti, C, Fanti, R., O'Dea, C.P., Schilizzi, R.T., p. 157.

[P91] Padrielli, L., Eastman, L., Gregorini, L., Mantovani, F., Spangler, S.R., 1991 *Astron. Astrophys.* **249**, 351.

[D93] Dallacasa, D., Fanti, C., Fanti, F., Schilizzi, R.T., Spencer, R.E., 1993 in preparation.

The Quasar 4C39.25 on A Scale of Tens of Mas

Shengyin Wu *   *Ivan I.K. Pauliny-Toth* and *Richard W. Porcas* [†]

Abstract

The preliminary results of an EVN–Shanghai VLBI observation of 4C39.25 at 18 cm are presented in this paper. They show a complicated morphology of the source on a scale of hundreds of parsecs and the possible existence of superluminal motion on this scale. A model of an uncoplanar, undulating jet is implied by the results.

A VLBI observation of 4C39.25 at 18 cm was conducted at epoch 1989.83 with 7 European telescopes and the Shanghai 25-m dish. The hybrid maps imaged with the data from European baselines alone and those from all baselines together, with corresponding restoring beams are shown in Figs. 1 and 2 respectively. The contour levels are -0.5, 0.3, 0.5, 1, 2, 5, 10, 20, ..., 90% of the peak in the figures. Fig. 1 indicates a much more complex morphology than that at 1984.26 (see [WPP87]). In addition to an obvious extension along P.A.=$-108°$ and $82°$ which is consistent with the direction of elongation found at the previous epoch, there is significant extension and curvature along P.A.=$-13°$—$-32°$. Fig. 2 shows even more complicated structure and some distinct features. The eastern features are distributed in about the same position angle as that of the compact center and the arcsec–jet (see [MAE90]). Faint features are still found and confirmed by both hybrid mapping and model fitting at distances of 10 and 21.3 mas and position angles of $-32°$ and $-20.5°$ respectively.

Our result, compared with those in [ZMA90,MAE90], could provide some indications of variation of the morphology on scales of hundreds of pc. If the stronger eastern component (5.2 mas from the compact center) is identified as that revealed at 6cm (1988.44) separated 4.3 mas from the compact centre, superluminal motion of this component with $\beta_{app} = 5.67$ could be deduced, taking account of the spectral dependence of the separation (16:18) [WPP87]. The position of the weak **NW** component found at 1984.26 was between the two **NW** features revealed this epoch, but it is hard to identify it as either **NW** feature. Considering that all **NW** components appear successively in a curvilineal trace, and that at least two possible components (b of [MAE90] and the eastern component mentioned) move superluminally eastwards on different scales, our observations lend support to the idea that the "core" in 4C39.25

---

*Beijing Astronomical Observatory, Chinese Academy of Sciences, Beijing 100080, P.R. China. This author was partially supported by an honorarium from the MPG in Germany and a grant # 1860621 from the NSFC in China.

[†]Max–Planck–Institut für Radioastronomie, Auf dem Hügel 69, 5300 Bonn 1, F.R. Germany.

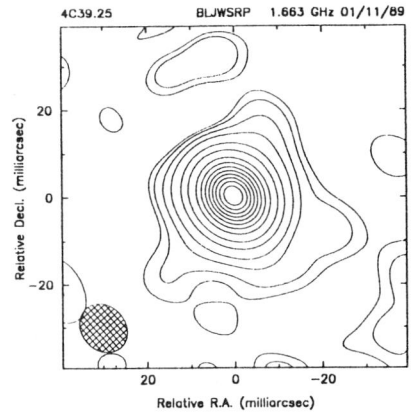 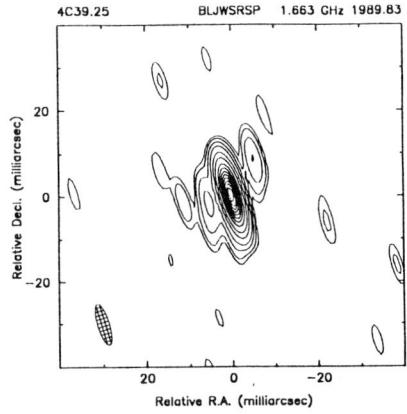

Fig. 1. The hybrid map of 4C39.25 from EVN data.

Fig. 2. The hybrid map of 4C39.25 from all baselines.

is invisible and located somewhere to the north–west of the compact triple [MAE90].

The variation of the position angle with respect to the compact centre with the distances of components is shown in Table 1 by combining the positions of components on different scales found at recent epochs (our VLBI and MERLIN observations and [ZMA90,MAE90]). The total curvature is 97° from the remote NW component to the most eastern, arcsecond hot spot (the latter will be published soon).

Table 1. Variation of the P.A. with the radii of components in 4C39.25.

	Eastern Features				a,b,c		NW Features		
R (mas)	3000	1000	11.3	5.2	1.89	2.64	10.1	18	21.3
P.A. (°)	77	83	98	110	-81	-73	-32	-26	-20.5

The morphology of the source and coexistence of several superluminal and stationary components can be explained using a spatial (or 3–dimensional) undulating jet with an invisible core and misaligned components, that involves a slight modification to the model suggested in [MAE90].

Acknowledgement. We are grateful to Dr. Ch. Hummel for his advice and helpful discussions regarding data and image processing, and to Dr. P. Wilkinson for his valuable suggestions as to reimaging the morphology.

References

[MAE90] Marcaide,J.M., Alberdi,A., Elósegui,P., et al., *Detection of a new component in the peculiar superluminal quasar 4C39.25*. Parsec–Scale Radio Jets (1990), pp. 59–65. ed. by J.A. Zensus and T.J. Pearson.

[WPP87] Wu,S.Y., Pauliny-Toth,I.I.K., and Porcas,R.W., *VLBI observations of 4C39.25 at 18 and 6 cm waves*. Chin. Astron. Astrophys. Vol. 11 (1987), pp. 201–206.

[ZMA90] Zhang,Y.F., et al., *4C39.25, A twisted compact jet?* Parsec–Scale Radio Jets (1990), pp. 66–70. ed. by J.A. Zensus and T.J. Pearson.

Twisted Pc–Scale Structure in the BL Lacertae Object Mrk 501

J.M. Wrobel and J.E. Conway *

Abstract

The BL Lacertae object Mrk 501 is hosted by the $m_{pg} = 13.7$ mag elliptical galaxy UGC 10599 (Abraham, McHardy, & Crawford 1991, MNRAS, 252, 482). The radio continuum structure of Mrk 501 is puzzling. It exhibits a one–sided "core–jet" whose elongation position angle (PA) appears to twist through about 79° between 2 and 100 milliarcsecond (mas), or 1 and 46 h^{-1} pc, as revealed in published global VLBI, EVN, and VLA images (van Breugel & Schilizzi 1986, ApJ, 301, 834; Pearson & Readhead 1988, ApJ, 328, 114). The origin of this dramatic structural twisting on pc scales is difficult to investigate with existing radio data, since no image adequately follows the twist on scales of tens of mas. To remedy this problem, 6 cm continuum data on Mrk 501 were acquired in 1991 with the enhanced MERLIN and with the partially constructed NRAO VLBA, augmented by one VLA antenna. Images from these data sets are shown in Fig. 1. These images are being used to follow the details of the structural twist, and to interpret both the twist and other properties of Mrk 501 within the context of three possible models for the pronounced observed curvature: (1) precession of mildly relativistic twin jets; (2) deflection and brightening by strong dynamical interaction with clumped ambient gases; and (3) helical distortions caused by Kelvin–Helmholtz instability of a light relativistic jet expanding in response to an external pressure gradient. NRAO is operated by Associated Universities, Inc., under cooperative agreement with the National Science Foundation.

Fig. 1. – Naturally weighted images of Mrk 501 (1652+398). Image contours are logarithmic, with 7 given per decade. Lowest positive (solid) contour is 0.15% of peak. Negative contours are dashed. Boxed ellipse indicates Gaussian restoring beam FWHM cross–section. Scale is 100 mas = 46 h^{-1} pc. (a) RCP image from 10–hour track with 6 MERLIN antennas. Center frequency and bandwidth are 4996 MHz and 7 MHz, respectively. Peak is 881 mJy beam^{-1}. (b) LCP image from 16–hour track with 4 VLBA antennas plus one VLA antenna. Center frequency and bandwidth are 4992 MHz and 2 MHz, respectively. Peak is 695 mJy beam^{-1}.

---

*NRAO, P.O. Box O, Socorro, New Mexico, USA 87801

Fig. 1a – Mrk 501 MERLIN

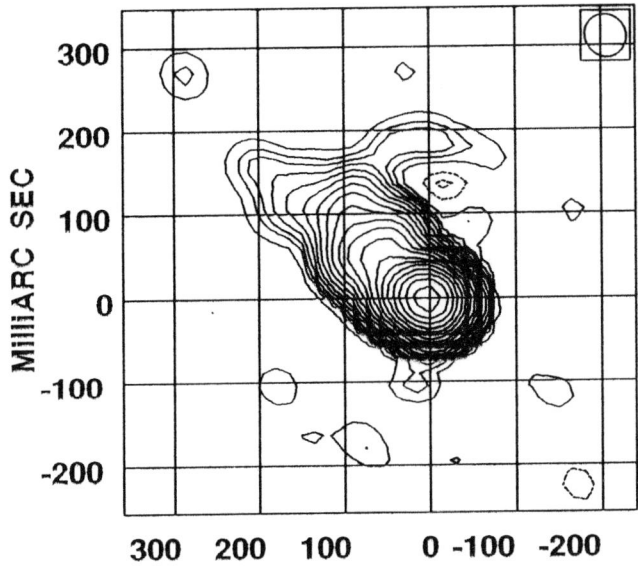

Fig. 1b – Mrk 501 VLBA

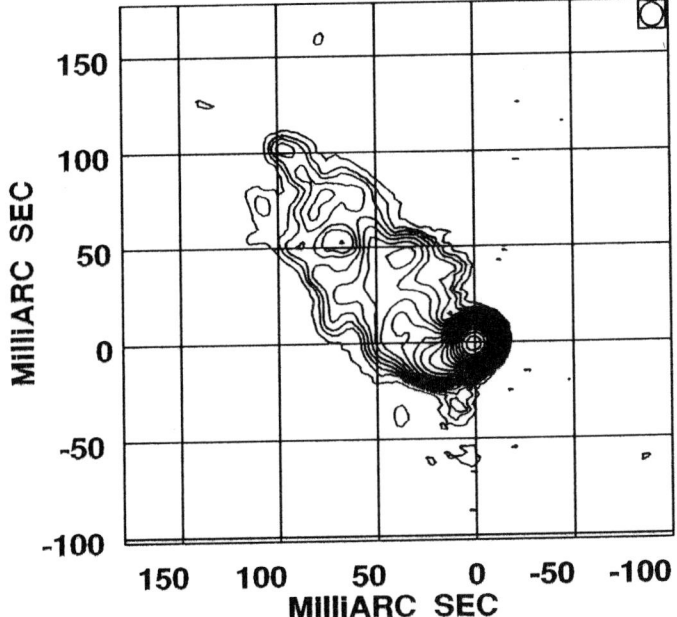

A New VLBI Image of the Archetypal CSS Source 3C286 at 5 GHz

F.J. Zhang [*]   R.E. Spencer [†]   R.T. Schilizzi [‡]   C. Fanti [§]

Abstract

Studies of CSS sources over the past decade have yielded considerable information on theirmorphologies, spectral characteristics, and statistics [1, 2, 3, 4]. Many of these objects show complex structures on a wide range of angular scales from a few mas to hundreds of mas. We have made a detailed investigation of 3C286 in order to understand further the detailed physics and to study possible interaction of the radio emission with the ambient medium.

The main characteristics of 3C286 are typical for the CSS QSO class. It is one of the most luminous CSS's, and has previously been observed with the Cambridge 5-km telescope, VLA, MERLIN, and VLBI. The morphology of 3C286 appears to be intermediate between the core-jet types (like 3C138) and the grossly distorted objects (e.g. 3C380). A few maps obtained from previous VLBI observations show that the structure of 3C286 is symmetric: the jet extends in the direction of about $-135°$ and is replicated on the opposite side; the core consists of two components oriented in the same position angle as the extended structure [5, 7]. With a VLBI array including EVN and Haystack, 3C286 was observed in October, 1989 (epoch of 1989.8). The results show a complicated structure in the 'core' component: the bifurcation of the central region is readily apparent, and the jet appears as a sinuous structure [9]. But the dynamic range is very low when a single 'outrigger' telescope is in the array. In order to make a detailed study of the 'core', and confirm the wiggle seen in the jet, a new observation with a large global MK2 array at 5 GHz was executed again at epoch of 1990.9 (in November, 1990). The array consisted of EVN, USVN, VLA (tied array), partial VLBA telescopes, and some Russian telescopes. 3C345 and OQ208 were used as fringe finder and calibrator repectively.

---

[*]Shanghai Observatory, Shanghai, China.
[†]Nuffield Radio Astronomy Laboratories, Jodrell Bank, UK.
[‡]NFRA, Dwingeloo, Netherlands.
[§]Instituto di Radioastronomia del CNR, Bologna, Italy.

When calibration was done we found that OQ208 is not an ideal calibrator because this source can be resolved on both intercontinental baselines and long U.S. transcontinental baselines. The initial map of 3C286 at epoch of 1990.9 is shown in Fig.1. The restoring beam and peak value of Fig.1 are 3.0×1.0 mas and 3.26×10^{-1} Jy/beam respectively. Contours are at levels: $(-1.0, 1.0, 3.0, 5.0, 7.0, 10.0, 15.0, 20.0, 30.0, 50.0, 70.0, 90.0) \times 3.26 \times 10^{-3}$ Jy/beam. The central region of 3C286 is clearly resolved into two components designed C1 and C2. Their dimensions are about the same. The separation between C1 and C2 is ~ 5.9 mas. The peak values of flux density for both C1 and C2 are 327 mJy and 310 mJy respectively. Component C2 is located on the north-east side of C1 and is a little weaker than C1. Figure 2 shows a one–dimensional slice which passes through the centers of both C1 and C2. It is difficult to determine which component is the real core of 3C286. New 1.67 GHz VLBI data obtained from an observation of a large global VLBI array is being processed. We should obtain much information for both soectral index and structure after its mapping is finished. This will help us tp put forward a reasonable model.

The jet extends in the direction of about $-135°$. Its sinuous structure is very clear. There is no counter–jet found on the north–east side at the level of our dynamic range on the map. The behaviour of the jet suggests that there could be a relativistic precession occurring in the center and fluid material with helical motion along the jet. The wiggles and bends in the source are intrinsically large. Relativistic beaming may not be important if the source is not aligned close to the line of sight. Another possibility is that the wiggles are apparent and due to relative limb brightening of a wider jet as seen in M87. If this is true, 3C286 is only the second source to show limb brightening.

References

[1] Fanti C., et al., 1985, *A&Ap*, **143**, 292

[2] Fanti C., et al., 1989, *A&Ap*, **217**, 44

[3] Fanti C., et al., 1990, *A&Ap*, **231**, 333

[4] Pearson T.J., Perley R.A., & Readhead A.C.S., 1985, *Astr. J.*, **90**, 738

[5] Phillips R.B., & Shaffer D.B., 1983, *Ap. J.*, **271**, 32

[6] Spencer R.E., et al., 1989, *MNRAS*, **240**, 657

[7] Spencer R.E., 1990, private communication

[8] van Breugel W.J.M., Miley G., & Heckman T., 1984, *Astr. J.*, **89**, 5

[9] Zhang F.J., et al., 1990, in C. Fanti, R. Fanti, C.P. O'Dea & R.T. Schilizzi, eds, CSS and GHz–Peaked Spectrum Radio Sources, p.107

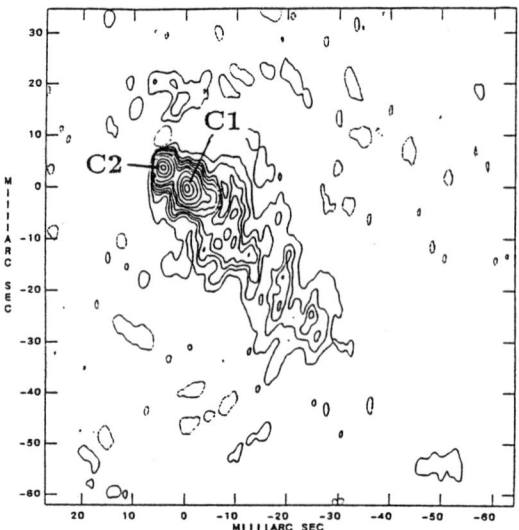

Figure 1. VLBI map of 3C286 at 5 GHz

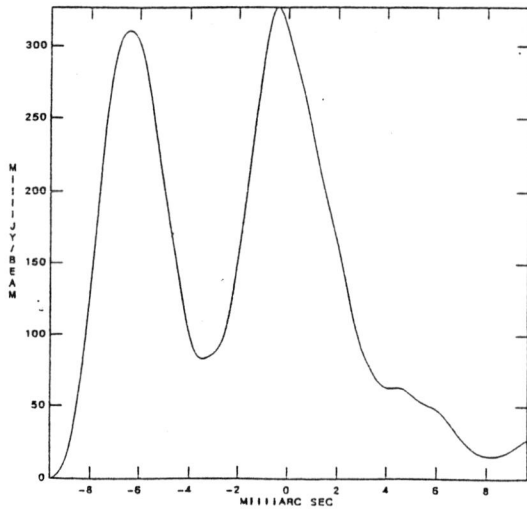

Figure 2. A one-dimension slice which passes through the centres of both C1 and C2

VLBI Structure of 3C395 at 2.3 and 8.4 GHz: Observations and Numerical Simulations

L. Lara [*]   A. Alberdi [*]   J.L. Gómez [*]   J.M. Marcaide [†]
T.W.B. Muxlow [‡]

Abstract

New maps of 3C395 at 2.3 and 8.4 GHz are presented. With the help of a jet emission numerical code and all available observational results, the physical parameters of the relativistic jet are constrained. The positional shift of the opaque core at 2.3 GHz with respect to 8.4 GHz is estimated, allowing us to obtain an spectral-index map of the source.

On November 1st 1990 the quasar 3C395 (m_v=17.5, z=0.635) was observed with a global array of 8 antennas. The observations were carried out simultaneously at 2.3 and 8.4 Ghz using the Mark III recording system in mode A, assigning half of the synthesized bandwidth to each frequency [2]. We present the resulting maps from each frequency data set. Our results confirm once more component 2 (see Fig. 1 and 2) as stationary relative to component 1, the core. Apart from undersampling in the monitoring of component 3, our maps seem to confirm the strong deceleration in the superluminal motion of this component as reported by Simon et al. [4]. Additionally, our map at 8.4 GHz also shows a new component emerging from the core.

We have used a relativistic jet emission numerical code [1] in order to model 3C395. Such code calculates the emission of bent shocked relativistic jets considering an input set of geometric and physical parameters. The model parameters can be iteratively improved to make the model emission match the observations. Component 3 can be reproduced with a plane shock wave travelling along a jet that is turning away from the line of sight. In order to simulate the increase in flux density associated to component 2 and its stationary character, we consider that in the region associated to this component the jet is almost aligned to the observer.

---

[*] Instituto de Astrofísica de Andalucía, Apdo. 3004, 18080 Granada, Spain.
[†] Dpto. de Matemática Aplicada y Astronomía, Universitat de València, 46100 Burjassot, Spain.
[‡] NRAL, Jodrell Bank, Macclesfield, Cheshire SK11 9DL, England.

Our numerical solution provides frequency dependent core positions as a result of changes in the optical depth. This dependence can only be measured through accurate astrometry with respect to an external reference, work which is underway. Based on the 0.7 mas shift between 2.3 and 8.4 GHz indicated by the numerical model and by a similar result measured by Marcaide and Shapiro [3] on another source, we have introduced such a shift in our observed maps and we have constructed the spectral-index map of the compact structure of 3C395 shown in Fig.3.

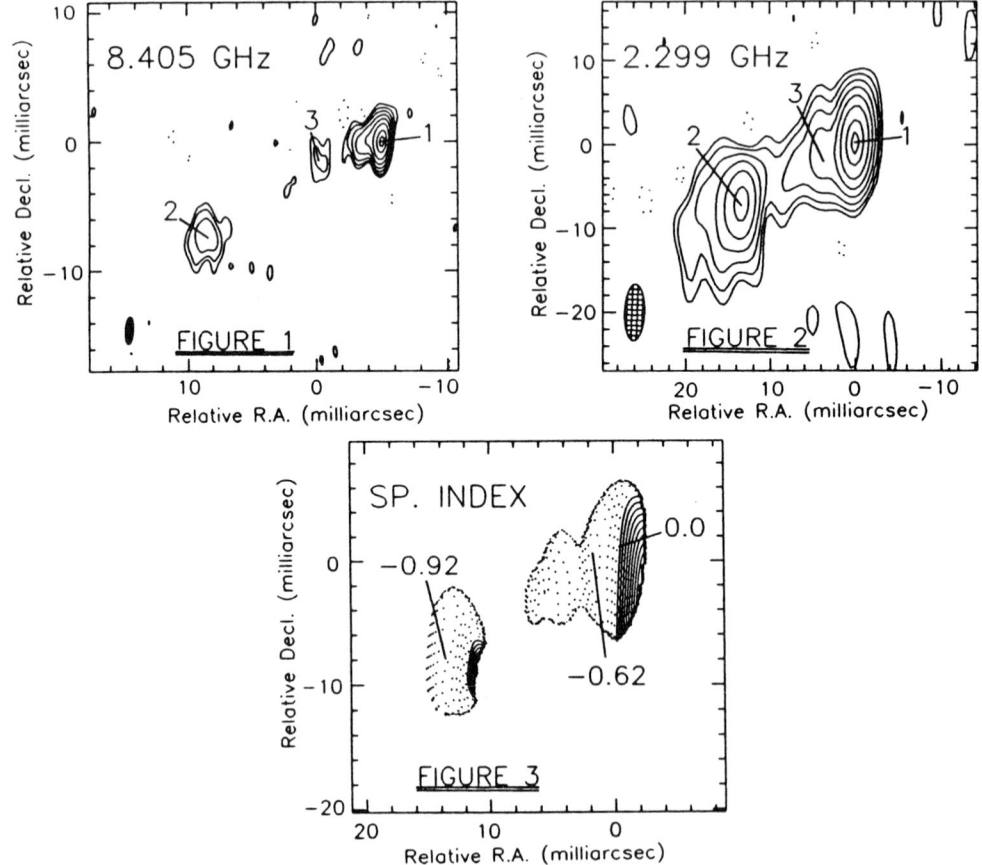

References

[1] Gómez J.L., Alberdi A., Marcaide J.M., *submitted to A.&A. (1993)*.

[2] Lara L., Alberdi A., Marcaide J.M., Muxlow T.W.B., *in preparation, (1993)*.

[3] Marcaide J.M. and Shapiro I.I., *Ap.J.*, 276 (1984), pp 56-59.

[4] Simon R.S., Johnston K.J., Spencer J.H., *The Impact of VLBI on Astrophysics and Geophysics*. M.J. Reid and J.M. Moran (eds.), Kluwer Academic Publishers (1988), pp 21-22.

2.3 GHz VLBI Images of Southern Hemisphere Radio Galaxies and Quasars

David W. Murphy * and the SHEVE team [†]

Abstract

The first images from VLBI data taken with Southern Hemisphere radio telescopes have recently been made as part of the SHEVE (Southern Hemisphere VLBI Experiment) collaboration. In this paper we present 2.3 GHz images of two quasars (PKS0237-233 and PKS0438-436) and two radio galaxies (PKS1549-790 and PKS1934-638).

In Figure 1(a) we show the image of the GPS radio galaxy PKS1934-638 which has a 'compact double' radio morphology (Tzioumis et al, 1989, AJ, 98, 36) with two radio components separated by 83 pc (42 mas). This image includes data taken from Hartebeesthoek as well as 6 Australian antennas. The extra resolution from the trans-Indian Oceanic baselines enables the morphology of each individual component to be examined. The E. component is elongated parallel to the line joining the two components whereas the W. component is elongated perpendicular to this line. These very different morphologies are at variance with those expected if 'compact doubles' are young lobe-dominated radio sources, which would predict that this 'compact double' source should have two equal VLBI components that are 'edge-brightened'.

The 2.3 GHz image of the flat spectrum radio galaxy PKS1549-790 is shown in Figure 1(b) and was made from data taken with 6 Australian antennas. This source has an interesting curved 'jet' with the 'jet' initially heading in a NE direction followed by a large bend. The W. component of this source is simply fit by a 2 Gaussian component model: one component which is unresolved and may be tentatively identified as the 'core' and other which is an elongated Gaussian of axial ratio 3:1 with a major axis (FWHM) of 18 pc (23 mas) and maybe identified as the start of a 'jet'. The elongated

---

*Jet Propulsion Laboratory, California Institute of Technology, Pasadena, CA. 91109, U.S.A. This research was carried out in part at the Jet Propulsion Laboratory, California Institute of Technology, under contract with the National Aeronautics and Space Administration.

[†]Members of the Southern Hemisphere VLBI Experiment (SHEVE) team are listed as co-authors on the paper by R. A. Preston et al. in these proceedings.

E. component is fit by a single Gaussian model of axial ratio 10:1 and major axis (FWHM) of 30 pc (38 mas).

In Figure 1(c) we show the 2.3 GHz image of the GPS quasar PKS0237-233 which was made with antennas in Australia and one in S. Africa. The source is just resolved with data from Australia only but the extra baselines across the Indian Ocean enable two jet components in addition to the core to be clearly detected.

The 2.3 GHz image of the flat spectrum quasar PKS0438-436 (Figure 1(d)) was made with data from Australian antennas only and has a double morphology. Both components are resolved. The SE component is almost circular but the NW component has an axial ratio of 3:1 and the higher brightness temperature of the two components.

This paper has shown it is possible to obtain images with the present SHEVE array. We are continuing to monitor these sources at 2.3 GHz and 8.4 GHz.

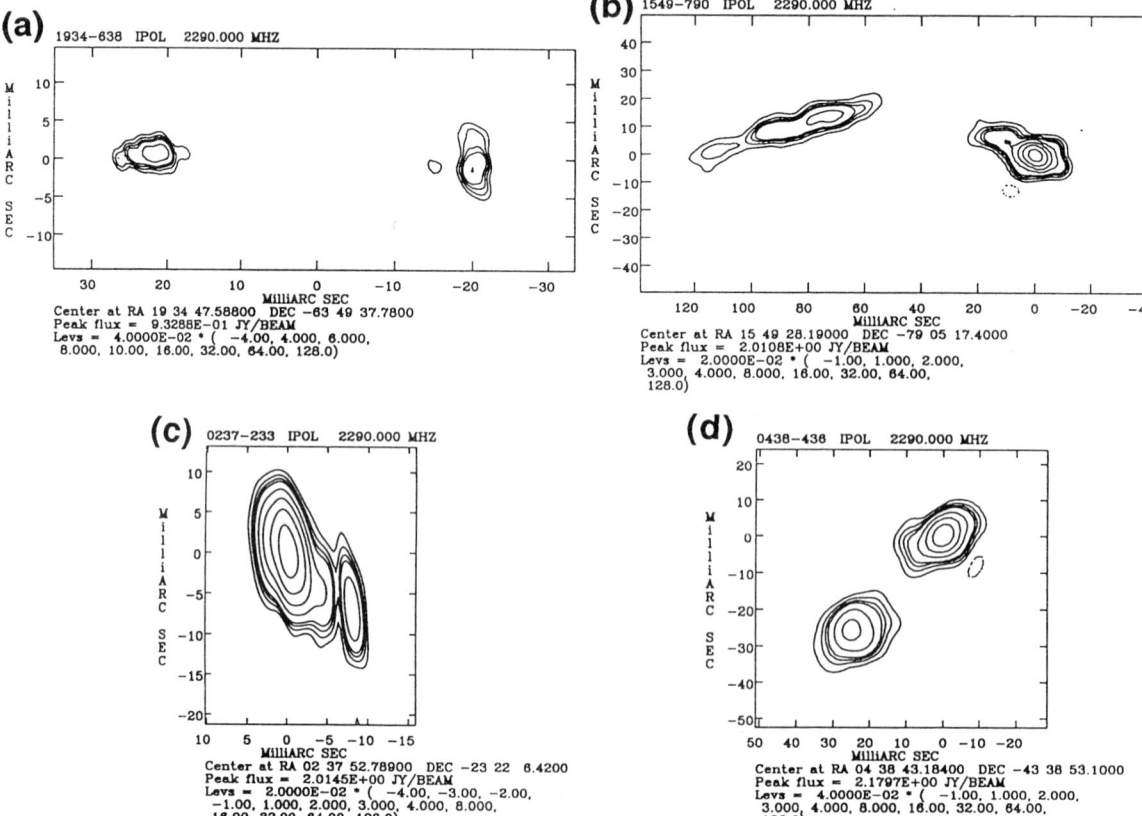

Figure 1.   2.3-GHz VLBI maps of (a) PKS1934-638 (b) PKS1549-790 (c) PKS0237-233 and (d) PKS0438-436

Observations of Radio Galaxies in the S/X Band with the Triangle Medicina - Noto - Madrid

Tiziana Venturi,[*] *Luigina Feretti,*[*] *Gabriele Giovannini,* [*]

Jon Marcaide, [†]   *Lucas Lara,*[‡] *Maria J. Rioja,* [‡]

Corrado Trigilio,[§] *Grazia Umana* [§]

1. Introduction

VLBI observations of a sample of 10 radio galaxies of total luminosity in the range 5×10^{23} - 2×10^{26} W/Hz at 0.4 GHz were obtained with the MarkIII, in the S/X band, using the telescopes at Medicina, Noto and Madrid. The observed galaxies belong to the complete sample defined in [GFC90] and were chosen among those which appeared to be resolved in the 6 cm snapshot data.

The aim of these observations is to obtain structural and spectral information at the parsec scale for radio galaxies characterized by low-intermediate power. Here we will present preliminary results on the morphology of the 10 radio galaxies under consideration.

2. Observations and Preliminary Results

The ten radio galaxies under study were observed with the triangle Medicina - Noto - Madrid (32 m), using the S/X receiver (S band = 2.3 GHz, X band = 8.4 GHz) in snapshot mode. The observations were carried out with the MarkIII/C recording system, with seven channels recording in the S band, and the remaining seven in the X band. Each source was observed for a total of ~2.5 hours. For two sources, i.e. 4C31.04 and 3C452, only the baseline Medicina-Noto was available. The resolution

---

[*]Istituto di Radioastronomia-CNR, Bologna, Italy
[†]Universidad de Valencia, Valencia, Spain
[‡]Instituto de Astrofisica de Andalucia, Granada, Spain
[§]Istituto di Radioastronomia-CNR, Noto, Italy

Table 1 - Details on the sources

Source	logP (408 MHz)	S band Model	S band Comp. Separ. mas	X Band Model	X Band Comp. Separ. mas
4C31.04	25.10	one-sided	12.4	undet.	-
NGC1167	24.05	undet.	-	undet.	-
0648+27	23.70	undet.	-	undet.	-
4C29.30	25.08	unresolv.	-	one-sided	10.0
3C264	24.85	one-sided	13.7	one-sided	5.7
1144+35	24.15	2 comp.	18.2	2 comp.	19.9
3C303	26.11	one-sided	14.3	one-sided	3.7
3C338	25.25	two-sided	-	two-sided	-
3C452	26.30	one-sided	14.0	undet.	-
3C465	25.30	one-sided	8.9	undet.	-

of these observations is of the order of 16×10 mas in the S band, and 4.5×2.5 mas in the X band.

The data were correlated in Bonn, and the data reduction was carried out with the Caltech package. Each source was first modelfitted, then for those sources observed with a sufficient uv-coverage a few cycles of self-cal and mapping were attempted.

On the basis of the modelfitting and/or best map, we classified the ten radio galaxies according to the derived morphology. In Table 1 we summarise the characteristics of the best model fitted to the visibilities and the component separation for the resolved sources.

At the present sensitivity and resolution, all sources, but 3C338 and 1144+35 are either point-like or one-sided. 3C338 has a two-sided jet in both bands, while 1144+35 resembles a compact double, with separation of $\sim$ 20 mas. For the three FR-1 sources with a VLBI core-jet morphology, i.e. 4C29.30, 3C264, 3C465, the parsec scale jet is on the same side of the nucleus as the strongest arc-sec scale jet, as it is found in the stronger FR-2 galaxies and in radio loud quasars.

References

[GFC90] Giovannini G., Feretti L., Comoretto G., *VLBI observations of a complete sample of radio galaxies. I. Snapshot observations.* ApJ 358 (1990), 159-163.

The Structure of the Hot Spots in 3C 295

Gregory B. Taylor [*]   *Richard A. Perley* [†]

Abstract

We present new observations of the hot spots of 3C 295 at 1.7 GHz with 20 milliarcsecond resolution using the newly constructed Very Long Baseline Array (VLBA) of the NRAO. Our observations reveal a remarkable contrast in morphology between the northwest and southeast hot spots. The northwest hot spot is very compact with a bright central component of only 40 × 60 mas (280 × 410 pc for $H_o = 50$ km s^{-1} Mpc^{-1} and $\Omega = 1$), while the southeast hot spot has an elongated structure. From VLA and MERLIN observations at 15, 8, 6, and 1.7 GHz we find that the spectra of the two hot spots are consistent with a steep power law with no spectral break. The spectral index of the northwest hot spot is 1.0, and that of the southeast hot spot is 0.89. The absence of a break in the spectra of both hot spots is unusual.

We have also analyzed the low frequency turnover in the integrated spectrum of 3C 295 in light of new observations made by Konstantin Sokolov with the UTR-2 telescope. We find that the low frequency turnover can be explained by thermal absorption from an inhomogeneous ISM.

1. The 18 cm VLBA Observations

We observed 3C 295 at 1664 MHz for 12.5 hours on May 20-21, 1991 with 6 VLBA antennas and a single VLA antenna. The VLBA antennas used were located at Pie Town, Kitt Peak, Los Alamos, Fort Davis, North Liberty, and Owens Valley. The initial clock offsets were found by performing fringe fitting to observations of 3C 84 and 3C 345 using the NRAO Mark II correlator. The complete correlation of the entire observing run was performed using the Block II correlator at Cal Tech.

2. Results

A contour plot of the northwest hot spot of 3C 295 at 20 mas resolution (Figure 1) shows a round hot-spot, with a bright elongated central component of size 40 mas

---

[*]Arcetri Observatory, Largo E. Fermi 5, I-50125, Firenze, Italy
[†]NRAO, Box 0, Socorro, NM 87801, USA

× 60 mas (280 × 410 pc). The diameter of the entire northwest hot spot is ∼ 150 mas (1.04 kpc). The southeast hot spot (Figure 1) has quite a different morphology. Our VLBA map reveals an 'S' shaped structure to the hot-spot, although the knotty appearance is most likely the result of incomplete u-v coverage. The overall shape and boundaries of the 18 cm VLBA map agree well with the 2 cm VLA map of Perley and Taylor [PT91]. The total flux densities of the two hot spots at 18 cm are similar – 3.6 ± 0.4 Jy in the northwest hot spot and 4.0 ± 0.4 Jy in the southeast hot spot.

From the VLA observations presented in Perley and Taylor [PT91] at 2, 3.6 and 6 cm, and MERLIN observations at 20 cm (Chidi Akujor - private communication) we have compiled flux density measurements at 0.3" resolution for the hot spots in 3C 295. The spectra of both hot spots are consistent with a straight power law fit between 1.7 and 15 GHz. A linear least squares fit to the spectrum of the northwest hot spot yields a spectral index of 1.02 ± 0.020 with a χ^2 value of 0.54. The spectrum of the southeast hot spot is not quite so steep with a best fit spectral index of 0.89 ± 0.024 with a χ^2 value of 0.66. These χ^2 values indicate that the data are consistent with a straight power law with a confidence of 99.9%.

Figure 1.

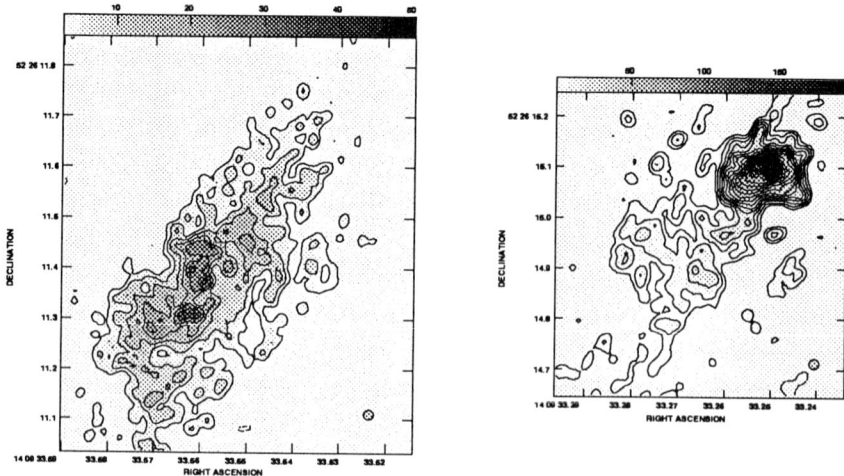

Further discussion of these results will be described in Taylor and Perley [TP92].

References

[PT91]   Perley, R.A., and Taylor, G.B., *VLA observations of 3C 295 - a young radio galaxy?*, AJ, Vol. 101 (1991), pp. 1623-1631.

[TP92]   Taylor, G.B., and Perley, R.A., *The structure of the hot spots in 3C 295*, Astr. and Astrophys., (1992) in press

Unusual Features in QSR 3C147 – A Unified Explanation of the Jets and BLR Clouds

H. S. Chu [*]   F. J. Zhang [†]   R. L. Mutel [‡]   L. I. Matveyenko [§]
R. E. Spencer [¶]

Abstract

We present a combined MERLIN - EVN map of the quasar 3C147 at 1.67 GHz together with a VLA polarization map at 15 GHZ. The images show that 3C147 has a bright core with a two sided asymmetrical jet structure and an unusual sideways ejection. We propose a model in which an ordered annular magnetic field present in the core region of 3C147 collimates the main jet by the pinch process. Relativistic beaming is also occurring but the jets are intrinsically different and asymmetric due to asymmetry in the magnetic fields on each side of an accretion disk. The unusual sideways ejection can be explained by a coronal mass ejection (CME) from an accretion disk corona, as can the isotropically distributed BLR clouds of this and other AGN.

A combined map (fig.1) of the QSR 3C147 was produced from data observed by 6 telescopes of the European VLBI Network and by MERLIN at 1.67 GHz (Zhang et al. 1991). Comparisons with a VLBI map at 329 MHz (Simon et al. 1990), a MERLIN map at 4995 GHz (Akujor et al. 1990) and a VLA polarization map at 15 GHz (fig. 2) (Kellermann and Crane 1990, private communication) confirm that 3C147 has a bright core with two–sided asymmetrical jet structure and an unusual sideways ejection.

The main jet is ~230 mas (2 pc) long, is well collimated and extends to the SW. The northern jet is diffuse and much longer (~700 mas) than the main jet. We suggest that these two jets are independent and do not constitute a twin jet since the brightness and length ratios do not fit the standard relativistic beaming model. The ~50 mas long feature to the SE also does not fit into the standard scheme. It seems

---

[*] Purple Mountain Observatory, Academy of Sciences, China.
[†] Shangai Observatory, Academy of Sciences, China.
[‡] Department of Physics and Astronomy, University of Iowa, USA.
[§] Institute for Space Research, Russian Academy of Sciences, Russia.
[¶] NRAL, University of Manchester, Manchester, UK.

unlikely that the northern jet is merely a diffuse structure surrounding a Doppler diminished counter–jet. A cocoon of extended emission surrounds the main jet. The jet is also edge brightened in our EVN alone map. These features, together with the high degree of collimation, can be explained if there is an ordered annular magnetic field in the core region and jet. The VLA map indeed shows strong polarization near the core indicating the presence of such a field. Magnetic pinch effects will keep matter flowing in the axial direction well collimated.

The N jet shows no connection with the core on the combined MERLIN–EVN map, only a generic relation to the cocoon surrounding the main jet. The MERLIN and VLA maps also show that the N jet is isolated from the core. This jet may not be formed by a continuous flow from the core but possibly due to an outburst from an accretion disk corona. Its diffuse, fragmented and interrupted structure in our combined 6 cm map (Zhang et al. 1991) is more consistent with this hypothesis than its origin as a jet from a double black hole. The most unusual feature is the sideways ejection, totally unexpected in the two–sided jet model. This feature may give an insight into the nature of 3C147, and we propose an accretion disk corona model (fig. 3), supported by the following points:

1. The sideways ejection is non–axial and may be due to a different ejection mechanism to that of the main jet.
2. The accretion disk is likely to form a corona somewhat similar to that on the Sun and in binary X-ray sources. Since there are basic qualitative similarities between the creation of solar and galactic magnetic fields by a MHD dynamo process (Parker 1990), we can expect such features as prominences and flares also to occur in a galactic disk corona. The sideways ejection could thus be of magnetic origin.
3. Due to differential rotation and magnetic bouyancy the disk magnetic fields may escape from the disk to form bipolar loops. Footpoint motion will force these magnetic loops together, forming current sheets at their interfaces. Reconnection of field lines can occur, releasing magnetic energy and producing flares. We can expect the ejection of coronal material into interstellar space to occur, with an energy source ultimately derived from the rotation of the galaxy. As on the Sun this coronal mass ejection (CME) will take the form of part of an arcade and could give rise to the observed SE feature.
4. Since the magnetic field structure is generally not symmetric about the accretion disk then the ejection may not be axial, as observed.
5. The origin of the isotropically distributed BLR clouds in AGN may be explained by the existence of many variously sized CME ejected in different directions from the accretion disk around the central black hole.

Thus CME with asymmetric jets can explain many features of the radio structure in 3C147, though we expect that some degree of Doppler boosting is also occurring. It will be interesting to see whether distorted features in other compact radio sources can be explained by this model.

References

[1] Akujor, C.E., Spencer, R.E., Wilkinson, P.N., 1990, *MNRAS*, **244**, 362.

[2] Parker, B.N., in *Galactic and Intergalactic Magnetic Fields.*, *IAU Symp.* 140, ed. R. Beck et al., (1990), p. 169.

[3] Simon, R.S., Readhead, A.C.S., Wilkinson, P.N., 1990, *ApJ*, **354**, 140.

[4] Zhang, F.J., Akujor, C.E., Chu H.S., Mutel, R.L., Spencer, R.E., Wilkinson, P.N., Alef, W., Matveyenko, L.I. & Preuss, E., 1991, *MNRAS*, **250**, 650.

Fig.1 EVN-MERLIN combined map of 3C147 (Zhang et al. 1991)

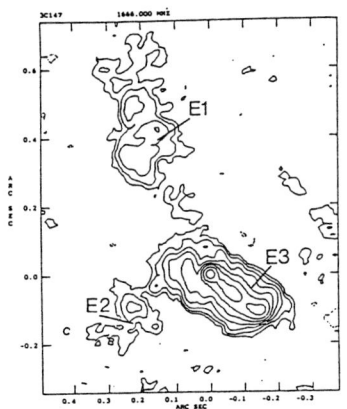
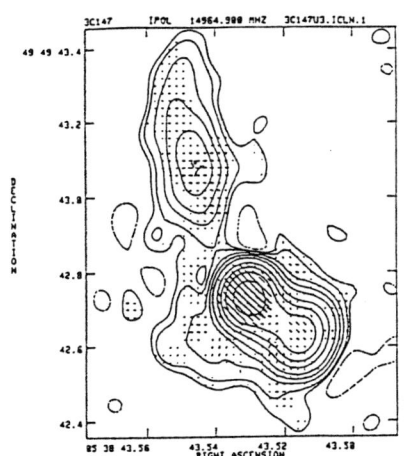

Fig.2 VLA map at 15 GHz from 1983 data (Kellermann and Crane, unpublished). Polarization line 1"=0.379 Jy/beam; contours at 10, 50, 20, 10, 5, 2, 1, 0.5, 0.2, 0.1, -0.1 percent of the peak brightness of 1.41 Jy per beam. Synthesized beam 0.091 × 0.082" P.A.−13.6°.

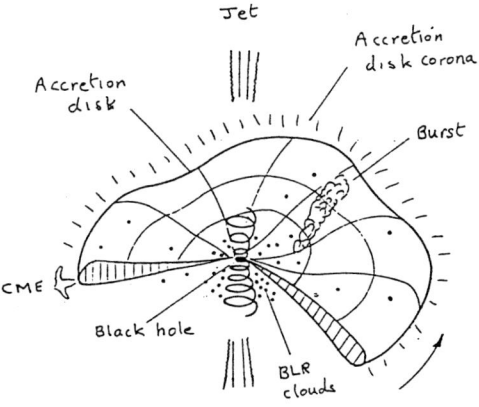

Fig.3 A model of jet origin, BLR clouds and activity in 3C147. The accretion disk may be warped. The two jets may be asymmetric and independent. The counter-jet may not be seen due to Doppler beaming.

Observations of Jets with MERLIN

T.W.B. Muxlow [*]

Abstract

Since it began operations in its original form in 1980, MERLIN has been an important instrument for the study of radio jets because it combines high angular resolution with sensitivity to steep spectrum emission. In addition, it remains the linking array between the lower angular resolution VLA and the extreme resolution of VLBI. I present an example of *matched resolution* imaging of the quasar 3C418 with array combinations of MERLIN with both the VLA and EVN in order to investigate the spectral evolution along the jet of this core-dominated source. On its own, the resolving power of MERLIN has been used to investigate apparent super-luminal motion on the kpc scale size in the radio galaxy 3C120 where observations separated by eight years show evidence of motion close to the core but not in a jet feature 4 arcsec from the nucleus. Recent improvements to MERLIN have transformed its imaging capability and many new results both here and elsewhere at this conference utilize the polarization imaging of the enhanced array. I present new MERLIN/VLBI data combinations in the investigation of the unusual quasar 3C395 where for the first time it is unambiguously shown that the one-sided radio jet bend through a projected angle of nearly 180 degress within 0.1 arcsec of the nucleus. New MERLIN has proved to be ideal for the imaging of the short length jets now found in the nuclei of Seyfert galaxies. I present recent images of the nuclear region of three nearby Seyferts NGC4151, Mkn3, and NGC1068.

1. MERLIN 1

In its original configuration from 1980, MERLIN interferometric data consisted of single polarization correlations of 8MHz bandwidth in real time from 6 antennas out to a maximum separation of 127 kms (although in the late 1980s this was supplemented by the addition of a single (18 m) element of the MRAO one mile telescope at Cambridge). The original antennas were all phase coherent with two-way L-Band locking links. Most images were made by self-calibration in a manner similar to VLBI.

The instrument was well-suited to the study of extra-galactic jets and many of the well-known examples were mapped. To illustrate the use of MERLIN with other arrays I include *matched resolution* images at 150 mas from MERLIN at 5GHz together with the VLA at 15GHz and MERLIN/EVN at 1.7GHz of the quasar 3C418 (z=1.698). Figure 1 shows the MERLIN 5GHz and VLA 15GHz images with the spectral index between 1.7 and 15GHz superposed in greyscale demonstrating spectral steepening along the jet. By fitting a Jaffe/Perola type spectral shape to the

---
[*]N.R.A.L., Jodrell Bank, Macclesfield, Cheshire, SK11 9DL, U.K.

images pixel by pixel along the jet with an initial electron energy spectrum derived from the observed jet spectral index close to the core, it is possible to investigate the spectral break frequency as a function of jet position. The knots are thus interpreted as sites of interaction where the jet brightens and the spectrum hardens although there is no evidence for major bulk reacceleration. Converting the break frequency to a spectral age via an equipartition magnetic field around $50\mu G$ derives an age for the material at the end of the jet close to 1.3 Myr. The total jet length is about 40 kpc implying a jet speed of 0.1c. Allowing for projection effects with the jet lying 10 degrees to the line of site increases the estimated speed to around 0.6c.

The high image fidelity of MERLIN has allowed very fine measurements of source structure to be made. To illustrate this I include results of MERLIN 1.7GHz observations of the jet in the radio galaxy 3C120 (Muxlow & Wilkinson, 1991). Here images from 1980 and 1988 observations have shown that apparent super-luminal motion as seen by VLBI in the core persists into the inner jet to about 0.15 arcsec from the core but that a knot in the jet 4 arcsec from the core is stationary. The inner jet material is found to have moved out by 19 mas in the eight years separating the images (See Figure 2).

2. MERLIN 2

MERLIN has recently been upgraded with the addition of improved cooled receiver systems, together with broad-band dual polarization telescope links and a new 32 m high frequency antenna at Cambridge. This has transformed its imaging capability with full polarization capability and major sensitivity enhancements. In addition, the array is now fully phase calibrated allowing astrometric position measurements to be made.

MERLIN combination with VLBI is increasingly common and recent work in collaboration with a group in Granada on the quasar 3C395 has confirmed the earlier suggestions of Saikia *et al* 1990 from MERLIN 1 images that the jet in this object turns through a projected angle of 180 degrees with 100 mas of the core before running into an extended component to the north west. Figure 3 shows a series of images of the jet in this source at differing angular resolutions from the global VLBI 5GHz image at 2 mas through the MERLIN/VLBI combination image at 9 mas to the MERLIN image at 50 mas . Work on these data and on an associated VLA 5GHz dataset is in continuing.

MERLIN has also recently been used for investigations of the small-scale jets in the nuclear regions of Seyfert galaxies. Collimated outflow in such objects has not,until now, been positively established; however MERLIN high resolution 5GHz observations have revealed the existence of jets in a number of these nuclei. I present new MERLIN 2 images of the nuclear emission in the Seyfert galaxies NGC4151,

Mkn3, (See Pedlar *et al.*, and Kukula *et al*, this meeting) and NGC1068. In NGC4151 MERLIN and MERLIN/VLA combinations have shown twin-sided jets over the inner 2.5 arcsecs.. In Mkn3 the MERLIN image reveals an unambiguous twin-sided jet. In addition, the new positional capability of MERLIN allows us to identify the true nucleus in this source for the first time. The MERLIN images of NGC4151 and Mkn3 are shown in Figures 4 and 5.

In NGC1068 (see Figure 6) we find a one-sided flaring jet strikingly similar to the VLBI jet found in the quasar 3C48 (Wilkinson *et al.*, 1991) although an order of magnitude smaller in physical size and radio luminosity. Comparison with an existing 22GHz VLA image by Ulvestad *et al.*, 1987 shows that the southernmost feature in the MERLIN image is flat spectrum and as one moves further along the jet to the North-East so the spectrum steepens. We thus identify the southernmost component in the MERLIN map as the true nucleus of NGC1068.

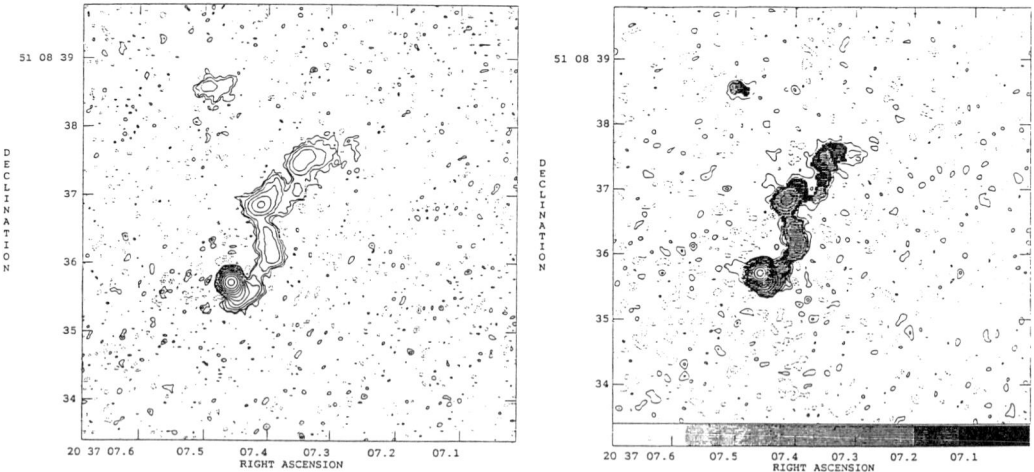

Figure 1. MERLIN image of the quasar 3C418 at 5GHz (left). VLA image of 3C418 at 15GHz (right). Both have angular resolution=150mas. At right the greyscale encodes the spectral index between 1.7 and 15GHz.

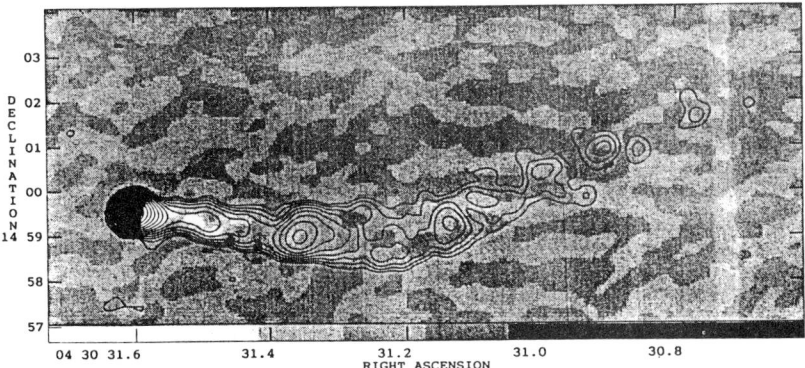

Figure 2. MERLIN 1.7GHz image of the radio galaxy 3C120 (epoch 1988.92). The difference image between this and an epoch 1980.98 image is shown in greyscale. Positive-going excursions in the difference image are shown black, negative-going ones white. An outward shift is characterized by a positive followed by a negative excursion when scanning westward along the jet.

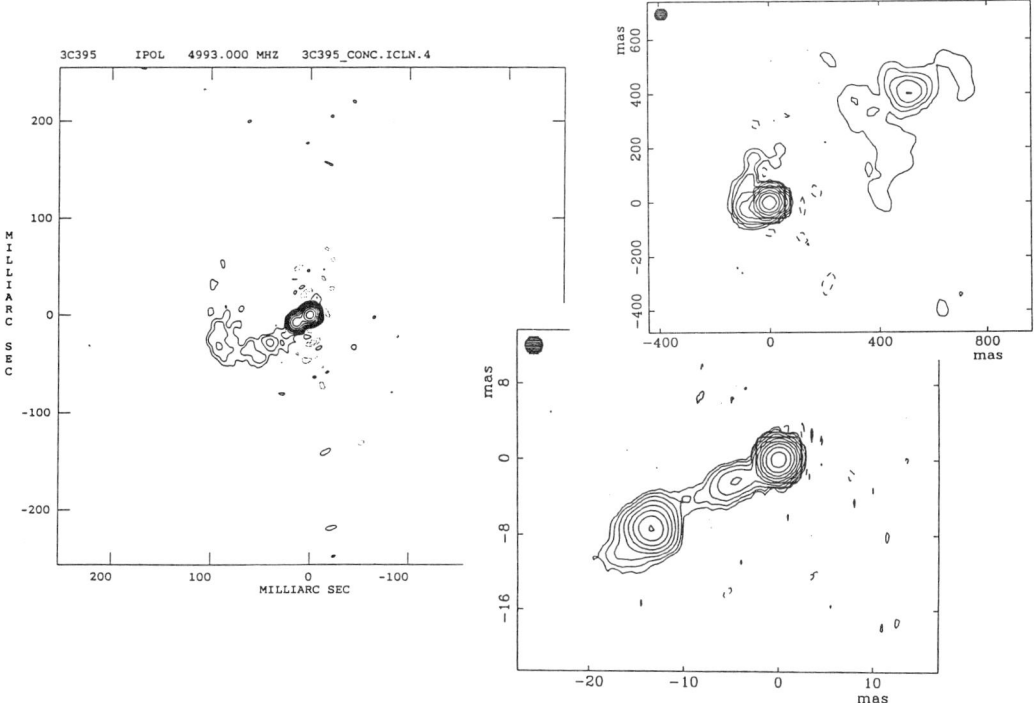

Figure 3. MERLIN2/VLBI 5GHz combination images of the quasar 3C395. MERLIN (upper right, beam 50 mas), MERLIN+VLBI (left, beam 9 mas), and Global VLBI (lower right, beam 2 mas)

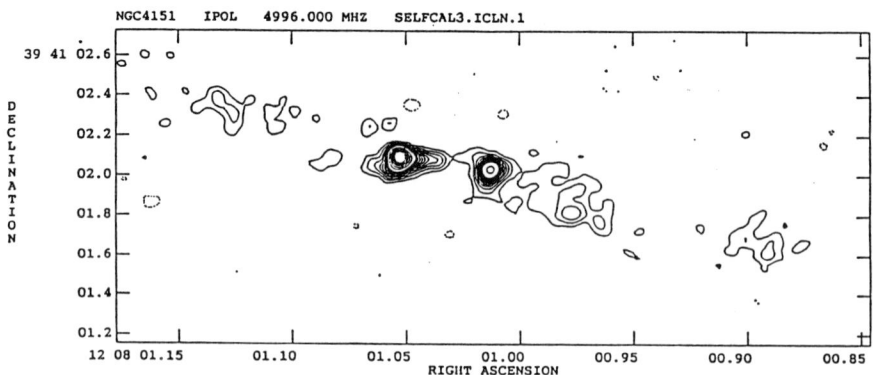

Figure 4. MERLIN 2 5GHz image of the nucleus of NGC4151 (After Pedlar *et al.*, this meeting)

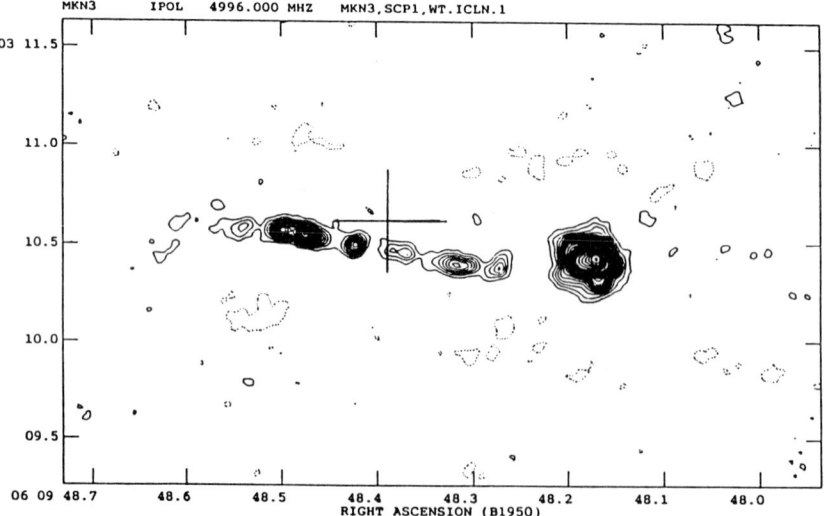

Figure 5. MERLIN 2 5GHz image of the nucleus of Mkn3 (After Kukula *et al.*, this meeting)

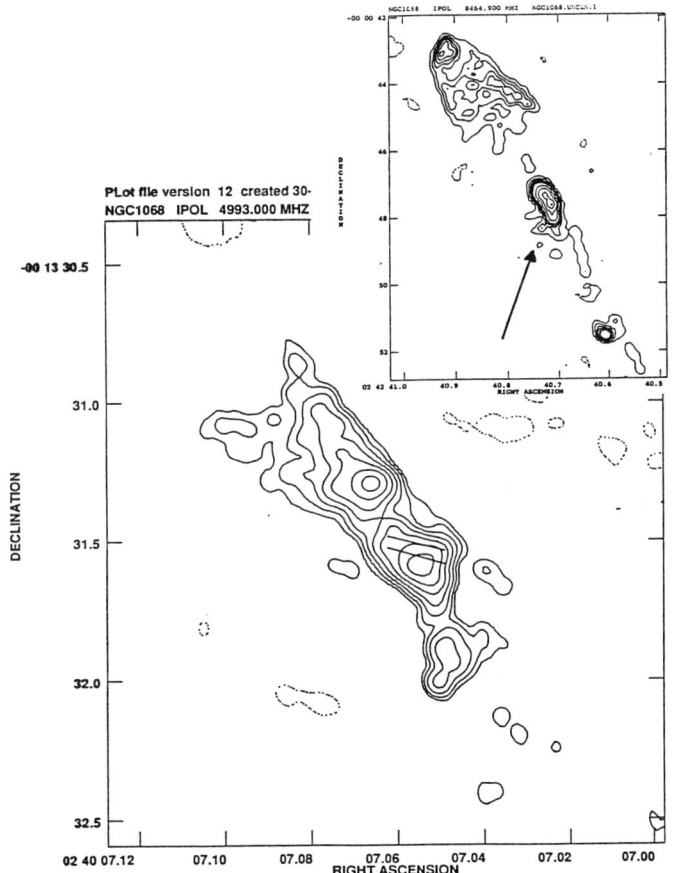

Figure 6. MERLIN 2 5GHz image of the nucleus of NGC1068. (Insert: Unpublished 8.4GHz VLA image of NGC1068, Kukula, private communication)

References

Muxlow T.W.B. & Wilkinson P.N., 1991, *MNRAS*, **251**, 54.

Saikia D.J., Muxlow T.W.B, & Junor W., 1990, *MNRAS*, **245**, 503.

Ulvestad J.S., Neff S.G., & Wilson A.S., 1987, *AJ*, **93**, 22.

Wilkinson P.N., Tziuomis A.K,, Benson J.M., Walker R.C., Simon R.S., & Kahn F.D., 1991, *Nature*, **352**, 313.

Fine Structure in the Jets of 3C 219

R.A. Perley [*], A.H. Bridle [†], D.A. Clarke [‡]

Abstract

VLA observations at 2 and 3.6 cm of the "partial" jet and counterjet in 3C219 at resolutions of 0.18 to 0.4 arcseconds have revealed fine structure with important implications for jet models. Both jets abruptly end at bright, compact features, dropping from the maximum to zero intensity in less than 100 pc. There is no evidence of bow shocks in advance of either jet. The jet tips are of similar size and brightness, and are aligned with the core with a bend angle of less than 0.2 degrees. Sinusoidal (and thus possibly helical) oscillations in the ridge lines of both jets are found, with inverted symmetry. The jet edges are remarkably linear, despite internal variations in structure, and are limb-brightened. There is evidence for a previously unknown nuclear jet of length ~ 2 kpc, pointed towards the main (SW) jet. The 2cm-3.6cm spectral index of the main jet is very high, exceeding 1.0 in most places.

1. Introduction

3C219 is a moderately luminous ($P_{1.4} = 3 \times 10^{26}$ W Hz^{-1}) FRII radio galaxy with a bright jet that extends for 18 arcsec to the SW from the nucleus, and a fainter counter-jet that extends to about 5 arc sec from the nucleus to the NW. The jet is of particular interest because it disappears abruptly despite the presence of a bright hotspot some 60 arcseconds from the nucleus which presumably signifies a continuation of an active jet somewhere beyond the apparent point of disappearance. Clarke et al. (Ap.J., **385**, 173, 1992) proposed two models to explain the observed structure of 3C219:

(a) Symmetric, relativistic, restarting jets, with the brightness asymmetry of the jets being due to Doppler favoritism and the length asymmetry being being due to the difference in light travel time to the observer from the approaching jet and receding counter-jet. In this model, outward-moving shocks at the leading edges of

---

[*]NRAO, P.O. Box O, Socorro, NM, USA 87801
[†]NRAO, Edgemont Road, Charlottesville, VA, USA 22903
[‡]Center for Astrophysics, 60 Garden St.,Cambridge MA, USA 02138

the jets decelerate them and make the counter-jet tip visible by reducing a previously unfavorable Doppler factor.

(b) A continuous, two sided jet, whose emissivity is enhanced close to the galaxy through lateral compression from X-shocks caused by the adjustment of the supersonic flow to the declining external pressure gradient. The emission is quenched when the jet comes into pressure equilibrium with the atmosphere. In this model, the brightness asymmetry is ascribed to a side-to-side aysmmetry in the pitch angle of the magnetic field (higher pitch angle in the counter-jet) and the geometrical asymmetry is random.

Both models can account for some aspects of the large scale structure, but both contain *ad hoc* assumptions. The sensitive high resolution observations reported here were made in an attempt to discriminate further between the models.

Two eight-hour sessions with the VLA were used to observe 3C219 at 3.6 and 2cm in the **A**-configuration in Sept. 1991. The image shown in Fig. 1 reveals the following new features: (1) The emission drops from the maximum of 12 mJy/sq. arcsec to zero within 100 pc of each jet tip. A wider-field image shows there is no compact emission anywhere between the jet tips and the hotspots. (2) The tips of the jet and counter-jet are remarkably similar to each other in size and brightness. (3) A line drawn from the jet tip through the nucleus passes through the counter-jet tip within 0.05 arcseconds, so the maximum bend angle is 0.2 degrees. (4) Sinusoidal structures are present in both jets, with inverted symmetry. (5) The edges of the main jet are linear, despite variations in its substructure, and are limb brightened in many places. In addition, a core-subtracted image reveals a 2 kpc nuclear jet which is aligned with the main jet and which points toward it. (The tip of this inner jet is just visible in Figure 1.) Comparison of 0.4 arcsecond images shows that the 3.6 to 2cm spectral index of the SW jet exceeds 1.0 in most places, while the jet tips have spectral indices of 0.5 to 0.7.

2. Discussion

The abrupt ends and bright tips of the jet and counter-jet pose a strong challenge to the continuous flow/field-compression model, which cannot readily reproduce such rapid quenching of the emissivity. In contrast, the rapid drop in emissivity and the similarity in structure and brightness of the tips of the jet and counter-jet are predicted by the restarting jet model, as is the inverted symmetry of the internal structures. Does this mean the restarting jet model is favored, and the continuous flow/field decompression model is now to be rejected? Not necessarily.

A supersonic restarting jet propagating through the lobes of a radio source must drive ahead of it a bow shock which travels in a radio-emitting plasma and so must significantly enhance the radio emissivity of the lobe medium. No traces of such bow-shock structures are evident in advance of either jet in 3C219, however. Furthermore, the great compactness of the features at the tips of the jets, and their strong alignment across the nucleus, can be explained in a restarting jet model only under the assumptions of constancy in the jet direction and steadiness of the shock structures in time. The line linking the jet and counter-jet tips and the nucleus is not the symmetry axis of the more diffuse jet emission, however, and there are hook-like features in both the jet and the counter-jet behind their bright tips. These new results question whether an assumption of strong symmetry is reasonable.

It remains to be seen whether the new details of the internal structure of 3C219 revealed by these observations can be accommodated in either of the models previously discussed by Clarke et al. A successful model should now account for the following: (1) the jet/counter-jet length asymmetry, including the apparent scale asymmetries between the hooks near their tips, (2) the strong asymmetry in brightness between all features except the very tips of the jets, (3) the remarkable compactness of the jet tips, (4) their strong alignment across the nucleus, and (5) the apparent lack of stand-off features marking bow shocks beyond the jet tips. While the "born-again" relativistic jet model predicts (1) and (2), (3) remains unexplained, the travel-time asymmetries make (4) coincidental unless the jet direction is very stable and (5) conflicts with a prediction of the model. The field-compression model predicts (3) and (5) but (1), (2) and (4) remain as coincidences or unexplained phenomena.

We conclude that neither existing model is fully satisfactory, and more imaginative, and complex, models may be needed to explain all the symmetries of the new fine structures revealed by these observations.

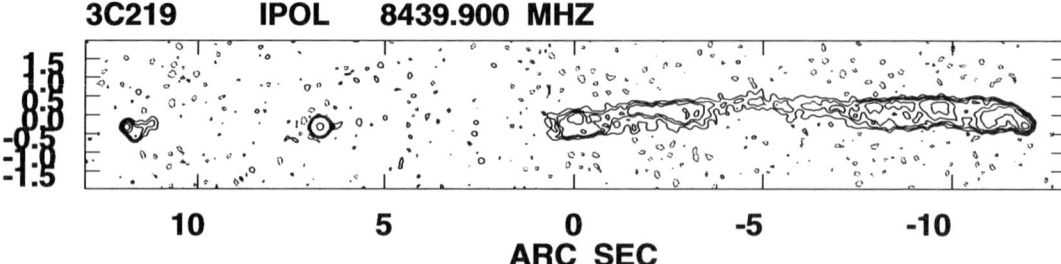

Fig. 1. The jets of 3C219 at 3.6cm wavelength and 0.18" resolution. The main, or SW, jet occupies the right half of the figure, the counter-jet is seen in the left side, while the nucleus, with the tip of the nuclear jet protruding on the right side, is in between. The jet has been rotated CCW by 50.5 degrees and the contours are at multiples of -1,1,2,3,5,9, and 15 times 35 μJy/beam. An extra contour at 40.9 mJy/beam is added to show the FWHP of the synthesized beam.

The Giant Radio Quasar 4C74.26: pc to Mpc Scale Structure

K.M. Blundell [*]   *P.J. Warner*[*]   *J.M. Riley*[*]   *T.J. Pearson* [†]
P. Alexander[*]

Abstract

We present data for the giant ($10' \Rightarrow 1.6\,\mathrm{Mpc}$) quasar 4C74.26 which show that it has a remarkable one-sided jet observed on scales from 12 pc to 400 kpc. First epoch VLBI observations show the pc-scale jet to be extended in almost exactly the same position angle as the kpc-scale jet seen in our VLA[1] images.

4C74.26 has a one-sided jet at 1.4 GHz which emanates from the core and extends for at least $150''$ (400 kpc) ($H_o = 50\,\mathrm{kms^{-1}Mpc^{-1}}$), in p.a. $\sim 160°$. A VLBI jet is detected at 5 GHz. The emission is highly elongated in p.a. $161°$. It extends for 4.5 mas (12 pc) and is unresolved transversely ($\leq 0.5\,\mathrm{mas}$). There is a limit of 0.6 mJy/beam on the surface brightness of any counter–jet to the north of the core, so that it is at least a factor of 70 fainter than the VLBI jet.

Model 1: The brightness asymmetry of the two hotspots may be explained simply if the jet is intrinsically one-sided, the brighter hotspot being the one which is currently 'fed' by the jet. Some kind of 'flip-flop' mechanism is assumed; the radiative lifetime of the hotpots and lobes would clearly be longer than the lifetime of the jet. The source does not exhibit the Liu-Pooley effect. This is the effect wherein the lobe on the jet side has a flatter spectral index than the other lobe. This effect can be interpreted, in a model assuming an intrinsically one-sided jet, by particle acceleration on the jet side with an older electron population on the non-jet side. This would be evidence against an intrinsic one-sided jet in 4C74.26. However, this effect has only been observed in small powerful sources which may therefore be considerably younger than 4C74.26. The Liu-Pooley effect could be assumed to be washed out in this source if the radiative lifetime of the lobes is indeed long. In this case the consequences of a lobe being energised by a jet would not be apparent.

---

[*]MRAO, Cavendish Laboratory, Cambridge, CB3 0HE, U.K.
[†]Astronomy Department, Caltech 105-24, Pasadena, 91125, U.S.A.
[1]The NRAO is operated by Associated Universities, Inc., under co-operative agreement with the National Science Foundation.

Model 2: Different densities, asymmetric entrainment etc. might account for the observed asymmetries in jet emission, hotspot and lobe brightness. However, it seems unlikely that asymmetries in the environment would persist over the scales of the VLBI jet (12 pc) up to that of the source (1.6 Mpc).

Model 3: If the asymmetry is due to Doppler Boosting, the difference in the brightness of the hotspots implies that some of the radiation from the hotspots must also be beamed. If they were boosted the spectral indices of the two hotspots might be expected to differ, calculations show them to be the same, however. Spectral index analysis using more frequencies is currently being undertaken and will clarify this. If the axis is at 45° to the line of sight then the de-projected size of the source is 2.3 Mpc. A jet-to-counterjet flux density ratio of ≤ 70 can be achieved at 45° by Doppler Boosting at relatively modest speeds ($\geq 0.91\,c$). If the jets are highly relativistic the angle to the line-of-sight could be as great as 50°, although this would be outside Barthel's quasar cone of 45°. [Barthel 1987 *Ap. J.* **336** 606]

From the above discussion, we reject asymmetry in the environment of the source (model 2) as a plausible explanation for jet–sidedness. It is not yet possible to say whether the source is described by an intrinsically one-sided jet (model 1) or whether the asymmetry arises from relativistic boosting (model 3), but further analysis of multi-frequency data may enable us to distinguish between them.

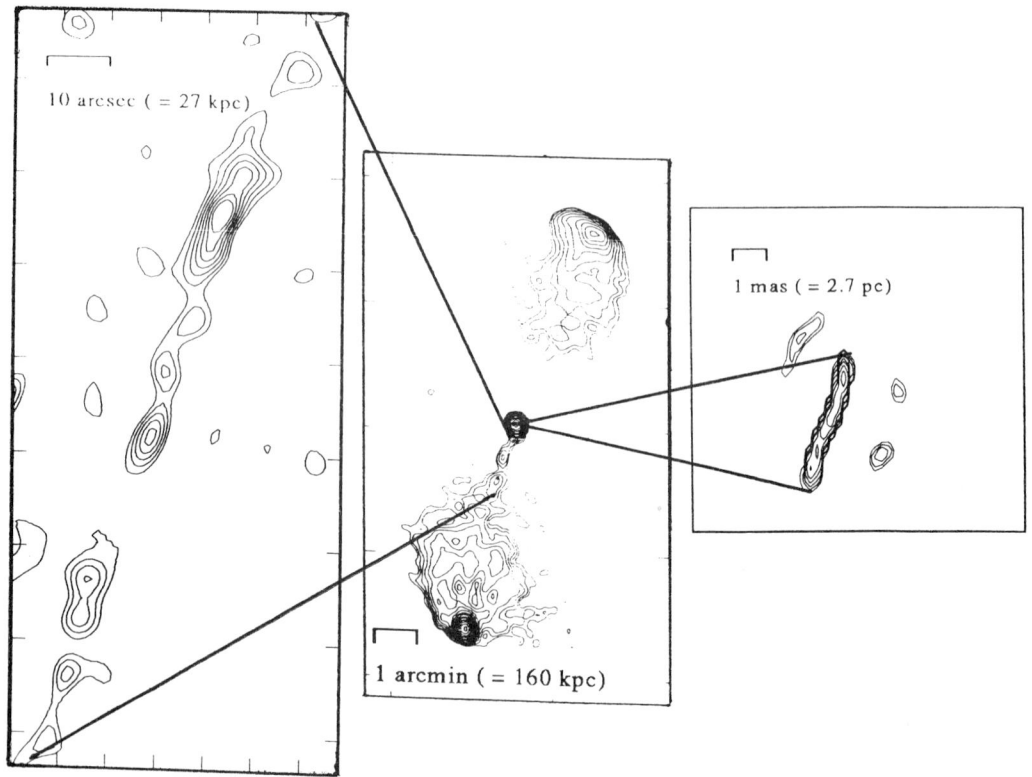

Extended Radio Jets without Hotspots

Chidi E. Akujor *   Simon T. Garrington †

Most jets in powerful radio sources terminate in bright hotspots at the outer edge of the radio lobe. These hotspots are thought to mark the locations of strong shocks where electrons are accelerated to synchrotron energies. However, in some high-luminosity (FR2) sources, the jets either fade gradually or contain several bright knots along their length, but not at the end of the jet. Such behaivour, though common in low-luminosity (FR1) sources, occurs in $\lesssim 5\%$ of FR2 sources. It has been suggested that the jets in these sources lose a substantial fraction of their bulk kinetic energy along their length (Swarup *et al.* 1982). Alternatively, the contrast between the jet and hotspot may be enhanced by relativistic beaming (Akujor and Garrington 1991).

We are currently investigating the radio structure and polarisation characteristics of a sample of about 25 hotspotless sources using multi-frequency (1.6, 5.0 and 8.4 GHz) VLA and MERLIN observations.

Our results so far suggest that:

> These sources often lack a distinct lobe on the jet–side; although there is often a diffuse halo surrounding the whole source, *e.g.* 1150+497, 3C200.

> The hotspotless phenomenon occurs in both lobe–dominated (*e.g.* 1857+566) and core dominated sources (*e.g.* 1642+690 and 3C418).

> Many of these jets are very knotty (1049+616), a few show strong bends associated with knots (e.g. 1857+56), while many have clear wiggly pattern (e.g 3C200).

> Several sources show a characteristic oscillatory offset between the peak polarisation and the peak intensity along the jet. This behaivour is most clearly seen in 0800+608 (Jackson *et al.* 1990) and has been interpreted in terms of the magnetic flux-tube mode of Königl and Choudhuri (1985).

---

*Onsala Space Observatory, Onsala, S–439 92, Sweden and Nuffield Radio Astronomy Laboratories, Jodrell Bank, Cheshire SK11 9DL, UK.
†Nuffield Radio Astronomy Laboratories, Jodrell Bank, Cheshire SK11 9DL, UK.

References

Akujor, C.E. and Garrington, S.T. 1991 *Mon. Not. R. astr. Soc.* **250**, 644

Jackson, N.J., Browne, I.W.A., Shone, D.L. and Lind, K.L. 1990 *Mon. Not. R. astr. Soc.* **244**, 750

Königl, A. and Choudhuri, A.R. 1985 *Astrophys. J.* **289**, 188

Swarup, G., Sinha, R.P. and Saikia, D.J 1982 *Mon. Not. R. astr. Soc.* **201**, 393

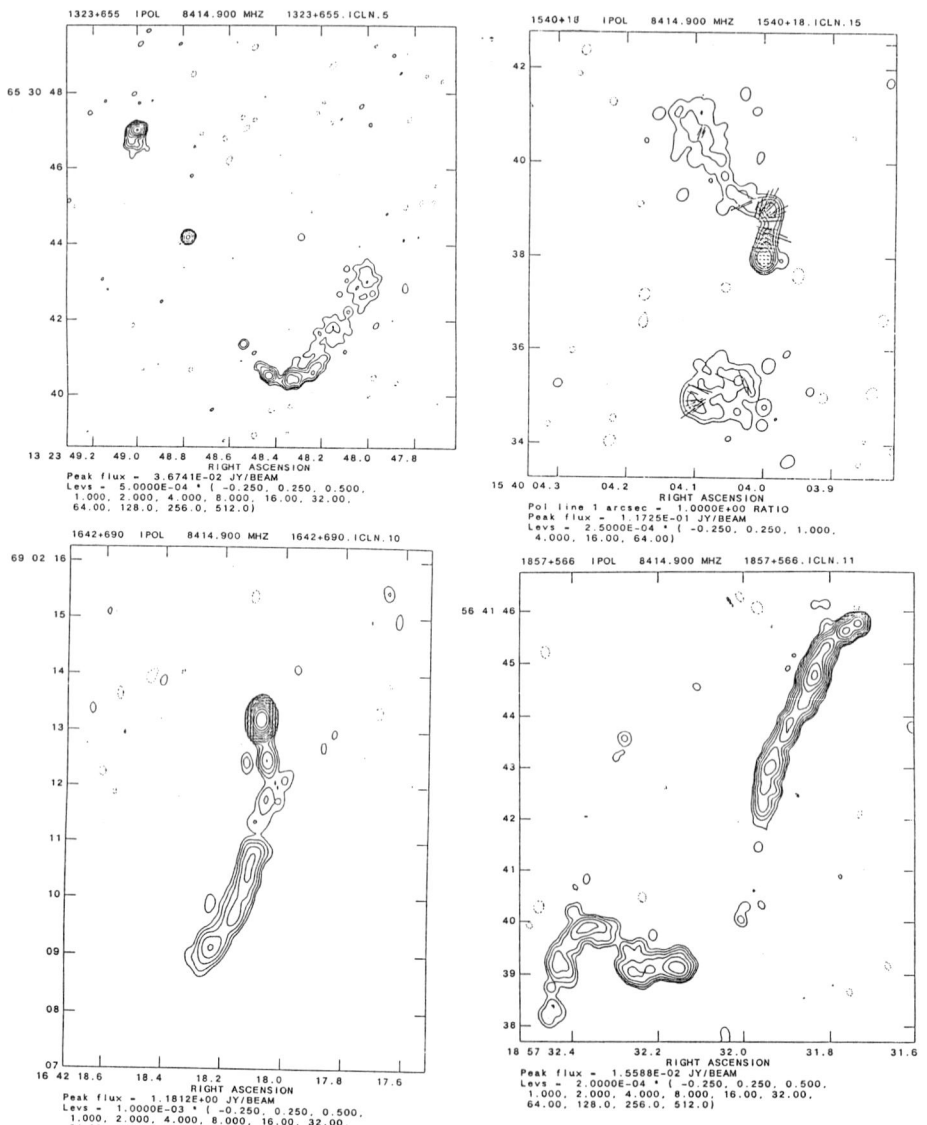

Figure 1. VLA images at 8.4 GHz of the quasar jets in 1323+655, 1540+18, 1642+690 and 1857+566

3C216: A Galactic–size powerful radio source?

Chidi E. Akujor [*]   *Richard W. Porcas* [†]
Everton Lüdke & David L. Shone [‡]

Abstract

Detailed imaging of the complex powerful source, 3C216 provides information on different scales suggesting that it is a galactic–size object.

3C216 is a complex radio source associated with a BSO (z=0.688) and has blazar properties – rapid optical variability and strong, variable optical polarisation. It is also classed as a compact steep–spectrum source since most of its flux density arises from a compact radio structure, $\theta \sim 2.5$ arcsec and on account of its steep radio spectrum, $\alpha = 0.9$ ($S \propto \nu^{-\alpha}$).

3C216 does not have the conventional double or triple structure usually associated with bright radio sources, but has some of the most complex and convoluted structure amongst active energetic galaxies. This is a report on our continuing investigation of the structure of 3C216 with the VLA, MERLIN and VLBI in order to answer the question: Is it a normal triple radio source seen end-on (*foreshortened by projection*) or is it really a galactic–size object (*dwarfed and distorted by interaction with surrounding plasma ?*).

The relatively bright core, the observed superluminal motion and the extended halo (8.0 arcsec) [BPR88], [FPA92] at low frequency (see map) suggest a source in which projection is important. The majority of CSSs have weak cores. If 3C216 was being viewed at say, 10-30° the corresponding deprojected size of 58 – 165 kpc is typical of double sources of the same intrinsic luminosity.

But other factors which favour an intrinsically small and distorted source are:

(i) gross distortion and misalignment in structure on all scales – apparent lack of connection between the jets on different scales and with the large–scale structure.

---

[*]Onsala Space Observatory, Onsala, S–439 92, Sweden and Nuffield Radio Astronomy Laboratories, Jodrell Bank, Cheshire SK11 9DL, UK.
[†]Max Planck Institute fur Radioastronomie, Auf dem Hugel 69, Bonn, F.R.G.
[‡]Nuffield Radio Astronomy Laboratories, Jodrell Bank, Cheshire SK11 9DL, UK.

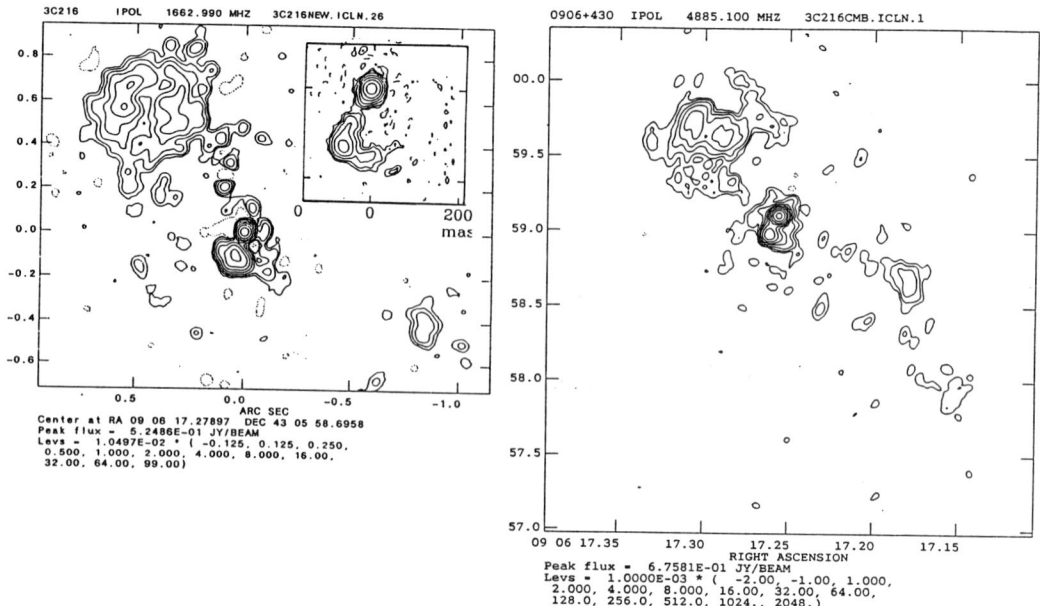

Figure 1. Combined MERLIN–EVN map (left) at 1.6 GHz of 3C216 (inset is the EVN only map) and combined MERLIN–VLA map at 5 GHz

(ii) the unusually sharp bend at a bright knot of the VLBI jet – the flat radio spectrum of the knot, $\alpha \sim 0.4$ is an indication of a physical interaction. This is supported by the high O[II] emission line luminosity of 3C216.

(iii) combined MERLIN–EVN map suggests that the jet may have turned round to connect the Northern lobe. Note that at high resolution often the jet is connected to the brighter lobe.

(iv) depolarisation asymmetry – comparison of our 5GHz map with high frequency data shows that the Northern component is depolarised, indicating high faraday depth (see Akujor & Garrington, this proceedings).

References

[BPR88] Barthel, P.B., Pearson, T.J. & Readhead, A.C.S., 1988, ApJ, 329, L51-55.

[FPA92] Fejes, I., Porcas, R.W. & Akujor, C.E. 1992, A & A, 257, 459-464.

The sub–arcsecond radio structure of 4C39.25

N. Jackson *   I.W.A. Browne †   A. Alberdi ‡   J. Marcaide §

Abstract

A new 5GHz 70mas resolution MERLIN image of the superluminal quasar 4C39.25 shows structures within the previously unresolved jet. The presence of a straight–line "jet" is now more obvious, and the structure of the terminal hotspot is very similar to that of larger sources. This provides further evidence for basic similarity between sources such as 4C39.25 and larger sources.

4C39.25 is a redshift 0.699 core–dominated quasar with arcsecond–scale extended emission. Browne et al. (1982) argued that such core–dominated quasars were intrinsically weak "classical double" lobe–dominated quasars, viewed at such an angle that their cores appeared brighter due to Doppler boosting (Orr and Browne 1982). The detection of superluminal motion (a component moving between two stationary components: Marcaide et al. 1985, Shaffer et al. 1987) in this object supported this idea and has been modelled by Marcaide et al. (1989) in terms of a relativistic jet bending close to the line of sight. >0.15" resolution VLA and MERLIN observations (Marcaide et al. 1989, Marscher et al. 1991) show limited high resolution structure, consisting of a bright terminal blob (2" E of core, "the hotspot"), emission trailing west from this and a knot at 0.7" E. There is a lower surface brightness lobe 2" W.

The new MERLIN 5GHz observations were performed on 1991 September 7: figure 1 shows the new map. Its limiting factor is the residuals due to closure errors from the extremely bright central point source, although the peak to lowest believable contour is still ≃10000:1. Full details of the data obtained will be given elsewhere (Jackson et al, in preparation). There are minor artefacts very close to the 10Jy core, the most important being the slight extension to the southwest of the core and the trail of emission at the level of the lowest contour to the northwest of the knot.

The new map shows all the features that have been seen at lower resolutions, and the western lobe is just visible: these features reinforce our conclusion that the map

*Sterrewacht Leiden, Postbus 9513, 2300RA Leiden, Netherlands
†NRAL, Jodrell Bank, Macclesfield, Cheshire SK11 9DL, England
‡Instituto de Astrofísica de Andalucia, Apt. Correos 2144, Granada E18080, Spain
§Universitat de Valencia, Valencia, Spain

is basically sound. In it we see further features within the hotspot, namely a condensation at the eastern end together with a subsidiary ridge to its west. The southern side of this structure is sharply defined, and a ridge of emission enters it from the west. This ridge is exactly in a line with the knot and the core. We suggest that for the first time in this object, there is now evidence for an arcsecond scale jet, although it does not fulfil all the criteria defined by Bridle and Perley (1984) for the use of the word "jet". This is one of the few small core–dominated radio quasars in which such evidence is available, and it appears to very similar to structures seen in more extended classical double sources. Evidence of this kind in small core–dominated sources promises to be an important consistency check on models which claim that the two classes of source are indeed the same types of source seen from a different angle.

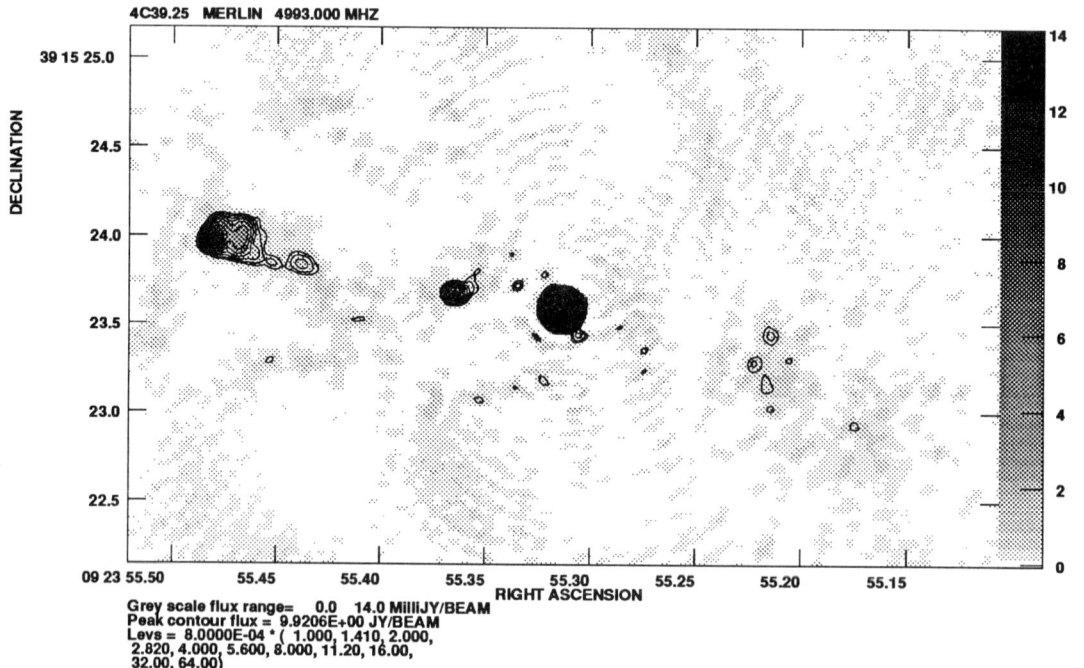

References
Bridle, A. H., and Perley, R. A., 1984, ARA&A, 22, 319
Browne, I. W. A., et al., 1982, MNRAS 198, 673
Marscher, A. P., et al., 1991, ApJ, 371, 491
Marcaide, J., et al., 1989. A&A 211, L23
Marcaide, J. M., et al., 1985. Nature, 314, 424
Orr, M. J. L. and Browne, I. W. A., 1982, MNRAS 200, 1067
Shaffer, D. B., et al., 1987, ApJL 314, L1

3C159: a peculiar radio galaxy

Pietro L. Cerchiara, Franco Mantovani [*]
Ian W.A. Browne, Tom W.B. Muxlow [†]

Abstract

We present a MERLIN observation of the total intensity and polarization of 3C159: an extended radio source that contains a prominent double hot spot in one of the two outer lobes.

We show that the low frequency variability of the source can be the result of a combined effect of the high linear polarization of the source with the ionospheric Faraday rotation during the variability observations.

1. Introduction

3C159 is a very peculiar object that shows variable radio flux density at 408-MHz ([F81]), a steep radio spectrum and an extended double radio structure with a double hot spot. It is a powerful radio source with a 408-MHz luminosity of 1.6×10^{27} W/Hz and an overall linear size of 115 Kpc (H_0=100 km/s$\times$Mpc, q_0=0) ([B85]).

Browne et al. suggested that the flux density variation, which shows a 6 months periodicity, could arise from the combined effect of a hypothetical high linear polarization of 3C159 and changes in the ionospheric Faraday rotation.

Here we present a MERLIN map of the total intensity and of the polarization of the radio source.

2. Discussion and conclusions

The absence of a very compact high-brightness component excludes the interpretation of the low frequency variability as a result of relativistic beaming or refractive scintillation. The high linear polarization of the source at 408 MHz (10%) (Fig. 1), leads us to believe that the low frequency variability can be the result of a combined effect of ionospheric Faraday rotation and high linear polarization. Indeed,

---

[*]Istituto di Radioastronomia, Consiglio Nazionale delle Ricerche, Bologna, Italy.
[†]Nuffield Radio Astronomy Laboratories, University of Manchester, Jodrell Bank, England.

the plane of polarization of the source emission relative to the linearly polarized E-W arm of the Northern Cross Bologna telescope can be swept by changes in ionospheric Faraday rotation during the observations. A source with a linear polarization > 6% could exhibit spurious variations of the order of magnitude reported in 3C159, if the ionospheric Faraday rotation changes by > 90^0 between observations. The 3C159 variability observations were made at transit, during the day in summer and at night in winter ([B85]). This could produce systematic changes in the measured flux density with time of the year because the ionospheric electron content changes by large amount between night and day.

References

[F81] Fanti R., Fanti C., Ficarra A., Mantovani F., Padrielli L., Weiler K.W., *Astron. Astrophys. Suppl. Ser.*, 1981, Vol. 45, p. 61.

[B85] Browne I.W.A., Mantovani F., Muxlow T.W.B., Padrielli L., Romney J.D., in *Active Galactic Nuclei, Proceeding of a workshop in Manchester*, 1985, p. 68.

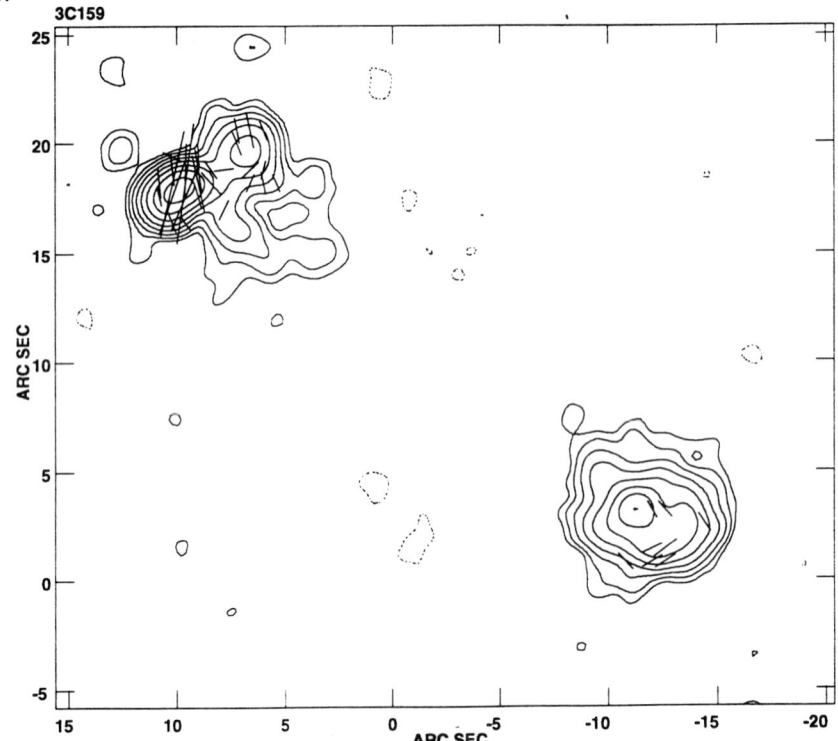

Figure 1: *408 MHz MERLIN map. The beam is 1.2×1.2 arcsec^2. Contours are at -5, 5, 10, 20, 40, 80, 160, 320, 640 mJy beam^{-1}. A vector 1 arc second long represents a polarized intensity of 12.5 mJy beam^{-1}. The noise is 3.5 mJy beam^{-1}.*

SUPERLUMINAL MOTION AT 4 ARCSECONDS IN 3C120?

R. C. Walker and J. M. Benson *

Abstract

Previous attempts to measure the apparent velocity of the 4 arcsecond knot in 3C120 with the VLA [WWB88] and MERLIN [MW91] gave conflicting results regarding the presence of superluminal motion. Both measurements were at the limit of what could be done with very careful observations with the instruments, so it was not clear which was right. We report here third epoch observations with the VLA that indicate no motion, in agreement with the previous MERLIN result.

1. Introduction

There are no direct ways to measure the velocity of the material in extragalactic radio jets. Perhaps the best indirect velocity indicator is the measurement of superluminal motions on parsec scales in many quasars. The standard model for apparent superluminal motions requires relativistic motions nearly along the line-of-sight so the observation of such motions is usually taken as good evidence for a relativistic jet. Until recently, measurements of motions of jet components have been confined to observations on parsec scales using VLBI. However, with very high dynamic range observations using the VLA or MERLIN, it is just possible to measure superluminal motions in the most favorable sources on time scales of a few years.

Of the superluminal sources normally observed with VLBI, 3C120 is the best candidate for such observations. It is the closest and therefore shows the largest angular rates of motion — about 2.5 mas per year. It also has a large scale jet with a pronounced knot at 4 arcseconds from the core. However the knot is weak relative to the core and resolved at about half an arcsecond. The observations are best made with MERLIN at 18 cm or the VLA at 6 cm and all known measures must be taken to obtain the highest possible dynamic range. Measurement of a motion of 2.5 mas per year over 4 years with the VLA requires detecting a change of separation 35 times smaller than the beam, of two features that differ in flux density by a factor of 15!

---

*National Radio Astronomy Observatory, Socorro, NM, USA. The NRAO is operated by Associated Universities Inc. under cooperative agreement with the National Science Foundation.

2. Results

Our first VLA observations were made in October 1983 and July 1987. A shift of 9 mas was measured using cross correlation of sections of the images [WWB88]. This was thought to be about a 3 sigma result. Using, as had we, existing data for the early epoch, Muxlow and Wilkinson attempted to confirm this result but found no motion at 18 cm in the 8 years between December 1980 and December 1988 [MW91] at significance level similar to ours. We report here third epoch observations with the VLA made in August 1991. Cross correlation of the new image with that from July 1987 (much better than Oct. 1983), gives a formal result of a motion toward the core of 2.4 mas, essentially a null result within the noise. A careful analysis will be required before a formal upper limit to the jet velocity can be derived. That upper limit may still be just over the speed of light, but it will be significantly smaller than the velocity seen with VLBI.

It now seems that the 4 arcsecond knot in 3C120 is not moving at the speed of the VLBI components. We still do not fully understand why our first measurement gave a positive result. We do not suspect that there was a technical problem with the earlier measurement but rather that changes in the source combined with noise contrived to give the result seen. Possible contributing effects include a shift in the centroid of the core plus the inner jet due to the 50% rise in core flux density between the epochs, a core shift due structural changes (VLBI observations at the same times suggest this should have caused the opposite effect), and a knot centroid shift due to brightness or other structural changes. The relatively low quality of the October 1983 image probably also contributed.

Unfortunately, a negative result does not say much about the jet velocity. Both standing shocks and geometric effects can give stationary bright spots in a relativistic jet.

References

[MW91] Muxlow T. W. B., and Wilkinson P. J., *MERLIN Observations of Superluminal Motion in the jet of 3C 120.* M.N.R.A.S. Vol. 251 (1991), pp. 54-62.

[WWB88] Walker R. C., Walker M. A., and Benson J. M., *Evidence for Superluminal Motion on Kiloparsec Scales in 3C 120.* Ap. J. Vol. 335 (1988), pp. 668-676.

Polarization Observations of CSS Sources With the Enhanced MERLIN

E. Lüdke, T.W.B. Muxlow, S.T. Garrington, R.E. Spencer *
C.E. Akujor [†]

Abstract

MERLIN phase II polarization observations at 5 GHz and 60 mas resolution of archetypical CSS sources are presented for the first time. The maps show high asymmetry in polarization in symmetric compact double sources where regions nearest to the core depolarize faster than distant ones. As it is seen in normal radio galaxies and quasars, compact jets in these sources ends on the brightest hotspot, which depolarizes faster than counter-jet side. Other observational features are briefly discussed.

The polarization study of compact steep-spectrum sources (CSSs) is a powerful tool for investigating the physics of beam-gas interaction and the physical properties of the dense intergalactic medium surrounding them.

Their small linear sizes ($\leq$ 4 arcsec) make them suitable for detailed studies with the enhanced MERLIN. In particular, MERLIN with a new 32m telescope, larger bandwidth (30 MHz), cooled receivers and polarization capabilities allowed us to map the CSSs at high resolution (60 mas) and with low noise levels ($\leq 500\mu$Jy/beam).

The observations comprises 12 hours source tracking and the phase-referencing technique to point sources within 5 degrees of the CSSs was used to correct for ionospheric fluctuations and to derive the feed crosstalks with 0.5 % error. 3C286 was used to set the position angles with an accuracy of 5°. The data were processed with the NRAO AIPS package and the maps have high dynamic range ($>$ 1000:1).

There is evidence for steep-spectrum galactic-scale bent jets ($\leq$3 Kpc extension) with knots and wiggles, linking cores to the brightest hotspot. There is also evidence for weak and unpolarized cores at 5GHz with peak flux $\leq$ 1 mJy/beam (e.g. 3C216, 3C190 and 3C318) and 8 % polarized flux appears in cores with unresolved VLBI jets.

---

*University of Manchester, NRAL-Jodrell Bank, Macclesfield, Cheshire, SK11 9DL, England
[†]Onsala Observatory, Onsala, S43992, Sweden

Comparing these maps with the published VLA maps at 15 GHz (van Bruegel et al, 1992), we infer that the dense environment surrounding these CSSs contributes significantly to the depolarization of the synchrotron radiation and large amounts of Faraday rotation have been found between 5 and 15 GHz. The extended structures near the core of 3C268.3 and 3C67 (figure 1) depolarize faster between 5 and 15 GHz than the hotspots farther away, which may be understood as if the surrounding gas is denser at the source center and the external lobes are seen through less depolarizing medium (Garrington, Conway and Leahy , 1991) with magnetic field irregularities smaller than 130 pc (Ho=75 Km/sMpc, qo=0.5).

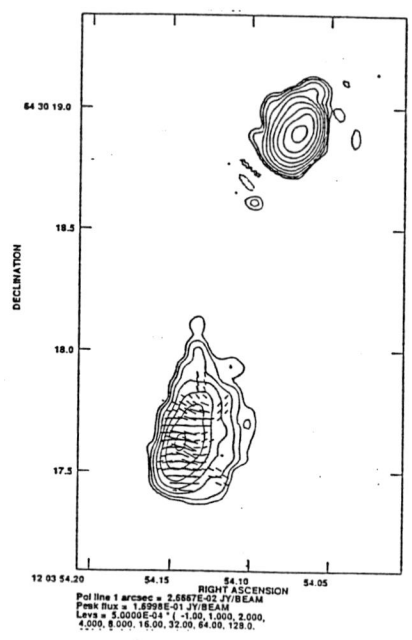
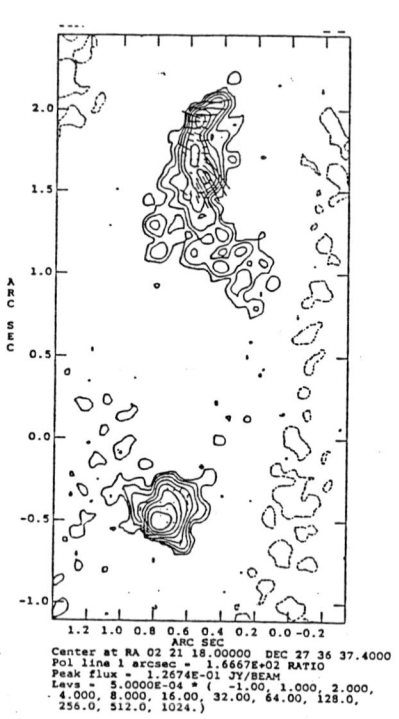

Figure 1. MERLIN II polarization maps of 3C268.3 and 3C67. The polarization degree ranges from 7 to 12 % and the sensitivity limit of polarized emission is $\approx 750\mu$ Jy/beam.

van Bruegel W.J.M., Fanti C., Fanti R., Stanguellini C., Schillizi R.T. and Spencer R.E., 1992, A & A, 256, 56

Garrington S.T., Conway R.G. and Leahy J.P.,1991, MNRAS, 250, 171

Polarization Observations of Compact Steep Spectrum Sources

Chidi E. Akujor *   *Simon T. Garrington* †

In studying compact steep-spectrum (CSS) radio sources, radio polarization observations provide an important diagnostic of the environments of these sources, while asymmetry in the degree of polarization of individual components may provide information on the orientation of the source to the line of sight (Garrington *et al.* 1990). The relative importance of these two factors – environment and orientation – is a key question in understanding the nature of CSS sources.

Previous studies (*e.g.* van Breugel *et al.* 1984) have suggested that Faraday effects may be strong in CSS sources, but to date no systematic study using observations at several wavelengths with comparable resolution has been made. We have started such a study with the aim of observing all steep-spectrum ($\alpha \geq 0.6$) 3C sources with linear sizes less than 15 kpc (for $H_0 = 75$ kms^{-1}Mpc^{-1}) and angular sizes larger than 0.8 arcsec. For sources with angular size larger than 2 arcsec we use the VLA at 1.3 - 1.6 and 5 GHz in A and B configurations at 1 arcsec resolution, while for the smaller sources we use MERLIN at 1.4 - 1.7 GHz and the VLA at 8.4 GHz at 0.2 arcsec resolution.

Figure 1 shows examples of our new 8.4 GHz VLA maps. We have calculated the polarization parameters for each source and comparison with data at 15 GHz given by van Breugel *et al.* (1984) shows that nearly all sources depolarize significantly between 15 and 8.4 GHz. While some of the depolarization may be due to the lower resolution of the 8.4 GHz data, the dominant effect is probably Faraday depolarization. We estimate that the observed depolarization wavelength $\lambda_{1/2}$ for these sources is typically 4 cm, compared to about 20 cm for 3C sources in general. This confirms other suggestions of high Faraday depths in CSS sources, but highlights the problem of obtaining matched resolution data at wavelengths near $\lambda_{1/2}$ for the smaller sources. Further progress may be made by comparing MERLIN 5 GHz and VLA 22 GHz maps at 75 milliarcsec resolution.

The 8.4 GHz VLA maps also show a marked asymmetry in the degreee of polarization of the two components, which we attribute to an asymmetry in depolarization. In

---

*Onsala Space Observatory, Onsala, S–439 92, Sweden and Nuffield Radio Astronomy Laboratories, Jodrell Bank, Cheshire SK11 9DL, UK.

†Nuffield Radio Astronomy Laboratories, Jodrell Bank, Cheshire SK11 9DL, UK.

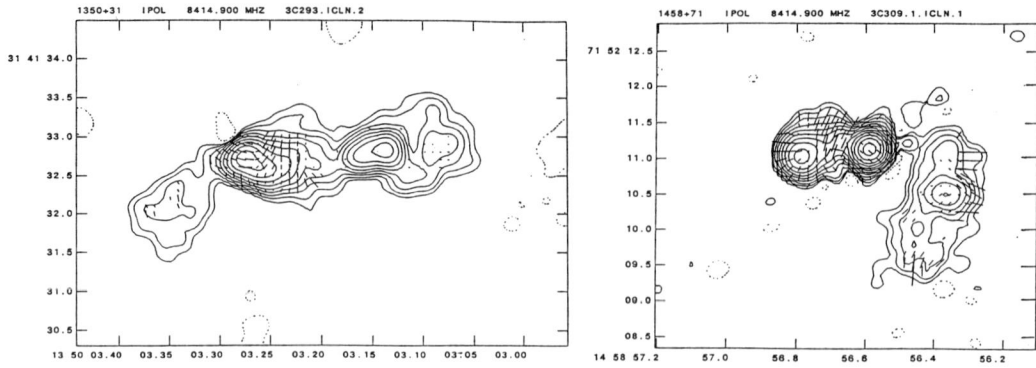

Figure 1. VLA maps at 8.4 GHz of the CSS sources 3C213.1 and 3C309.1

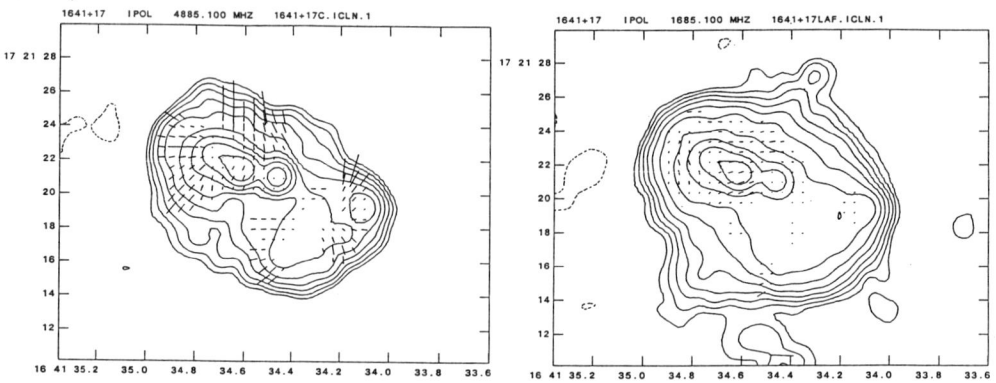

Figure 2. VLA maps at 5 and 1.7 GHz of 3C346, showing asymmetric depolarization in the lobes

all sources with jets, the lobe opposite the jet depolarizes more rapidly as found by Garrington et al. (1990) for larger sources. Our lower resolution maps at 5 and 1.5 GHz confirm this trend in the larger CSS sources, and an example is shown in Fig. 2.

References

Garrington, S.T. et al. 1991, *Mon. Not. R. astr. Soc.* **250**, 171.

Garrington, S.T. and Conway, R.G. 1991. *MNRAS* **250**, 198.

van Breugel, W, Miley, G. and Heckman, T.1984. *Astron. J.* **89**, 5.

Radio Polarisation Observations of 807 Flat Spectrum Sources

Chidi E. Akujor *   *Alok R. Patnaik* [†]

We have observed 807 flat spectrum radio sources with the VLA at 8.4 GHz. These are snapshot observations only for 1 to 1.5 minutes and therefore not sensitive to the extended emission ([Pa92]). Hence the results presented are the polarisation properties of radio cores. The sources were selected from the Green Bank catalogues at 1.4 and 5 GHz using the following criteria. (i) S(6cm) $\geq$ 200 mJy, (ii) Spectral index flatter than 0.5, (iii) $35° \leq \delta \leq 75°$ and, (iv) $|b| \geq 2.5°$.

Spectral index of these sources have been derived using flux densities at 365 MHz (Texas) and 8.4 GHz. Optical identifications of the sources are from POSS prints, supplemented by APM scans. Nearly 70% are identified with stellar objects, 20% have no optical counterparts, and 6% are associated with galaxies.

We find that the median fractional polarisation for stellar objects (mostly quasars) (2.5±0.1 %) is higher than that for galaxies (2.0±0.2 %). The values for BL Lacs is 3.0± 0.3 %, however, we have only 10 known BL Lacs in the sample. The empty fields (2.4±0.1 %) have similar distribution as quasars, suggesting that flat spectrum empty fields are quasars.

The mean spectral index of the sources between 365 MHz and 8.4 GHz are: stellar objects $\sim$ 0.12; galaxies $\sim$ 0.19; empty fields $\sim$ 0.22 and BL Lacs $\sim$ 0.15. There are no obvious differences in the spectral index distributions. Also there is no correlation between spectral index and either fractional polarisation or redshift.

The distribution of PA(VLBI structure) – PA(Core polarisation, E–vector) appears to be uniform (see [ZP85]; [AP92]). However, PA(VLA structure) – PA(Core polarisation) has an obvious peak about 90°. This suggests that the magnetic field at the core is well aligned with the VLA–scale structure but not so for VLBI–scale structure.

---

*Onsala Space Observatory, Onsala, S–439 92, Sweden and Nuffield Radio Astronomy Laboratories, Jodrell Bank, Cheshire SK11 9DL, UK.

[†]Nuffield Radio Astronomy Laboratories, Jodrell Bank, Cheshire SK11 9DL, UK.

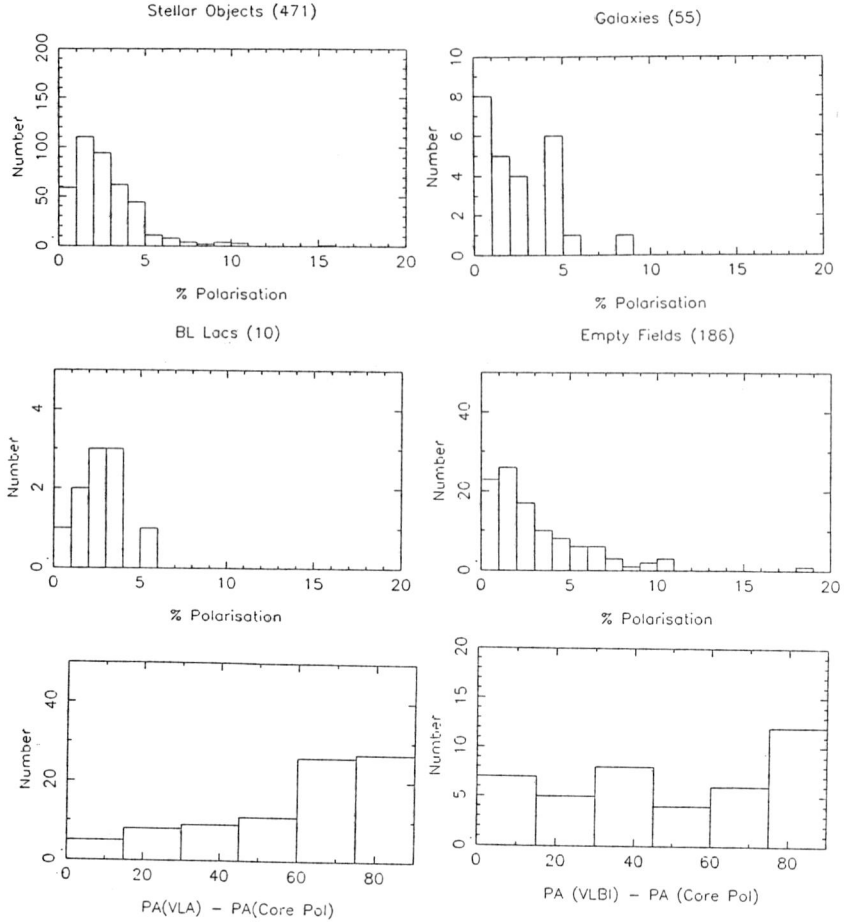

Figure 1. A histogram of core % polarisation for different classes (top and middle) and position angle differences (bottom)

References

[AP92] Akujor C.E., Porcas R.W., 1992, in 'Extragalactic Radio Sources: From Beams to Jets', Roland J., Sol H., Pelletier G., eds., Cambridge Univ. Press, p134.

[Pa92] Patnaik A.R. et al. 1992, MNRAS, 254, 655

[ZP85] Zensus A., Porcas R.W., 1985, in 'Active Galactic Nuclei', Dyson, J., ed. Manchester Univ. Press, p54.

Low Frequency Variability and Structure of the Quasar 3C 345

L.I.Matveenko [*]

The activity of the nucleus of the quasar 3C 345 is accompanied by strong radio outbursts, with a typical lifetime of a few months. The outbursts appear first, and are strongest, at short wavelengths, and then propagate to longer wavelengths while decreasing in amplitude. However the variability of the decimeter radio emission is again increased. This contradicts the model for synchrotron emission. The brightness temperature of the emission region calculated from the timescale and amplitude of the variability is about $T_b \sim 10^{16}$K.

It has been suggested that the low frequency variability arises in the interstellar medium, [Ri86]. If the variations are intrinsic to the source, the Lorentz factor ≥ 100 required is much larger than that derived from the observations. In both cases, the source would be compact and the brightness temperature would be very high and variable. Another possibility is that the low frequency variability is determined by absorption.

We observed 3C 345 in 1981.2 at λ 18 cm, and between 1983 and 1990 at λ49 cm with a global VLBI array [MGP92]. The core at λ18 cm has a size $\sim 0.3 mas$, flux density $F = 0.2\ Jy$ and brightness temperature $T_b \sim 10^{12}$ K. The core spectrum has a low frequency cut-off ($\alpha \geq 3$), and the core is not visible at λ49 cm.

The compact part of the jet at $\lambda 49 cm$ had: size Q=5.9x3.6mas, flux density $F = 1.94\ Jy$ and brightness temperature $T_b = 0.4 \cdot 10^{12}$ K at epoch 1983.9; Q=3.6x4.7 $mas, F = 2.2\ Jy, T_b = 0.5 \cdot 10^{12}$ K (epoch 1986.8); Q=5x8 $mas, F = 3.7\ Jy, T_b = 0.4 \cdot 10^{12}$ K (epoch 1988.5) [PEG91]; Q=6.5x6.5 $mas, F = 4.2 Jy, T_b = 0.4 \cdot 10^{12}$ K (epoch 1990.8). At 18 cm (epoch 1981.1) the component had $F = 1.2 Jy$, Q=0.3 mas and $T_b \sim 6.8 \cdot 10^{12}$ K.

The source size is enlarged by interstellar scattering. The scattering angle is:

$$Q_{sc} \sim 10^{-3} \lambda^2 |\sin b|^{-0.5}\ [mas] \tag{1}$$

which for 3C 345 is $Q_{sc} = 3\ mas$ at λ 49cm. After correction of the measured angular sizes of the source for scattering the brightness temperature becomes $T_b \sim (0.5 - 1.3) \cdot 10^{12}$ K.

The spectrum of an outburst between epochs 1981.5 - 1982.5 had a deep low frequency cut-off and a spectral index $\alpha \sim 3.2$, [BGH86], [MGP92]. The spectral index of the compact component was $\alpha \geq 3$. This suggests a distribution of relativistic electrons

---

[*]Space Research Institute, Moscow

in an ionized medium. The spectrum before and after the active period was flat, but the radio emission was higher than earlier. It shows that the UV emission decreased and the transparency of the ionized medium increased.

Optical studies of quasars, including 3C 345, show that the nuclei are surrounded by an ionized medium which radiates emission lines. The size of the emission region is $\sim 10^{21}L_{-46}cm$ with an electron density $N_e \sim 10^{4-6}$ cm^{-3} and a temperature $T_e \sim 10^4$ K, [Ne87]. The 3C 345 optical–UV luminosity is $\sim 10^{45} erg\ s^{-1}$, [BGH86], and the region size $\sim 30pc$ (8 mas). The UV emission changes the ionization of the medium and the timescale of variations ranges from a few days to a few years.

The low-frequency variability can be characterized by "negative" outbursts, anti-correlated with those at high frequencies [AA82] and uncorrelated for quasars located closely together in the sky [SC81]. These results can be explained in terms of absorption of the relativistic electrons responsible for the synchrotron emission by an ionized medium surrounding the nucleus.

The core size is determined by the distribution of the magnetic field, energy and density of relativistic electrons and ionized medium. The optical depth of the ionized medium is:

$$\tau = 0.08 \cdot T_e^{-1.35} \cdot f^{-2.1} \cdot \int N_e^2(l)dl \qquad (2)$$

VLBI measurements at different frequencies, [MGP92], show that the radius of the core depends on the frequency as: $R(f) = 1.05 f^{-0.54}$ pc. Assuming that the electron density distribution is $N_e(l) \sim l^{-\beta} cm^{-3}$ and $\tau \sim 1$, then $R(f) \sim f^{-2.1/(2\beta-1)} pc$, and $N_e(l) \sim 7 \cdot 10^6 l^{-2.44} cm^{-3}$. The recombination time of an ionized medium is $t \sim 10^5 \cdot N_e^{-1} yr$, [Ne87] and so at the frequency 0.4 GHz we have $R \sim 1.7pc$, $N_e \sim 10^6 cm^{-3}$ and $t \sim 0.1yr$, which does not limit the timescale of the low-frequency variability.

The 49 cm flux density and size of the compact component has increased by ~ 2 times during the observing period, but the millimeter emission has decreased by ~ 2 times, [Va92], which corresponds to a decrease of the nuclear activity. The increase in the 49 cm observations may be explained by an increase in transparency of the ionising medium.

References

[AA82] Aller H.D., Aller M.F., *Proc. NRAO Workshop on "Low Frequency Variability of Extragalactic Radio Sources", Cotton and Spangler (eds.), 1982, pp. 105.*

[BGH86] Bregman J.N., Glasgold A.E., Huggins P.J., et al., *Multifrequency observations of the superluminal quasar 3C 345. Ap.J., Vol.301 (1986), pp.708-726.*

[MPS86] Matveenko L.I., Pauliny-Toth I.I.K., Scherwood W.A., Bååth L.B., Kus A.J., *Decimeter-wavelength structure in quasar 3C 345. Pisma Astron. Zh., Vol. 12 (1986), pp. 156-167.,(Sov. Astron. Lett., Vol. 12, pp.63-67.)*

[MGP92] Matveenko L.I., Graham, D., Pauliny-Toth I.I.K., Scherwood W.A., Bååth L.,B., Kus A.:(1992). *The structure of the quasar 3C 345 at 49 cm.* in preparation.

[Ne87] Netzer H., *The physics and structure of AGN.* "Astrophysical Jets and Their Engines", Series C: Mathem. and Phys. Science, Kundt (ed.), Reidel Publishing Company, Vol. 208 (1987), pp. 103-124.

[PEG91] Padrielli L., Eastman W., Gregorini L., Mantovani F., Spangler S., A&A, Vol. 249 (1991), pp. 351-357.

[Ri86] Rickett B.J., *Refractive interstellar scintillation of radio sources.* ApJ, Vol. 307 (1986), pp. 564-574.

[SC81] Spangler S.R., Cotton W.D., *Broadband radio observations of low-frequency variable sources.* Astron.J.,Vol. 86 (1981), pp. 730-746.

[Va92] Valtaoja E.:1992., *private communication.*

On Angular Structure Studies of Very-Steep Spectrum Sources Found at Decametric Wavelenghths

Konstantin P. Sokolov *

1. Introduction

Previous studies of extended radio sources with low surface brightness and very steep ($\alpha \geq 1.0$, where $S \sim \nu^{-\alpha}$) spectra were mainly based on strong 3CR-sources or sources selected at frequencies above 178 MHz. We intend to study the angular stucture of the sources from a much weaker sample which has been selected at a frequency below 100 MHz.

2. Why select the sources found at very low frequencies ?

Studying the structure of extended sources with low surface brightness and very steep spectra that have been found at very low ($\nu < 100$ MHz) frequencies is of considerable astrophysical interest because:

1) these objects represent the final stage in the evolution of extragalactic radio sources, i.e. the objects in which the energy supply process to their extended components has stopped;

2) the effect of the IGM on the overall structure and spectra of the sources should manifest itself in a most pronounced way for their extended components, so the data obtained can be used to probe the surrounding IGM parameters;

3) these objects have a lifetime of $\sim 3 \times 10^9$ yrs. comparable with the age of the Universe, i.e. they could be used for estimating the IGM's parameters for earlier cosmological epochs;

4) these objects constitute a subsample of the "parent population" of extragalactic radio sources which is not biased by alignment effects;

5) these sources are characterized by the least scatter in their intrinsic luminosities, so they are the most promising objects to be used as a "standard candle".

---

*Institute of Radio Astronomy, Kharkov 310002, Ukraine

3. The UTR-2 very low-frequency sample of radio sources

The source sample needed for such a study has been selected from the Very Low-Frequency Sky Survey made at 10, 12.6, 14.7, 16.7, 20 and 25 MHz with the UTR-2 radio telescope (Braude et al. 1985).

The UTR-2 sample contains 265 sources selected by the following criteria:
1) flux density at frequency of 16.7 MHz is $S \geq 29$ Jy;
2) the spectral indices in the range 16.7 - 178 MHz are $\alpha \geq 1.0$;
3) ranges of declinations are $-13^o \leq \delta \leq 20^o$ and $41^o \leq \delta \leq 60^o$;
4) galactic latitudes $\mid b \mid \geq 15^o$.

The flux-density limit of the sample corresponds to a value lower than 2.8 Jy at 178 MHz, i.e this sample is much weaker than the well-studied 3CR-sample. On the other hand, the accepted spectral-index limit has proven to be an efficient criterion for searching the most distant objects. The sample contains radio galaxies, quasars and sources which have not been identified with previously known objects.

4. Conclusions

The main goals of the angular structure study of the UTR-2 sources are:
1) to obtain radio images at two or more frequencies in the range of 100 - 1000 MHz for sources with very steep low-frequency spectra;
2) to obtain the spectra of their extended components which can be used for estimating ages and expansion velocities of the sources;
3) to obtain accurate positions for the unidentified sources;
4) analysis of the morphology, angular sizes and spectra of the extended components of unidentified objects, radio galaxies and quasars, as well as search for their dependence on the IGM's parameters and the redshift;
5) search for very old "fossil" and distant radio sources;

To reach these aims, radio images in the range of 100 - 1000 MHz with resolution not worse than 20 beams across the sources are needed. The mean angular source dimensions are expected to be $\sim$ 100 arcseconds. Therefore, we need maps obtained with a resolution of $\simeq$ 5 arcseconds within the range. The VLA and MERLIN meet these requirements.

So, results of the study can provide us with new information on physical conditions in the extended components of extragalactic radio sources and their environments.

References

Braude, S.Ya., Sharykin, N.K., Sokolov, K.P., Zakharenko, S.M., 1985. *Astrophys. and Space Science, 111,* 1.

Optical and HST observations of jets

William B. Sparks *

Abstract

The Hubble Space Telescope for the first time enables optical and UV images of jets to be obtained with spatial resolution comparable to radio interferometric techniques. A detailed comparison of HST and VLA images of the M87 jet is presented, showing similaritites between the two, but also identifying differences. The optical and UV data are more localised than the radio and the jet is narrower. An optical hot-spot indicates the presence of an active counterjet. The discovery of a new optical jet in 3C 264 is described. High resolution HST images have revealed previously unsuspected fine structure in the jets of 3C 66B and 3C 273.

1. Introduction

This review is concerned with *Hubble Space Telescope* (HST) observations of optical non-thermal synchrotron jets in radio galaxies. These form a diverse sample ranging over many orders of magnitude in size and luminosity and with very different characteristics, both regarding their internal morphology, polarization and the relationship between optical and radio emission. The FWHM of the HST point spread function is typically ≈ 60 millarcsec, which for the first time gives astronomers a chance to compare optical and UV imaging data directly to the excellent high spatial resolution VLA and MERLIN data at radio wavelengths. The data to be described were obtained using the European Space Agency's Faint Object Camera (FOC) on-board the HST. This camera offers the highest spatial resolution available on the HST and is sensitive across a broad wavelength range from 6000Å in the red to 1200Å in the far-UV. Macchetto (1991) and (1992) reviewed recent HST observations of jets. For an inventory of optical jets, and summary of up-to-date thinking on optical ground-based observations, see Meisenheimer (1991). Other comprehensive review material includes the proceedings of the 1992 STScI Annual Workshop *"Astrophysical Jets"*, and Hughes (1991).

Highly collimated outflows are one of Nature's most favoured means of energy transport, yet one of the least understood. For non-thermal synchrotron jets, optical and

---

*Space Telescope Science Institute, 3700 San Martin Drive, Baltimore, MD 21218, U.S.A.

UV observations offer the prospect of identifying sites of particle acceleration and constraining magnetic field properties because of the very short synchrotron lifetimes of optically emitting electrons. The optical and radio images of a synchrotron jet probe quite different electron populations — Lorentz factors for 'optical' electrons are $\sim 10^6$, compared to ~ 1000 for radio. The effect of the jet on its environment and vice versa may be studied effectively in the optical, and data may be obtained on the host galaxy. Optical polarization observations are free of Faraday rotation effects, and the high spatial resolution and blue sensitivity of the HST/FOC will enable improved detection frequencies to be estimated.

2. The Jet of M87

HST has observed the most famous optical synchrotron jet, that of M87, discovered by Curtis (1918). The synchrotron nature of the ~ 2 kpc long jet has been demonstrated over the years through ground-based polarization and spectral index studies, and the overall correspondence between optical and radio morphology has been well-established (e.g. Baade 1956, Warren-Smith et al. 1984, Schlötelberg et al. 1988, Fraix-Burnet et al. 1989, Stiavelli et al. 1991).

Fig. 1 shows an FOC image of the jet alongside a VLA image at the same scale and resolution. The FOC image is a composite derived by scaling and mosaicing many separate images taken at different times and pointings. The effective wavelength is ≈ 3700Å and the deconvolution method is the Lucy (1974) iterative method. (Results are described in detail in Sparks, Biretta & Macchetto 1992 in prep.) Immediately obvious is the filamentary structure familiar from the ~ 0.1 arcsec resolution VLA radio images of Owen et al. (1989).

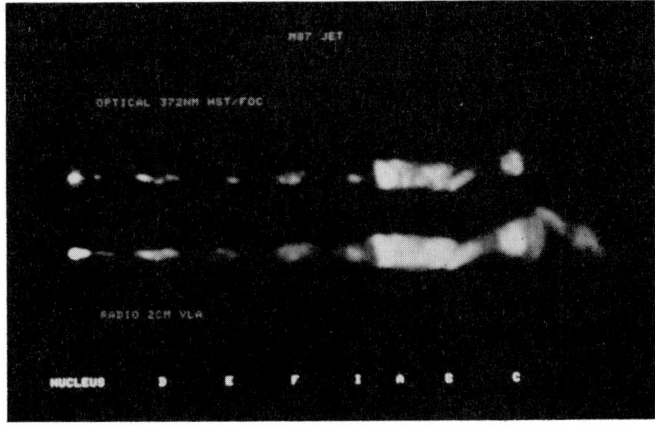

Figure 1. FOC and VLA images of the jet of M87, courtesy J. Biretta.

Both ground-based data (Stiavelli *et al.* 1991), and early analysis HST observations (Boksenberg *et al.* 1992) have stressed the similarity between optical and radio morphology. Fig. 1 shows that indeed the optical/UV and radio emission are remarkably similar, given the difference in electron populations. Nevertheless, they are *not* identical, and to summarize the *differences*: the optical data are more contrasty than the radio data; the faint inter-knot radio emission is not present in the optical with the present S/N; the innermost jet between knot D and the nucleus is quite different; knot D itself has a different structure, as does knot E; the jet is narrower in the optical than in the radio, seen most easily at knot A which is also convex to the nucleus rather than concave.

We may be witnessing the effects of differences in the strength or topology of the magnetic field; localized particle acceleration at shock sites defined by the more compact optical knots; or secular variations in the output of the jet, e.g. Rees (1978), Morabito *et al.* (1988). A simple boundary layer model is contradictory to these data, in that the optical do *not* show prominent limb brightening.

The question of whether the jet is intrinsically two-sided has been revitalized by the discovery of an optical synchrotron hot-spot opposite to the bright jet (Stiavelli *et al.* 1992, Sparks *et al.* 1992). The short optical synchrotron lifetime strongly suggests that energy is being supplied to the SE lobe at the present time and that the jet has an invisible but otherwise roughly symmetric counterpart.

3. Fine Structure in Other Radio Galaxies

Other jets studied by the Faint Object Camera Team include those in PKS0521−36, and 3C 66B, see Macchetto *et al.* (1991a,b). The jet of PKS0521−36 is fully resolved having a width typically 0.4 arcsec, corresponding to a width of several hundred pc, length ∼ 10 kpc. Only smoothly distributed emission is seen. 3C 66B on the other hand, with similar observational material and intrinsic dimensions about half the size of the PKS0521 − 36 jet, shows fine strands in the optical that had not previously been seen in 0.3 arcsec resolution radio data, Leahy *et al.* (1989). Jackson *et al.* (1992, and this conference) present new radio images that reveal fine structure in the radio data as well as in the optical images.

4. Discovery of a new Synchrotron Jet

Several radio galaxies have been observed by the FOC and in one of them, 3C 264 (NGC3862, see Baum *et al.* 1988), a new optical jet was discovered, Crane *et al.* (1992). Fig. 2 shows the Lucy-deconvolved image of the jet. The length is only 0.6 arcsec, and it is barely resolved, if at all, across its width. The corresponding metric length is only of order 400 pc which makes it the shortest optical/radio jet known. The optical spectral index is consistent with those of other synchrotron jets.

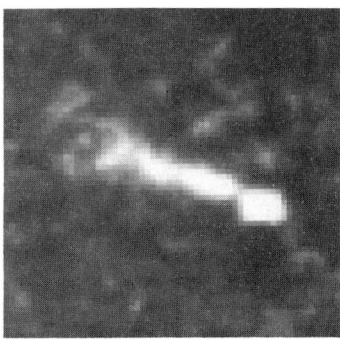

Figure 2. FOC deconvolved image of the jet in NGC3862, Crane et al. (1992).

The synchrotron nature of the feature has yet to be unambiguously established through polarimetry and detailed comparison to high resolution radio data, however the qualitative appearence and quantitative colours are strongly suggestive that this is indeed a new optical synchrotron jet.

5. 3C 273

From the smallest optical jet to the largest — 3C 273, with an extent $\sim 50-100$ kpc, shown in Fig. 3. The images show a wealth of detail, confirming in large part and improving upon the analysis of Evans et al. (1989). These data utilized the imaging polarimetric capability of the FOC and it will be possible to derive magnetic field directions. Here the radio morphology, Conway et al. (1992), Muxlow (this conference), is completely different to the optical. The radio peaks beyond the end of the

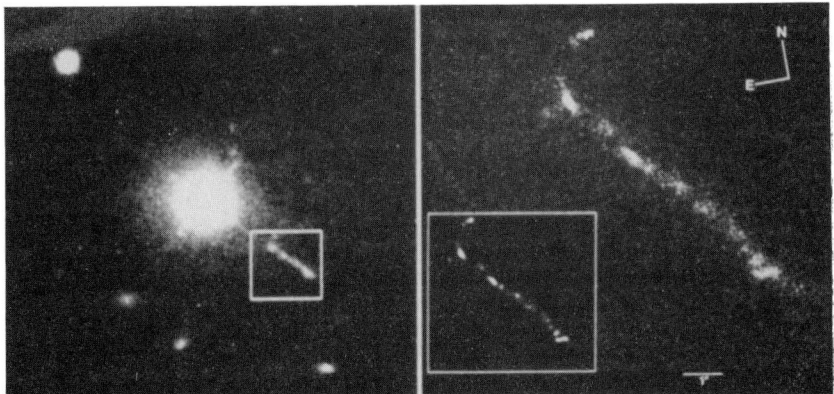

Figure 3. The jet of 3C 273. On the left, a ground-based CFHT image, and on the right, an FOC image without deconvolution. The inset shows an MEM deconvolved version of the FOC image, see Thomson, Wright & Disney (1992).

optical jet and the radio width is of order 1 arcsec compared to the 0.2 to 0.3 arcsec of the optical data.

6. Summary

In conclusion, new optical and UV studies at the highest spatial resolution possible, acquired with the FOC on the HST, have offered astronomers the chance to compare in great detail the different structures of jets seen at wavelengths corresponding to very distinct electron populations. Filamentary structures appear to be common, but not ubiquitous. In M87 there are subtle differences between optical and radio images that may be attributable to short electron lifetimes and diffusion lengths within the optical band, or to different local conditions such as magnetic field strength. The data are inconsistent with the simplest boundary layer models in which all emission arises on the surface of a non-radiating cone. Secular variations may also play a role.

Jets may be significantly more common than presently realized since the HST has already serendipitously discovered a new, previously unsuspected jet.

Future optical/UV spectral index maps together with similar improvements in spatial resolution at X-ray wavelengths with AXAF will allow us to make great progress on understanding the nature of the jet phenomenon.

7. Acknowledgements

The author is indebted to all members of the Faint Object Camera Investigation Definition Team, particularly F. Macchetto. Also, G. Miley, N. Jackson and J. Biretta helped with the analysis of many aspects of the data. In addition, I thank Bob Thomson for the 3C 273 data.

References:

Baade, W., 1956, *Astrophys. J.*, **123**,550.

Baum, S., Heckman, T., Bridle, A., van Breugel, W. & Miley, G., 1988, *Astrophys. J. Suppl.*, **68**,643.

Boksenberg, A. et al., 1992, *Astr. Astrophys.*, **261**,393.

Crane, P. et al., 1992, *Astrophys. J. (Letters)*, in press.

Conway, R.G., Garrington, S.T. & Perley, R.A., 1992, *Astr. Astrophys.*, in press.

Curtis, H.D., 1918, *Publ. Lick Obs.*, **13**,11.

Evans, I.N., Ford, H.C. & Hui, X., 1989, *Astrophys. J.*, **347**,68.

Fraix-Burnet, D., Le Borgne, J.-F. and Nieto, J.-L., 1989, *Astr. Astrophys.*, **224**,17.

Hughes, P.A., 1991, *Beams & Jets in Astrophysics*, C.U.P..

Jackson, N., Sparks, W.B, Miley, G.K. & Macchetto, F., 1992, *Astr. Astrophys.*, in press.

Leahy,J.P., Jägers, W. & Pooley, G.G., 1989, *Astr. Astrophys.*, **156**,251.

Lucy, L.B., 1974, *Astron. J.*, **79**,745.

Macchetto, F., 1991, In *Testing the AGN Paradigm*, eds. S.S. Holt, S.G. Neff, C.M. Urry: A.I.P. Conference Proceedings 254, p 409.

Macchetto, F., 1992, In *Science with the Hubble Space Telescope*, Sardinia, eds. P. Benvenuti & E. Schreier.

Macchetto, F. et al., 1991 a, *Astrophys. J.*, **369**,L55.

Macchetto, F. et al., 1991 b, *Astrophys. J.*, **373**,L55.

Meisenheimer, K., 1991, In *Physics of Active Galactic Nuclei*, Heidelberg, eds. W. Duschl & S. Wagner.

Morabito, D.D., Preston, R.A. & Jauncy, D.L., 1988, *Astron. J.*, **95**,1037.

Owen, F.N., Hardee, P.E., and Cornwell, T.J., 1989, *Astrophys. J.*, **340**,698.

Rees, M.J., 1978, *Mon. Not. R. astr. Soc.*, **184**,61P.

Schlötelberg, M., Meisenheimer, K. and Röser, H.-J., 1988, *Astr. Astrophys.*, **202**,L23.

Sparks, W.B., Fraix-Burnet, D., Macchetto, F. & Owen, F.N., 1992, *Nature*, **355**,804.

Stiavelli, M., Møller, P. & Zeilinger, W.W., 1991, *Nature*, **354**,132.

Stiavelli, M., Biretta, J., Møller, P. & Zeilinger, W.W., 1992, *Nature*, **355**,802.

Thomson, R.C., Wright, A.E. & Disney, M.J., 1992, In *Science with the Hubble Space Telescope*, Sardinia, eds. P. Benvenuti & E. Schreier.

Warren-Smith, R.F., King, D.J. and Scarrott, S.M, 1984, *Mon. Not. R. astr. Soc.*, **210**,415.

The Optical Counterpart of the East Lobe of M87

J.I. González-Serrano [*]   *I. Pérez-Fournon* [†]   *W. Junor* [‡]

Abstract

We report on the results from optical imaging of the East lobe of M87. Our CCD images allow us to resolve it into several peaks not previously reported in either radio maps or recent optical images. New $UBVRI$ photometry is used to discuss its overall radio-to-optical spectrum. Although the optical emission of the lobe is associated with the radio emission, our new data reveal that the brightest radio and optical features do not match well.

1. Introduction

Recently it has been reported the detection of an optical counterpart to the East radio lobe of M87 ([1], [2]). In the most recent radio data at $\lambda 6$ cm with $0''.4$ resolution ([3]), the brightest feature, called θ, is almost diametrically opposite the jet. The detection of polarization at this position ([1]) identifies the optical emission as synchrotron radiation. The interpretation is that the feature is a hot spot fed by an invisible counterjet ([1], [2]). In this work we present deep CCD imaging of M87 from which we have found substructure in the East lobe not reported in previous optical or radio observations.

2. Observations and data reduction

M87 was observed at the INT and WHT telescopes at La Palma. Several exposures in the broad-band filters U, B, V, R, I, and z were obtained. At the WHT the seeing was $0''.9$ but conditions were not photometric. We observed also the quasar pair 1038+528 A+B in order to obtain very accurately the plate scales since the relative position of both quasars is known with microarsecond precision from VLBI observations ([4],

---

[*]Departamento de Física Moderna, Facultad de Ciencias, Universidad de Cantabria, 39005 Santander, Spain.
[†]Instituto de Astrofísica de Canarias, 38200 La Laguna, Tenerife, Spain.
[‡]VLA, National Radio Astronomy Observatory, P.O. Box 0, Socorro, NM 87801, USA.

[5]). The plate scale obtained in this way was $0.''2744 \pm 0.''0002$ pixel^{-1} for the WHT GEC camera.

The M87 galaxy light distribution was modelled and subtracted by fitting ellipses to the isophotes by the procedure described in [6]. Photometric calibration was done using aperture photometry of M87 ([7]). We have produced a total image by adding the residual frames in all the bands. In Fig. 1 we show a contour map, corresponding to this total image, of the brightest region of the East lobe that coincides with feature θ in [3].

3. Results and discussion

Our I, R, V, and B magnitudes are consistent, within the errors, with those given in [1] and [2]. In the later work, the feature is marginally detected at the 2σ level in the U band. We detect emission in this band at the 6σ level, having a magnitude of $m_U = 22.22 \pm 0.21$. The radio-to-optical spectrum is consistent with a power law of spectral index $\alpha = 0.86$ ($S_\nu \propto \nu^{-\alpha}$) from 2 cm to the I band and shows a strong curvature towards the optical range. In the optical, excluding the U band, we measure a spectral index of 2.5 ± 0.6, implying either a gradual steepening of the spectrum or a break at some frequency. The U band flux, on the other hand, implies that the high energy spectrum does not follow a single power law, presenting a further steepening between V and U frequencies. The spectral properties of feature θ are very similar to that of the final knots of the Western jet ([8]).

The centroid of the optical emission ($\alpha(1950) = 12^h 28^m 19\overset{s}{.}07, \delta(1950) = 12°39'50\overset{''}{.}6$) is located at $24\overset{''}{.}2$ from the nucleus in position angle $117\overset{\circ}{.}3$. When comparing our optical data (Fig. 1) with the radio map at $\lambda 6$cm ([3]) it is evident that, although both emissions are similar, in the optical there exists structure at the arcsec scale. The northern peak, which seems to be resolved and appears elongated, coincides with the maximum of feature θ in the radio map. Towards the South of the optical peak some substructure is resolved and three other peaks are observed. In the radio map a secondary peak is resolved while the remaining lobe emission appears to be flat. The secondary radio peak is not coincident with any optical maximum in our data. This fact suggests that there is a strong gradient in the spectral index along feature θ from North to South.

The result of an imperfect match of the peaks at radio and optical frequencies may have important implications on the theoretical interpretation of the physical nature of radio lobes and for models of synchrotron emission and particle acceleration in jets. The result also may imply that the acceleration mechanisms, and possibly also the hydrodynamical structure of the shocks produced in the lobe, favour the emission of photons at optical frequencies from positions close, but not exactly coincident with, those corresponding to the local maxima of the radio emission.

References

[1] Sparks W.B., Fraix-Burnet D., Machetto F., Owen F.N., *Nature (1992), 355, 804.*

[2] Stiavelli M., Biretta J., Møller P., Zeilinger W.W., *Nature (1992), 355, 802.*

[3] Hines D.C., Owen, F.N., Eilek J.A., *Astrophys. J. (1989), 347, 713.*

[4] Marcaide J.M., Shapiro, I.I., *Astron. J. (1983), 88, 1133.*

[5] Elósegui, P., Marcaide, J.M., *private comunication.*

[6] González-Serrano, J.I., Pérez-Fournon, I., *Astron. Astrophys. (1991), 249, 75.*

[7] Longo, G., de Vaucouleurs, A., *A general catalogue of photoelectric magnitudes and colors in the U, B, V system (1983), Univ. of Texas.*

[8] Pérez-Fournon, I., Colina, L., González-Serrano, J.I., Biermann, P.L., *Astrophys. J. (Letters) (1988), 329, L81.*

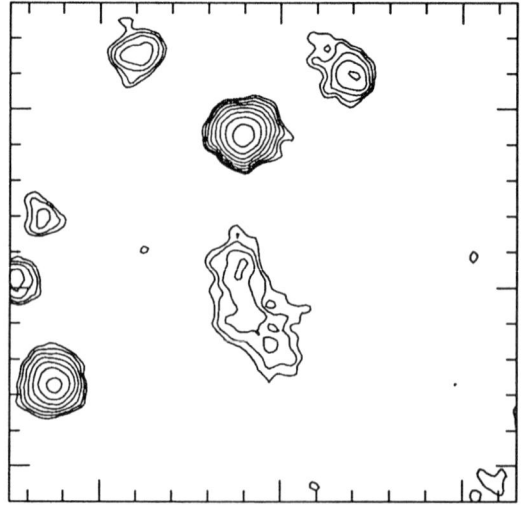

Figure 1.- Contour map of the optical emission detected at the position of the brightest part of the East radio lobe. The separation between tick marks is 1".

Optical Observations of Compact Steep Spectrum Radio Sources

Raffaella Morganti [*] *Clive N. Tadhunter* [†]

1. The Sample & the Observations

Compact Steep Spectrum (CSS) radio sources are that subset of powerful radio sources which have sub-galactic apparent dimensions. These sources are likely to be intrinsically small objects ($D < 30$ kpc) as suggested by [1] from statistical considerations (see also Dallacasa et al. these procedings).
The group of CSS considered here has been selected among the sources of the [2] 2.7 GHz sample (complete down to a flux density of 2Jy) with $\delta < 10°$ and $z < 0.7$ (for details see [3]).
This **complete sample** contains 87 radio sources and among them there are 8 CSS sources (≤ 4 arcsec ~ 30 kpc — $H_o = 50$ km s^{-1}Mpc^{-1}). For all objects we obtained spectra at low dispersion (covering, in the majority of the cases, the region [OII]λ3727 – [OIII]λ5007) and radio maps at 6cm (~ 4 arcsec resolution) using the VLA, the Australian Telescope or collected from literature.

2. Results

In Fig. 1 are plotted the "diagnostics" obtained using [OIII]λ5007 luminosity, line ratio and radio power for both CSS sources and the other objects in the sample. It is evident from these diagrams that there are **no gross differences or systematic trends which might suggest more extreme conditions in the Narrow Line Region of the CSS sources.** As pointed out by [4] the trend visible in the diagnostic [OII]/[OIII] vs [OIII]/Hβ is likely to be mainly determined by changes in the *ionization parameter* (i.e. number of ionizing photons). The lines in the plot show the results for photoionization models calculated for three different continuum spectra and different values of the ionization parameter.
However, the diagnostic diagrams involving the strong emission lines are rather crude and a more complete analysis involving a wide range of emission lines, par-

---
[*]Istituto di Radioastronomia, via Irnerio 46, 40126 Bologna, Italy.
[†]Department of Physics, University of Sheffield, Sheffield S3 7RH, England

ticularly those sensitive to extreme densities, should be made. In fact, in the case of 1934-63, [5] noticed that the spectrum was quite unusual among the narrow-line radio galaxies because of the strong [OI]λ6300 and [SII]λ4070 emission. They claimed that this, coupled with the strength of [OIII] and the relative weakness of [OII]λ3727, indicates a high electron density in the narrow line region. To check this we measured the [SII]λ4070 in some of the CSS sources in our sample and compared the results with what obtained for other radio galaxies (collected from literature). Apart from the extreme case of 1934+63, (note that at the dispersion of their data this line was blended with Hδ) the values found for the CSS do not differ, on average, from the ones of the classical radio galaxies. Then, no evidence for a density systematically higher has been found.

If this can be confirmed (from more detailed optical studies), the most likely explanation for the compact nature of the sources is that they **are young objects but with already formed AGN nuclei, in which the radio emission did not manage yet to go through the ISM and grow like in extended radio sources.**

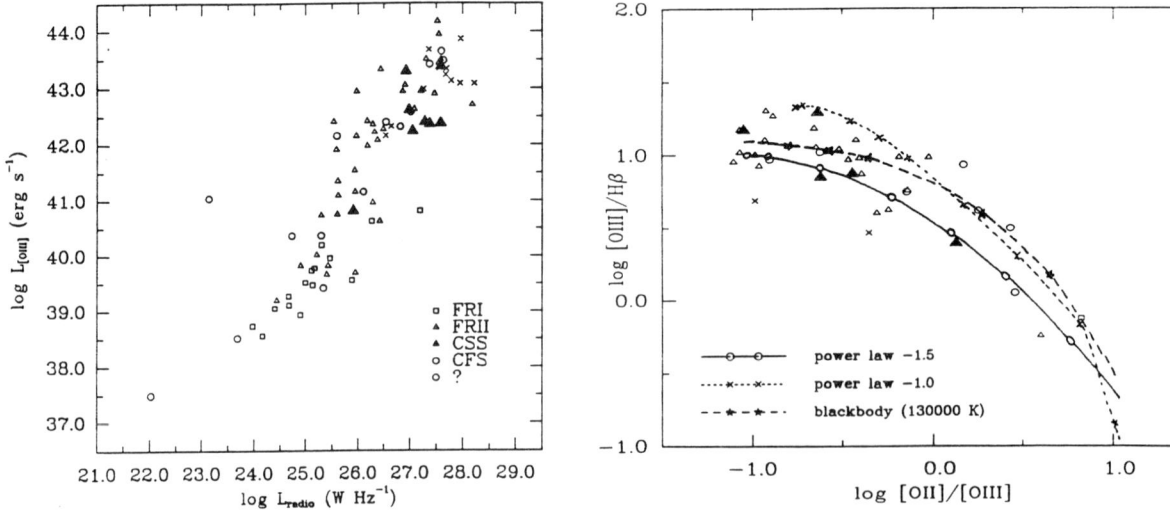

References

[1] Fanti, R. *et al.* 1990, *Astron. Astrophys.*, 231, 333.

[2] Wall & Peacock 1985. *Mon. Not. R. astr. Soc.* **216**,173

[3] Tadhunter *et al. Mon. Not. R. astr. Soc.* submitted

[4] Robinson *et al.* 1987. *Mon. Not. R. astr. Soc.* **227**,97.

[5] Fosbury *et al.* 1987. *Mon. Not. R. astr. Soc.* **225**,761.

Optical polarization observations revealing the emergence of new radio components in blazars

Leena Valtaoja *

The optical polarization of blazars is very variable in timescales differing from source to source as well as in a given object. It is also common that the polarization shows wavelength dependence, which may change its character from time to time, again in different timescales. The position angle varies as well, but it is observed to depend on the wavelength less often than the degree of polarization. We are attempting to explain these seemingly random variations by connections to radio events.

In parsec-scale VLBI maps of blazars one is able to resolve nonstationary components, which are commonly identified as shocks moving along the jet. These shocks are very probably causing the variability of the radio flux in the sources (Marscher and Gear 1985, Aller et al. 1985, Hughes et al. 1989). We propose that the high frequency tails of the same shocks affect the optical polarization as well.

When the evolving shock is first seen in the optical region its high frequency spectrum is flatter than that of the jet component. Since the compressed shock is also more polarized than the underlying jet continuum, one should observe the degree of polarization to decrease towards the red. While the radio outburst is reaching its peak, the high frequency tail of the shock spectrum steepens by energy losses and the longer wavebands should have the largest flux and polarization, meaning that the degree of polarization should increase to the red. Moving further down in frequency, the shock does not have any effect in the optical region any more.

The subtraction of the quiescent spectrum from the radio data allows us to follow the evolution of the shock during the radio flare (Valtaoja E. et al. 1988). By comparing the radio flux measurements at Metsahovi with the coinciding optical observations of the same objects we have been able to fit the behavior of the frequency dependent optical polarization with growing radio shocks (Valtaoja L. et al. 1991a).

The same shock can also explain the observed similarity in the degree and in the position angle of polarization in the variable optical and radio components (Valtaoja L. et al. 1991b), since in the model both the radio and the optical flux originate in the same component.

---

*NORDITA, Blegdamsvej 17, DK-2100 Copenhagen, Denmark

In summary, this means that by following the behavior of the wavelength dependence of the optical polarization one can predict the appearance of the radio flare, which also means the emergence of a new component in the VLBI-map. Of course there are all the complications: the difficulty in excluding those optical events which are not due to shocks, the interpretation of possible overlapping shocks, and the fact that different sources have different basic wavelength dependences in their optical polarization. But in principle...

References

Aller, H.D., et al., 1985, ApJ, 298, 296

Hughes, P.A., Aller, H.D., and Aller, M.V., 1989, ApJ, 341, 54

Marscher, A.P., and Gear, W.K., 1985, ApJ, 298, 114

Valtaoja, E., et al., 1988, A&A 203, 1

Valtaoja, L., Valtaoja, E., Shakhovskoy, N.M., Efimov, Yu., and Sillanpä ä,ØA., 1991a, AJ, 101, 78

Valtaoja, L., et al., 1991b, AJ, 102, 1946

Interpretation of Multiwavelength Observations of Nonthermal Extragalactic Radio Sources

Alan P. Marscher *

Abstract

Radio observations of compact extragalactic radio sources study the emission at parsec scales, and are generally interpreted in terms of a relativistic jet model. The "inner jet" region that connects the parsec-scale jet with the central engine can only be observed at wavelengths shorter than those in the radio regime, namely the submillimeter to γ-ray portions of the spectrum. Multiwavelength observations demonstrate that there is a strong relationship between the nonthermal emission at lower and higher frequencies. Future multiwavelength monitoring holds the key to understanding the geometry and physics of the inner jet.

1. Introduction

At the sub-milliarcsecond resolution of VLBI, we find that a majority of strong, compact, extragalactic radio sources have core-jet structure [PeR88]. The almost universally adopted interpretation is that the emission arises from well-collimated jets of nonthermal plasma flowing out from the nucleus at relativistic speeds (see [Mar93] for a recent review of this model and the observations that support it). The "core" is a very compact, stationary component at one end of the jet. Nevertheless, its size is 2–3 orders of magnitude larger than the dimension of the central engine according to the favored accreting black hole paradigm. High-frequency VLBI is crucial for exploring the properties of the core and comparing these with the expectations of the jet model. Still, it is likely that the connection between the radio jet and the central engine can be studied fully only by observing at frequencies much higher than is possible with VLBI.

In general, one expects high-frequency emission to arise mainly from the region closest to the ultimate energy source, since electrons that radiate at these frequencies lose

---

*Department of Astronomy, Boston University, 725 Commonwealth Ave., Boston, MA 02215, USA. This work was supported in part by US National Science Foundation grant AST-9116525 and NASA grants NAGW-1068 and NAG 5-1566.

energy rapidly once they exit the site of energy injection. Nevertheless, rejuvenation of these electrons can occur far downstream, for example in shock waves. In addition, inverse Compton scattering can produce X-rays and γ-rays wherever high-energy electrons and a strong photon field are found. It is therefore important to find ways to determine where the high-frequency emission arises so that the details of the observed time variations, spectra, etc., can be related to the physics and geometry of the source. Since sub-milliarcsecond resolution is not currently possible at wavelengths shorter than 1.3 mm, we must rely on comparison of multiwavelength observations with theoretical models to establish where the emission arises and how this relates to the physics and structure of the jet.

2. A Quick Review of Multiwavelength Observations

Multiwavelength spectra of compact extragalacitc sources exhibit flat radio spectra (known to be caused mainly by the superposition of self-absorbed components in the jet), which turn over and become optically thin at frequencies above a few 10^{10} to $\sim 10^{12}$ Hz. Above the turnover frequency, [Lan86] find that the spectra become monotically steeper with frequency, although the spectra measured by [Bro89] are in many cases well described by power laws over several decades of frequency.

BL Lac objects tend to be highly variable at X-ray energies and to have steeper X-ray spectra than do quasars [Mri92]. The spectra of some BL Lac objects appear to be continuous from radio to X-ray frequencies, indicating a common emission mechanism — presumably synchrotron radiation — in the compact jet. Quasars are also variable at X-ray energies, although not many cases are well documented. The X-ray spectra of quasars are rather flat [WiE87], too much so to be explained as continuations of the optical-uv emission. The recent detections of a number of quasars and BL Lac objects containing compact jets by the *Compton Observatory* at hard γ-ray energies demonstrate that a significant, in some cases dominant, fraction of the nonthermal luminosity is emitted at extremely high energies ([Har92]; [DeS92]). The γ-ray flux from the quasar 3C 279 was reported to be variable by a factor of three over a 4-month period and also to be significantly variable on a timescale of a few days [Kan92].

There are a number of studies that have demonstrated that flares in nonthermal sources are often broadband in nature. For example, a direct correspondence between X-ray and radio–infrared variability has been found in 3C 279 [Mak89]. In the case of BL Lac, [Kaw91] found that the X-ray flux was correlated with the submillimeter-wave flux. [Bre90] and [HuB92] find that, in BL Lac and several other blazars, there is no significant time delay between features in the optical and infrared light curves, but that there is a delay of about 1 year between the weakly correlated optical and radio variations. The optical variations can be characterized by a combination of shot noise and flicker, whereas the radio variations have power spectra similar to shot noise. The conclusion is that flickering is a high-frequency phenomenon in the sources studied, and that the radio emitting region is larger than, but connected to, the site of the optical emission.

3. A Quick Review of Models for the Inner Jet

Two possible models linking the parsec-scale jets with the central engine are tapered, accelerating jets and highly relativistic particle beams (see Fig. 1). For the beam model, [Phi87] and [MeK89] have shown that up-scattering of the uv photons emitted by the accretion disk decelerates the electron-positron stream to a terminal bulk Lorentz factor ~ 10. The scattering, along with plasma instabilities, can also randomize the pitch angles such that the beam becomes a flowing plasma by this point, which corresponds to the core of the radio jet. A model intermediate between these two describes the inner jet as a beam of relativistic plasma flowing along essentially straight field lines [Bak88]. At some distance from the central engine, plasma instabilities introduce a substantial random component to the magnetic field such that the source emits incoherent synchrotron radiation downstream of this point.

If the acceleration of the relativistic electrons occurs only in the region closest to the central engine, the synchrotron emission at uv, optical, and IR frequencies is confined to this region as well, which is opaque to radio emission ([Mar80]; [MGC92]). The dependence of the magnetic field, relativistic electron density and maximum energy, and bulk Lorentz factor on distance down the jet leads to a frequency-dependent size of the emission region and a steep spectral index. The timescale of variability is therefore shorter at higher frequencies and time delays are expected as a flare propagates from high to low frequencies. Outside of the UVOIR region, the highest energy electrons emit only at lower frequencies, having suffered from radiative and adiabatic losses. However, if the inner jet does not open too abruptly, the maximum radio emission occurs where the Lorentz factor, and hence the Doppler boosting, is strongest. This region is then identified as the radio core, with the radio jet visible on the downstream side. Substantial self-Compton γ-ray and X-ray emission can occur either in the UVOIR region or the radio core (see [MGC92]). In addition, (inverse) Compton relection of optical and uv photons from the accretion disk can take place in the UVOIR region, producing X-rays and γ rays [DSM92].

Compton reflection of the optical and uv photons from the accretion disk (or ambient photons from emission-line clouds or regions of electron scattering) off a highly relativistic stream of electrons [MeK89] emits γ rays from the deepest part of the inner jet and X-rays somewhat downstream of this, with the nonthermal optical to radio emission occurring farther out.

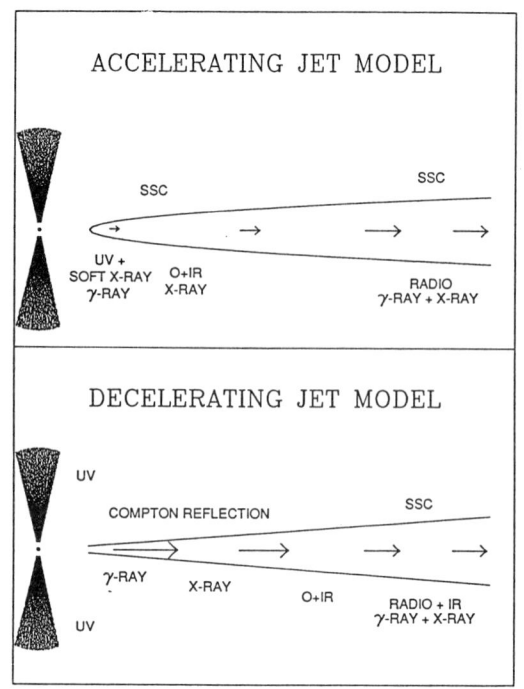

Figure 1. *Two basic models (not drawn to scale) for the inner jet that connects the compact radio jet with the central engine, here depicted as a massive black hole with an accretion disk. The lengths of the arrows inside each jet correspond to the magnitude of the bulk Lorentz factor of the jet flow on a logarithmic scale. The primary emission mechanism of each region is indicated above the jet, with "SSC" corresponding to synchrotron self-Compton emission. The main frequency bands of the nonthermal emission are indicated below the jet, with synchrotron or Compton reflection on top and self-Compton on bottom.*

4. Tests of the Models through Multiwavelength Observations

There are three aspects of multiwavelength observations that can be used to test models for the emission mechanisms and source structure and physics: the value of the spectral index, the relative flux densities at different wavebands if the spectra do not connect smoothly, and correlated (or uncorrelated) variability of brightness.

The spectral index of a given uniform synchrotron-emitting source reflects the energy distribution of the relativistic electrons. Any spectral steepening corresponds to radiative losses competing with energy gains. In the accelerating jet model, however, the source is nonuniform, with gradients in magnetic field, etc. The value of the spectral slope therefore results from a combination of the electron energy distribution and the gradients of the physical quantities, which in turn depends on the geometry of the jet (see [Mar80] and [GMT85] for the relevant relations).

While it appears well established that the nonthermal radio to optical emission is synchrotron radiation, it is not so clear what causes the X-ray and γ-ray emission. For quasars and some BL Lac objects, the flat (spectral index $\lesssim 1.0$) X-ray spectra suggest that inverse Compton emission might be the main mechanism. If this is self-Compton scattering, the ratio of γ-ray to X-ray luminosity cannot exceed that of X-ray to infrared luminosity (see [MaB92] and [BlM92]). In addition, the value of the spectral index across one waveband is directly related to that of another, although not as trivially as usually asserted.

In each of the models for the inner jet, the emission regions at different wavebands are connected but lie at different distances from the central engine. One can therefore potentially use multifrequency observations to discriminate among the models. Disturbances propagating down the jet are time-delayed at different wavebands, depending on the location of the primary emission region at each frequency. Measurement of such time delays would therefore reveal the jet geometry, the location of the particle acceleration, and possibly the speed of the flow as a function of distance along the jet.

Fluctuations in jet flow can also cause shocks to form. [BlK79] proposed that such shocks correspond to the apparently superluminal knots found in VLBI images of compact jets. [MaG85] showed that, at high frequencies, the emission behind the shock has frequency-dependent structure, with the highest frequency radiation confined to the region immediately behind the shock front where the emitting electrons have not yet suffered significant radiative energy losses. Such shocks can occur in any portion of the jet between the base and the region near the radio core. (Radiative losses are probably not important far downstream of the core.) [MGT92] have explored the variations in brightness and spectrum expected from a shock propagating down a jet containing hydromagnetically turbulent plasma. The overall evolution of the flare spectrum is a rather abrupt rise, followed by a decrease in turnover frequency as the peak flux remains roughly constant, and eventually a decline. As the shock encounters turbulent eddies, simultaneous brightness and polarization fluctuations occur at higher frequencies with slightly time-delayed and less pronounced variations at lower frequencies (but still above the self-absorption turnover).

5. Conclusions

The region between the radio core and the central engine is the great unexplored region of compact jets, yet that is where the most interesting physics is likely to occur. As millimeter and (let's hope!) submillimeter wave VLBI progresses over the next several years, we may eventually be able to image the core and perhaps part of the inner jet. During the same time period, the availability of the *Compton* Gamma Ray

Observatory, *ROSAT* and *ASTRO-D*, as well as ground-based radio, submillimeter, infrared, and optical observatories provides an outstanding opportunity to undertake coordinated multiwavelength observations. Such campaigns, combined with theoretical models, promise to unlock the secrets of the inner jet.

References

[Bak88] Baker, D.N., Borovsky, J.E., Benford, G., Eilek, J.A. 1988, ApJ, 326, 110
[BlK79] Blandford, R.D., Königl, A. 1979, ApJ, 232, 34.
[BlM92] Bloom, S.D., Marscher, A.P. 1992, in The Compton Observatory Science Workshop, ed. C.R. Shrader, N. Gehrels, B. Dennis (NASA Conf. Publ. 3137), 339.
[Bre90] Bregman, J.N., et al. 1990, ApJ, 352, 574.
[Bro89] Brown, L.M.J., et al. 1989, ApJ, 340, 129.
[DeS92] Dermer, C.D., Schlickeiser, R. 1992, Science, 257, 1642.
[DSM92] Dermer, C.D., Schlickeiser, R., Mastichiadis, A. 1992, A&A, 256, L27.
[GMT85] Ghisellini, G., Maraschi, L., Treves, A. 1985, A&A, 146, 204.
[Har92] Hartman, R.C., et al. 1992, ApJ, 385, L1.
[HuB92] Hufnagel, B.R., Bregman, J.N. 1992, ApJ, 386, 473.
[Kan92] Kanbach, G., et al. 1992, IAU Circular no. 5431.
[Kaw91] Kawai, N., et al. 1991, ApJ, 382, 508.
[Lan86] Landau, R., et al. 1986, ApJ, 308, 78.
[Mak89] Makino, F., et al. 1989, ApJ, 347, L9.
[Msi92] Maraschi, L. 1992, in Variability of Blazars, ed. E. Valtaoja M. Valtonen (Cambridge Univ. Press), 447.
[MGC92] Maraschi, L., Ghisellini, G., Celotti, A. 1992, ApJ, 397, L5.
[Mar80] Marscher, A.P. 1980, ApJ, 235, 386.
[Mar92] Marscher, A.P. 1992, in Physics of Active Galactic Nuclei, ed. S.J. Wagner W.J. Duschl (Heidelberg: Springer-Verlag), in press.
[Mar93] Marscher, A.P. 1993, in Astrophysical Jets, STScI Symposium Series, 6, ed. D. Burgarella, M. Livio, C. O'Dea. (Cambridge Univ. Press), in press.
[MaB92] Marscher, A.P., Bloom, S.D. 1992, in The Compton Observatory Science Workshop, ed. C.R. Shrader, N. Gehrels, B. Dennis (NASA Conf. Publ. 3137), 346.
[MaG85] Marscher, A.P., Gear, W.K. 1985, ApJ, 298, 114.
[MGT92] Marscher, A.P., Gear, W.K., Travis, J.P. 1992, in Variability of Blazars, ed. E. Valtaoja M. Valtonen (Cambridge Univ. Press), 85.
[MeK89] Melia, F., Königl, A. 1989, ApJ, 340, 162.
[PeR88] Pearson, T.J., Readhead, A.C.S. 1988, ApJ, 328, 114.
[Phi87] Phinney, E.S. 1987, in Superluminal Radio Sources, ed. J.A. Zensus T.J. Pearson (Cambridge Univ. Press), 301.
[WiE87] Wilkes, B.J., Elvis, M. 1987, ApJ, 323, 243.

New X-ray Observations of Extragalactic VLBI Sources

D.M. Worrall [*]   M. Birkinshaw [†]   C.R. Gwinn [‡]

1. Introduction

ROSAT [T92] is providing the best opportunity so far to study both the structure and spectrum of X-ray emission from the vicinity of extragalactic VLBI sources. Whereas the X-ray emission from radio-loud quasars and BL Lac objects is commonly believed to be dominated by non-thermal radiation from a compact jet boosted by relativistic beaming, some degree of jet misalignment reveals other X-ray components of physical significance. The earlier *Einstein* Observatory mission could separate X-ray components only for nearby radio galaxies such as M87 [BSH91]. With ROSAT's improved angular and spectral resolution, such studies can be extended to higher redshifts and more objects. Here we report some new results from a series of investigations we are undertaking with ROSAT to probe the gas structure and pressure surrounding radio jets, to separate and identify emission components using the measured X-ray spectra, and to test models of relativistic beaming. Some of these studies which are not discussed further in this report are in collaboration with groups from CalTech, the University of Tübingen, and STScI.

2. The Radio Galaxy NGC 6251

NGC 6251 is noted for its large one-sided jet [PBW84] and slightly misaligned VLBI core-jet structure [J86]. Whereas we anticipated measuring X-ray emission from gas confining the jet and possibly the jet itself, the detected X-rays were limited to a region consistent with the Point Response Function (PRF) of the ROSAT PSPC detector (fig. 1). From our derived upper limit to any spatial X-ray extent we conclude that the kpc-scale jet (at least beyond ∼ 25 arcsec from the core) and the VLBI jet *are not confined* by the pressure of X-ray emitting gas.

---

[*]Harvard-Smithsonian Center for Astrophysics, Cambridge, MA, U.S.A. This author was supported by NASA grant NAG5-1724.

[†]Harvard-Smithsonian Center for Astrophysics, Cambridge, MA, U.S.A. This author was supported by NASA grant NAG5-1648.

[‡]University of California, Santa Barbara, CA, U.S.A.

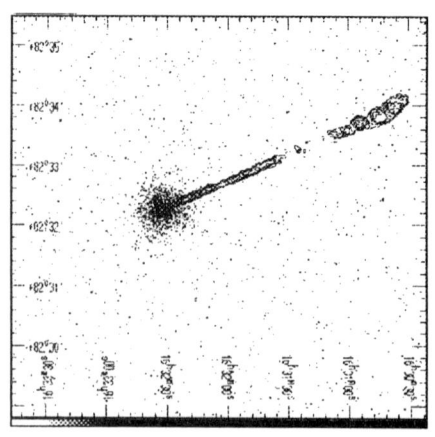

Figure 1. Contour plot of a 330 MHz radio map of NGC 6251 [B92] superimposed on the ROSAT PSPC 0.2-2.4 keV X-ray image. The X-ray emission is well confined to a small region around the core of the radio source (and the nucleus of the elliptical galaxy).

The spectrum of the X-ray core emission, when fitted to a power-law model, requires absorption far in excess of that measured for our Galaxy [S92] and NGC 6251 itself [vG89]. A more realistic interpretation appears to be that the emission is comprised of two (or more) components; a power law and 0.5 keV thermal emission together give a good fit to the data and a self-consistent picture for the energy balance in NGC 6251 is beginning to emerge.

We find that the 0.5 keV gas is unstable to cooling and falls into the nucleus on a time scale comparable to the lifetime of the radio jet ($\sim 10^8$ yr) and with a mass infall rate ($\sim 0.4 M_\odot$/yr) typical of that required to fuel a medium-power AGN. Our decomposition of the X-ray core into beamed (non-thermal) and unbeamed (thermal) emission fits a unification scheme that relates BL Lac objects and radio galaxies [PU90] and implies that NGC 6251's VLBI jet lies at $\sim 25°$ to the line of sight. The derived relativistic Doppler factor of ~ 2 is consistent with production of the non-thermal X-rays by simple spherically symmetric self-Compton scattering from the VLBI core. Our identification of the X-ray thermal emission with the unbeamed component required by unification models will be tested by searching for such a component in other low-luminosity radio galaxies. More details are reported by [BW92].

3. Two Scintillating Radio Sources/BL Lac Objects

Little is known about the X-ray properties of sources showing unusually strong refractive interstellar scattering [F87]. We have observed two, the BL Lac Objects 0954+658 and 1749+091, with the ROSAT PSPC to search for structural evidence of scattering sites and anomalies in the AGN spectra or light curves. 0954+658 has exhibited additional intraday radio and optical variability which has been argued to be of an intrinsic origin [W90].

Our X-ray observations showed the sources to be 'normal' X-ray-emitting BL Lac

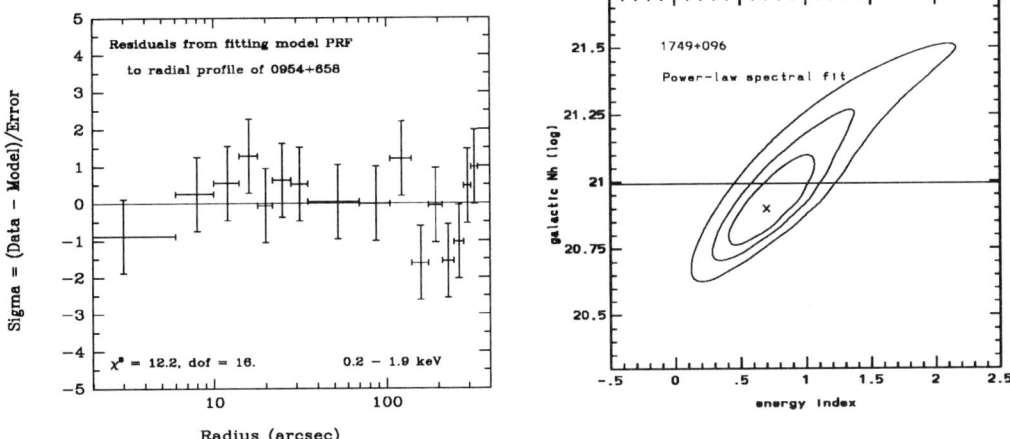

Figure 2. (a) Both sources have spatial distributions consistent with the PRF; example shows 0954+658. (b) Both sources fit power-law spectra. 1749+096 is the best constrained and shown are 68%, 90% and 99% joint-confidence contours for energy index and hydrogen column density. The Galactic value of $\log N_H = 20.99$ is acceptable and implies a spectral index of $0.84 \pm 0.14 (1\sigma)$, consistent with the average found previously for BL Lac objects [WW90].

objects. Neither showed evidence for spatial extent beyond the PRF (fig. 2a) or anomalies in their X-ray spectra (fig. 2b). The ~ 6.8 ksec total exposure for 0954+658 was spread over 8 days with an average count rate of ~ 3 counts/min; 1749+096 gave ~ 5 counts/min in ~ 9.5 ksec split not only over days but in two intervals a year apart. No significant X-ray variability has been found so far in these data. We now plan to check whether weak X-ray features detected within a few arcmin of each BL Lac object have counterparts on VLA maps of the region, and to evaluate the likelihood that these X-ray features might be extended as opposed to weak point sources.

References

[BSH91] Biretta, J.A., Stern, C.P. & Harris, D.E. 1991, *AJ*, 101, 1632.
[B92] Birkinshaw, M., Zheng, X.W., Ho, P.T.P. & Reid, M.J. 1992, in preparation.
[BW92] Birkinshaw, M. & Worrall, D.M. 1992, *ApJ*, submitted.
[F87] Fiedler, R.L. *et al.* 1987, *Nature*, 326, 675.
[J86] Jones, D.L. *et al.* 1986, *ApJ*, 305, 684.
[PU90] Padovani, P. & Urry, C.M. 1990, *ApJ*, 356, 75.
[PBW84] Perley, R.A., Bridle, A.H. & Willis, A.G. 1984, *ApJS*, 54, 291.
[S92] Stark, A.A. *et al.* 1992, *ApJS*, 79, 77.
[T92] Trümper, J. 1992, *QJRAS*, 33, 165
[vG89] van Gorkom, J.H. *et al* 1989, *AJ*, 97, 708.
[W90] Wagner, S. *et al.* 1990, *A&A*, 235, L1.
[WW90] Worrall, D.M. & Wilkes, B.J. 1990, *ApJ*, 360, 396.

High Resolution UV/Optical/IR Imaging of Active Galaxies

Martin J. Ward [*]

Abstract

For nearby Seyferts high resolution images are capable of revealing the dusty molecular torus, the existence of which is one of the central tenants of the AGN paradigm. Arcsec and sub-arcsecond scales correspond to regions in which the effects of the anisotropic continuum emission on the narrow-line gas are clearly visible, in the form of ionization cones. A number of Seyfert nuclei are surrounded by circum-nuclear starbursts, and mid-IR imaging is a powerful means of identifying their presence. High resolution, mostly IR, imaging of luminous IRAS selected galaxies has yielded a wealth of sub-arcsecond detail. In some cases this strengthens the link with starformation, supernovae and their remnants, but the central issue of whether this class of object conceals a hidden Seyfert/quasar remains ambiguous.

1. Introduction

It is now more than ten years since the ESO conference on *The Scientific Importance of High Angular Resolution at Infrared and Optical Wavelengths* [ULK81]. Since then the field of AGN research has undergone something of a revolution. The impetus for this came from the recognition of the importance of directionally selective absorption of radiation from the nucleus, leading to the development of the so-called *Unified Schemes*. It is of historical interest to recall that the foundational observations that led to the recognition of the significance of collimation, namely polarized broad emission line components in narrow-line AGN (Seyfert 2s), and the position angle dependence of the extended narrow-line morphologies, did *not* require high-spatial resolution measurements. However, it was subsequently deduced that the primary cause of this collimation should be associated with a component of physical size a few, up to a few tens of parsecs, which for nearby AGN is in the range of interest for sub-arcsecond observations.

---

[*] Astrophysics, Nuclear Physics Laboratory, Oxford University, 1 Keble Rd, Oxford, OX1 3RH, UK.

In the following discussion it will be useful to bear in mind some typical dimensions associated with the activity at the centres of galaxies. As an example consider an AGN with a redshift of 0.01. Taking $H_o = 50$, 1 arcsecond corresponds to 300 parsecs. For a 10^8 solar mass black hole the gravitational diameter is 10^{-5} parsecs ($3\ 10^{-8}$ arcsec). The diameter of the accretion disc region in which much of the UV/soft X-ray emission is liberated, is around 100 times larger. Assuming that the central engine radiates at its Eddington limit, for an AGN of this luminosity the broad-line region (BLR) would be about 0.05 parsec (0.2 milli-arcseconds). The size of the putative dusty molecular torus that is believed to hide the BLR in Seyfert 2s, is uncertain, but could lie in the range 5–50 parsecs (0.02–0.2 arcsecs). Further out the narrow-line region would be visible above and below the poles of the torus, and can extend hundreds of parsecs, eventually merging with the gas rotating with the host galaxy. It is clear that the components of interest for high resolution imaging are; the dusty torus, the morphology of the NLR gas close to the collimation site, and the circum-nuclear starburst if present.

2. Some Technical Considerations

In principle the true diffraction limit of a telescope can be achieved by the use of speckle imaging techniques. Whilst resolutions of a few hundredths of an arcsecond have been achieved on bright stars, the results on much fainter sources like AGN have been less spectacular. For extragalactic sources typically a resolution of a few tenths of an arcsecond is obtained. Pioneering studies like those of Meaburn et al [MMV82], gave indications of the compactness of the nucleus, but the inferred multi-component structure is less certain. More recent near infrared speckle observations of the bright Seyferts NGC 1068 [CPC87] and NGC 4151 [ABC90], have set upper limits on the extent of the unresolved component at about 0.2 arcsecond (20 parsecs) in both cases. However, limitations in the technique again made it difficult to say much about the nature of the emission surrounding the compact source. Speckle observations have also been made of the same two Seyferts in the light of the strong emission line [OIII] 5007Å, [ECP89]. In this case we now have a direct check of the speckle results by comparison with HST images [EFK91]. The position angles of structures seen in the speckle maps are confirmed, but in general there is little exact correspondence between individual components on the sub-arcsecond scale.

The uncertainty in the component structures resulting from speckle observations, can of course be eliminated by direct imaging. Instruments that take out tilts and piston deformations of the wavefront are capable of achieving about a factor 2 improvement under good seeing conditions using a small telescope. Early results from such techniques were obtained on the radio galaxy Cygnus A [Tho84], showing a dusty morphology bisecting the image. Recent observations using a high resolution camera (HRCAM) on the CFHT have achieved 0.4 arcsecond resolution images of the quasar 3C273 [HuN91]. It is interesting to compare those images with observations of 3C273 taken with the HST [TWD92]. Clearly the resolution gap is narrowing between ground based observations and those taken with the HST in its aberrated state.

For the remainder of this review I will concentrate on ground based imaging of AGN, mostly at infrared wavelengths. The HST optical and ultraviolet imaging results on AGN are only now, at the time of writing, becoming available (at this time the Sardinia conference on results from the HST, is the most up to date reference for this work, see [TBC92] and [Bok92]).

3. Seyfert Nuclei

The highest resolution images are naturally of the brightest Seyferts, NGC 1068 and NGC 4151. At 11.2 microns the nucleus of NGC 4151 is resolved at 0.16 arcseconds (15 parsecs), [NGS90]. This immediately rules out non-thermal models, although there may still be some contribution from an unresolved nuclear source. These observations are in good agreement with thermal models, in which dust grains are heated by the central source. On the other hand mid-infrared observations of NGC 1068 [Cam92] show that it is extended on a scale of about 2 arcseconds (200 parsecs) in the direction of the radio and optical emission line morphology. This suggests a link between the narrow-line region and emission from warm dust grains at a few hundred degrees. The Seyfert 1.5 NGC 7469 has been imaged at 11.3 microns [KBA92a], showing an elongated structure 0.7x0.4 arcseconds (450x220 parsecs). In all three cases the results suggest a thermal origin for the mid-IR emission. However the detailed explanations do differ. In NGC 4151 the size is consistent with the central Seyfert nucleus heating circum-nuclear dust. In NGC 1068 the 10 micron emission may result from dust in the NLR with some heating *in situ*. For NGC 7469 the extended mid-IR is very likely associated with the kiloparsec scale starburst surrounding the Seyfert nucleus. Despite having cast some doubt on the reality of some of the multi-component structures seen in the speckle images of NGC 4151 and NGC 1068, optical speckle images of the nucleus of NGC 7469 [HMW89] do show sub-arcsecond structures in a circum-nuclear ring, that coincide with radio continuum peaks [WHH91]. The mid-infrared image is more elongated than either the radio or the optical speckle image structure, but this may simply result from the lower sensitivity detecting only the brightest components.

Turning to the optical/ultraviolet region, some recent HST results are summarized in [Bok92]. Emission line images in [OIII] 5007Å show that bi-conicial structures aligned with the radio jet or compact lobe components, are common in Seyfert 2s, and is also seen in the Seyfert 1 nucleus of NGC 4151. Another interesting feature is that the ultraviolet continuum is unresolved ie. less than about 0.1 arcsecond, in Seyfert 1s, whilst it is resolved in Seyfert 2s [LFG91]. Ground based images in the near-UV (3600Å) [PoR92] show extended blue continuum emission aligned with, but extending beyond, the radio morphology. Their interpretation is that the extended continuum is due to electron or dust scattering of the nuclear light, the same 'mirrors' that reflect the hidden BLR.

Finally, what of the molecular torus itself? Maps of NGC 1068 taken in the molecular hydrogen line at 2.12 microns [Cam92] show two *clouds*, one of which is close to the nucleus. It is currently unclear whether this *is* the torus, or whether it is just a molecular cloud seen in projection against the nucleus. Perhaps more convincing evidence is the structure seen in the HST image of M51, taken in visible light [TBC92]. A dark lane indicating absorption, is seen across the nucleus. If this is the torus seen edge-on, then its diameter would be about 20 parsecs.

4. IRAS Galaxies and Starbursts

Galaxies with high mid-far infrared luminosities ($L>10^{11}$ L(solar)) have been the subject of intensive study since their discovery by IRAS. Put simply the question is, do these high luminosities arise from super-starbursts, or the absorbed optical/UV continuum from a hidden quasar re-radiated in the infrared? It was hoped that high resolution infrared imaging, which can penetrate the dusty nucleus, would help answer this question. A near IR survey showed a significant excess of double nuclei [CGM90], suggesting mergers. But the fact that the nuclei are more compact at longer wavelengths lends support to the hidden quasar hypothesis. For a somewhat lower luminosity sample than [CGM90], it is found that the near IR emission is resolved, and is typically extended over 0.5 kiloparsecs, [EBH90] and [FWP92]. Mid infrared observations are more difficult, but the limited data that currently exists, [KBA92a] and [KBA92b], show that the bulk of the emission comes from regions smaller than 1 kiloparsec. Ten micron observations of the nearby starburst in M82 [TeG92], demonstrate the complexity of the situation, since in this case the near and mid-IR emission originates from different components.

Two of the best studied luminous IRAS galaxies NGC 6240 and ARP 220, both have double nuclei separated at near IR wavelengths by about 0.5 kiloparsecs, [EBH90] and [GCM90]. In these galaxies the double radio continuum components, [EBH90] and [BeW87], are cospatial with the IR nuclei. Although hidden quasars are not ruled out, in these cases the data are consistent with a pure starburst interpretation. The quest should continue to resolve the starburst/quasar enigma. Encouragement may be derived from the success of high resolution IR imaging in revealing the hidden quasar in Cygnus A [DWM91].

References

[ABC90] Ayers G.R., Benson J., Carels K., Dyck H.M., Spillar E., *Speckle Observations of the Central Region of NGC 4151. Ap.J.*, Vol. 360 (1990), pp. 471-473.

[BeW87] Becklin E.E., Wynn-Williams C.G., *Ground Based 1-32 micron Observations of ARP 220. Star Formation in Galaxies*, CP-2466, NASA. Ed. C.J. Lonsdale., (1987), pp. 643-649.

[Bok92] Boksenberg A., *Imaging of AGN. Science with the Hubble Space Telescope*. Proceedings Published by ESO. (1992) Ed. P. Benvenuti.

[CGM90] Carico P., Graham J.R., Matthews K., Wilson T.D., Soifer B.T., Neugebauer G., Sanders D.B., *Near Infrared Morphology of Ultraluminous Infrared Galaxies*. Ap.J. Letters., Vol. 349 (1990), pp. L39-L42.

[CPC87] Chelli A., Perrier C., Cruz-Gonzalez I., Carrasco L., *High Spatial Resolution IR Observations and Variability of the Nuclear Region of NGC 1068*. Astron. and Astrophys., Vol. 177 (1987), pp. 51-62.

[Cam92] Cameron et al., *High Resolution 1-20 micron Imaging of NGC 1068*. Proceedings of the Third Teton Summer School on Evolution of Galaxies., (1992), pp. 297-298.

[DWM91] Djorgovski S., Weir N., Matthews K., Graham J.R., *Discovery of an Infrared Nucleus in Cygnus A: An Obscured Quasar Revealed?*. Ap.J. Letters., Vol. 372 (1991), pp. L67-L00

[EBH90] Eales S.A., Becklin E.E., Hodapp K-W., Simons D.A., Wynn-Williams C.G., *Subarcsecond Infrared Structures at the Centres of Infrared-Luminous Galaxies*. Ap.J., Vol. 365 (1990), pp. 478-486.

[ECP89] Ebstein S.M., Carleton N.P., Papaliolios C., *Speckle Imaging of NGC 1068 and NGC 4151*. Ap.J., Vol. 336 (1989), pp. 103-111.

[EFK91] Evans I.N., Ford H.C., Kinney A.L., Antonucci R.R.J., Armus L., Caganoff S., *HST Imaging of the Inner 3 Arcseconds of NGC 1068*. Ap.J. Letters., Vol. 369 (1991), pp. L27-L30.

[FWP92] Forbes D.A., Ward M.J., DePoy D.L., Boisson C., Smith M.G., *Near-Infrared Images of LINER and Starburst Galaxies*. M.N.R.A.S., Vol. 254 (1992), pp. 509-524.

[GCM90] Graham J.R., Carico D.P., Matthews K., Neugebauer G., Soifer B.T., Wilson T.D., *The Double Nucleus of ARP 220 Unveiled*. Ap.J. Letters., Vol. 354 (1990), pp. L5-L8.

[HMW89] Hofmann K-H., Mauder W., Weigelt G., *Speckle Observations of NGC 7469*. Proceedings of ESO Workshop, (1989), Extranuclear Activity in Galaxies, ed. E.J.A. Meurs., R.A.E. Fosbury., pp. 35-37

[HuN91] Hutchings J.B., Neff S.G., *0.4-Arc-Second Images of 3C273*. P.A.S.P., Vol. 103 (1991), pp 26-31.

[KBA92a] Keto E., Ball R., Arens J., Jernigan G., Meixner M., *Subarcsecond Mid-Infrared Imaging of NGC 1614 and NGC 7469*. Ap.J., Vol. 389 (1992), pp. 223-226.

[KBA92b] Keto E., Ball R., Arens J., Jernigan G., Meixner M., *Mid-Infrared Imaging of MKN 231 and ARP 220*. Ap.J. Letters., Vol. 387 (1992), pp. L17-L19.

[LFG91] Lynds et al., *NGC 1068: Resolution of Nuclear Structure in the Optical Continuum*. Ap.J. Letters., Vol. 369 (1991), pp. L31-L34.

[MMV82] Meaburn J., Morgan B.L., Vine H., Pedlar A., Spencer R., *Speckle Observations of the Nucleus of NGC 1068*. Nature., Vol. 296 (1982), pp 331-334.

[NGS90] Neugebauer G., Graham J.R., Soifer B.T., Matthews K., *The Size of NGC 4151 at 11.2 microns*. A.J., Vol. 99 (1990), pp. 1456-1460.

[PoR92] Pogge R.W., DeRobertis M.M., *Extended Near-UV Continuum Emission in Seyfert 2s*. Preprint, Ap.J., accepted.

[TeG92] Telesco C.M., Gezari D.Y., *High-Resolution 12.4 micron Images of the Starburst Region in M 82*. Ap.J., Vol. 395 (1992), pp 461-465.

[Tho84] Thompson L.A., *High-Resolution Imaging from Mauna Kea: Cygnus A*. Ap.J. Letters., Vol. 279 (1984), pp L47-L49.

[TBC92] Tsvetanov Z., Bohlin R., Caganoff S., Evans I., Ford H., Hartig G., Kriss G., Kinney A., Uomoto A., *Observations of AGN within the FOS GTO Program*. Science with the Hubble Space Telescope. Proceedings Published by ESO. (1992) Ed. P. Benvenuti.

[TWD92] Thomson R.C., Wright A.E., Disney M.J., *FOC Images of the Optical Jet in 3C273*. Science with the Hubble Space telescope. Proceedings Published by ESO. (1992) Ed. P. Benvenuti.

[ULK81] Ulrich M.H., Kjar K., *Scientific Importance of High Angular Resolution at Infrared and Optical Wavelengths*. Proceedings of the ESO Conference - Garching (1981).

[WHH91] Wilson A.S., Helfer T.T., Haniff C.A., Ward M.J., *The Starburst Ring Around the Seyfert Nucleus in NGC 7469*. Ap.J., Vol. 381 (1991), pp. 79-84.

MERLIN 5GHz Observations of Supernova Remnants in M82

T.W.B. Muxlow*   A. Pedlar*   P.N. Wilkinson*   D. Axon*
A.G. deBruyn [†]

The strong 60-100 μ emission observed from the nuclear regions of many spiral galaxies is generally interpreted as thermal emission from dust grains heated by a population of early-type stars created in a burst of star formation (Rieke & Lebofsky 1979). The star formation rates required to power the IR emission imply supernovae rates of 0.5 to 10 per year (Scoville & Young 1983) unless the IMF is biassed to low mass stars. However the extinction to the majority of starburst regions is >10-20 mag. in V and hence most of the supernovae in these regions will not be detected optically.

Radio observations of the nearby starburst galaxy M82 revealed $\sim$30 compact objects (Unger et al 1984, Kronberg, Biermann & Schwab 1985) which were tentatively identified as either radio supernovae (RSNe) or young supernova remnants (SNRs). The strongest source (41.9+58) has a luminosity several hundred times that of Cassiopeia A, nevertheless its monotonically decreasing flux density and shell-like structure determined by VLBI are consistent with it being an unusual SNR about 50 years old (Wilkinson & deBruyn, 1990). Nearly all of the stronger compact objects in M82 also have decreasing flux densities (Kronberg and Sramek 1985) with half-lives of tens of years implying that that the "refreshment rate" of the population should be one new source every few years.

The compact components are ideal targets for MERLIN Phase 2 and hence in July 1992 M82, was observed for 38 hours at 4.993 GHz with a bandwidth of 15 MHz in each hand of circular polarisation. The flux density scale was calibrated using 3C286 while the telescope phases were determined using the compact source 0955+697 which is only 0°.32 away from M82. Because the components are distributed over a region 50 × 10 arcsec in extent the data were taken in spectral-line mode in order to avoid bandwidth smearing. After applying the phase corrections determined on 0955+697 three adjoining 1024 × 512 fields were imaged with the AIPS task MX, after first applying a rotation of 15 degrees to align the major axis of M82 with the horizontal. The rms noise obtained was $\sim$ 60μJy/beam. Two preliminary sub-images showing individual components are presented in Figure 1. It will be possible to improve the images by self calibration and deeper cleaning.

Our results, although preliminary, are consistent with the compact objects in M82 being SNRs rather than young RSNe. We have positive detections of at least 13 sources almost all of which are extended compared with our 50 mas (0.8 pc) beam. One of them, 40.7+550, has a clear shell-like structure $\sim$ 3pc in extent. Almost all

---

*N.R.A.L., Jodrell Bank, Macclesfield, Cheshire, SK11 9DL, U.K.
[†]Radiosterrenwacht, Postbus 2, 7900 AA Dwingeloo, The Netherlands.

the sources not detected by us are unresolved with the 0.35 arcsec beam of the VLA at 6 cm and have flux densities greater than 1 mJy; our failure to detect them implies either that they have decayed rapidly or, more likely, that they are extended—an unresolved 1 mJy source would have a signal-to-noise ratio greater than 16 in our image. The evidence suggests, therefore, that the sizes of our non-detected sources are 200 ± 100 mas i.e. 3.2 ± 1.6 pc. Assuming a typical expansion rate of ~ 5000 km s^{-1} the largest sources in M82 must be several hundred years old.

Although a full analysis has yet to be carried out, it seems clear that the exceptionally high supernovae rates implied both by the infra-red luminosity and the flux density decay of the brightest sources, are not supported by these new radio data. Rather the lack of any new sources in the 1992 MERLIN image as compared with a 1981 VLA image (Kronberg and Sramek 1985), together with the estimated ages of the largest remnants, suggests a rate nearer to 0.1 per year.

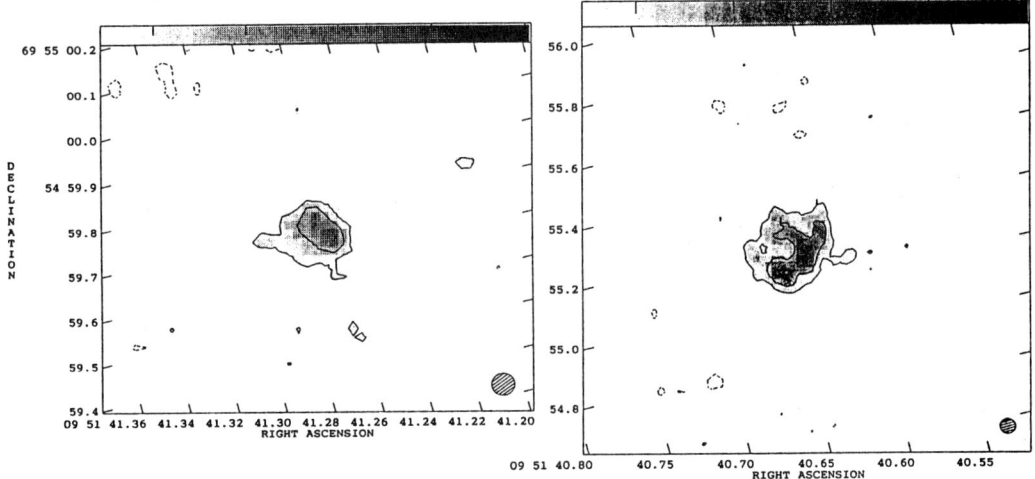

Figure 1. MERLIN 5GHz maps of SN remnants 41.3+596(left), and 40.7+550(right). Contours are at -1, 1, 2, 4, 8 × 3σ (3σ = 0.061 mJy/beam). Beam=50mas circular.

References

Kronberg P.P, Biermann P., & Schwab F.R., 1985, *Ap.J.*, **291**, 693.

Kronberg P.P, & Sramek R.A., 1985, *Science* **227**, 28.

Rieke G.H., & Lebòvsky M.J., 1979, *Ann.Rev.Atron.Astrophys.* Vol. **17**, 477.

Scoville N, & Young J.S., 1983, *Ap.J.*, **265**, 148.

Unger S.W.,Pedlar A.,Axon D.J.,Wilkinson P.N.,& Appleton P.J., 1984, *MNRAS*, **211**, 783.

Wilkinson P.N., & deBruyn A.G., 1990, *MNRAS*, **242**, 529.

Subarcsecond observations of the radio jet in NGC4151

A. Pedlar, M.J. Kukula, T.B. Muxlow, D.J. Axon [*]
S. Baum, C. O'Dea [†]   S.W. Unger [‡]

Abstract

We present MERLIN and VLA images of the radio nucleus of NGC4151. We discuss the misalignment between the radio jet and Extended Narrow Line Region, and conclude the simplest explanation is density bounding of a the UV cone by gas in the disk of the galaxy.

1. Introduction & Observations

Many of the earlier radio studies of Seyfert nuclei were carried out before the recent MERLIN upgrade (resulting in higher sensitivity and angular resolution) or in snapshot mode with the incomplete VLA. Hence many of these earlier radio images can be significantly improved both in sensitivity and resolution, which is essential if they are to be compared in detail with HST and other subarcsecond optical images.

In Figure 1 we show the high angular resolution (75mas) 5GHz observations of NGC4151 using data from the new MERLIN. In Figure 2 we present a high sensitivity 8Ghz image of NGC4151 produced from VLA A & B array observations which are sensitive to low brightness emission, and in addition can be used to constrain thermal free-free contributions from both the NLR and ENLR components (Pedlar et al. 1991).

2. Evidence for collimated ejection

The present observations support the conclusions of the earlier work, based on the elongated structure within 150pc of the nucleus, that collimated ejection is taking

---
[*]NRAL, University of Manchester, Jodrell Bank, Macclesfield, Cheshire, UK
[†]STScI, 3700 San Martin Dr., Baltimore, USA.
[‡]RGO, Madingley Rd., Cambridge, UK

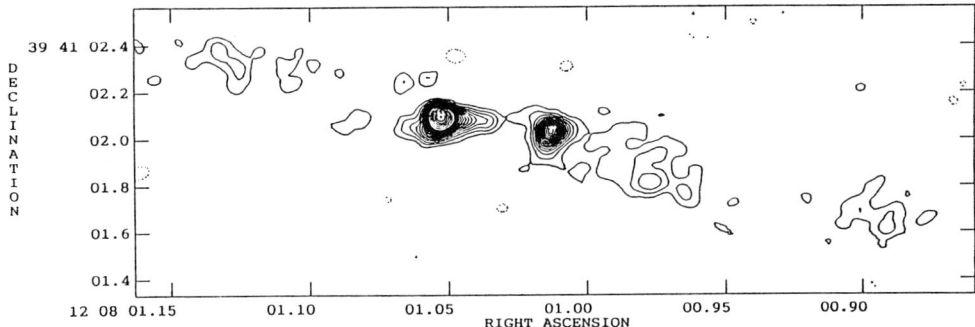

Figure 1. Contour map of the naturally weighted MERLIN 5GHz image. The resolution is 75 mas and the contour intervals are 0.5 mJy beam^{-1} to 8 mJy beam^{-1} and then 2 mJy beam^{-1}.

place along PA $\sim 77°$ and $257°$ giving rise to a two sided radio jet. The detection of a component 4.5 arcsec (300pc) to the east in PA $\sim 77°$ reinforces this picture.

The flatter spectral index of the eastern component of the central 0.4 arcsec double ('C4' following the naming scheme of Carral et al. 1990) is consistent with it containing a synchrotron self-absorbed core and hence being the component associated with the Seyfert nucleus. All the strongest components in the 'jet' lie close to the PA $77°$ line, although there are deviations of several degrees between individual components. In particular the central 460mas double (C3 & C4) is aligned along PA $83°$, and there is marginal evidence for a 'wiggle' of period ~ 1 arcsec along the jet.

The MERLIN 5GHz observations show a small (150mas) jet pointing away from component C4 directly towards component C3, further evidence, in addition to the flatter spectral index, that C4 is the nucleus and C3 is a compact knot in the jet. This 'jet' will also be included most earlier flux determinations of C4. The only other observations which resolve it are the 1.6GHz and 5GHz VLBI measurements of Harrison et al. (1986). Hence the compact core, excluding the 150 mas jet, has a spectral index of $\alpha = -0.3$ between 1.6 and 5GHz. A two dimensional gaussian fit to the core component shows it to have a size of 33×17 mas and to be elongated in PA 62 ± 5 degrees, in approximate agreement with the VLBI results. The PA of the nuclear component differs from that of the arcsecond jet ($77°$), and is closer to the alignment of much of the ionised NLR and ENLR gas associated with the nucleus.

3. Comparison with Optical Narrow Line Emission

NGC4151 is one of the few Seyferts in which the NLR can be resolved by ground based optical techniques. Ulrich (1973) showed the NLR to be several arcsec in

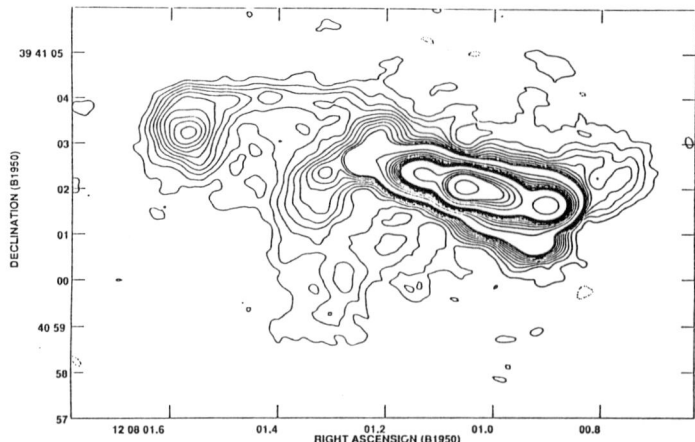

Figure 2. Contour map of the VLA 8GHz natural weighted image. The angular resolution 0.52 x 0.44 arcsec in PA 88° and contours are shown in three ranges - 0.02 to 0.2, 0.2 to 2, 2 to 10 & 10 to 30 mJy beam^{-1}, with intervals 0.02, 0.2, 2 & 10 mJy beam^{-1} respectively.

extent and to consist of a least 4 distinct clouds. The NLR appears to be extended in PA $\sim 45°$, and is clearly not aligned with the radio collimation axis in PA $\sim 77°$. A superposition of our 8GHz radio image and the [OIII] image of Perez et al. (1989) shows the radio emission to be associated with the part of the ENLR to the east along PA $\sim 77°$. The predominant elongation of the ENLR, however, is in PA $\sim 228°$ ($48°$) from the nucleus and like the NLR is not aligned with the collimation axis of the radio jet.

The most reasonable possibility is that the misalignment is due the combination of a wide UV cone, co-axial with the radio jet, and density bounding of the ENLR. As the active nucleus is in the center of a disk of neutral hydrogen, if the collimation axis of the UV cone and radio jet is at an arbitrary angle to the plane of the galaxy, then simple geometrical arguments can account for the apparent misalignment between the collimation axis and the ionised gas in the ENLR (Figure 3). Hence we might only expect to see the ENLR only where the UV cone intercepts the neutral gas close to the plane of the galaxy. This interpretation is supported by the similarity of ENLR and neutral hydrogen velocity measurements (Pedlar et al. 1992) consistent with the ENLR being close to the plane of the galaxy. The lack of a clear cone shaped structure could be consistent with the edge of the UV cone is 'grazing' the disk. If this is the case then the brighter SW component of the ENLR is pointing $\sim 8°$ away from a plane perpendicular to our line of sight. In this model the ENLR, rotation axis of the galaxy and the collimation axis (of UV and Radio) would be expected in the same plane, perpendicular to the plane of the galaxy which has an angle of inclination of $\sim 21°$. Thus if these assumptions are correct it is possible to determine the true collimation axis of the radio jet and UV cone. Simple geometry shows the

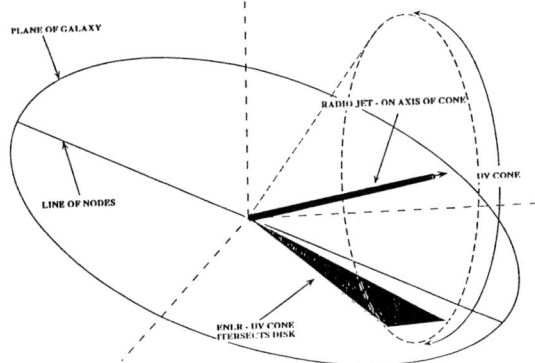

Figure 3. A sketch to show to show how the UV cone can be density bounded by neutral gas in the disk of the galaxy. The radio jet is assumed to lie on the axis of the UV cone. The ENLR is formed where the UV cone intercepts the neutral disk and in general need not align with the collimation axis of the nucleus.

collimation to be aligned approximately 25° from the galaxy rotation axis and hence 65° from the ENLR and 40° from the line of sight. This implies a large opening angle for the UV cone $\sim 120°$, approximately twice that seen in other ENLR and inferred from statistical arguments on numbers of Seyfert 1 and Seyfert 2 nuclei. However as the ENLR is in the plane of an almost face on galaxy (i=21°) then clearly an opening angle of this order is required if we are to see the BLR as required in 'unified schemes' of Seyferts.

If the above interpretation is correct then it will enable us to determine the true angle to the line of sight of the radio and UV collimation in galaxies in which the UV cone intercepts the planar gas distribution of the host galaxy. In general we would expect the radio jet and ENLR to be misaligned except when the collimation axis is close to the plane of the galaxy or we are seeing the phenomenon from a special direction.

References

[1] Carral, P., Turner, J.L. & Ho, P.T.P., 1990, *Ap.J.*, **362**, 434.

[2] Harrison, B., et al., 1986, *Mon.Not.R.astr.Soc.*, **218**, 775.

[3] Pedlar, A., et al., 1991, *Proc Heidelberg meeting on 'Physics of AGN'*.

[4] Pedlar, A., et al., 1992, *Mon.Not.R.astr.Soc.*, (in press).

[5] Perez, E., et al., 1989, *Mon.Not.R.astr.Soc.*, **241**, 31p.

[6] Ulrich, M-H., 1973, *Astrophys.J*, **227**, L55.

A VLBI Survey of Luminous IRAS Galaxies

Colin J. Lonsdale [*]   *Harding E. Smith* [†]   *Carol J. Lonsdale* [‡]

Abstract

The results of a new VLBI survey of 31 bright IRAS galaxies are reported. These galaxies, thought to be dominated by compact nuclear starbursts despite Seyfert 2-like excitation in some of them, constitute the upper end of the IR luminosity distribution, and are arguably among the best candidates for the presence of buried AGN hidden from optical view by starburst-related dust.

We have detected high brightness temperature emission from roughly half the observed sample, and many of the detected sources show evidence for non-pointlike or complex structure. Although the possibility of a new class of exceptionally powerful radio supernovae must be considered, the most probable explanation for the results is that an AGN indeed lies shrouded by dust in the nuclei of many, and perhaps all of these IR luminous galaxies.

1. Introduction

There has been increasing interest in recent years in the relationship between active star formation in galaxies, as evinced by high infrared luminosity, and AGN characteristics. This interest has been stimulated by schemes which suggest a causal or evolutionary relationship between the AGN and starburst phenomena. Ideally, we wish to determine whether nuclear starbursts and AGN typically, or only occasionally, coexist, and whether typical starburst and AGN properties are correlated. The principal difficulty in such a determination is the large optical depth to the nucleus presented by the dusty starburst environment at most observing wavelengths. However, the optical depth at cm radio wavelengths is expected to be small, and by observing such galaxies with a sensitive VLBI array designed to detect high brightness-temperature AGN-related emission, it is possible to conduct a search for AGN buried in dusty nuclear starbursts. We report here such a survey, which was successful in detecting numerous cases of high T_b emission.

---

[*]Haystack Observatory, Westford MA 01886, USA.

[†]Center for Astrophysics and Space Sciences, and Department of Physics, University of California, San Diego, La Holla CA 92093-0111, USA, also Infrared Processing and Analysis Center, California Institute of Technology 100-22, Pasadena CA 91125, USA

[‡]Infrared Processing and Analysis Center, California Institute of Technology 100-22, and Jet Propulsion Laboratory, Pasadena CA 91125, USA

2. Sample Selection and Observations

Condon et al. (1991) presented 8.44GHz VLA images of the 40 most luminous members of the IRAS Bright Galaxy Sample (BGS) with $\log[L_{FIR}/L_\odot] \geq 11.25$. Most of the images show nuclear features of angular extent comparable to or smaller than the 0.25 arcsecond restoring beam, but which are resolved in almost all cases. Nevertheless, these data did not exclude the possible presence of milliarcsecond-scale emission at levels detectable with current VLBI techniques. We constructed a subsample of 31 potentially detectable objects, by using what amounts to a radio flux density limit (there is a minimum flux density at which it is possible to hide a 1–2mJy unresolved point source in the 8.4GHz VLA images).

The VLBI observations were performed, using MkIII mode-A at a wavelength of 18cm, on September 29, 1991. In addition to the most sensitive antennas available, namely Effelsberg, phased VLA, Greenbank and Arecibo, we used the south-western U.S. VLBA antennas at Pietown, Los Alamos, Kitt Peak, and Fort Davis, together with the Westerbork tied array and the Jodrell Bank MkIA telescope to provide a number of relatively short baselines. The correlation was done at Haystack Observatory, and the calibration of the resulting correlation coefficients was performed in standard fashion.

By examination of the variation of correlated flux density with projected baseline length for detected sources, we were able to classify each object as "simple" or "complex", where "complex" can be taken to imply a lack of circular symmetry, and the strong possibility of multiple peaks in the brightness distribution.

3. Results and Interpretation

Of the 31 observed galaxies, 21 exhibited brightness temperatures in excess of 10^5K, and 17 had $T_b > 10^6$K. Of the latter 17, at least 10 contain "complex" structure as defined above. In many cases, lower limits on the brightness temperature exceed 10^8K.

Condon et al. (1991) modelled their galaxies with ultracompact starbursts so dense as to be optically thick to free-free absorption at 1.6GHz and to dust extinction at $25\mu m$. These authors derive an upper limit of $\sim 10^5$K for the brightness temperature of purely starburst-related emission. Assuming that this starburst model is substantially correct (it enjoys strong support from the VLA data), our results imply that compact components *not of starburst origin* are common in ultraluminous IRAS galaxies. The question as to what *is* the origin of these compact components now arises.

Two possibilities are evident. Most obviously, we could be detecting AGN radio cores, the original goal of the project. In this case, the lack of other clear AGN indicators in these galaxies (e.g. broad emission lines, variability, and other QSO characteristics) would be attributed to obscuration of the nuclear regions by starburst-related dust.

Alternatively, we must consider radio supernovae (RSNs) of unprecedented radio luminosity. In fact, evidence has been presented for examples of such ultraluminous RSNs (Wilkinson and de Bruyn 1990, Yin and Heeschen 1991, but see Lonsdale, Lonsdale and Smith 1992), and this possibility must be carefully explored.

The limited structural information contained in our VLBI detections (employed in our classifications of "simple" and "complex" above) allows us to place stringent constraints on any RSN model of these compact components. The detected emission typically covers an angular extent of many milliarcseconds up to tens of milliarcseconds. The corresponding linear extents of many parsecs far exceed those of supernovae in the high T_b RSN stage, dictating the adoption of a model involving multiple supernovae per galaxy. Unless there are tens to hundreds of RSNs per galaxy, each RSN must belong to the hypothetical new class of ultraluminous RSN. Also, the characteristics of our VLBI detections typically require all the compact emission to originate within the central 1% or less of the volume of the nuclear starbursting region. While this is possible, it is a severe constraint for a supernova interpretation of our results, and we prefer to postulate instead that buried AGNs are responsible.

A striking aspect of our results is the lack of correlation of VLBI core properties with many other characteristics, some of which are considered reliable indicators of AGN activity. Of particular note is the fact that the emission line spectrum yields few, if any, clues to the existence of VLBI-scale radio emission. If high excitation and compact radio core were both reliable indicators of the presence of a true AGN, our results would indicate that low-excitation (H II) galaxies differ from their Seyfert brethren only in the quantity or patchiness of dusty material obscuring the high-excitation region from our view.

In summary, our results suggest that AGN-like cores are common in luminous infrared galaxies, but the lack of correlations with other properties complicates simplistic scenarios in which there is a monotonic evolution from starburst to AGN. Nevertheless, however indirect the relationship between the compact core source and the infrared emission may be, a relationship of some type does appear to exist.

REFERENCES

Condon, J. J., Huang, Z.-P., Yin, Q. F., and Thuan, T. X. 1991, *Ap.J.* **378**, 65.

Lonsdale, C. J., Lonsdale, C. J., and Smith, H. E. 1992, *Ap.J.* **391**, 629.

Wilkinson, P. N., and de Bruyn, A. G. 1990, *Mon. Not. R. Astr. Soc.* **242**, 529.

Yin, Q. F., and Heeschen, D. S. 1991, *Nature* **354**, 130.

Radio Observations of the Nuclear Environment of Seyfert Galaxies

Gregorini L. [*]   *Marziani P.* [†]   *Padrielli L.* [*]   *Rafanelli P.* [‡]

1. Introduction

In recent times deep optical observations of the circumnuclear regions of Seyfert galaxies have revealed the presence of extended (up to some kpc) and compact (some 100 pc) emitting regions either separated or connected to the nuclei by bridges of matter. The two-dimensional spectroscopic investigation of such structures has revealed that their physical features may be typical of high and low ionization supergiant HII regions or of regions ionized by the nucleus itself.

It is evident that the radio properties of such regions are also strongly dependent on the nature of the ionizing sources and that the relations between the nuclear radio powers and those of the radio structures around them could be helpful in understanding the mechanisms necessary to describe the source of their emission. In addition the high resolution reachable in the radio allows to disentangle complex structures not resolved in the optical.

We have then observed at 20 cm and 6 cm with the Very Large Array (A array) an optically selected sample of 13 Seyfert galaxies which show extended extranuclear and circumnuclear emitting regions.

2. Results

The high resolutions of the observations ($\simeq$ 1.2 arcsec and 0.3 arcsec at 20 cm and 6 cm respectively) permit to study radio emission from regions with linear dimensions of the order of 0.5 - 0.1 kpc.

NGC 1144, Mkn 463, Mkn 477, Mkn 1395, NGC 5953, Mkn 700 show radio structures at both these frequencies. Two other Seyfert galaxies (Mkn 298, Mkn 314) have been detected by other authors, but the lack of compact components did not

---

[*]Istituto di Radioastronomia, CNR, Bologna, Italy
[†]SISSA, Trieste, Italy
[‡]Dipartimento di Astronomia, Università di Padova, Italy

allow us to detect them. E0116+317, Mkn 673, Mkn 885, Kaz163, Mkn 309 have no radioemission at a level greater than 0.15 mJy/beam.

Here we present some preliminary results of two interesting systems: NGC 1143/1144 and NGC 5953/5954. Their peculiar properties are almost certainly the result of a collisional encounter.

NGC 1143/1144: The circumnuclear region of NGC 1144 is characterized by a complex radio structure formed by at least three not thermal compact sources plunged in a spiral arm detectable only at 20 cm (fig.1). At 6 cm the higher resolution permits to have more detailed structure information of the central region. The source E, which is a possible starburst, has no emission at 6 cm. The morphology of the sources B and C is typical of that of a double source: a core and two bright lobes. The spectral index ($S \propto \nu^{-\alpha}$) computed between 6 and 20 cm is steep ($\alpha \simeq 1.0$) with a steepening in the outer regions ($\alpha \simeq 1.3$). The source D has a spectral index of $\simeq 1.5$. The characteristics (morphology and spectral index) of the source A suggest that it could be the nuclear component. In it is a compact source slightly extended in the direction of B and C sources and the spectral index is flat ($\alpha \simeq 0.4$). No radioemission has been detected in NGC 1143.

NGC 5953/5954: In this system the radioemission is only present in NGC 5953. The whole galaxy is a diffuse and low brightness radio emitter at 20 cm, its nucleus is elongated along the NE–SW direction and a faint tail extends up to about 5 arcsec to the west of the nucleus (fig.2). At 6 cm only a faint nucleus elongated in north-south direction has been detected.

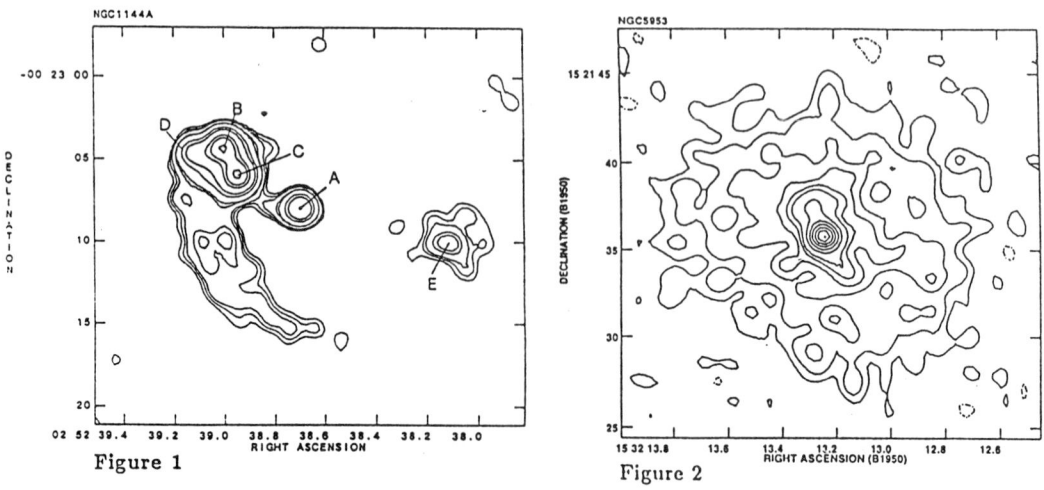

Figure 1

Figure 2

The Nuclear Ring of the Barred Spiral Galaxy NGC 1326

J.A. García-Barreto [*]   R.-J. Dettmar [†]   F. Combes[‡],   M. Gerin [‡]

B. Koribalski[†] [§]

Abstract

High resolution observations of the radio continuum and of the Hα emissions from the Barred Spiral Galaxy NGC 1326 indicate the presence of a circumnuclear structure within the central 20 arc-sec (or a radius of 670 pc). There is no evidence of the role of the compact nucleus in the formation of such structure since we did not detect any radio continuum emission from it at a level of few hundred μJy. The circumnuclear structure is a region of a on-going star formation and its origin is most likely related to galaxy dynamics as discussed by Binney and Tremaine (1987 in Galactic Dynamics, Princeton Univ. Press. New Jersey).

We interpret the presence of the nuclear ring and the outer ring in NGC 1326 as due to the gravitational perturbation of the bar: the nuclear ring to be found at the Inner Lindblad Resonance and the outer ring to be found at the Outer Lindblad Resonance.

A detailed analysis of the observations and interpretation is reported by García-Barreto et al. 1991 Rev. Mex. Astron. Astrof. **22**, 197.

---

[*]Instituto de Astronomía, Universidad Nacional Autónoma de México, México. JAG-B acknowledges partial financial support from the local organizing committee of the Sub-Arcsecond Radio Astronomy conference and from CONACYT (Mexico) that allowed him to attend this meeting.
[†]Radioastronomisches Institute der Universitat Bonn, Germany.
[‡]DEMIRM Observatoire de Meudon, France.
[§]Max Planck Institute für Radioastronomie, Bonn, Germany.

Extragalactic Masers - A Status Report

Willem A. Baan *

Abstract

The characteristics of OH, CH, H_2CO, and H_2O megamasers will be reviewed and what is known about the nature of the nuclear sources. High resolution studies of megamasers will be considered with regard to molecular diagnostics and dynamical studies of the nuclear region.

1. Introduction

Powerful extragalactic masers are known for OH, H_2O, H_2CO, and CH. Unusual physical circumstances in nuclear regions of galaxies at various stages of activity conspire to produce a beamed emission pattern of spectral line emission. A working model of the extragalactic (mega-) maser action is the *exponential amplification of background radio continuum by inverted foreground molecular material* [Baa85, Baa89, HaB85, HeW91]. The extragalactic amplifying regions are different from those for saturated masers within the Galaxy and the physical conditions for the relatively low-gain amplification process are more relaxed.

At present fifty-three OH megamasers are known, nine water masers, eleven formaldehyde masers and nine CH sources. The physics of the OH and H_2O megamasers is best understood because of the relatively large number of sources and because their strength allows for interferometric studies. The apparent gains are largest for H_2O sources (7 or more) and OH (0.05 to 3.4), while the gains for H_2CO and CH (0.001 to 0.02) are much lower. The IR pumping process for OH and the shock/collisional pumping for H_2O appears very effective. The beaming angles for the H_2O are likely to be very small, while for the other three molecules amplification occurs over large solid angles along paths with high gain or along the plane of the molecular disk. The weakness of formaldehyde and CH sources may point at a less efficient pump and the non-availability of molecular gas at the right density and radial distance. Furthermore, many such molecular line sources are below the present detection limits. Reviews of extragalactic maser sources have been published elsewhere [Baa85, Baa89, Baa91, HeW91, HBM91].

---

*Arecibo Observatory, P.O.Box 995, Arecibo, PR 00613. Arecibo Observatory is part of the National Astronomy and Ionosphere Center and is operated by Cornell University under contract with the National Science Foundation.

2. OH Megamasers

The understanding of the physical characteristics of OH megamasers is primarily based on the tight correlation of OH and FIR properties of the galaxies [Baa89, MBD89, HeW91]. Since the main infrared pumping lines for OH are at 35 and 53 μm, the quadratic L_{OH} -L_{IR} correlation becomes much tighter with the use of the 60 μm luminosity [BRF92]. There is (yet) no evidence for a non-linear L_{OH} -L_{IR} behavior.

Spectroscopic results of OH megamasers show that the division between Seyfert and starburst nuclei is almost reversed from that of the whole IRAS galaxy sample [BaS92]. Of the 35 megamaser galaxies that could be classified from the 42 sources observed, 20 (57%) are found to be Seyfert 2 or Liner, and 13 (37%) are Starburst nuclei. Two sources are Seyfert 1's. The predominance of Seyfert nuclei among the OH megamasers suggests a relation between megamaser activity and the *compactness of Seyferts nuclei* as compared with the SBN's (see [UlW84, NAS90]). Although interacting systems could have molecular gas in many directions, the simplest configuration of the OH amplifying gas would be an edge-on molecular disk. The predominance of Sy2's in the OH megamaser sample confirm earlier Seyfert scenarios with Sy2's being more edge-on than Sy1's [dZG85, Bar89].

The molecular properties of FIR galaxies have been found to be exceptional (for reviews see [HBM91, YoS91]). The OH megamaser are most extreme among the prominent FIR/molecular galaxies. The CO(1-0) properties of 32 megamasers show large molecular concentrations and an enhanced star formation efficiency (SFE) ratio $F_{FIR}/M(H_2)$ as compared with the Bright Galaxy Sample in the same luminosity range [BFH92, SSS91]. Interferometric data for a limited number of prominent molecular sources indicate that OH megamasers have the highest surface densities [ScS91]. Furthermore, the L_{HCN}/L_{CO} ratio for a limited sample of prominent sources shows that the mass ratio for high- and low-density gas is highest for those galaxies with OH megamaser activity [SDR92].

3. H$_2$CO and CH Megamasers

After the initial detection of formaldehyde megamaser emission in IC 4553 (Arp 220) [BGH86], recent detections have raised the number of sources to eleven [BHU92]. The formaldehyde emission lines are typically only a few milliJansky and have been seen among OH emitters but also among OH absorbers. A quadratic L_{H_2CO} - L_{6cm} relation has already been seen for the limited sample of formaldehyde megamasers. The inversion of the lowest K-doublet of formaldehyde may be achieved with a radio continuum [BGH86]. Just like the amplifying OH clouds, the formaldehyde clouds are likely to trace high density molecular material with density of 10^4 -10^6 cm^{-3}.

CH emission has been detected in a total of nine galaxies [WGH80, BGG91]. CH is a known low-density tracer and all CH emission has been found in strong OH absorbers. Two of these sources are H_2O emitters and two others are H_2CO emitters.

4. H_2O Megamasers

The powerful H_2O megamasers observed in some nearby galaxies have allowed detailed studies of the maser structure. Although superposition of line emission and continuum has not been shown in detail for individual extragalactic maser components, the notion of amplification within narrow cones by foreground pockets of gas appears viable [HBS90], but there are alternative interpretations [Gre92]. High resolution studies show that the extragalactic maser features occur in clusters as found in HII-regions in the Galaxy. Evidence exists from a number of megamaser sources that collisional pumping in shocked molecular regions can provide the necessary inversion of the water molecules [EHM89]. Long term monitoring studies reveal time variability of the velocity, the linewidth, and the strength of individual extragalactic maser features [HaB90, BaH92]. This variability is quite consistent with unsaturated maser amplification and can been explained in terms of variable pumping of foreground material. H_2O masers require higher amplifying optical depths than the other megamasers. Although these gains may not anymore represent low-gain amplification, the necessary gains are much lower than those required for saturated masers. Because collisional pumping is very density dependent only the densest clumps are inverted, clusters of maser spots are to be expected in extragalactic sources as well. However, 10^6 Jy maser features equivalent to one 0.3 Jy feature in the prominent source NGC 3079 are not observed in the Galactic Center. The extraordinary strength of extragalactic masers must indeed be connected with the background radio continuum and is not easily explained in terms of Galactic maser phenomenology.

5. Megamaser Family Ties

Unusual conditions prevail in luminous FIR/megamaser galaxies resulting from the large nuclear concentrations of molecular gas and the extreme nuclear activity. The occurence of megamaser activity specifically among the most extreme molecular sources confirms the presence of some threshold for the surface density and the radiation density for the radiative inversion of the molecules.

The observational differences of the various molecular megamasers result from satisfying the different physical requirements for pumping and amplification. The integrated flux of the *nuclear radio source* may be correlated with the FIR flux but its compactness is related to the nature of the nuclear activity. The tendency of OH megamasers to associate with Seyfert nuclei rather than starburst nuclei could be related not only to the compactness of the radio source but also to the depth of the potential well experienced by the molecular gas. The *powerful central radio/FIR pump source* must provide an inversion in the molecular OH and H_2CO gas in a significant part of the central molecular structure. The collisional pump required for water vapor requires strong shocks passing through the molecular disk. Local pumping in foreground H_2O regions may thus be uncorrelated with the nuclear activity, although the nuclear radio source is needed for amplification. The *availability of molecular gas at the right density* along the line of sight is correlated with the nature of the central source and the depth of the nuclear potential well. Mergers and interactions facilitate the buildup of the large molecular column densities in the nuclear region through accretion and direct deposits. The *combination of radial density and pumping conditions* is critically dictated by *evolutionary stage in the nuclear activity*. The OH and H_2CO megamasers are confined to luminous FIR/radio galaxies during early stages of evolution for the nuclear activity, where much of the molecular structure is still intact. On the other hand, the CH and H_2O amplification occurs in galaxies with more evolved nuclear activity, which may exhibit population inversions and amplification in non-nuclear regions.

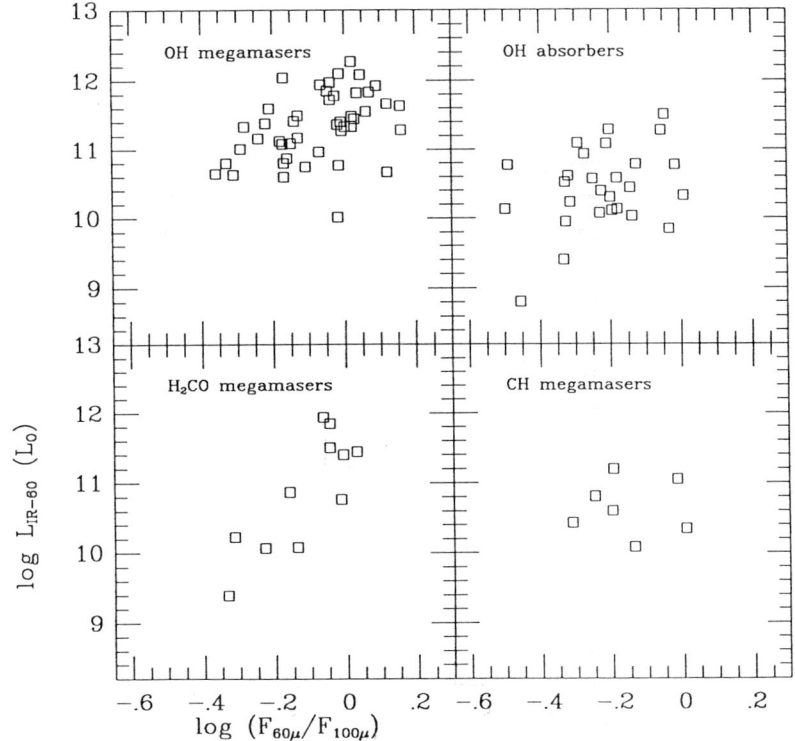

Figure 1. The $60\mu m$ IR luminosity versus the $\log(F_{60\mu m}/F_{100\mu m})$ color distribution of OH emitters and absorbers and of formaldehyde and CH emitters.

The correlation of the various source populations is illustrated in Figure 1. The $L_{IR-60\mu m}$ versus $\log(F_{60\mu m}/F_{100\mu m})$ color diagram show that the OH emitter and absorbers have a complementary distribution. The formaldehyde sources lie along the lower L_{IR} edge of the OH distributions, while the CH sources agree with the OH absorber distribution. Water sources are not clearly correlated with the other sources and require localized foreground pockets of shocked high-density gas.

6. High Resolution Studies

The line and continuum strength and compact structures of the megamasers makes sub-arcsecond observations possible and necessary. The observable universe for molecular emissions is very large in particular for the OH and CO lines. Detections of OH at $\zeta = 0.265$ [BRF92] and of CO at $\zeta = 2.286$ [Brv91] confirm the presence of previously unrecognized populations of sources. The line width of the OH gigamaser of close to 2400 km s^{-1} signifies the unusual nature of these sources. The combined signatures of molecular lines sampling different density ranges and different excitation conditions will allow detailed studies of the extreme circumstances in nuclear regions.

High resolution studies have been done for a small number of the *most powerful OH megamasers*. Only the following systems have been investigated: IC 694 / NGC 3690, IC 4553 (Arp 220), 3Zw35, Mkn 273, and IR17208-0014 (see [MoC92]). Except for IC 4553 (with Merlin and EVN) and another OH megamaser IR12112+0305 (with VLA-A) all OH emitters are unresolved at 18 cm with non-VLBI instruments. The three nearest of these galaxies show a velocity gradient in the line emission, which is related to the rotation of the nuclear molecular disk. The Seyfert 2 identification of the majority of OH galaxies would suggest that many of these nuclei are indeed compact and have a well-developed nuclear molecular disk. At lower resolution the line emission is found to be superposed on the continuum emission but an exact superposition at the highest resolution has not yet been shown. The general model of low-gain will work as long as the line emission falls within the continuum contours of the nuclear source, except that the gain needs to be slightly larger. The double pronged emission profiles of many of the strong OH megamaser sources appear related to the presence of a compact double source in the nucleus as for IC 4553, 3Zw35 and IR12112+0305. The double peaked absorption spectra in nearby sources NGC 660, 3628 and 3079 is due to a nuclear double structure as well. The double radio and NIR source of IC 4553 is due to two merging nuclei.

The *extragalactic H_2O masers* occur within the nuclear structure of the galaxies but they need not coincide with the strongest components of the radio source. At low resolution (VLA-A) the maser and the continuum are superposed. VLBI studies of the NGC 3079 and M33/IC 133 show multiple maser features within the confines of the radio source and making up the maser lines [HBS90, GMR90, GMR92]. These multiple features resemble Galactic-sized HII-regions.

Many thanks are due to Aubrey Haschick and our many collaborators for these ten years of megamaser work.

References

[Baa85]   Baan W.A., *Nature*, Vol. 315 (1985), p. 26.

[Baa89]   Baan W.A., *Astroph. J.*, Vol. 338 (1989), p. 804.

[Baa91]   Baan W.A., *Skylines, the Third Haystack Conf.*, P.A.S.P. Conf. Proc., eds. A.D. Haschick, P.T. Ho (1991), p. 45.

[BFH92]   Baan W.A., Freund, R.W., Haschick A.D., *(1992), in preparation*

[BGH86]   Baan, W.A., Güsten, R., Haschick, A.D., *Astroph.J.*, Vol. 305 (1986), p. 830.

[BaH92]   Baan, W.A., Haschick, A.D., *Astroph. J. (1992), submitted.*

[BHU92]   Baan, W.A., Haschick, A.D., Uglesich, R., *Astroph. J. (1992), submitted.*

[BaS92]   Baan, W.A., Salzer, J., *(1992), in preparation.*

[BRF92]   Baan, W.A., Rhoads, J., Fisher, K, Haschick, A., Altschuler, D., *Astroph. J. (Letters)*, Vol. 396 (1992), p. L99.

[Bar89]   Barthel, P.D., *Astroph. J.*, Vol. 336 (1989), p. 606.

[BGG91]   Bottinelli, L., Gouguenheim, L., Gérard, E., Le Squeren, A.M., Martin, J.M., Dennefeld, M., *Dynamics of Galaxies and their Cloud Distributions*, (1991), IAU Symp. 146, eds. F. Combes and F. Casoli, (Kluwer: Dordrecht), p. 442.

[Brv91]   Brown, R.L., vandenBout, P.A., *Astron. J.*, Vol. 102 (1991), p. 1956.

[dZG85]   de Zotti, G., Gaskell, C.M., *Astron. Ap.*, Vol. 147 (1985), p. 1.

[EHM89]   Elitzur, M., Hollenbach, D., McKee, C., *Astroph. J.*, Vol. 367 (1989), p. 333.

[Gre92]   Greenhill, L., *Astrophysical Masers*, eds. A. Clegg and G. Nedoluha (1992), in press.

[GMR90]   Greenhill, L., Moran, J.M., Reid, M.J., Gwinn, C.R., Menten, K.M., Eckart, A., Hirabayashi, H., *Astroph. J.*, Vol. 364 (1990), p. 513.

[GMR92]   Greenhill, L., Moran, J.M., Reid, M.J, Haschick, A.D., Hirabayashi, H., Baan, W.A., *Astroph. J. (1992), in press.*

[HaB85]   Haschick, A.D., Baan, W.A., *Nature*, Vol. 314 (1985). p. 144.

[HaB90]   Haschick, A.D., Baan, W.A., *Astroph. J. (Letters)*, Vol. 355 (1990), p. L23.

[HBS90]   Haschick, A.D., Baan, W.A, Schneps, M.H., Reid, M.J., Moram, J.M., Güsten, R., *Astroph. J.*, Vol. 356 (1990), p. 149.

[HBM91]   Henkel, C., Baan, W.A., Mauersberger, R., *Astron. Astroph. Review*, Vol. 3 (1991), p. 47.

[HeW91]   Henkel, C., Wilson, T.L., *Astron. Ap.*, Vol. 229 (1991), p. 431.

[MBD89]   Martin, J.M., Bottinelli, L., Dennefeld, M., Gouguenheim, L., Le Squeren, A.M., Paturel, G., *C.R. Acad. Sci. Paris*, Vol. 308(II) (1989), p. 287.

[MoC92]   Montgomery, Cohen. R.J., *Mon. Not. R. Astron. Soc.*, Vol. 254 (1992), p. 23P.

[NAS90]   Norris, R.P., Allen, D.A., Sramek, R.A., Kesteven, M.J., Troup, E.R., *Astroph. J.*, Vol. 359 (1990), p. 291.

[SSS91]   Sanders, D.B., Scoville, N.Z., Soifer, B.T.,, *Astroph. J*, Vol. 370 (1991), p. 158.

[ScS91]   Scoville, N.Z., Sargent, A.I., Sanders, D.B., Soifer, B.T., *Astroph. J.(Letters)*, Vol. 366 (1991), p. L5.

[SDR92]   Solomon, P., Downes, D., Radford, D., *Astroph. J. (Letters)*, Vol. 387 (1992), p. L55.

[UlW84]   Ulvestad, J.S., Wilson, A.S., *Astroph. J.*, Vol. 278 (1984), p. 544.

[WGH80]   Whiteoak, J., Gardner, F.F., Höglund, B., *Mon. Not. R. Astron. Soc.*, Vol. 190 (1980), p. 17P.

[YoS91]   Young, J.S., Scoville, N.Z., *Ann. Rev. Astron. Astroph.*, Vol. 29 (1991), p. 581.

Toward Weighing a Hidden Seyfert Nucleus

C.R. Gwinn [*]   R.J. Antonucci [†]   R. Barvainis [‡]   J. Ulvestad [§]
S. Neff [¶]

Abstract

We present preliminary results of VLBI observations of the H_2O megamaser in the nucleus of NGC 1068. Using the most sensitive antennas and a new search technique, we detect the maser with VLBI and set a lower limit on its gradient in Doppler velocity with position. The direction of the gradient is perpendicular to the jet seen in VLA maps and HST images of the nucleus. Under the assumption that the maser is in Keplerian rotation about the nucleus, our lower limit on the magnitude of the gradient yields a preliminary upper limit on the mass of the nucleus of 4×10^9 solar masses.

Nuclei of Seyfert galaxies are relatively nearby, relatively less luminous examples of the active galactic nuclei present in radio-quiet quasars. Seyfert galaxies are divided into two classes, with Seyfert 1 showing broad permitted lines. A fruitful hypothesis has been that Seyfert 2s are identical to Seyfert 1s, except that the active nucleus is obscured by a dust torus.

The Seyfert 2 NGC1068 is something of a Rosetta Stone to the dust-torus picture, because polarized emission from its nucleus shows the broad lines characteristic of Seyfert 1s. This suggests that the polarized light originates in the nuclear region but is reflected into the line of sight, perhaps by free electrons at the base of the kpc-scale jet [AnM85]. HST, VLA and MERLIN observations indicate that the southernmost of the three major components within the arcsecond-scale nucleus is the "true" nucleus [Mux92]. The jet is seen as an approximately conical region of optical and radio emission, extending ~ 20" northest of the nucleus.

An H_2O megamaser lies in the nucleus of NGC1068 [CHL84]. The brightest emission extends over ~ 100 km/s, and is redshifted relative to the galaxy by 300 km/s.

---

[*]University of California, Santa Barbara, California 93106, USA.
[†]University of California, Santa Barbara, California 93106, USA.
[‡]Haystack Observatory, Westford, Massachusetts 01886, USA.
[§]Jet Propulsion Laboratory, California Institute of Technology, Pasadena, California 91109, USA.
[¶]Goddard Space Flight Center, Greenbelt, Maryland 02771, USA.

We observed the megamaser in NGC1068 on 22 June 1991 with antennas at Bonn (100 m), Madrid (70 m), Haystack (38 m), Green Bank (42 m), VLA (27×25 m), and Goldstone (70 m). Data were recorded with the Mk III VLBI system and processed with the Haystack Mk IIIA processor.

We processed the data with a novel technique. A linear gradient of Doppler velocity with position on the sky will produce a linear gradient of interferometric phase with observing frequency. Traditionally, such a gradient of phase is interpreted as a delay offset, for a continuum source. With knowlege of the true source position, and an accurate *a priori* estimate of the delay from nearby calibration sources, we can place limits on the gradient of Doppler velocity on the sky.

We detected the maser with relatively low SNR, but at the predicted position, so that statistical significance is high. We place an upper limit of (30 km/s)/(2700 AU) on any gradient of Doppler velocity with position on the sky, along a position angle of 25°. The masers cannot lie closer to the Seyfert nucleus than 1 pc, the distance at which dust sublimates. From this relatively model-independent lower bound on radial distance, and our observational lower limit on velocity gradient, we place an preliminary upper limit on mass within the maser's orbit of 4×10^9 solar masses, under the assumption of Keplerian rotation.

References

[Mux92] Muxlow T., *These Proceedings*, 1992.

[AnM85] Antonucci R.R.J., Miller J.E., *Spectrapolarimetry and the Nature of NGC1068*. Ap. J. Vol. 297, pp. 621-632.

[CHL84] Claussen M.J., Lo K.-Y., *Circumnuclear Water Masers in Active Galaxies*. Ap. J. Vol. 308, pp. 592-599.

Unified Schemes for Active Galactic Nuclei

C. Megan Urry [*]

Abstract

The basic tenet of unified schemes for active galactic nuclei (AGN) is that their appearance is strongly orientation dependent, so that our current classification is dominated by random pointing directions rather than physical properties. Two kinds of strongly anisotropic radiation, obscuration by optically thick matter and relativistic beaming of radio emission, are important in many AGN. I outline possible mini-unifications of different classes of AGN which involve these two forms of anisotropy, including recent work on the statistics of the radio-loud schemes. While we do not yet understand all the subtleties of the unified schemes, it is clear that to first order, some of these ideas have to be right.

1. Theoretical Expectations and Observational Facts

The prevailing (but not necessarily correct) picture of the physical structure of AGN includes a central supermassive black hole, whose gravitational potential energy is the ultimate source of the AGN luminosity. Matter pulled toward the black hole loses angular momentum through viscous or turbulent processes in an accretion disk, which is heated and glows brightly at optical, ultraviolet, and perhaps even soft X-ray wavelengths. Some of this light may be obscured by a torus or warped disk of gas and dust, on much larger scales than the accretion disk. Outflows of energetic particles occur along the poles of the disk or torus, escaping and forming collimated jets and sometimes giant radio sources when the host galaxy is an elliptical, but forming only very weak radio sources when the host is a gas-rich spiral. The plasma in the radio jets, at least on the smallest scales, streams outward at a velocity near the speed of light, beaming radiation in the forward direction. This picture of AGN is inherently asymmetric, a fact that inspires the unified schemes discussed below.

Observationally, AGN can be broadly characterized by their optical spectra (broad, narrow, or extremely weak line emission) and their radio-loudness. Most AGN are

---

[*]Space Telescope Science Institute, 3700 San Martin Drive, Baltimore, MD 21218, USA. This work was supported in part by NASA grant NAG5-1034.

radio-quiet. Type 1 AGN, called Seyfert 1 galaxies or quasars depending on luminosity, have bright continua and broad emission lines. Seyfert 2 galaxies have weak continua and only narrow emission lines, meaning either that they have no high velocity gas or, as we now believe, the line of sight to such gas is blocked by absorbing material. The high-luminosity analogs of Seyfert 2 may be infrared-luminous galaxies [S88][W92][Ho91] and/or narrow-line X-ray galaxies [M82].

Roughly 10% of all AGN are radio-loud. These include Broad-Line Radio Galaxies (BLRGs) and Steep-Spectrum Radio Quasars (SSRQs), which differ only in luminosity. Radio-loud AGN with only narrow emission lines are known as Narrow-Line Radio Galaxies (NLRGs). The low-luminosity NLRGs are Fanaroff-Riley type I objects [FR74], characterized by diffuse, symmetric radio jets and weakly defined radio lobes. The high-luminosity NLRGs are Fanaroff-Riley type II, with sharp jet and lobe boundaries and bright hot spots. The hosts galaxies also differ, with FRIs tending to lie in optically bright central galaxies in clusters. Radio-loud AGN with very weak or no detectable emission lines are called BL Lac objects; as far as we know, there are no truly radio-quiet BL Lacs.

A small fraction of radio-loud AGN evidence extreme activity: very rapid variability, unusually high and variable polarization, high brightness temperatures, and superluminal velocities of compact radio cores. These are collectively called blazars, and can be separated into broad-line (type 1) objects, like Optically Violently Variable quasars (OVVs are essentially the same as Highly Polarized Quasars, Flat-Spectrum Radio Quasars, or superluminal quasars; [F89][ILT91][V92][W92]), or BL Lac objects. The overall radio-through-X-ray continuum emission of OVVs and BL Lacs is very similar, suggesting that the fundamental processes of energy generation are related. (They do have significant differences in polarization properties, notably on VLBI scales; Gabuzda, this volume.)

The empirical division of AGN into types 1 and 2, and into blazars and non-blazars, is almost certainly aspect-dependent. Unified schemes suggest that current AGN classification depends primarily on the angle between the principal axis of symmetry and the line of sight. The ultimate hope is that these schemes will allow us to understand fundamental properties of active galaxies like whether all AGN contain central massive black holes.

2. Anisotropy: Obscuration and Relativistic Beaming

Although the innermost regions of most AGN are too small to be imaged, considerable evidence exists for strongly anisotropic emission even on the smallest scales. This anisotropy takes two principal forms: relativistic beaming of radio jets (important only in radio-loud objects) and obscuration of IR-UV light by optically thick gas and dust.

Over the past fifteen years or so, a number of arguments have suggested the presence of obscuring material in AGN, particularly in the type 2s. The most direct argument comes from optical spectropolarimetry of type 2 objects, particularly observations of a NLRG, 3C 234 [A84], and the nearby Seyfert 2, NGC 1068 [AM85]. Spectra of their polarized light have broad lines, like type 1 rather than type 2 objects, probably caused by electron scattering of a luminous, type 1 nucleus that is hidden from direct view by a thick torus of gas and dust (whose axis is the radio jet axis in the case of 3C234). Many other polarization and scattering observations of AGN support this picture of scattered light from a luminous, hidden type 1 continuum source.

Additional evidence comes from observations at other wavelengths. Compact, bright, infrared cores and/or wavelength-independent perpendicular polarization have been found in about a dozen NLRGs, of both high and low luminosities. In a few cases, broad wings have been detected for the infrared Paschen lines [F86][W91], indicating high-velocity gas that is obscured at optical wavelengths. The weakness of the X-ray continuum from type 2 AGN relative to type 1s is also consistent with the idea of an obscured nucleus [LE82]. The variable hard X-ray source in Cygnus A, a classic high-luminosity NLRG, argues for a compact nuclear source [A87]. Finally, light cones seen in many nearby type 2 AGN, notably Cen A [M91] and NGC 1068 [E91][P88], suggest ionization of the extended narrow-line gas by a hidden continuum source with type 1-like luminosity.

Relativistic beaming is important when the emitting plasma has a bulk relativistic velocity with respect to the observer. An observer located in or near the path of the plasma sees a greatly enhanced intensity and shorter time scales relative to a plasma at rest. The extremely rapid variations and high luminosities seen in blazars are most likely explained in this way [BR78]. In some blazars, the rapid variations observed correspond to inferred brightness temperatures greatly in excess of the Compton limit, $T \sim 10^{12}$ K [Q89]. A number of blazars have recently turned out to be strong γ-ray sources, and in the case of 3C279, variable γ-ray emission can dominate the bolometric luminosity [Ha91], leading to a model-independent argument for relativistic beaming of the γ-rays [MGC92]. Perhaps the most direct indication of relativistic beaming is superluminal motion on the smallest (VLBI) scales, with typical speeds in the range $1 - 10c$, easily explained if by a closely aligned, oncoming, relativistic jet.

Radio jets imaged on the sky are often one-sided, almost exclusively so in the high-luminosity AGN. On small scales, the one-sidedness is likely due to relativistic beaming, especially in those sources exhibiting superluminal motion. Relativistic jets on larger scales are problematic, as they would have enormous kinetic energies, but since the sidedness is generally the same on large and small scales, it follows that both may result from relativistic beaming. Perhaps the most compelling argument for relativistic speeds on the larger scales stems from the large study of depolarization asymmetry in radio lobes [GC91][T92]. In 49 of 69 extended radio sources, the lobe on the counterjet side is more depolarized than the one on the jet side, just as if it were further away (and had more depolarizing material along the line of sight). Some details of interpretation remain to be worked out, but it seems plausible that most jets are intrinsically two-sided and appear one-sided due to relativistic beaming on both small and large scales.

Evidence for the preponderance of relativistic jets comes from large surveys of radio-loud AGN [HR89][ILT91]. These studies show that, just as one would expect if all jets were relativistic, the ratio of core radio emission (presumably relativistically beamed) to extended (clearly unbeamed) radio emission is correlated with optical polarization, optical power-law fraction, degree and rapidity of variability, jet curvature, and superluminal motion, and is inversely correlated with linear size.

3. Unified Schemes for AGN

If we accept the abundant evidence for strongly anisotropic radiation, then AGN oriented differently with respect to us will look very different, and we have no choice but to investigate unification schemes. For radio-quiet objects, unified schemes are still not thoroughly worked out. It is clear there are some problems with the simplest picture, where type 1 and type 2 classifications result from lines of sight that miss or intercept obscuring material, respectively, at a minimum suggesting a change in the geometry of the obscuring material with luminosity and/or redshift [L91].

For radio-loud objects, the picture is a bit better defined, with blazars being the aligned version of NLRGs. The usual indications for any particular unified scheme come from isotropic or quasi-isotropic properties. For example, the extended radio emission around the unresolved cores of BL Lac objects is quite similar to the radio lobe power in FRI galaxies [P92][WMA84][AU85]. The host galaxies of OVV quasars and BL Lacs correspond to those of high- and low-luminosity NLRGs, respectively [SH90][VW90] [U89][Br89]. The environments also appear to match [SH90][S91], although more observations of BL Lac fields are needed.

The prevailing low- and high-luminosity unified schemes for radio- loud AGN are: (1) BL Lacs are aligned FRIs [B83], and (2) either OVVs are aligned steep-spectrum radio quasars [OB82] or aligned FRII galaxies (with BLRG/SSRQs being at intermediate angles; [P87][Ba89]). Because the fundamental assumption that the AGN orientations are random with respect to us, these unified schemes make clear predictions for the relative numbers of each class of AGN as a function of orientation. This situation is complicated by the strong selection effects introduced by anisotropic radiation patterns. Since unbiased selection of useful samples is not currently viable, the only alternative is to take the selection effects into account in a model-dependent way. The parameters derived from fitting the luminosity functions then lead to predictions about other properties, like the expected distributions of superluminal velocities and core-to-extended radio emission.

Radio-quiet schemes have not been tested this way, because the radiation pattern is still very uncertain, but the radio-loud unified schemes have been explored in greater detail. At high luminosity, the luminosity functions and observed ratios of core-to-extended radio flux for the 2 Jy radio sample of AGN (which contains 83 FRIIs, 34 SSRQs, and 50 FSRQs, most with known redshifts; [WP85]) is consistent with the larger scheme wherein radio galaxies constitute the parent population, with a mean bulk Lorentz factor of $\gamma \sim 10$ [PU92]. This means FSRQs have beams within $\theta \sim 14°$ of the line of sight, while SSRQs lie in the range $14° \lesssim \theta \lesssim 40°$. The SSRQ-FSRQ-only scheme (ignoring the radio galaxies) can not be ruled out, but is much harder to reconcile with the available data.

The available data are also in quantitative agreement with the low-luminosity (FRI-BL Lac) unified scheme. For X-ray selected BL Lacs, the inferred Lorentz factor is $\gamma \sim 3$ [PU90], while for the higher quality radio data, a range is required, with $\langle\gamma\rangle \sim 7$ [UPS91]. The apparent dependence of the Lorentz factor on wavelength, can be explained in some inhomogeneous jet models (e.g., [GM89]), and is underscored by the fact that X-ray-selected BL Lac objects are 1-2 orders of magnitude more numerous than radio-selected BL Lac objects [UPS91].

Despite notable progress, a number of significant problems remain with these unified schemes, including the large deprojected sizes of radio quasars; slight mismatches in infrared properties, host galaxies, and environments of radio galaxies and quasars; complicated structure of jets [LB85], usually ignored in the statistical tests; wavelength dependence of beaming (at least in BL Lacs; [UPS91]); the fact that large-scale FRI jets are subluminal; and evolution, which due to poor statistics has so far been finessed with models.

4. Conclusions

Strong anisotropy in AGN is well established. This means some unified scheme(s) must be right. In radio-quiet objects there is some connection between narrow-line and broad-line objects via orientation, but a detailed scheme has not been tested quantitatively. For radio-loud objects, relativistic beaming is likely to be very important, on large as well as small scales. Two schemes, one unifying low-luminosity (FRI) radio galaxies with weak-lined blazars (BL Lac objects), and the other unifying high-luminosity (FRII) radio galaxies with type 1 blazars (OVV/HPQ/FSRQs) are both consistent with available data. In the latter case, a scheme omitting the NLRGs is not ruled out but is less likely [PU92][Ba89]. If these unified schemes are correct, all quasars are seen from a preferred angle (since the broad-line region is visible). This may explain why Lyα edges are not seen.

References

[A84] Antonucci, R.R.J. 1984, ApJ, 278, 299

[AM85]   Antonucci, R.R.J., Miller, J.S. 1985, ApJ, 297, 621

[AU85]   Antonucci, R., Ulvestad, J. 1985, ApJ, 294, 158

[A87]   Arnaud, K.A., Johnstone, R.M., Fabian, A.C., Crawford, C.S., Nulsen, P.E.J., Shafer, R.A., Mushotzky, R.F. 1987, MNRAS, 227, 241

[Ba89]   Barthel, P.D. 1989, ApJ, 336, 606

[BR78]   Blandford, R.D., Rees, M.J. 1978, in Pittsburgh Conference on BL Lac Objects, Ed. A.N. Wolfe (Pittsburgh: Univ. of Pitt. Press), p. 328

[B83]   Browne, I.W.A. 1983, MNRAS, 204, 23p

[Br89]   Browne, I.W.A. 1989, in BL Lac Objects, ed. L. Maraschi, (Heidelberg: Springer Verlag), p. 401

[E91]   Evans, I.N., Ford, H.C., Kinney, A.L., Antonucci, R.R.J., Armus, L., Caganoff, S. 1991, ApJ, 369, L27

[F86]   Fabbiano, G., Willner, S.P., Carleton, N.P., Elvis, M. 1986, ApJ, 304, L37

[FR74]   Fanaroff, B.L., Riley, J.M. 1974, MNRAS, 167, 31p

[F89]   Fugmann, W. 1989, A&A, 205, 86

[GC91]   Garrington, S.T., Conway, R.G. 1991, MNRAS, 250, 198

[GM89]   Ghisellini, G., Maraschi, L. 1989, ApJ, 340, 181

[Ha91]   Hartman, R.C., *et al.* 1991, ApJ, 385, L1

[Ho91]   Hough, J.H., Brindle, C.B., Wills, B.J., Wills, D., Bailey, J. 1991, ApJ, 372, 478

[HR89]   Hough, D.H., Readhead, A.C.S. 1989, AJ, 98, 1208

[ILT91]   Impey, C.D., Lawrence, C.R., Tapia, S. 1991, ApJ, 375, 46

[L91]   Lawrence, A. 1991, MNRAS, 252, 586

[LE82]   Lawrence, A., Elvis, M. 1982, ApJ, 256, 410

[LB85]   Lind, K.R., Blandford, R.D. 1985, ApJ, 295, 358

[MGC92]   Maraschi, L., Ghisellini, G., Celotti, A. 1992, in Testing the AGN Paradigm, eds. S.S. Holt, S.G. Neff, C.M. Urry, (New York: AIP), p. 439

[M91]   Morganti, R., Robinson, A., Fosbury, R.A.E., di Serego Alighieri, S., Tadhunter, C.N., Malin, D.F. 1991, MNRAS, 249, 91

[M82]   Mushotzky, R.F. 1982, ApJ, 256, 92

[OB82]   Orr, M.J.W., Browne, I.W.A. 1982, MNRAS, 200, 1067

[P92]   Padovani, P. 1992, A&A, 256, 399

[PU90] Padovani, P., Urry, C.M. 1990, ApJ, 356, 75

[PU92] Padovani, P., Urry, C.M. 1992, ApJ, 387, 449

[P87]   Peacock, J.A. 1987, in Astrophysical Jets and Their Engines, ed. W. Kundt, (Dordrecht: Reidel), p. 185

[PK92] Pier, E., Krolik, J.H. 1992, ApJ, 399, L23

[P88]   Pogge, R. 1988, ApJ, 328, 519

[Q89]   Quirrenbach, A., Witzel, A., Krichbaum, T., Hummel, C.A., Alberdi, A., Schalinski, C. 1989, Nature, 337, 442

[S88]   Sanders, D.B., Soifer, B.T., Elias, J.H., Madore, B.F., Matthews, K., Neugebauer, G., Scoville, N.Z. 1988, ApJ, 325, 74

[SRT90] Scarrott, S.M., Rolph, C.D., Tadhunter, C.N. 1990, MNRAS, 243, 5p

[SH90] Smith, E.P., Heckman, T.M. 1990, ApJ, 348, 38

[S91]   Stickel, M., Padovani, P., Urry, C.M., Fried, J., Kühr, H. 1991, ApJ, 374, 431

[T92]   Tribble, P.C. 1992, MNRAS, 256, 281

[U89]   Ulrich, M.-H. 1989, in BL Lac Objects, ed. L. Maraschi (Heidelberg: Springer-Verlag), p. 45

[UMC91] Urry, C.M., Marziani, P., Calvani, M. 1991, ApJ, 371, 510

[UP91] Urry, C.M., Padovani, P. 1991, ApJ, 371, 60

[UPS91] Urry, C.M., Padovani, P., Stickel, M. 1991, ApJ, 382, 501

[US84] Urry, C.M., Shafer, R.A. 1984, ApJ, 280, 569

[V92]   Valtaoja, E., Terasranta, H., Urpo, S., Nesterov, N.S., Lainela, M., Valtonen, M. 1992, A&A, 254, 80

[VW90] Véron-Cetty, M.-P., Woltjer, L. 1990, A&A, 236, 69

[WP85] Wall, J.V., Peacock, J.A. 1985, MNRAS, 216, 173

[W91]   Ward, M.J., Blanco, P.R., Wilson, A.S., Nishida, M. 1991, ApJ, 382, 115

[WMA84] Wardle, J.F.C., Moore, R.L., Angel, J.R.P. 1984, ApJ, 279, 93

[W92]   Wills, B.J., Wills, D., Breger, M., Antonucci, R.R.J., Barvainis, R. 1992, ApJ, 398, 454

The Effect of Turbulence on the Large Scale Structure of Radio Jets

S.A.E.G. Falle [*]

Abstract

A $k - \epsilon$ turbulence model is used in combination with an adaptive grid code to study the effect of turbulence on the propagation of hypersonic jets. The results seem to suggest that such jets tend to become turbulent at quite short distances from the source unless the external density decreases rapidly with distance from the soure.

1. Introduction

It has long been fashionable to assume that the radio jets in FRI sources are subsonic and turbulent, while those in the more powerful FRII sources are highly supersonic and therefore largely unaffected by turbulent entrainment of surrounding material (Begelman, Blandford & Rees 1984). If radio jets are indeed fluid like, then this must be true to some extent since it would otherwise be impossible for FRII jets to propagate for many jet radii without dissipating most of their kinetic energy. However, this does not mean that turbulence does not play a crucial role in the dynamics of such jets. Even if the jet can escape becoming turbulent before it reaches the working surface, it will then encounter strongly curved shocks which generate vorticity and thus enhance the level of turbulence. It is therefore quite likely that the backflow into the cocoon is turbulent.

In a previous paper (Falle 1991) it was shown that one would generally expect an inviscid jet to become roughly self-similar if the external density decreases like $1/R^\alpha$ where R is the distance from the source and $0 < \alpha < 2$. The self-similarity is only approximate because the flow near the working surface is dominated by periodic vortex shedding which cannot be properly modelled without including viscous effects. At these Reynolds' numbers it is not the laminar viscosity that matters, but the turbulent one and there is clearly no really good way of dealing with this. It is, however, possible to get some idea of what might happen by using a turbulence model of the kind that are routinely used in engineering calculations.

---

[*]Department of Applied Mathematical Studies, The University, Leeds, LS2 9JT, England.

There are many models of this sort, but the uncertainties are such that there is no point in using anything too sophisticated, particularly since there is very little experience of applying these models to supersonic flow. Fortunately, a modified form of the $k - \epsilon$ model has been applied to steady supersonic jets and to gives results which are in reasonable agreement with experiment (Dash & Wolf 1983). Of course, this does not mean that it is any good for unsteady jets with vortices, but it should at least give us some idea of the effect of turbulence on these jets.

2. Turbulence Model

It is well known that in a high Reynolds' number turbulent flow the non-linearity of the Navier-Stokes' equations drives an energy cascade from large scale eddies down to eddy sizes which are small enough for the ordinary viscosity to be important. The scale of the largest eddies is typically that of the whole turbulent region, while the eddies in which the energy dissipation takes place are a factor $1/R_e^{3/4}$ smaller, where R_e is the Reynolds' number. Since turbulence is essentially three dimensional and R_e is extremely large, there is obviously no way in which such flows can be computed directly on currently available machines. Instead one has to try and simulate the effect that the turbulent eddies have on the mean flow by increasing the transport coefficients, particularly the viscosity, in regions where there is a lot of turbulence. The crudest way of doing this is to multiply the laminar viscosity by the Reynolds' number everywhere, but this takes no account of the huge variation in turbulent intensity that occurs in all but the simplest turbulent flows.

It is much better to make some attempt to calculate the properties of the turbulence and to use these to estimate the turbulent viscosity etc. A $k - \epsilon$ model does this by introducing two variables to describe the turbulent eddies, the turbulent energy per unit mass k and the turbulent dissipation rate ϵ. Since the turbulent energy resides mainly in the large eddies, while the d issipation occurs in the small ones, it is possible to think of k and ϵ as describing the large and small scale turbulence respectively. One can then use dimensional arguments and experimental results to get expressions for the transport properties of the turbulence in terms of k and ϵ.

This still leaves the problem of calculating k and ϵ and, as it is not possible to derive appropriate equations rigorously from the Navier-Stokes equations, one has to rely upon heuristic arguments. Since turbulence is advected by the mean flow and in some sense diffused by the eddies, we clearly need to include both advective and diffusive terms our equations. For k the other terms are fairly obvious. Turbulent energy is generated by the action of the turbulent viscosity on the mean flow and it is converted to heat at the dissipation rate ϵ. The k equation therefore takes the form

$$\frac{\partial}{\partial t}\rho k + \nabla.\rho k \mathbf{v} = \nabla.\mu_t \nabla k + P - \rho.\epsilon$$

where

$$\mu_t = \frac{0.09}{(1+19M_t^2)} \frac{\rho k^2}{\epsilon} \qquad M_t = \left(\frac{\rho k}{\gamma p}\right)^{1/2}$$

are the turbulent viscosity and Mach number respectively. Here P is the turbulent production rate which is equal to the rate of working of the turbulent stresses.

The ϵ equation is much harder to justify, but the usual form is

$$\frac{\partial}{\partial t}\rho\epsilon + \nabla.\rho\epsilon\mathbf{v} = 0.77\nabla.\mu_t\nabla\epsilon + (1.4P - 1.94\rho\epsilon)\frac{\epsilon}{k}$$

The dimensionless coefficients in these equations are derived from experiments on free turbulent shear layers.

3. Hypersonic Jets

Any fluid jet at high Reynolds' number will be bounded by a turbulent shear layer. Experiments on supersonic jets suggest that the width of this layer grows linearly with distance down the jet and that the angle θ_t that the inner edge of the layer makes with the jet boundary behaves like

$$\theta_t \sim \frac{1}{M_j}$$

if the mach number M_j of the jet is sufficiently large. This means that such a jet can avoid becoming turbulent provided its opening angle θ_j is larger than θ_t. However, a jet can only have constant opening angle as long as its ram pressure is much greater than the external pressure and since the ram pressure in such a jet decreases like $1/R^2$ where R is the distance from the source, the jet will eventually reconfine and its opening angle decrease if the external pressure decreases more slowly than this. Most FRII jets do seem to have a constant opening angle near the source, but then appear to reconfine before they reach the radio lobes. It is therefore not at all obvious that they can avoid becoming turbulent even if they are hypersonic when they emerge from the source.

Whether the reconfinement is brutal enough to make the jet turbulent depends on the pressure just outside the jet and this is not likely to be the same as in that in the galactic atmosphere. The jet material fills a hot cavity and the pressure in this cavity can be substantially larger than the undisturbed pressure and much more uniform (Falle 1991). In fact the jet must reconfine and will probably become turbulent unless the external density decreases more rapidly than $1/R^2$.

4. Numerical Calculations

Since our purpose is to look at the effect of turbulence on the large scale structure, it makes sense to consider the same problem as that computed by Falle (1991) without any physical viscosity. In that paper it was assumed that two symmetric, hypersonic, conical jets emerge from their parent galaxy with power P, mass flux Q and opening angle θ_j into an environment with zero pressure and constant density. Figure 1 shows pressure and turbulent intensity k for a jet with $\theta_j = 13^0$ and the results should be compared with those in Falle (1991). The most striking difference is that the jet becomes fully turbulent and subsonic long before it reaches the working surface and there is therefore no pronounced high pressure region at the end of the jet i.e. no hot spot. This is very different from the essentially inviscid jets described in Falle (1991). The main reason why the jet becomes turbulent so quickly is that even if it is not fully turbulent at the hot spot, the backflow is and this makes the pressure in the cavity more uniform than it would otherwise be. The result is that reconfinement occurs closer to the source than it does in the inviscid case.

Figure 2 shows what happens when we reduce the opening angle to 5^0 and allow the external density to decrease like $1/R$. In this case the jet goes further before it becomes turbulent, but it is still subsonic by the time it reaches the working surface. At earlier times the jets remain supersonic all the way out to the working surface in both cases, but their length to width ratio is then much smaller than that of typical FRII jets.

5. Conclusions

This paper has described a first attempt to include the effect of turbulence in a numerical simulation of the flow in a hypersonic jet. Our results show that such jets must become turbulent unless the density in the environment decreases so rapidly that there is no significant reconfinement before the jet reaches the working surface. This basically means that the external density must decrease faster than $1/R^2$ for much of the length of the jet, but it must become more uniform at some point, otherwise the jet would not be decelerated and there would be no working surface.

These conclusions do, of course, depend upon our somewhat dubious treatment of the turbulence, but they are not likely to be too wrong unless relativistic and magnetohydrodynamic effects can keep the turbulence under control. Whether or not they can do so is hard to say, but our results do at least suggest that there is not much point in doing purely inviscid calculations.

References

[1] Begelman, M.C., Blandford, R.D., & Rees, M.J., 1984, *Rev. Mod. Phys.*, **56**, 255.

[2] Dash, S.M., & Wolf, D.E., 1983, *AIAA paper 83-0704*.

[3] Falle, S.A.E.G., 1991, *Mon. Not. R. astr. Soc.*, **250**, 581.

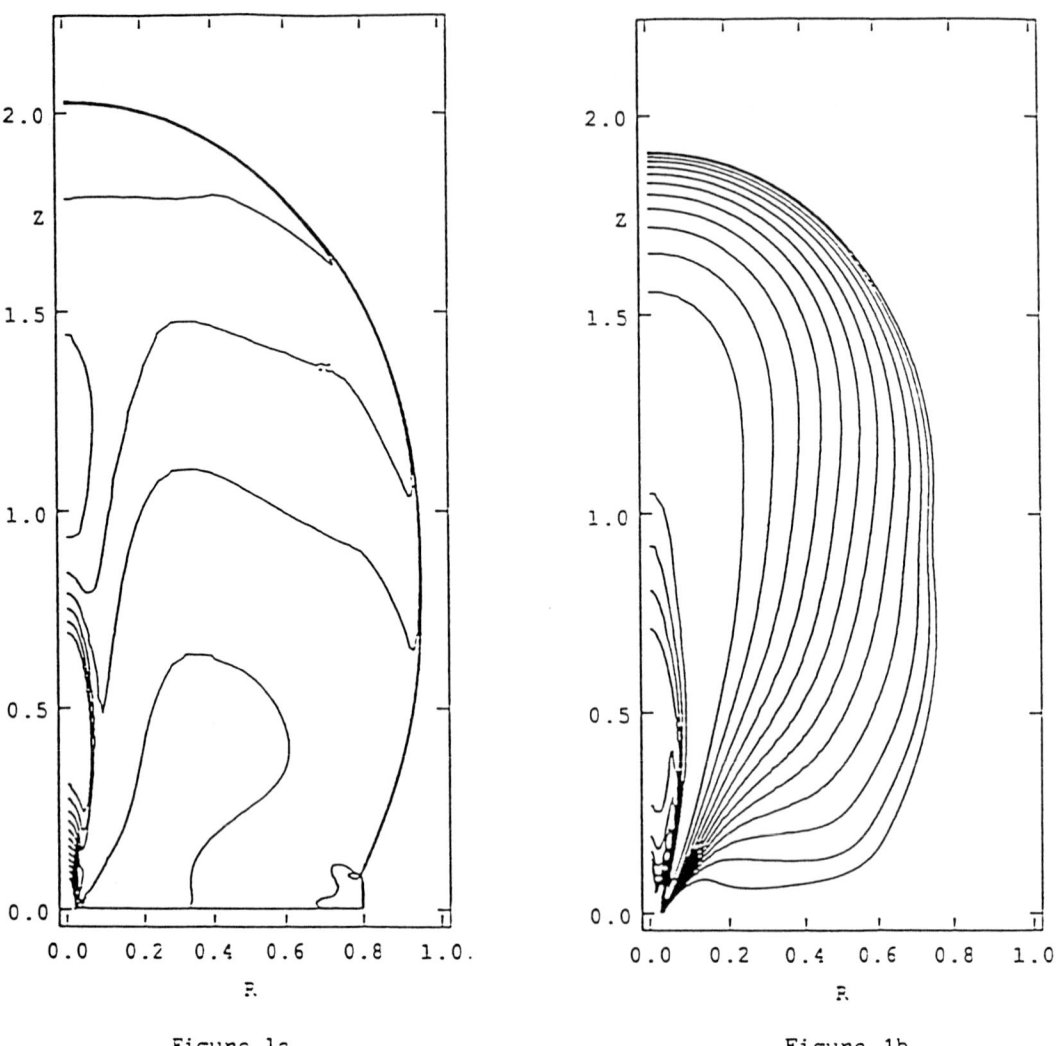

Figure 1a.   Figure 1b.

Figure Captions

Figure 1a. Logarithmic contours of pressure for the same parameters as in Falle (1991) at $t = 2.33$. There are 11 equally spaced contours between 0.1 and 1.0.

Figure 1b. Logarithmic contours of turbulent intensity k at $t = 2.33$. There are 11 equally spaced contours between 0.01 and 2.0.

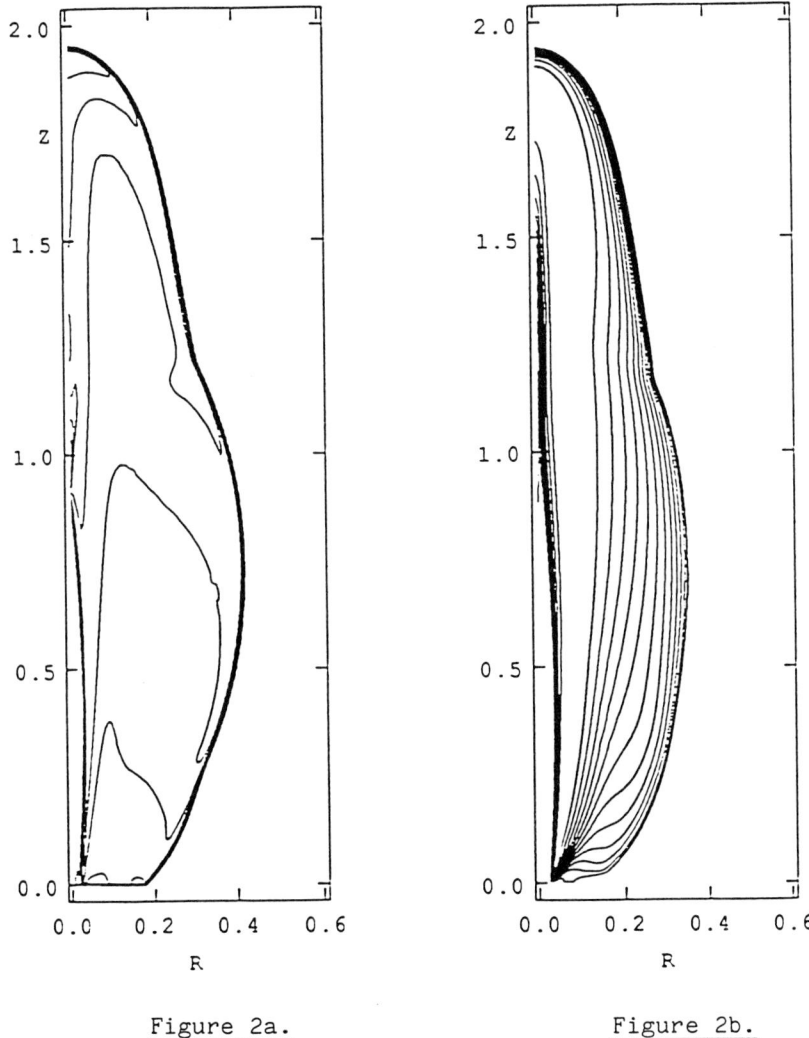

Figure 2a.   Figure 2b.

Figure 2a. Logarithmic contours of pressure for a jet with an initial opening angle of 5^0 in an external density that decreases like $1/R$. There are 11 equally spaced contours between 10^{-3} and 1.62. This is the latest time at which there is a detectable hot spot at the end of the jet.

Figure 2b. Logarithmic contours of turbulent intensity k for a jet with an initial opening angle of 5^0 in an external density that decreases like $1/R$. There are 11 equally spaced contours between 0.1 and 7.6.

Relativistic Flow in Low-Luminosity Radio Jets

Robert A. Laing *

Abstract

Several lines of evidence suggest that jets in FRI radio galaxies may be relativistic on small scales, decelerating to non-relativistic speeds at large distances from the nucleus. The consequences of relativistic aberration for the appearance of polarized radiation from a two-component, decelerating jet consisting of a perpendicular-field core surrounded by a slower parallel field shear layer are considered. For a range of angles to the line of sight between 30° and 90°, several observed phenomena are reproduced.

1. The Case for Relativistic Flow in FRI Jet Bases

Jets in weak (FRI) radio galaxies are predominantly two-sided and the flow velocities in their outer regions cannot, therefore, be significantly affected by Doppler boosting. The jet bases, however, are one-sided [BP84] and the hypothesis that these are relativistic has been given more credibility by recent observations of motions in the M87 jet [BO91] and depolarization asymmetry [PMCFR92]. The idea that BL Lac objects are FRI radio galaxies with relativistic jets on pc scales [UPS91] also has the consequence that the flow must decelerate before reaching the symmetrical outer structure.

Deceleration of a relativistic jet causes a large fraction of its kinetic energy to be dissipated as heat. Phinney [PH83] discussed the deceleration of a jet from relativistic to subrelativistic speeds and showed that, if its luminosity is sufficiently low (as expected for FRI sources), the jet can be recollimated by the pressure gradient of the surrounding galactic atmosphere. The present study therefore adopts the hypothesis that FRI jet bases are relativistic and attempts to devise a model which explains the result that one-sided jet bases almost always have a longitudinal apparent field.

---

*Royal Greenwich Observatory, Madingley Road, Cambridge CB3 0EZ.

2. The model

The velocity profile and field structure assumed in the model have been suggested previously to explain the observed polarization structure of FRI radio jets. Two antiparallel, but otherwise identical conical jets propagate away from a galactic nucleus along a direction which makes an angle θ to the line of sight. The jet cores have velocity $\beta_c c$ and contain a magnetic field which has no longitudinal component but is otherwise random [L80]. The surrounding shear layer has velocity $\beta_s c$ ($\beta_s < \beta_c$). Its field is entirely longitudinal. The shear layer contributes a fraction $f \approx 0.1$ of the total intensity of the core for a static jet in the plane of the sky. The model is a gross oversimplification of the structure of a real jet, but does have the crucial feature that there are two components with *different* velocities.

3. Observational consequences

If FRI jets have field and velocity structures of this type and decelerate away from the nucleus, the observable consequences are as follows:

- Jets tilted by a moderate angle to the line of sight ($30° < \theta < 90°$) undergo a field flip from longitudinal to transverse as they slow down, since the shear layer is slower than the core and suffers less beaming away from the line of sight.

- The emission from both jet components is heavily suppressed close to the nucleus unless $\theta < 30°$, giving a possible explanation for the observed "gaps" [BP84]. The rest-frame surface-brightness falls more rapidly with distance from the nucleus than appears from observation.

- Jets in the plane of the sky are precisely symmetrical. They have weak (Doppler-dimmed) $\mathbf{B}_\parallel$ bases on both sides of the nucleus.

- Jets tilted by a moderate angle are one-sided close to the nucleus. Both jets have $\mathbf{B}_\parallel$ bases, but the approaching one is brighter. Part of the $\mathbf{B}_\perp$ region of the approaching jet is also Doppler boosted, but the least symmetrical emission is the $\mathbf{B}_\parallel$ base.

- Jets very close to the line of sight (within the beaming cone of the core) are one-sided, but have $\mathbf{B}_\perp$. These are BL Lac objects in the unified models.

- $\mathbf{B}_\perp$ jets are always centre-brightened but $\mathbf{B}_\parallel$ jets are limb-brightened, since the emission from the centre must be suppressed for the longitudinal component to dominate. This is observed in M87 [OHC89].

- The approaching jet has a more centrally peaked brightness distribution, especially near the field transition.

- Differences in collimation between the two jets are not expected if the pattern speed of the flow is small.

- Depolarization asymmetry is expected, as the lobe with the brighter jet base is seen through less magnetoionic material in the galaxy halo.

4. Conclusions

The jet model discussed here and, in more detail, by Laing (*in preparation*) is clearly very crude, but does predict a number of the observed properties of FRI radio jets. A key test is the observation of transverse apparent velocities $\approx$ c in representative sources of this type.

References

[BO91] Biretta, J.A. Owen, F.N., *Velocity Structure of the M87 Jet: Preliminary Results. Parsec-scale radio jets*, Cambridge University Press (1991), pp. 125-128.

[BP84] Bridle, A.H., Perley, R.A. *Extragalactic Radio Jets*. Ann. Rev. Astr. Astrophys., 22 (1984), 319-358.

[PMCFR92] Parma, P., Morganti, R., Capetti, A., Fanti, R., de Ruiter, H.R., *Polarization properties at 1.4 GHz of low luminosity radio galaxies*. Astron. Astrophys., in press (1992).

[L80] Laing, R.A., 1980. *A model for the magnetic-field structure in extended radio sources*. Mon. Not. R. astr. Soc., 193 (1980), 439-449.

[L88] Laing, R.A., *The sidedness of jets and depolarization in powerful extragalactic radio sources*. Nature 331 (1988), 149-151.

[OHC89] Owen, F.N., Hardee, P.E. Cornwell, T.J., *High-resolution, high dynamic range VLA images of the M87 jet at 2 centimeters*. Astrophys. J., 340 (1989), 698-707.

[PH83] Phinney, E.S. *Ph.D. Thesis. University of Cambridge (1983)*.

[UPS91] Urry, C.M., Padovani, P., Stickel, M., *Fanaroff-Riley I Galaxies as the Parent Population of BL Lacertae Objects. III. Radio Constraints*. Astrophys. J., 382 (1991), 501-520.

Deceleration of Relativistic Jets in FR–I Radio Galaxies

Komissarov S. Sergey *

Abstract

The interaction between stellar winds and relativistic jets is considered. Relativistic fluid dynamics with mass injection is used to study the importance of the interaction for the jet dynamics. It concluded that the low power extragalactic jets (FR–I) should effectively decelerate to subrelativistic velocities at typical scales of the King core of the parent galaxy.

The attractive idea of unifying BL Lac objects with radio galaxies of the FR–I class ([1] [2]) requires *relativistic velocities* for FR–I jets at parsec scales ($\gamma = 4$ to 10). The observed properties of FR–I jets at kiloparsec scales, on the contrary, suggest subrelativistic velocities ([3]). Moreover, at kiloparsec scales FR–I jets are most probably *supersonic* flows with $M_j \geq 1$. Thus, the problem is to find a reasonable way *to decelerate initially relativistic FR–I jets to subrelativistic but supersonic velocities*. One possibility is the entrainment of mass lost by stars in the form of a stellar wind inside the jet.

The consideration of the interaction between a jet and winds typical for ellipticals shows that it can be described by fluid dynamics. Typical Reynolds numbers are high enough to assume long turbulent tails downstream of stars and effective mixing. This makes it possible to use relativistic fluid dynamics with mass injection to study the global effect of the interaction on the jet dynamics.

A quasi–one–dimensional approximation was used for investigating in more detail pressure–confined stationary jets. I accept a King distribution for the volume source of mass–energy and an isothermal stationary distribution for the intragalactic gas. The behaviour of the solutions is determined by the value of the following dimensionless parameter

$$\kappa = \frac{c q_0 r_c}{P_0 \gamma_0^2} \qquad (1)$$

---

*Astro–Space Centre, Lebedev Physical Institute, Leninsky Prospect 53, 177924, Moscow B–333, Russia.

where c is the speed of light, r_c is the radius of the King core, q_0 and P_0 are the volume density of the mass source and the pressure in the core of the galaxy and γ_0 is the intial Lorentz factor of the jet ($\gamma_0 = 6$ in the calculations). For $\kappa \sim 1$, effective deceleration of the jets to subrelativistic and subsonic velocities followed by a quite rapid increase in the jet radius is observed at a scale $\sim r_c$ (see Fig. 1). At higher distances the jets recollimate and accelerate again to supersonic ($M \geq 1$) velocities. For $\kappa > 1$ dramatic deceleration and unreasonably high spreading rates of jets are found. Most likely, the approximation of pressure balance across the jet becomes incorrect due to strong dissipation. For $\kappa \ll 1$ the effect of mass entrainment is negligible. A value of $q_0 \sim 10^{-41} g \ cm^{-1} s^{-1}$ corresponding to the critical $\kappa = 1$ agrees with the mean rate of mass loss by stars $\frac{\dot{M}}{M} \sim 10^{-12}$ to $10^{-13} yr^{-1}$ expected for ellipticals ([4] [5]). Thus pressure confined jets with $\gamma \sim 6$ should effectively decelerate. However, only FR–I jets having low kinetic luminosity ($\leq 10^{42} erg \ s^{-1}$) can be pressure confined inside the typical King core.

It is easy to show that in the general case the effect of mass entrainment is significant if the following parameter is not too small:

$$\xi = \frac{\dot{M}_g}{\dot{M}_j^0} \qquad (2)$$

where $\dot{M}_j^0$ is the initial mass flux of the jet and $\dot{M}_g$ is the rate of mass aquisition by a jet in a galaxy. For a conical jet with a top angle $\theta = 0.1 \ radians$ and kinetic luminosity L_j, and a galaxy with a mass of stars $M_g \sim 10^{11} M_\odot$ and mean rate of mass loss by stars $\sim 10^{-13}$ to $10^{-12} \ yr^{-1}$

$$\xi \sim 0.1 \div 1 \times (L_j/10^{43} erg \ s^{-1}). \qquad (3)$$

One can conclude that the considered effect should be weak for the most powerful FR–II jets and can be strong for FR–I jets.

References

[1] Browne I. W. A., 1983, *MNRAS*, **204**, 23.

[2] Urry C. M., Padovani P., *to be published in the Proceedings of the Georgia State University Conference on the Variability of Active Galactic Nuclei, May 1990.*

[3] Bridle A. H., Perley R. A., 1984, *Ann. Rev. Astron. Astrophys*, **22**, 319.

[4] Faber S. M., Gallagher J., 1976, *ApJ*, **204**, 365.

[5] Gisler G. R., 1976, *Astron. Astrophys.*, **51**, 137.

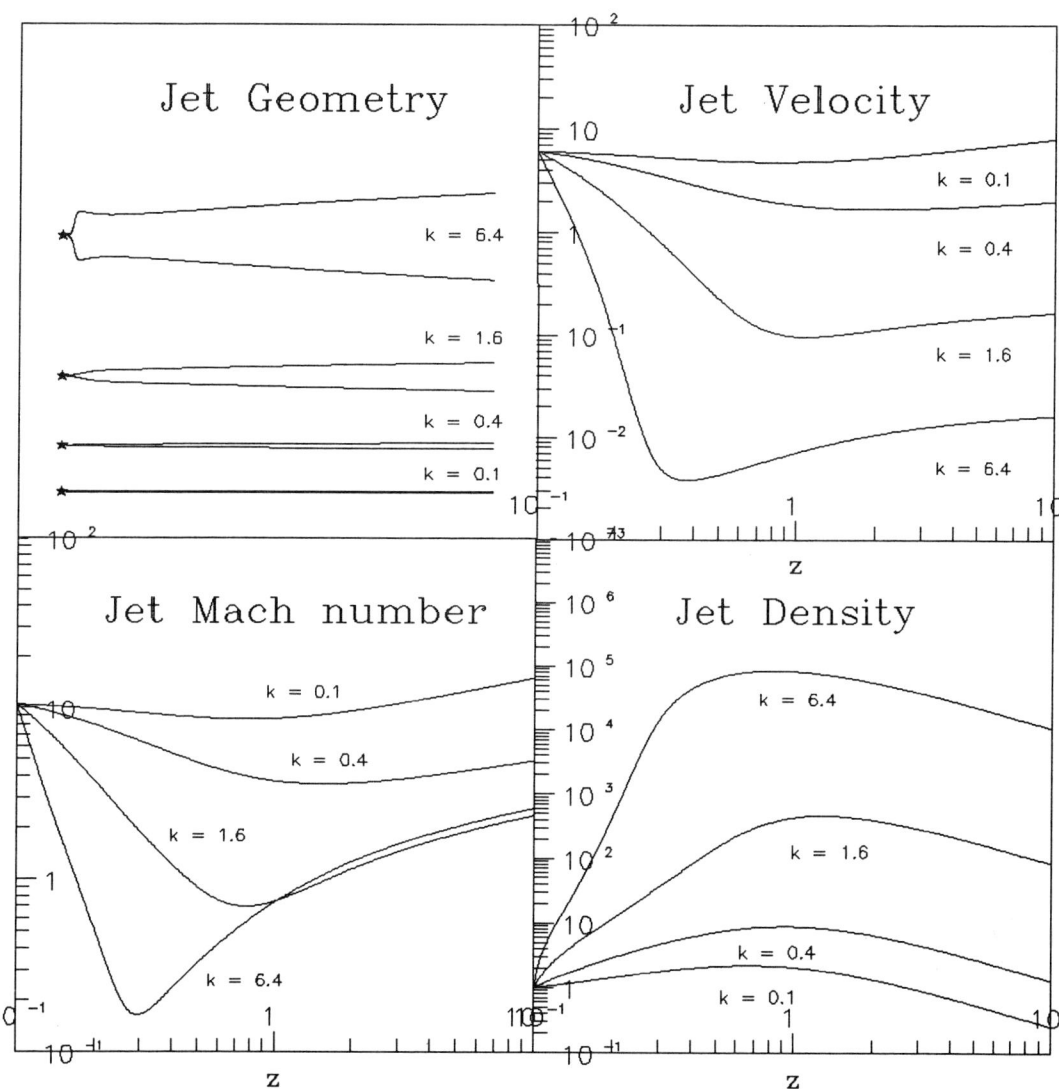

Figure 1. The variations of jet radius, four-velocity, Mach number and density of thermal plasma with distance from centre of galaxy. The unit of length along the z-axis corresponds to the radius of the X-ray core; the King radius, $r_c = 0.5$.

Models of Jets with Multiscale Structures of Knots

A C. Raga[*]

Abstract

Stellar and extragalactic jets often show structures of well aligned knots. This paper describes a model in which such knots are the direct product of a random time-variability of the source. The occurrence of many knot-merging processes leads to the formation of a self-similar knot structure in which the average knot spacing is proportional to a power of the distance from the source. It appears that knot structures in some astrophysical jets might show such characteristics.

1. Jets from time-dependent sources

The aligned knots observed in astrophysical jets can be interpreted as the result of a time-variability of the jet source. For example, a jet ejected with a time-independent velocity, but with a highly time-dependent density will be formed by a number of high density "blobs" separated by low density regions.

A different structure is obtained for the case of a jet ejected with constant density, but with a time-dependent velocity. If the amplitude of the velocity variability is larger than the sound speed, the steepening of these velocity variabilities (as they are advected downstream by the flow) results in the formation of internal shock waves in the jet beam ([R78]). These shock waves can form in pairs, each pair forming an "internal working surface" that travels down the flow ([R90], [KR92], [RK92]).

It is clear that as a result of a strictly periodic source variability, a high Mach number jet will develop a periodic structure of knots (either "blobs" or "internal working surfaces"), see above. On the other hand, for a non-periodic variability the spatial structure is not periodic. In this case, subsequent knots will not have identical velocities, so that "knot merging processes" will occur. These merging processes will have

---

[*]School of Mathematics, The University of Leeds, Leeds LS2 9JT and Department of Astronomy, The University of Manchester, Manchester M13 9PL, UK. The author acknowledges the support of a SERC fellowship.

the direct effect of increasing the average separation Δx between successive knots as a function of distance x from the source.

It is intuitively clear that after many knot merging processes (i.e. far away from the source) the "memory" of the details of the source variability is lost, and the velocity and position distribution of the knots along the jet reaches a "self-similar" form, which only depends on the nature of the "knot merging law". Such a solution has no preferred spatial scale, so that one would suspect that the mean separation between knots follows a $\Delta x \propto x^p$ law (where x is the distance from the source).

This can be shown indeed to be the case by carrying out numerical simulations of jets from sources with a random time-variability. [R92] finds that a source with a random velocity variability produces a distribution of knots in the jet that closely follows a $\Delta x \propto x^p$ law, with $p = 2/3$. This value of p can also be determined from a simple analysis of the scaling of the conservation laws (describing the knot merging processes) and the Boltzmann equation (which describes the evolution of the statistical distribution of knots). [R92] also shows that a source that ejects momentum conserving "blobs" of identical masses, but with different velocities, produces an asymptotic structure of knots in the jet which follows a $\Delta x \propto x^p$ power law with $p = 1$. A third possibility is a jet which has successive outflow events with highly variable velocity and mass. Such a source variability will result in a value of $p \approx 1/2$.

2. Comparisons with observations

The model described above suggests the intriguing possibility of being able to determine some of the physical characteristics of the knots with a simple observation of the knot separation vs. distance dependence in astrophysical jets. We illustrate this in figure 1, where we show the knot separation vs. distance relationship obtained for the stellar jet HH 30 (taken from an optical image of [M90]), and the radio jets 3C 111 (from [LP84], and [L87]) and NGC 6251 (from the radio map of [PBW84]).

The problems with this kind of measurement are clear from the results presented in figure 1. For the case of HH 30, the total number of knots is so small that it is really impossible to find a statistically significant Δx vs. x dependence. The measurements that we show actually are separations between pairs of successive knots, and not average knot separations. A power law $\Delta x \propto x^p$ dependence, with $2/3 < p < 1$ might be consistent with the HH 30 measurements.

For the case of 3C 111, we find that a power law with $p \approx 1$ best fits the measurements. However, this is probably a result of the fact that the two first points (closest to the source) and the two last points (at distances $\sim 10''$ from the source) were obtained from two different maps, with very different spatial resolutions ([LP84], and [L87]). Our $p \approx 1$ value probably indicates that in both maps we are seeing knots of about the same size as the beam, and that the total number of beamsizes covering each map are more or less similar. Such an observational bias will be present in any Δx vs. x measurement caried out in maps obtained with different beam sizes.

The results for NGC 6251 have been obtained from a single radio map ([PBW84]). This jet shows a large number of knots, so that in principle a better determination of a possible Δx vs. x relationship can be carried out. However, the results are highly uncertain due to the subjective nature of the definition of what one calls a "knot" in the intensity map of the jet. For this object, we find that the data are consistent with a $\Delta x \propto x^p$ dependence, with $p \approx 0.5$ (see Figure 1). The implications of this result are discussed in §1.

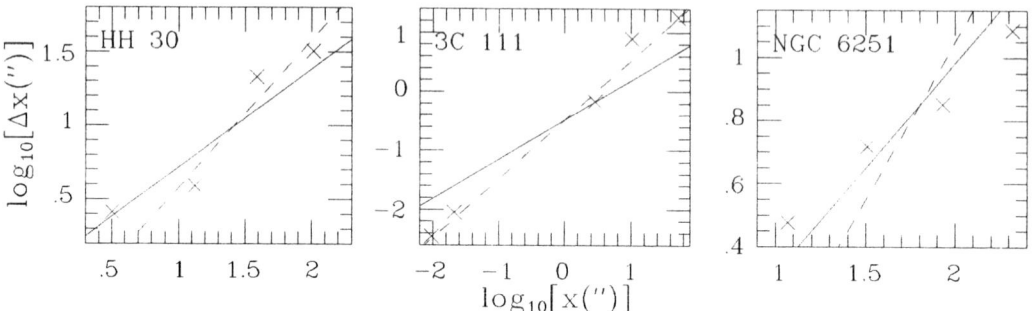

Figure 1— Logarithmic Δx vs. x dependence for HH 30 (left), 3C 111 (center) and NGC 6251 (right). The solid line corresponds to a power law of index 2/3, and the dashed line to a power law index of 1.

References

[KR92] Kofman, L., and Raga, A. C. 1992, *Ap. J.* **390**, 359.

[L87] Linfield, R. 1987, *Ap. J.* **317**, 121.

[LP84] Linfield, R., and Perley, R. 1984, *Ap. J.* **279**, 60.

[M90] Mundt, R., Ray, T. P., Bührke, T., Raga, A. C., and Solf, J. 1990, *Astron. Astrophys.* **232**, 37.

[PBW84] Perley, R. A., Bridle, A. H., and Willis, A. G. 1984, *Ap. J. Suppl.* **54**, 291.

[R92] Raga, A. C. 1992, *M.N.R.A.S.* (in press).

[R90] Raga, A. C., Cantó, J., Binette, L, and Calvet, N. 1990, *Ap. J.* **364**, 601.

[RK92] Raga, A. C., and Kofman, L. 1992, *Ap. J.* **386**, 222.

[R78] Rees, M. 1978, *M.N.R.A.S.* **184**, 61p.

Interpretation of Superluminal Radio Sources as Bent Shocked Relativistic Jets

A. Alberdi [*] *J. L. Gómez* [*] *J. M. Marcaide* [†]

Abstract

The evolution of parsec-scale radio structures of compact sources can be explained in terms of shock waves propagating along bent jets. These jets consist of narrow-angle cones of plasma flowing at bulk relativistic velocities, within tangled magnetic fields, emitting synchrotron radiation.

We have developed a numerical code which solves the synchrotron radiation transfer equations to compute the total and polarized emission of these bent shocked relativistic free jets. We have considered the inhomogeneities in the jet physical parameters, as particle density, flow velocity, and magnetic field along the line of integration when the jet is not seen side-on.

1. Introduction

The majority of compact extragalactic radio sources observed at milliarcsecond resolution consist of a compact core with an one-sided knotty jet. The standard working hypothesis explains this morphology in terms of a relativistic plasma flowing in a narrow-angle cone (a "relativistic jet") with travelling shock waves producing the moving knots (MGT91). According to this model, the predominant emission in the knots comes from the shocked gas.

We have developed a numerical code which solves the synchrotron radiation transfer equations to compute the total and polarized emission of bent shocked relativistic jets (GAM93). Similar calculation had previously been made for rectilinear jets by our group and by other authors (e.g., HAA89). However, our code is not restricted to rectilinear shocked relativistic jets, but it can be best used for jets with bends present in their trajectories (AMM93, GAM93). The method of integration in our code is substantially more elaborate than previous ones.

---

[*]Instituto de Astrofísica de Andalucía, P.O. Box 3004, 18080-Granada, Spain
[†]Dpto. de Matemática Aplicada y Astronomía, Universitat de València, 46100-Burjassot, Spain

Our code divides the jet in cells, and these cells are grouped together in integration columns parallel to the line of sight. We consider the differential Doppler boosting factor between consecutive cells due to a change in the velocity vector and/or in the orientation of the jet with respect to the observer, associated to the presence of shock waves in the jet structure and/or bends in the jet geometry.

2. Results

In this contribution we consider the effects of both bends and shock waves in the jet. For a quiescent rectilinear relativistic jet, the total flux density profile presents a maximum associated to the position at which the jet becomes optically thin. At the very same position, the polarized flux density presents a deep minimum and the polarization angle changes by 90 degrees. This minimum of the polarized flux is associated with the change in the predominancy of the two mutually perpendicular polarized flux components. In Fig. 1 we show the total and polarized flux density maps for a quiescent rectilinear jet forming an angle of 20 deg with respect to the observer.

a) Effects of a bend

If the jet bends towards the observer, the total flux density increases due to the beaming effect and to the lengthening of the integration column. However, the polarized flux density will increase or decrease depending on the opacity of the bent region. If the bend produces a transition from optically thin to optically thick in the bent region, the polarized intensity would diminish and the polarization angle would rotate by 90 degrees. In Fig. 2 we show the total and polarized flux density of the above mentioned jet in which we have introduced a bend of 5 deg towards the observer for a certain jet length. The bent-region manifests itself as a new component (a flux increase) in the total flux map, and as a decrease in the polarized flux map.

b) Effects of a shock wave

A shock wave produces a strong increase in the total flux density due to the enhancement of the particle density, the amplification of the magnetic field, and the energization of the individual particles, while the polarized flux density will increase or decrease depending on the resulting optical depth of the region heated by the shock wave. When the shock wave is propagating close to the core region, the shocked area is optically thick and there is correlation between the total flux density flare and the polarized flux density decay. There is a double rotation of 90 degrees of the polarization angle, one due to the optically thin-thick transition of the shocked region and the other due to the enhancement of the component of the **magnetic field parallel to the shock front**. In Fig. 3 we present the total and polarized flux maps for the above mentioned jet in which we have introduced a shock wave travelling along it. The shocked region appears in the total flux map as a (moving) component while a minimum in the polarized flux map can be seen at the same position. However, when the shock wave advances to positions further along the jet, the shocked region becomes optically thinner and increasingly a correlation between the total flux density and the polarized flux density takes place. In this case, a component will also be found in the polarized flux map.

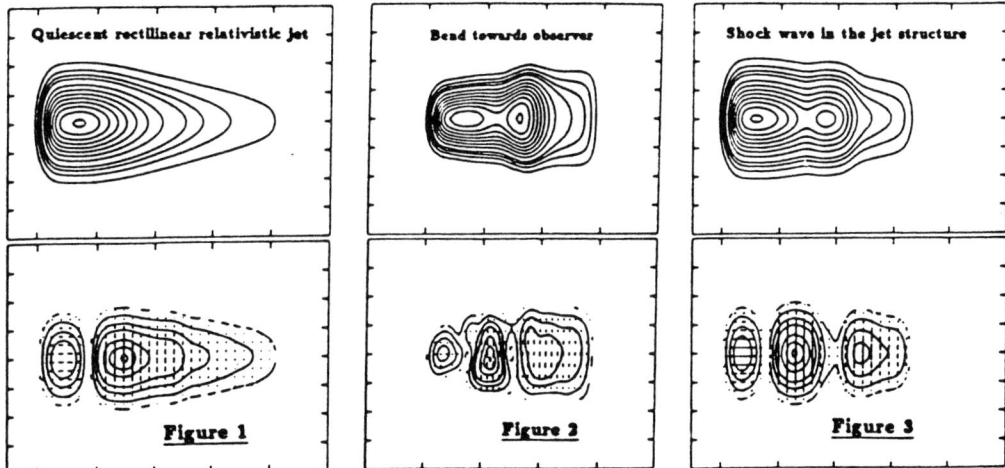

We have applied our code to reproduce the compact structure, kinematic evolution and time flux density evolution of 4C 39.25, and to obtain its jet physical parameters (AMM93), as well as to determine the spectral behaviour of the quasars 1038+528 A and B (this conference), the structure of 3C 395 (this conference), and the total and polarized flux density maps of the quasar 3C 273B (GAM92).

References

[AMM93] Alberdi, A., Marcaide, J.M., Marscher, A.P. et al. *Ap. J.* (1993), Jan. 1st issue

[GAM93] Gómez, J.L., Alberdi, A., Marcaide, J.M., *A. & A.* (1993), submitted

[GAM92] Gómez, J.L., Alberdi, A., Marcaide, J.M., *Proceedings of the Symp. "Astrophysical Jets"*, 1992

[HAA89] Hughes, P.A., Aller, H.D., Aller, M.F., *Ap. J.* 341 (1989) 54-67

[MGT91] Marscher, A.P., Gear, W.K., Travis, J.P., *Proceedings of the Symp. "Variability of Blazars"*, ed. E. Valtaoja (Cambridge Univ. Press), 1991

Interpreting VLBI Sources As Twisted Jets

P. A. Hughes, M. F. Aller, H. D. Aller and A. Rosen [*]

Abstract

We present a simple model for visualizing the structure and evolution of shocked, parsec scale relativistic jets that runs on a PC. The simulation reproduces commonly observed features.

1. Introduction

Recent VLBI observations have shown that bending is a common feature of parsec scale jets, while work in progress by the authors suggests that the overall morphology of such sources may change perceptibly over a time scale of years. The temporal and spectral characteristics of the integrated flux, the fractional polarization, and the VLBI structure are sensitively dependent on observer orientation because of the relativistic nature of these flows. Because detailed modelling (that performs radiation transfer calculations through flows of arbitrary geometry) is too computer intensive to allow an extensive exploration of parameter space, we have built a simple model that allows the visualization of such flows, in preparation for more detailed study.

2. The Model

We specify the flow morphology by parameterizing the locus of the jet axis; on this we "thread" circular cylinders of increasing radius, each of which is taken to be a simple synchrotron emitter. In our simulation, the "observer" specifies an orientation, and parameters of the quiescent flow: the Lorentz factor, rate of expansion and magnetic field decline, and electron spectral index. A "component" (a shock) may be propagated along the flow, and is characterized by whether it moves 'forwards' or 'backwards', its length and its strength (compression). We compute the total and polarized flux from each "pill", allowing for Doppler frequency shift, boost, aberration and time delays in general, and for the compression-induced polarization for shocked

---
[*] Astronomy Department, University of Michigan, Ann Arbor MI 48109-1090 USA. Anyone interested in obtaining a copy of this utility is invited to send a message to hughes@astro.lsa.umich.edu. This work was supported in part by grants NSF AST-8815678 and NASA NAGW-2135.

regions. We have implemented this code for 387 and 486 PCs, using MS Fortran 5.1: this renders the code portable and facilitates its use for demonstration, teaching and research. Different intensities are mapped into different densities of pixels within an 8 × 8 random mask, and local fractional polarization is color coded.

3. Results

The moving component between two stationary features seen in 3C 395, for example, and the changing, complex structure seen in 2134+004 are easily reproduced in generic form; we have made no attempt so far to make detailed models of these sources. The figure shows a "screen dump" illustrating the propagation of a single superluminal component and the corresponding "light curve". A stationary component appears where the local flow is almost along the line of sight. Note the complex structure of the light curve, suggesting that some structure, *e.g.* the double peaks sometimes seen in monitoring data, might not result from discrete events, but rather from the modulation of a single event by a curved flow. A Mark II version of this code will add the facility to produce maps allowing for a) obscuration of some parts of the flow by other, opaque segments, and b) convolution with a gaussian beam. Furthermore, the user will be able to specify arbitrary geometry.

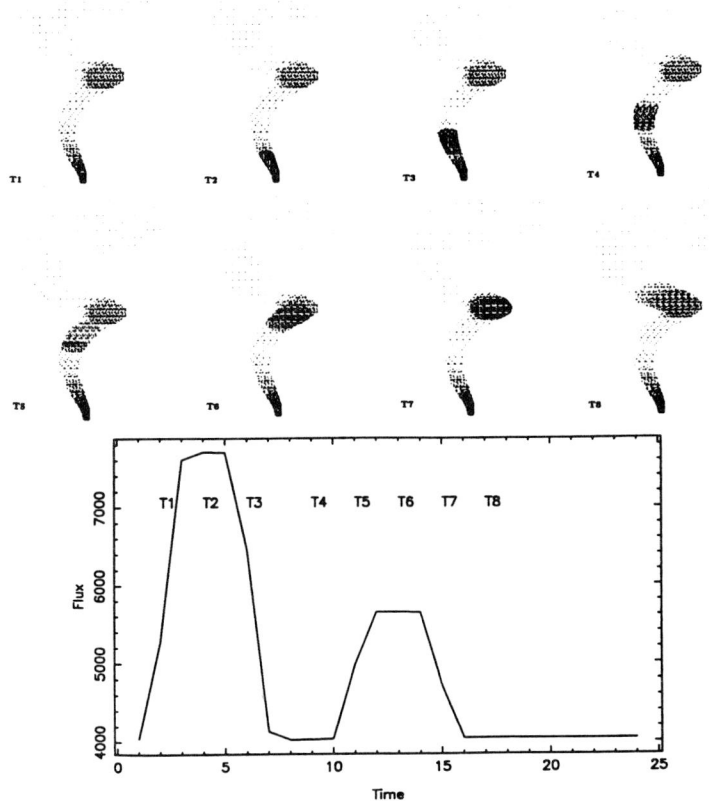

Helical Jets and the Misalignment Distribution for Core Dominated Radio Sources

J.E. Conway*and D.W.Murphy [†]

Abstract

We describe how helically distorted jets can explain the large population of core dominated sources which show apparently orthogonal misalignments between their parsec and kiloparsec scale jets.

1. Misalignments between Parsec/Kiloparsec Jets

In their sample of core-dominated radio sources Pearson and Readhead 1988 (PR survey ApJ 328,114) found that the distribution of apparent misalignments between the position angles of parsec and kiloparsec scale jets showed an unusual and unexpected form. Their misalignment distribution had two peaks, one centered close to 0° (aligned sources) and another near 90° (misaligned or orthogonal sources). A histogram (see Figure 1) formed by adding new data from the PR survey and from the VSS (Variable Sources Sample, Wehrle et al 1992, ApJ 391,589) confirms this result. Attempts at fitting this distribution with simple models based on small intrinsic bends between parsec and kiloparsec scales have not been successful, there is always an excess of orthogonal sources. The best such model, invoking two populations of sources with different intrinsic bend angles and core γ's, can be rejected at the 98% confidence level (Conway and Murphy 1993, ApJ submitted).

2. The Helical Jet Model

Although the aligned population can be explained by a model in which there are small bends between parsec and kiloparsec scales such a model cannot explain the misaligned sources. We have systematically searched (see Conway and Murphy 1993) for geometries which explain these sources and below we can only briefly describe our successful model.

---

*NRAO, P.O. Box O, Socorro, NM 87801, USA
[†]Jet Propulsion Laboratory, 4800 Oak Grove Drive, Pasadena, CA 91109, USA

Consider (see Figure 2a) a helically distorted VLBI jet with a low pitch angle. This jet initially lies on the surface of a cone whose opening angle equals that of the helix opening angle ζ. Further out assume that the amplitudes of the helical distortions eventually saturate so that on kiloparsec scales the jet moves inside the initial helix cone; perhaps as shown in Figure 2 reaching a maximum amplitude and so being confined to the surface of a cylinder. Due to relativistic beaming, helices with a velocity vector at the point of maximum jet emissivity (the 'radio core' region) close to the line of sight will be preferentially selected. For a low pitch angle helix this beamed emission is almost outward away from the apex of the helix cone along the cone surface, hence objects in which the line of sight passes close to the cone surface as shown will be preferentially selected. Given such a geometry, the jet emerges from the core and moves in azimuth around the helix cone and we see in projection (see Figure 2b) a VLBI jet orientated approximately orthogonally to the projected cone axis of the helix. Due to the saturation of the helical distortions the kiloparsec jet will lie in the direction of this projected cone axis, hence approximately orthogonal parsec and kiloparsec scale jets will be obtained.

Our model does provide a reasonably good fit (see Figure 3) to the orthogonal population as shown in Figure 1. We find that acceptable fits require that $\gamma > 20(15°/\zeta)$ where ζ is the half-cone opening angle of the helix. Based on parsec/kiloparsec misalignments in *lobe-dominated* sources (e.g. Browne 1987, Superluminal Sources,(CUP) p 129.) we estimate $\zeta \leq 15°$ and hence $\gamma \geq 20$. Such large values of γ for the misaligned sources are consistent with their extreme blazar properties. These properties include high ($> 15\%$) optical polarization (Impey, Lawrence and Tapia 1992, ApJ 375,46) and large optical continuum dominance (the misaligned population contains all 8 sources that have at one time been classified as 'BL Lacs'). Once the effects of jet curvature are taken into account such large values of γ are also consistent with the observed proper motions in the jets provided the Hubble constant is at the lower end of its plausible range (i.e. $H_o = 40 kms^{-1} Mpc^{-1}$). Larger values of the Hubble constant would require the pattern speed in the VLBI jet to be smaller than the bulk velocity in the core.

3. Physical Mechanisms

The helical distortions, required by our model, could be generated either by Kelvin-Helmholtz (K-H) instabilites (Hardee 1987, ApJ 318,78) or by jet precession. The advantage of the former mechanism is that, under certain conditions, the amplitudes of the helical distortions are predicted to saturate as required by our model. However it appears hard to obtain helical distortions that are initially accurately confined to a conical surface, as our model requires. Simulations of *ballistic* precession models show that the required precession periods are inconsistent with the kiloparsec jet morphology. However precession models in which the jet forms a coherent rigid channel may be consistent. The dynamics of such systems are unknown and should be investigated to see whether a saturation of the helix is predicted. Alternatively a precessing jet might be brought closer to the precession cone axis via an interaction with structures in the radio lobes (e.g. Cox, Gull and Scheuer 1991, MNRAS 252, 558). There has been much discussion at this meeting about helical structures within VLBI maps. Although these distortions generally have much smaller wavelengths than the distortions required to explain the pc/kpc misalignments (whose wavelengths are measured in kpc's) there may be a connection between the two types of distortion. The breaking of axisymmetry due to precession may allow small scale K-H instabilites to be generated. In addition the orbital period associated with a precessing binary black hole system could resonate with the frequency for maximum growth of K-H instabilites and so drive such instabilites. Such mechanisms may be occurring in the misaligned source 3C345 and in 1928+738 (Hummel et al 1992, AA 257,489) where two scales of helical distortion apparently coexist.

Fig. 1 – The Misalignment Distribution for sources in the PR survey (black), VSS sample (white) and in both samples (grey).

Fig. 2 – The Helical Jet model. a) Plan. b) View along LOS.

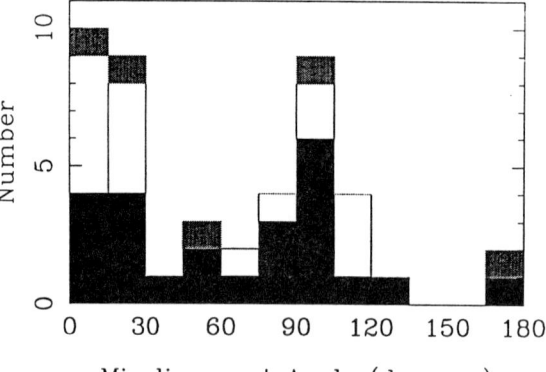

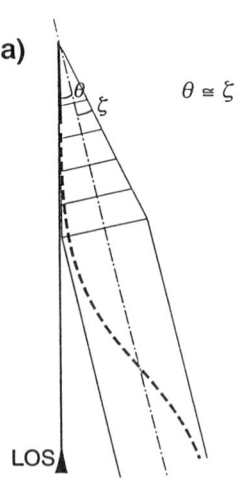

Fig. 3 – Model misalignment distribution assuming $\gamma = 20$, helix opening angle $\zeta = 15°$ and helix pitch $p = 1.5°$.

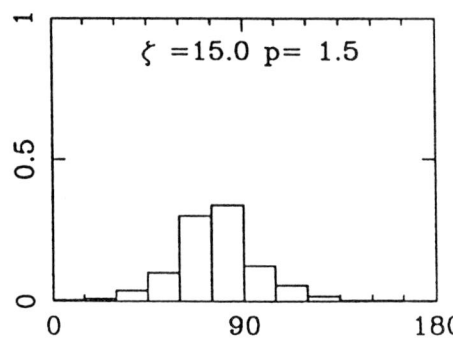

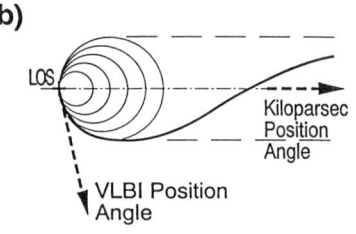

A magnetized helix in 3C345 ?

W. Steffen, T.P. Krichbaum, A. Witzel* and A. Zensus [†]

We analyzed the complex kinematics observed in the milli-arcsecond jet of the quasar 3C345. Components C4 (see e.g. B83) and C5 move on different curved trajectories and show variations in their superluminal speeds (Z92). We investigated helical jet-models with adiabatic expansion as explanations for the motion and flux density evolution of the components. The components were supposed to be constrained to a conical surface, which represented the underlying jet. The motion was described by three conservation quantities out of four: the opening-angle of the jet, the kinetic energy, the angular momentum and the momentum along the jet axis (C86). Out of these quantities we constructed four different (relativistic) models (fig.1). Some of these simplified models (No.1&4) predict the same kinematics as the more sophisticated approaches by Hardee (1987) and Camenzind (1986). Thus, we can use them as indicators to the overall physics in the parsec-scale jet.

Model 3 was ruled out as an explanation to the observed kinematics, since it cannot predict a complete revolution about the jet-axis. Such a revolution in contrast has been observed for the component C4 (fig.2).

We found reasonable fits for models showing angular momentum conservation. We obtained a mean angle to the line of sight $i = 7.5°$ and a Lorentz-factor $\gamma = 5.7$. These are the values for model 4. This model is equivalent to the kinematics of the magneto-hydrodynamical model by Camenzind (1986), which is characterized by helical magnetic field lines. The kinematics of model 1 (no angular momentum conservation) is similar to the one of the isothermal hydrodynamical model with helical instabilities by Hardee (1987). We found that this model was not able to fit the data with reasonable parameters, since it required opposite senses of revolution about the jet-axis for C4 and C5. It also predicts self-similar trajectories in the jet, which are not observed in 3C345.

Model 4 yields the best representation of the data. We therefore conclude that angular momentum, as well as helical magnetic fields, rather than helical instabilities, might dominate the motion observed in the milli-arcsecond jet of 3C345.

The flux density evolution of the component C4 was modelled considering differential Doppler-boosting and adiabatic isotropic expansion. Using our kinematic models the flux density evolution could not be explained by differential Doppler-boosting alone. Additional consideration of adiabatic expansion of a spherical synchrotron radiator yielded an acceptable fit to the data [S92]. We found that C4 could be modelled only with a flat spectrum. This is consistent with the observed spectrum and the complex structure of C4. Such a spectrum could be produced by an inhomogeneous structure of the jet-component.

---

*Max-Planck-Institut für Radioastronomie, Bonn, Germany.
[†]NRAO, Socorro, USA.

References

[B83] Biretta, J.A., et al., (1983), Nature, **306**, pp. 42-46.
[C86] Camenzind, M., (1986) *A&A,* **156**, *pp. 136-151* .
[H87] Hardee, P.E., (1987) *ApJ,* **318**, *pp. 78-92* .
[S92] Steffen, W., *Diploma-Thesis 1992, Univ. Bonn., and references therein.*
[Z92] Zensus, J.A., et al., *in preparation.*

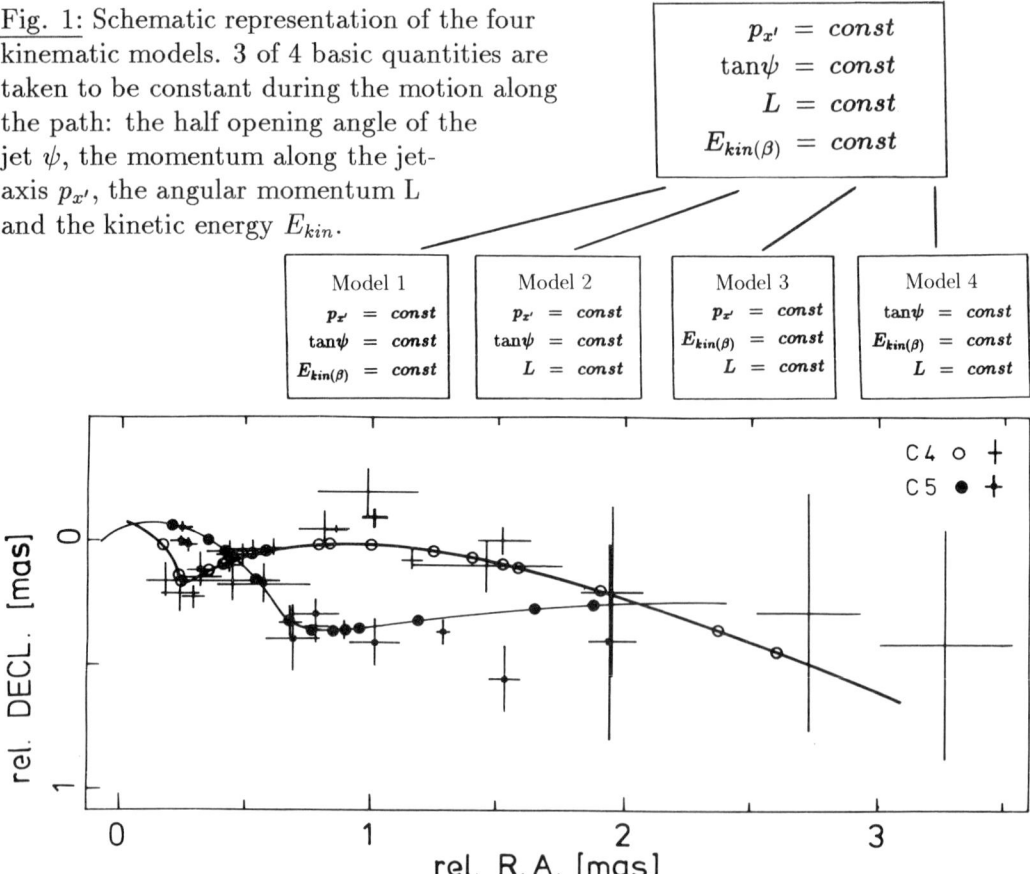

Fig. 1: Schematic representation of the four kinematic models. 3 of 4 basic quantities are taken to be constant during the motion along the path: the half opening angle of the jet ψ, the momentum along the jet-axis $p_{x'}$, the angular momentum L and the kinetic energy E_{kin}.

Fig. 2: Best fits to the motion of component C4 and C5 obtained with model 4 using essentially the same parameter-set. This model is characterized by a constant opening-angle of the jet, constant speed and angular momentum. The most important fit-parameters are discussed in the text. Crosses indicate positions at different epochs (freq.: 10.7, 22, 43 GHz; [S92]). Circles mark predicted positions for the observational epochs.

Models of Helical Jets

A.J. Kus - TRAO, Nicolaus Copernicus University, Torun, Poland. [*]

Abstract

In this paper a simple kinematic model of helical flow in radio jets is presented. The description of the model together with the basic information on possible solutions are given. The verification of the model is done by a comparison with the observed brightness distribution, measured apparent motion and PA of the polarized emission. Results obtained for subkpc jets in 3C309.1 and 3C380 are presented and compared with VLBI maps.

A helical structure in jets seems to be a common feature in a large number of objects. Detailed physical reasons for the emergence of such structures are given in some theoretical papers. One of the possible scenarios, suggested for radio jets, is a K-H instability triggered by precession in a dense gaseous environment. The model described here assumes : a bipolar continuous or discrete flow from a central engine along a fixed (in space) helical path (the same sense of winding is preserved on both sides); the velocity V along the trajectory is constant and its direction tangential to the line of the helix. The geometry (Fig.1) is described by : ψ - opening angle, ϕ - initial phase, τ - period of the first winding, θ - angle between source axis and the line of sight (LOS). Three elementary modes of the flow are considered : (a) constant wavelength λ, (b) constant Vz/V or constant pitch angle, (c) variable wavelength $\lambda(t)$. The brightness of an element in the flow is given by $Lo * D^{(2+\alpha)}$, where Lo is the luminosity in the flow frame, D is the Doppler factor, α - spectral index.

For a given set of parameters the model produces : (i) brightness distribution convolved with a circular gaussian beam, (ii) apparent velocity distribution and, if requested, (iii) polarization map. The direction of B is assumed to be frozen into the flow parallel to V. In the centre a small isotropic core flux is added to mark the engine position. Fig 2 shows the comparision of ballistic and helical (λ=const and Vz=const) flows. Due to a large variation in the position of V to the LOS there are big contrasts in the brightness produced by the helical flow.

High frequency studies of CSS quasars have uncovered characteristic features in subkpc structure which suggest the possibility of complex flow with helical signature. Very often strong components remain stationary, over many years, despite the evidence for BRM from IC or direct measurements of SLM. Some suggestion of SLM within the strongest, stationary component in 3C309.1 also supports the possibility of a fixed trajectory flow. The plausibility of such a model has been tested for 3C309.1 and 3C380. Fig3. shows a 5 GHz VLBI map and the model map. The following set of parameters were found to give one of the best fits : γ=5 , $\psi = 2°$, $\theta = 20°$, $\phi = -45°$,

---

[*]The author was partially supported by a Visiting Scientist Fellowship received from Onsala Space Observatory and from the Swedish Institute.

$\tau=-3000$y, left hand helix with $\lambda(t)$. The observed direction and the magnitude of apparent velocity in the core, as well as the derived γ, are used to constrain θ for a fixed value of ψ/θ ratio. There is a good correspondence of positions and shapes as well as relative brightness of major components as compared with the VLBI map. Thus the idea of helical flow in 3C309.1 seems to be well argued. To fit a similar model for 3C380 (Fig 4) has been a much more difficult task. The parameters found are: $\gamma = 5$, $\psi = 1.67°$, $\theta = 2.5°$, $\phi = 100°$, $\tau = +2500$y. A right hand helix with small θ is required to reproduce the observed radio emission. It has not been possible to make a suitable ballistic model to obtain high flux for component "B". A much higher efficiency of beaming in this area, which happens if V is closer to the LOS, is present in the helix proposed above.

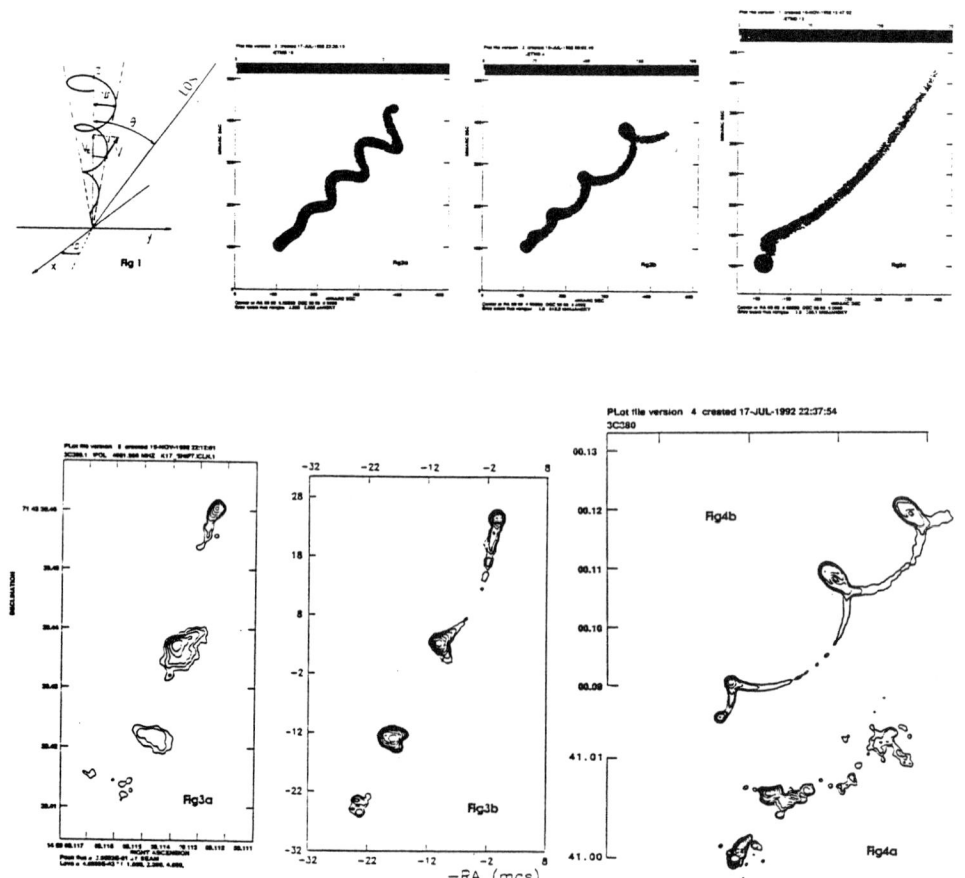

CSS sources and projection effects

Hardip S. Sanghera [*]   *Ralph E. Spencer* [†]

Abstract

Although the steep spectral indices of CSS sources indicate the absence of strong projection effects, the detection of superluminal motion and more misaligned radio structure in CSS quasars show that orientation effects may be important. Using a number of asymmetry parameters [1], we conclude that beaming and projection effects are more important in CSS quasars than radio galaxies, but the asymmetries in both types of source are predominantly due to differences in the density of the ambient medium and that lobe/gas interactions are important.

1. Introduction

Compact steep spectrum sources (CSSs) are a class of intrinsically small ($\leq 2''$, ≤ 10 kpc), high luminosity extragalactic radio sources, with steep high–frequency radio spectra ($\alpha > 0.5$). Their steep spectral indices imply a lack of strong projection effects, as the absence of a strong flat spectrum core discriminates against strong Doppler boosting and small viewing angles. However, projection and beaming effects could account for the greater number of misaligned quasars [1], the stronger radio cores and detection of superluminal motion in CSS quasars [2] and would indicate that orientation effects are in fact significant in such sources. We have investigated the strength of projection effects in a 54 source sample of CSSs [7] using several well known asymmetry parameters [3,4,5].

2. Results and Conclusions

Several aspects of the extended radio structure may be expected to depend on orientation and therefore provide useful asymmetry parameters *e.g.* largest angular size, Θ and projected linear size, L; the misalignment or bend angle, Ψ; the core to extended

---

[*] NRAL, Jodrell Bank, Macclesfield, SK11 9DL, U.K. This author was supported by a Research Studentship from the SERC.
[†] NRAL, Jodrell Bank, Macclesfield, SK11 9DL, U.K.

flux density ratio, R [8]; lobe distance (Q) and lobe flux density (F) ratios. However, we have as yet only detected a core in 48% of sources in the sample.

The results are consistent with a small degree of non–uniformity in the orientation of CSSs in the sky [2]. The distributions of R and Ψ showed a significant difference between radio galaxies and quasars, which is in contrast with the comparable degree of asymmetry in Q and F in both types of source. No significant correlation was found between any of the asymmetry parameters. This would imply that although there may be a small difference in the distributions of orientation angle, with quasars closer to the line of sight, lobe asymmetries are not strongly affected by orientation and are probably intrinsic in origin. In particular the lack of any significant correlation between linear size and any of the asymmetry parameters argues against the kinematic interpretation of lobe distance asymmetry.

CSS radio structures also appear to be more misaligned and asymmetric than that of the extended sources, consistent with either strong interaction with a dense ISM or orientation effects. The strong unified scheme [6] is derived from observations of extended sources in the narrow redshift range $0.5<z<1.0$ in the 3CRR sample, where there are twice as many radio galaxies as quasars. From the distribution of the ratio of space densities of CSS radio galaxies and quasars, over a number of redshift ranges, we find that such a scheme does not readily apply to CSSs. Although we find that projection effects are likely to be more important in CSS quasars than radio galaxies, asymmetries caused by differences in the properties of the ISM are likely to dominate. There is striking evidence that there is more emission line gas on the side closest to the nucleus [9] suggesting that the lobe distance asymmetries are the result of asymmetries in the distribution of gas around radio sources. The greater number of CSSs which exhibit enhanced emission in the closer lobe could then be explained by confinement by a denser environment on that side of the source.

References:

1) Sanghera,H. *et al*, 1991, in preparation.

2) Fanti,R. *et al*, 1990, Astr. Ap., 231, 333.

3) Macklin,J. T., 1981, Mon. Not. R. astr. Soc, 196, 967.

4) Hough,D.H. and Readhead,A.C.S., 1989, An. J., 98, 1208.

5) Kapahi,V.K., Parsec–scale radio jets, 1990, eds Zensus,J. and Pearson,T., CUP, 304.

6) Barthel,P.D., 1989, Ap. J., 336, 606.

7) Spencer,R.E. *et al*, 1989, Mon. Not. R. astr. Soc, 240, 657.

8) Orr,M.J.L. and Browne,I.W.A., 1982, Mon. Not. R. astr. Soc, 200, 1067.

9) McCarthy,P.J. and van Breugel,W.J.M. and Kapahi,V.K., 1991, Ap. J., 371, 478.

Can the filaments of the Radio Sources be really due to Synchrotron Thermal Instabilities?

Elisabete M. de Gouveia Dal Pino [*,†]   *Reuven Opher* [‡]

Abstract

Previous analyses of the evolution of synchrotron thermal (ST) instabililties in radio sources have shown that these sources are unstable to the development of ST modes whose final product is the formation of dense filaments aligned with the magnetic field. A question, however, has caused some controversy in the literature: are the ST filaments darker than the surrounding medium or brighter as indicated by the observations? In the present work, we address this question. Instead of the monoenergetic relativistic distribution of electrons of the previous analysis, we assume a power-law distribution, as observed, and investigate the brightness of the filaments relative to the ambient medium at each frequency. We find that: (i) the ST filaments are brighter than the ambient medium, in agreement with the observations and (ii) the spectral index is almost the same in both the filaments and in the ambient medium.

1. Introduction

VLA maps of galactic and extragalactic radio sources (e.g., Crab nebula, CygA, M87, and CenA) show the presence of inhomogeneous filamentary structures which are generally aligned with the magnetic field direction. Recently, we investigated the formation of these structures relating them to the development of nonlinear thermal instabilities mainly driven by the synchrotron emission of the relativistic electrons (DO89a, DO89b, DO91). In those analyses, we assumed a relativistic Maxwellian distribution of the electrons. In the present work, we consider a power-law distribution, in order to determine the brightness of the filaments produced by synchrotron

---

[*] Harvard- Smithsonian Center for Astrophysics, 60 Garden St., Cambridge, MA 02138, USA.

[†] Instituto Astronômico e Geofísico, Universidade de São Paulo, Av. Miguel Stefano, 2400, São Paulo, SP 01050, Brazil. This author was partially supported by the brazilian foundations BID/USP and FAPESP.

[‡] Instituto Astronômico e Geofísico, Universidade de São Paulo, Av. Miguel Stefano, 2400, São Paulo, SP 01050, Brazil.

thermal instablility (STI) relative to the unperturbed medium, at each frequency (see DO92 for a detailed description of the model).

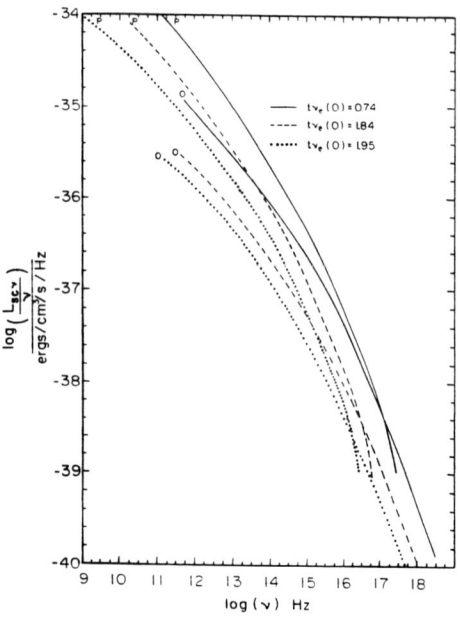

Figure 1. Log of the synchrotron volume emissivity per frequency (versus log of the frequency) calculated for the perturbed region and the adjacent unperturbed medium at different times $t\nu_e(0)$ (where $\nu_e(0) = 10^{-6} yr^{-1}$ is the initial expansion rate of the unperturbed medium). The curves corresponding to the perturbed region are labeled with "P" and the curves corresponding to the unperturbed region are labeled with "O".

In Fig.1, as initial equilibrium conditions, we considered the ambient parameters of the inner 40" radio jet of CenA where a complex filamentary structure has been detected (e.g., CBF86). The volume emissivity of the unperturbed system, $L_{sc\nu o}$, is determined from the observed spectrum from radio to x-ray (see DO92). These curves show that the condensed regions formed by STI are effectively brighter than the adjacent unperturbed regions for almost the entire frequency range, in agreement with the observations which indicate emissivity contrasts $\approx 4 - 10$ (e.g., DO92). We also found that the the spectral index of both the perturbed and the equilibrium regions are approximately the same and are $\alpha_E \approx 0.5$.

References

[CBF86]   Clarke, D.A., Burns, J.O., Feigelson, E.B., *Ap. J. (Lett.)*, Vol. 300 (1986), p. L41-L46.

[DO89a]   Gouveia Dal Pino, E. M., Opher, R., *Ap. J.*, Vol. 342 (1989a), p. 686-699.

[DO89b]   Gouveia Dal Pino, E. M., Opher, R., *MNRAS*, Vol. 240 (1989b), p. 573-590.

[DO91]   Gouveia Dal Pino, E. M., Opher, R., *Astr. Ap.*, Vol. 242 (1991), p. 319-333.

[DO92]   Gouveia Dal Pino, E.M., Opher, R., *MNRAS*, 1992 (submitted).

MODELING JETS ON MILLI- ARCSECOND SCALES: POLARIZED STRUCTURE AND FLUX SPECTRA

Yuri A. Kovalev *   Yuri Yu. Kovalev †

Abstract

Numerical results for the narrow straight line jets in the radial magnetic field of the active galactic nucleus are given. The typical normalized synchrotron flux density spectra has three main regions: increased (A), quasi flat (B) and decreased (C). Polarization and integral flux characteristics are different in these regions. If jet is stable, the polarized structure is well regulated and the integral degree of linear polarization will increase with the high frequency. These may be complicated by beam injection variability if the fixed frequency of an observation will be shifted to the other normalized spectra regions (from the region B to A or to C, as an example). Modification of this model are needed for interpretations of the curve jet observations.

The model was suggested by [Kar69] and developed by [Kur72], [OzU74], [KoM80]). We calculate numerically emission spectra and polarized intensity distribution under fixed angle θ between observer and jet directions ($\theta \ll 1$) and fixed power degree γ in electron energy distribution ($\gamma = 2.5$).

Polarization degree along the jet at a fixed normalized frequency of the spectra region B is decreased from approximately 0.1 in the start part ($R < R_0$) to a zero at a distance $R = R_0$ and is increased to about 0.7 to the end of the jet ($R > R_0$). This structure is polarized along and across the jet in the regions $R < R_0$ and $R > R_0$, respectively. The distance R_0 is shifted to the end (start) part with decreasing (increasing) frequency and, as a result, all structure is polarized along (across) jet direction at the sufficiently low (high) frequencies.

The intensity distribution for the stable jet gives, that the one- dimensional structure of the narrow jet can be divided by two components, as a rule: narrow (bright) and extent (background, weak). These component brightness ratio increases with the

---

*Astro Space Center, Lebedev Physical Institute, 117810 Moscow, Russia.
†Physical Department, Moscow State University, 119899 Moscow, Russia.

frequency. The linear polarization degree for the extent weak component is greater then it for the narrow bright component if frequency is not very high.

[GCR92] are reported that VLBI observed polarized structure of BL Lac objects and quasars are different, and the inferred magnetic fields in the jets in this sources are perpendicular and parallel to the jet axes,respectively (see also paper by D.H.Roberts and J.F.C.Wardle in this Conference Proceedings). In the above model this fact can be interpreted by the dominant emission from the part of the structure with $R < R_0$ for BL Lacs and with $R > R_0$ for quasars. It gives for these objects that normalized frequencies have to be different but physical nature of the emission may be common, and the magnetic field is parallel to the jet axes in both type of sources.

Nevertheless, polarization degree for the extent component may be too high to be in consistent with VLBI observations. These may indicate that: or 1) modifications of the straight line jet model are needed (to the curve jet model, as an example), or 2) the strong polarized extent component exist, but it is not observed (because it is weak), or 3) it is not enough time for forming the stable jet because of the jet variability, and the flares remnants are superposed.

The model is in agreement with many year spectra evolution observations for BL Lac and 3C 345. This simultaneous multi frequency spectra was observed at RATAN-600 ([BKK91]). New results at this model fitting to the flare in 0237+16 will be published by KL.

We thank organizer and Interferometrics Inc. for financial support that makes possible participating at this Conference for one of us.

References

[BKK91] Berlin A.B., Kovalev Yu.A., Kovalev Yu.Yu., Larionov G.M., Nidgelski N.A., Soglasnov V.A., *Proceedings of the Conference 'Variability of Blazars', Turku, 1991.*

[GCR92] Gabuzda D.C., Cawthorne T.V., Roberts D.H., Wardle J.F.C., *Astroph. J.* 388 (1992) 40- 54.

[Kar69] Kardashev N.S., *epilogue to Russian edition of:* Burbidge G.R., Burbidge E.M., *Quasi stellar Objects, Freeman (1967)* Mir, Moscow, 1969.

[KoM80] Kovalev Yu.A., Mikhailutsa V.P., *Sov.Astronomy* 24 (1980) 400-406.

[KoL93] Kovalev Yu.Yu., Larionov G.M., *in preparation (1993).*

[Kur72] Kurilchik V.N., *Astrophys. Lett.* 10 (1972) 115-119.

[OzU74] Ozernoy L.M., Ulanowsky L.E., *Sov. Astronomy* 18 (1974) 4-10.

A Problem of Different Kind of Asymmetries in Extended Double Radio Sources

Boris.V. Komberg *

The question of different kinds of asymmetries in extended double radio sources (db RS) is considered in terms of:

position of radio components;

direction of radio jets;

averaged depolarization (Garrington-Laing effect) [Nature ,**331**, 147, 149, 1988];

extended emission line regions;

averaged spectral indexes.

Some published possibilities to explain the observed asymmetries are discussed as follows:

1) inhomogeneous "Faraday" screen with randomly oriented cells of homogeneous field in hot gas of clusters;

2) inhomogeneous distribution of matter around the active nucleus, resulting in different dissipation of jet and counter-jet;

3) real difference in properties (in energy, velocity, duration of life) of jet and counter-jet.

It is shown that the observed asymmetries can be explained in a frame of a "flip-flop" model under the following assumptions: $\Delta t_{jet} = 10^6 yrs$, $\Delta t_{flip-flop} = 10^7 yrs$, and $\tau_{dbRS} = 10^8 yrs$. Here the asymmetry of the extended components is related to their different ages, and, hence, with different amounts of carried away gas.

Attention is given to the possible role of cooling flows in the formation of asymmetries of db RS in central regions of rich clusters. In particular, the distinction between db RS of FR II and FR I types on "$L_r - l_r$" plane might be connected with such an influence; "giant sequence" ($L_r \sim l_r^{-4.8}$) and "main sequence" ($L_r \sim l_r^{2.5}$) in the terminology of Shklovsky [Astron. J. (USSR),**39**, 591, 1962]

---

*Astro Space Center P.N. Lebedev Physical Institute, Moscow, Russia.

A model of a massive close binary system with opposite rotation of components is proposed. In this model an alteration of the jet direction could be achieved in a time interval $\Delta t_{flip-flop} = P/2$ if the binary system is immersed in a thick disc, whose plane coincides with the direction of component rotation and is perpendicular to their rotational plane.

As a development of Shklovsky's idea [Nature,**315**, 386, 1985] on recoil momentum in powerful one-side jet an oscillating motion of active galactic nuclei with period of about 10 yrs on scales of few kpc is proposed. Some specific features of QSS 3C 48 are linked with such a phenomenon [Wilkinson *et al.*, Nature,**313**, 312, 1991].

3-D simulations of continuous and recurrent cooling Jets

Elisabete M. de Gouveia Dal Pino [*][†]   *Willy Benz* [‡]

Abstract

We present fully three-dimensional hydrodynamical simulations of radiatively cooling, supersonic, heavy jets with both continuous and noncontinuous but periodic injection mechanism. The development of dynamical and thermal instabilities is also discussed.

1. Introduction

Recently, Raga (Ra88), Tenorio-Tagle (TCR88), and Blondin, Fryxell, and Konigl (BFK90), concerned with the dynamics of protostellar jets, have performed numerical simulations of two-dimensional, axisymmetric, continuous, supersonic jets including the effects of radiative cooling, neglected in previous analyses. In this work, we present the results of 3-D simulations of radiative cooling jets performed with a 3-D cartesian Smoothed Particle Hydrodynamics (SPH) code in order to explore a more realistic picture of the protostellar jets and associated HH objects (see DB92 and references therein for a detailed description of the model). The radiative cooling (due to collisional excitation and recombination) is implicitly calculated using the cooling function for a gas of cosmic abundances, cooling from $T = 10^6$ K (K89). A collimated, supersonic jet of radius R_j is injected in the bottom of the *ambient box* which has dimensions up to $40 R_j$ in the z direction and 12 R_j in the transverse x and y directions.

Figure 1 shows the central density contours of a cooling jet, with continuous injection, and an initial jet to ambient density ratio $\eta = n_j/n_a = 3$, $n_j = 60 cm^{-3}$, a pressure ratio $k = p_j/p_a = 1$, $R_j = 2 X 10^{16} cm$, a jet velocity $v_j = 400 km/s$, and jet and

---

[*]Harvard-Smithsonian Center for Astrophysics, 60 Garden St., Cambridge, MA 02138, USA.

[†]Instituto Astronômico e Geofisico, Universidade de São Paulo, Av. Miguel Stefano, 2400, São Paulo, SP 01050, Brazil. This author was partially supported by the brazilian foundations BID/USP and FAPESP.

[‡]University of Arizona, Steward Observatory and Lunar and Plan. Lab., University of Arizona, Tucson, AZ 85721, USA.

ambient Mach numbers $M_j = 20$ and $M_a = 11.55$. The cooling length parameter (which is defined by the ratio between the cooling length behind a shock and the jet radius) is given by $q_{bs} \simeq 10.5$ for the bow-shock and $q_{js} \simeq 0.39$ for the jet shock, implying that a radiative jet ($q_{js} < 1$) is propagating into an adiabatic ambient medium ($q_{bs} >> 1$). The entire evolution corresponds to an age $t \simeq 950 yrs$. Our results qualitatively agree with the 2-D simulations of BFK90. However, the removal of the axisymmetry has resulted in some relevant structural differences (see DB92). At the jet head, a dense shell develops, formed by the cooling of the shock-heated gas.

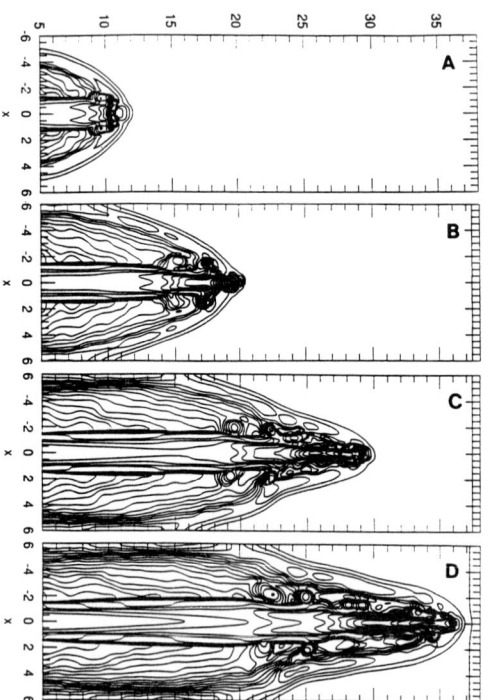

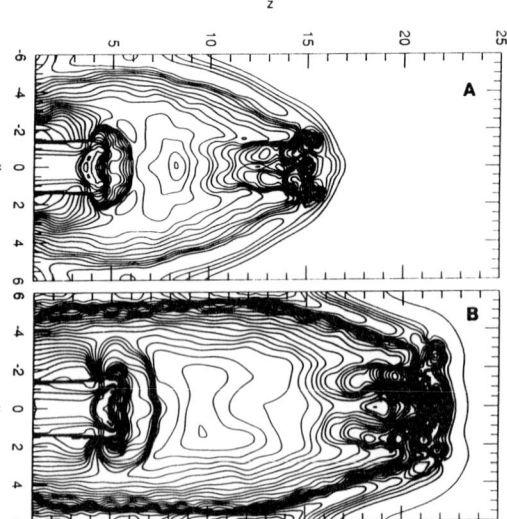

Figure 1. Density contours of a continuous, cooling jet with $\eta = 3$, and $M_j = 20$. The contour lines are separated by a factor of 1.3. The z, and x coordinates are in units of R_j.

Figure 2. Density contours of a recurrent jet with turning-on and turning-off periods $\simeq 100$ and $200 yrs$, respectivelly. The initial conditions are the same of Fig. 1. Fig.1a: $t \simeq 411 yrs$; Fig. 2b: $t \simeq 762 yrs$.

The radiative cooling reduces the thermal pressure which is deposited in the cocoon. As a result, the cocoon is smaller than in an adiabatic jet and has less pressure to collimate, drive (pinch and helical) Kelvin-Helmholtz instabilities, and reflect internal shocks in the beam (DB92). This is consistent with the absence of emission from the entire length of some observed stellar jets. The shell becomes dynamically unstable and disrupts into clumps, which eventually spill out to the cocoon forming an elongated plug of cold gas in the head (Figs. 1b-d). The disruption is caused by the combined effect of nonuniform cooling and the Rayleigh-Taylor instability (e.g., BFK90).

As the shell disrupts, its density undergoes time oscillations with a period $\simeq$ twice the cooling time behind the jet shock and a maximum density $\simeq 3.6 \times 10^4 cm^{-3}$. The oscillations are due to the development of global thermal instabilities at the radiative shock (e.g., DB92 and references therein). The clumpy structure of the dense shell and its variable density suggest that the knotty and variable emission pattern of some HH objects (e.g., HH1 and HH2) is a consequence of the thermal and dynamical instabilities in the dense shells at the heads of the associated jets.

Finally, it has been observed that some jets have a multiple bow shock structure (e.g., HH111 jet (Re89)). Assuming that such structures could be due to noncontinuous but periodic injections of the jet material into the ambient (probably associated with eruptive accretion phenomena at the central source; cf., Re89), we performed the numerical experiment of Fig. 2. In Fig. 2, the beam is periodically turned on and turned off. The turning-on period is $\simeq 100 yrs$ and the turning-off period is twice this amount. The initial conditions are the same of the Fig. 1. In Fig. 2a, there is a bow shock at the edge of the jet and a second one is being formed at the head of the emerging beam. After $\simeq 350 yrs$, the second bow shock has already reached the first one (Fig. 2b), forming with it a very large and knotty bow shock in accordance with the observations. (Notice that the leading bow shock is much larger than in the continuous jet of Fig. 1.) A third bow shock is emerging with the new beam. The tail which appears behind the bow shock at the edge is also seen on the observed intensity maps.

References

[BFK89] Blondin, J.M., Fryxell, B.A., Konigl, A., *Ap. J., Vol. 360 (1990), p. 370-386.*

[DB92]   Gouveia Dal Pino, E. M., and Benz, W., *Ap. J., 1992 (submitted).*

[K89]   Katz, J., *Princeton Thesis, 1989.*

[Ra88]   Raga, A., *Ap. J., Vol. 335 (1988), p. 820-828.*

[Re89]   Reipurth, B., *Nature, Vol. 340 (1989), p. 42-45.*

[TCR88] Tenorio-Tagle, G., Canto, J., Rozyczka, M., *Astr. Ap., Vol. 202 (1988), p. 256-266.*

Simulations Of Relativistic Jets.

Mark Bowman *

In order to maintain electrical neutrality in a plasma, two species of oppositely charged particles must be present. If the two components are at the same temperature, then we can derive a set of fluid dynamical equations in terms of a combined pressure, (Synge [1]). This is equally true if the species are of different rest masses, as is the case in an electron-proton mixture. Using a simple change of variables, these equations can be expressed in forms which are analogous to the classical conservation equations. This is an extension of the method devised by Chiu, [2]. Once the fluid equations have been reduced to this more familiar form, several classical parameters can be defined; such as a classical sound speed (hence a Mach number) and two dimensionless parameters,

$$\Gamma = \frac{a^2 \rho}{P} \quad and \quad \gamma = \frac{P}{\rho e} + 1 \ .$$

In the above a is the sound speed, P pressure, ρ density, and e is the internal energy. For an ideal fluid Γ is the ratio of the specific heat capacities, and in the case of a polytropic gas $\gamma = \Gamma$.

The conservative equations are solved using Godunov's method, which uses a Riemann solver to iteratively calculate the interaction of waves. This method employs the Rankine-Hugonoit jump conditions to calculate the change in fluid parameters across a shock. In the polytropic case only one post shock parameter is needed to define all the others, but now that the fluid parameters γ and Γ have a temperature dependency, the situation becomes more complicated. The modified jump condition leads to a relation for the deflection of a streamline by a shock that requires the values of γ for both the upstream and downstream flow.

The temperature dependency of the jump conditions requires multiple evaluations of the equation of state during the Riemann iterations. To avoid this, a local parametisation of the variable γ is employed during the calculation, (Colella and Glaz [3]). The information provided by the Riemann solver is used to update the fluxes in the conservative equations from which the fluid parameters are obtained.

References

SYNGE, J.L., 'The relativistic gas', North-Holland Publishing Company, Amsterdam, 1957.

CHIU, H.H., 1973. The Phys. of Fluids, 16, 825.

COLELLA, P. and GLAZ, H.M., 1985. J. Comput. Phys., 59, 264

Acknowledgment I would like to thank Dr S. A. E. G. Falle and Dr M. J. Wilson for providing me with the fluid code on which this work is based.

---

*University of Manchester, Nuffield Radio Astronomy Laboratories, Jodrell Bank, Macclesfield, Cheshire, SK11 9DL, U.K.

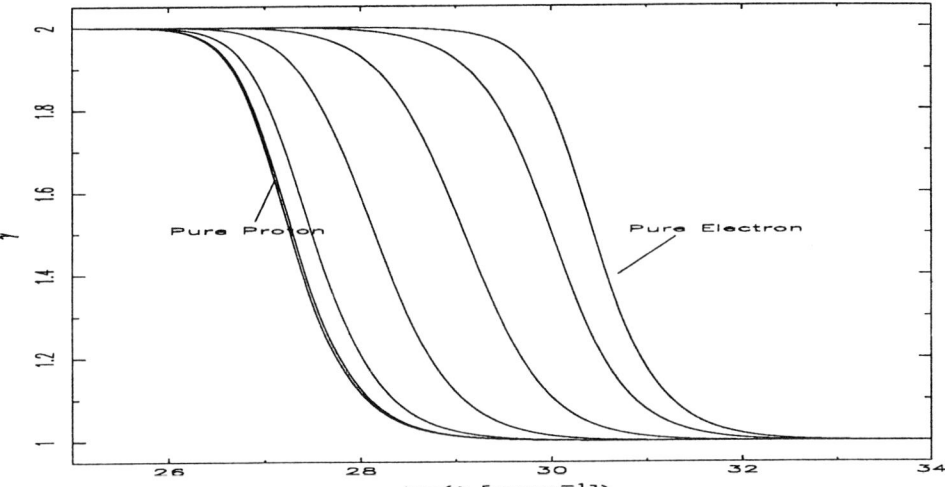

Figure 1. Variation in γ with inverse temperature, ξ, where $\xi = c^2/k\Theta$ (c is the speed of light, k is Boltzman's constant, and Θ is the absolute temperature). The various curves show the dependency for different ratios of proton to electron number densities.

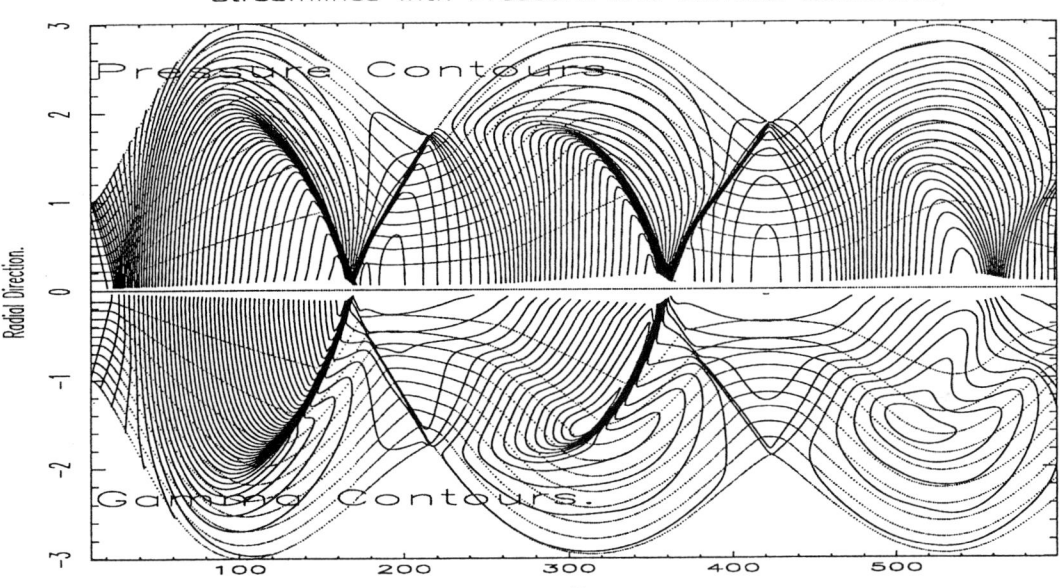

Figure 2. In this axisymmetric simulation, the jet can be seen emerging from the left with an initial Mach of 15 and absolute temperature of order 10^{13}K. The contours in the top half of the jet are pressure, whilst the contours on the lower half are of γ, which is a monotonic function of temperature, (see fig. 1). Initially the external atmosphere decays as a power law, after which it remains at a constant pressure. This causes the jet to over-expand with respect to its environment and a series of radial oscillations are created, mediated by recollimation shocks and expansion fans.

MilliArcsecond Structures of AGN at Different Redshifts

Leonid I. Gurvits *†

Abstract

The VLBI technique provides a unique possibility to investigate astrophysical objects with resolutions of milliarcsecond (mas) and higher. Due to the "long range" ability of radio astronomy this technique facilitates studies of mas structure properties of objects distributed over the widest range of cosmological distances, *i.e.* cosmological redshifts. Such a combination of the highest angular resolution and the coverage of the longest range of cosmological distances gives an opportunity to investigate both evolutionary evidences of discrete sources and properties of the Universe as a whole.

The only population of discrete sources known to be distributed over the widest range of redshifts from zero to about 5 is composed of active galaxy nuclei (AGN), mostly – quasars. Using the term AGN I will follow the wide classification given by Woltjer [Wol90]. The first question arising here is: do AGN show any dependence of their properties, particularly, structural properties, as a function of their redshift? Such a question was answered negatively about decade ago by Preston *et al.* [PMJ83]. But since that publication some new observational data became available.

1. Peterson *et al.* [PSJ82], Wright [Wri83], O'Dea [ODe90], Savage *et al.* [SJW90] reported that at higher redshift "humped" continuum radio spectra are more typical than at lower redshifts. Sources with such kind of spectra are classified as Giga-Hertz Peaked Sources (GPS). This type of spectra might have some reflection in the structural properties of sources.

2. The higher redshift objects tend to have closer directions of elongation of radio, infrared, and optical structures than their lower redshift counterparts (Chambers *et al.* [CMJ88], [CMv87]). Some explanation of this effect is proposed by Eales [Eal92].

---

*National Astronomy and Ionosphere Center, Arecibo Observatory, Arecibo, P.O. Box 995, Puerto Rico 00613. Operated by Cornell University under Contract with the National Science Foundation.

†Astro Space Center of P.N.Lebedev Physical Institute, Leninsky Pr. 53, Moscow, 117924, Russia.

3. An analysis of VLA maps of a few hundred sources allows us to conclude that the distortion of extended radio structures tends to increase at higher redshifts (Barthel and Miley [BaM88] and references therein). Some doubts on the application of the above conclusion for all kinds of extragalactic sources are argued by Kapahi [Kap90] and McCarthy *et al.* [MvK91]. Nevertheless, Kapahi [Kap90] confirms the conclusion for lobe dominated sources.

Some indications of increased bending on mas scale at higher redshifts are found by Wehrle *et al.* [WCU89]. This result is in agreement with VLBI maps of a few extremely high redshift quasars published by Gurvits *et al.* [GKP92].

All above mentioned tendencies are of marginal or indirect character. Can one apply any direct experimental evidences in searches for *"mas structure–redshift"* dependence?

Such a possibility can be realized using data from a large 2.3 GHz VLBI survey of 1398 predominantly – extragalactic sources (Preston *et al.* [PMW85]). The typical resolution of the survey is about 3 mas. A total of 917 sources were detected, *i.e.* correlated radio flux density associated with a compact structure was measured. At the time of publication of this survey in 1985, 478 extragalactic sources had measured redshifts. During the last few years, an additional 189 sources from the survey have had their redshifts determined, and 50 already known redshifts have been specified [VCV91]. These new optical measurements approach the "critical mass" needed for the declared application of the survey data since there are many high–redshift objects ($z > 2$) among the recently identified radio sources.

Let us concentrate our attention on those 337 sources (exclusively AGN) from the entire survey list which have known redshift z as well as measured total flux density S_{tot} and correlated flux density S_{corr}. The ratio $\Gamma = S_{corr}/S_{tot}$, so called a visibility amplitude, is a measure of source compactness.

An informative way to explore source structure is to examine the behavior of Γ as a function of an interferometer baseline length B. We note that the visibility amplitude never can exceed unity, *i.e.* $\Gamma \leq 1$, while $\Gamma = 1$ at $B = 0$. Γ remains equal to unity as the baseline increases if the source stays unresolved at all baselines, *i.e.* it is effectively a point source (Fig. 1, curve 1). Of course, an ideal point source is nothing more than an abstraction, and if one could increase the baseline indefinitely then at some baseline length the source would become resolved. Increasing the baseline further would result in a decrease of the visibility amplitude and asymptotical approach to zero (we do not discuss here practical matters of increasing of an interferometric baseline which is limited by telescopes sensitivity and propagation phenomena).

One may enquire if we can formulate any general properties of the $\Gamma(B)$ curves for AGN? This is not easy because of the wide diversity of possible source structures and the typical angular sizes of AGN at mas scales (*e.g.* see Pearson [Pea91] and

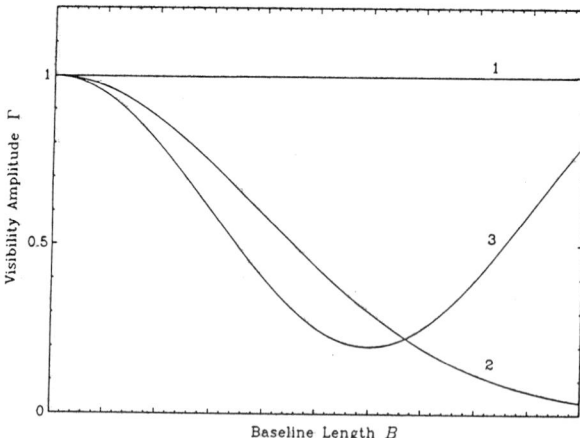

Figure 1. Visibility amplitude Γ as a function of interferometer baseline length B: 1 – unresolved source; 2 – typical curve for AGN structure averaged over large enough number of source; 3 – an example of a "prohibit" curve. The scale of the B-axis is arbitrary.

references therein). However, if we consider averaging a large enough sample of $\Gamma(B)$ curves obtained at the same frequency, we arrive at the remarkable conclusion that the averaged $\Gamma(B)$ curves tend to decrease monotonically with baseline length (Fig. 1, curve 2). It is important to note that there are two kinds of averaging. One is the direct averaging over the set of sources. Another kind of averaging is the "hidden" one corresponding to averaging over the different position angles of the interferometer baselines.

The statement above can be logically "proved by contradiction". Let us assume that the averaged $\Gamma(B)$ curve does not decrease monotonically. Then, together with the above–mentioned property for the visibility amplitude that $0 < \Gamma \leq 1$, this implies that there must be at least one local extremum for some finite baseline length (e.g. Fig. 1, curve 3). Physically such an extremum would determine a characteristic angular scale for fine structure of AGN. However, no such characteristic structural scale has yet been found with VLBI. Hence, following "Occam's razor", one must conclude that there is no other possible choice for the averaged $\Gamma(B)$ curve, except for it to be monotonically descending. To proceed, we will accept this statement as an *a priori* fact.

Fig. 2 shows values of Γ, averaged in redshift bins with $\Delta z = 0.2$, from the described above 2.3–GHz VLBI survey sample of 337 AGN. The apparent growth of Γ with redshift up to $z \simeq 1.0$ seems to be as expected, as this part of the Universe is well–described by Euclidean geometry and, in average, the further the sources, the smaller their apparent angular size, and hence the greater their measured Γ. The more interesting part of the diagram seems to be for redshifts $z > 1.0$. The apparent constancy of $\Gamma \approx 0.4$ looks quite compelling. To explore possible explanations, one needs to interpret measurements of Γ for the expanding Universe.

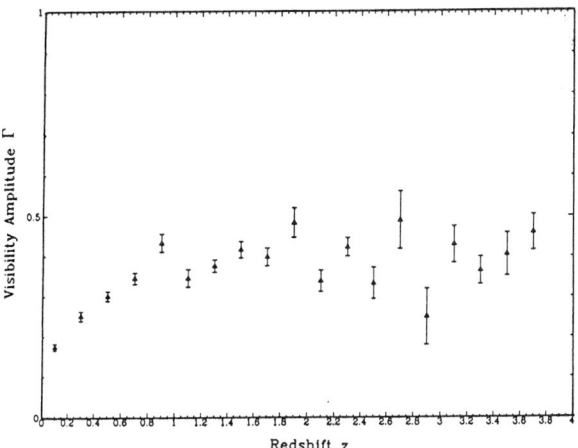

Figure 2. Visibility amplitude Γ as a function of source redshift z. Γ measurements are averaged over intervals $\Delta z = 0.2$. Total number of sources – 337. Bars correspond to $\pm 3\sigma$ errors.

Consider a source of a linear size l at a redshift z. It's corresponding angular size is $\vartheta = l(1+z)^2/D_L(z)$ where $D_L(z)$ is a luminosity distance (Weinberg [Wei72]). Now assume a VLBI baseline B to be optimized to an apparent angular source size $\vartheta = \lambda/B$, where λ is an observational frequency. Suppose we could move the source from the redshift z to some fixed reference redshift z_0. Now, if we can change the VLBI baseline such that the angular resolution will correspond to the same linear size l but at the reference redshift z_0, such a procedure will give the same visibility amplitude for the source at both redshifts (we imply that both the baseline and the source keep their orientation in image plane). The new baseline B_0 can be easily calculated as $B_0 = A \cdot B$, where the baseline modification coefficient A depends only on the variable z, the reference redshift z_0 and the cosmological deceleration parameter q_o

$$A = \frac{(1+z)^2 \cdot [z_0 q_o + (q_o - 1)(\sqrt{2q_o z_0 + 1} - 1)]}{(1+z_0)^2 \cdot [z q_o + (q_o - 1)(\sqrt{2q_o z + 1} - 1)]}$$

Thus we can modify the baselines for all sources in the sample, "moving" them to some fixed reference redshift. As we already noted, such a procedure does not change the visibility amplitude of a source. So, applying this procedure to the sample we can transfer $\Gamma(z)$ curve to $\Gamma(B)$ curve. This transform gives a clue to compare the experimental diagram $\Gamma(z)$ (Fig. 2) and the "theoretical" curve $\Gamma(B)$ (marked as 2 on the Fig. 1).

Fig. 3 shows a family of $A(z)$ curves for different values of q_o. Calculations were carried out for the reference value of $z_0 = 0.5$. Now we can enquire which curve for $A(z)$ agrees best with the experimental data.

This question should be answered within the uncertainty constraints of statistical

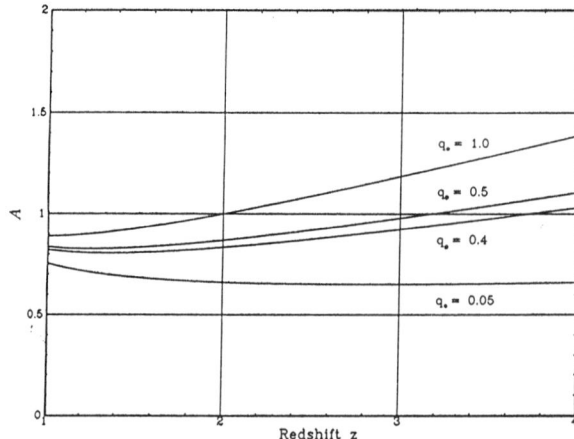

Figure 3. The baseline modification coefficient A as a function of redshift, $z_0 = 0.5$.

analysis. However, such an analysis would be difficult because of the large number of formal uncertainties, caused by the possible non-completeness of the source sample, and the statistically uncertain status of the basic statement on $\Gamma(B)$ curves. Nevertheless, one broad conclusion seems to be rather clear. The $\Gamma(z)$ curve derived from the observations will fit the "theoretical" $\Gamma(B)$ curve without any additional assumption on the nature of the source only if the value of $A(z)$ does not suffer significant variations in the range of investigated redshift, i.e. $1 \leq z \leq 4$. $A(z) = const$ would be apparently the best choice. As it is clear the Fig. 3 predicts a much higher probability for $q_o \leq 0.5$ than for greater values of q_o. This conclusion strains the flatness problem of cosmology (Peebles [Pee86]) since it narrows the range of possible values of q_o near 0.5. Note, that the latest value corresponds to the critical density, distinguishing between open and closed models of the Universe.

The estimation of q_o given here is in agreement with results of Kellermann [Kel92] and Vermeulen and Cohen [VeC92] also based on the analysis of VLBI data. All together mentioned and described above approaches show a high productivity of applications of VLBI for cosmology.

I appreciate discussions and advices of W.A.Baan, M.H.Cohen, N.S.Kardashev, K.I.Kellermann, B.M.Lewis, R.A.Preston, C.J.Salter, V.I.Slysh. I thank also R.A.Preston for providing VLBI survey data.

References

[BaM88]   Barthel P.D., Miley G.K., *Nature, Vol. 333, (1988), 319.*

[CMJ88]   Chambers K.C., Miley G.K., Joyce R.R., *Astrophys. J. Lett., Vol. 329, (1988), L75.*

[CMv87]   Chambers K.C., Miley G.K., van Breugel W., *Nature, Vol. 329, (1987), 604.*

[Eal92]   Eales S.A., *Astrophys. J., Vol. 397, (1992), 49.*

[GKP92]   Gurvits L.I., Kardashev N.S., Popov M.V., Schilizzi R.T., Barthel P.D., Pauliny-Toth I.I.K., Kellermann K.I., *Astron. Astrophys., Vol. 260, (1992), 82.*

[Kap90]   Kapahi V.K., *Current Science, Vol. 59, (1990), 561.*

[Kel92]   Kellermann K.I., *These Proceedings.*

[MvK91]   McCarthy P.J., van Breugel W., Kapahi V.K., *Astrophys. J., Vol. 371, (1991), 478.*

[ODe90]   O'Dea C.P., *MNRAS Vol. 245 (1990), 20p.*

[Pea91]   Pearson T.J. *Variability of active galactic nuclei*, eds. Miller H.R. & Wiita P.J., Cambridge University Press, Cambridge (1991), 134.

[Pee86]   Peebles P.J.E., *Nature, Vol. 321, (1986), 27.*

[PSJ82]   Peterson B.A., Savage A., Jauncey D.L., Wright A.E., *Astrophys. J Lett., Vol. 260, (1982), L27.*

[PMJ83]   Preston R.A. Morabito D.D., Jauncey D.L., *Astrophys. J., Vol. 269, (1983), 387.*

[PMW85]   Preston R.A. Morabito D.D., Williams J.G., Faulkner J., Jauncey D.L., Nicolson G.D., *Astronomical J., Vol. 90 (1985), 1599.*

[SJW90]   Savage A., Jauncey D.L., White G.L., Peterson B.A., Peters W.L., Gulkis S., Condon J.J., *Aust. J. Physics, Vol. 43, (1990), 241.*

[VeC92]   Vermeulen R.C., Cohen M.H., *These Procedings.*

[VCV91]   Véron-Cetty M.-P., Véron P., *A Catalogue of Quasars and Active Nuclei, 5th Edition,* ESO Sci. Report No. 10, 1991.

[WCU89]   Wehrle A.E., Cohen M.H., Unwin S.C., *Bull. American Astron. Soc., Vol. 21, (1989), 1094.*

[Wei72]   Weinberg S., *Gravitation and Cosmology*, J.Wiley & Sons, New York, 1972.

[Wol90]   Woltjer L., *Active Galactic Nuclei*, Saas-Fee Advanced Course 20, eds. Courvoisier T.J.-L. & Mayor M., Springer-Verlag, Berlin (1990), 1.

[Wri83]   Wright A.E., *Quasars and Gravitational Lens.* Proceedings of the 24th Liege Astrophys. Col., Universite de Liege (1983), 53.

The Angular Size—Redshift Relation for Compact Radio Sources

K. I. Kellermann[*]

Abstract

The angular size redshift dependence for compact radio sources is found to be consistent with the predictions of standard Friedmann cosmologies with a deceleration constant q_0 close to 1/2 corresponding to a density parameter, Ω near unity characteristic of inflationary cosmologies.

At the 1958 Paris Symposium on Radio Astronomy, Fred Hoyle (1959) emphasized the potential of using the angular size—redshift relation to distinguish among different world models. However, in practice observational attempts to use the θ - z relation suffer at optical wavelengths by the difficulty of making precision measurements of angular diameter and at radio wavelengths by the apparent change in their linear dimensions with redshift and luminosity.

By the early 1970's, sufficient resolution was being obtained with the radio interferometers at Cambridge, Caltech, and Green Bank. Miley (1970), and later Kapahi (1987), showed that the observed θ - z relation follows a simple $1/z$ law out to redshifts up to 2 where the angular separation of the lobes of double radio sources was about an order of magnitude smaller than expected from standard cosmological models. In order to save standard cosmologies, either the component separation must decrease with cosmic epoch or there must be an inverse relation between linear size and radio power. Oort *et al.* (1987), Singal (1988), and Kapahi (1989) have been able to fit the observed θ - z relation for double lobe sources within standard Friedmann cosmologies by

$$\ell \sim 400\ (1+z)^{-3}\ (P/10^{27})^{1/3}\ \text{kpc}$$

with H_0 = 50 km sec^{-1} Mpc^{-1}.

However, the apparent coincidence of the systematic changes in component separation with redshift by just the amount needed to cancel the expected geometric effects of standard world models raises what is perhaps the most serious question of whether the

---

[*]National Radio Astronomy Observatory, 520 Edgemont Road, Charlottesville, VA 22903. The NRAO is operated by Associated Universities, Inc., under cooperative agreement with the National Science Foundation.

redshift of galaxies and quasars is really due to the expansion of the universe (Sandage 1988).

Compact radio sources are likely to be free of these systematic evolutionary effects. Their characteristic lifetimes of only a few tens or hundreds of years are small compared with the age of the universe, even at the early cosmic epochs corresponding to redshifts of three. Moreover, compact radio sources are much smaller than any host galaxy so the radio structure should be independent of the intergalactic or intercluster medium which is systematically different at high z. Finally, unlike the extended radio sources, compact sources are not randomly oriented in the sky, but due to relativistic beaming, they are aligned close to the line of sight. Flux limited samples are expected to lie within a narrow range of projection angles centered on an angle $\theta \sim (1/2\ \gamma)$, where $\gamma = (1 - v^2/c^2)^{-1/2}$ (Cohen 1989). Thus projection effects for compact sources may be less important than for the randomly oriented extended sources. On the other hand, since relativistically beamed sources are aligned close to the line of sight, small changes in θ correspond to relatively large changes in projected size.

Probably, the most serious problem in using compact sources is that unlike the extended double lobe radio sources the size of compact core-jet radio sources is not unambiguously defined. In general the most distant jet components are the weakest, the most diffuse, and have the steepest spectra (e.g., Pearson 1990), so the apparent angular size of a compact radio source measured with VLBI will depend on sensitivity, angular resolution, and frequency. Since observations made at a frequency v_0 refer to an emitted frequency, $(1+z)v_0$, where the jets are smaller, high redshift sources may appear to be systematically smaller than low redshift sources.

A sample of 82 compact radio sources taken from the literature, e.g., Ekhart *et al.* (1987), Pearson and Readhead (1988), Wilkinson *et al.* (1992), Gurvits *et al.* (1992), and Wehrle *et al.* (1992), have been used to study the θ - z relation. The angular extent of each source was defined as the distance between the core and the most distant component whose peak brightness exceeds two percent of the core brightness. These dimensions were determined from published contour diagrams using only images made with global arrays having a resolution of about 1.5 milli arcseconds and a dynamic range of at least 100:1. Only sources with a 5 GHz luminosity greater than 10^{24} W/Hz were included. This is roughly the luminosity which separates the radio loud from the radio quiet quasars (e.g., Kellermann *et al.* 1989). Known BL Lac objects which are thought to be systematically smaller than quasars because they are oriented nearly along the line of sight, and compact steep spectrum sources, which are much larger than core-jet quasars, were excluded from the sample.

Figure 1 shows the observed mean angular size vs. redshift. The data include three unresolved sources and another five sources, where the resolution was inadequate to clearly identify a secondary component. The unresolved and slightly resolved sources do not appear to be concentrated near any particular redshift interval so they are unlikely to influence the results. Moreover, a similar analysis using median instead of mean sizes gives similar results.

In principal, it would be desirable to use an image for each source made at a frequency $v_0/(1+z)$ to insure that there is not a systematic change in apparent size with redshift due to the $(1+z)$ shift in frequency. However, with an array of fixed antennas, the change in resolution and sensitivity to extended lower surface brightness features would vary with redshift introducing even larger systematic errors. With the method used here of defining the angular size as the separation of bright components, the effect of observing at a fixed frequency, is thought to be small and is in the sense of an apparent reduction the measured size of the high redshift sources, or a decrease in the apparent value of q_0.

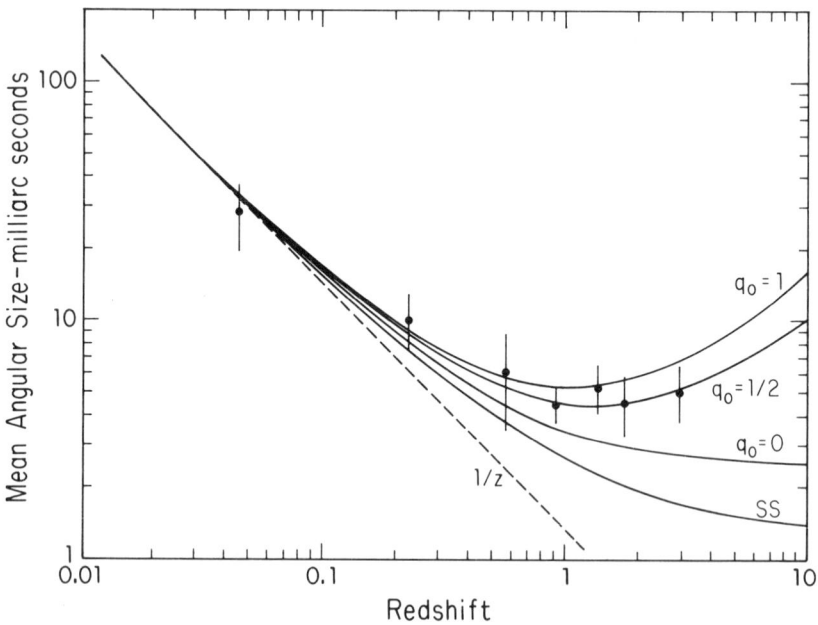

Figure 1. Mean angular size vs. redshift for 82 compact sources. The solid curves represent the expected dependence for Friedmann world models with various values of the deceleration parameter q_0 along with the Steady State model and the $1/z$ law observed for the separation of double lobed extended sources.

As expected from Friedmann world models with $q_0 > 0$, the observed mean angular size is essentially independent of redshift in the range $0.5 < z < 3$ and at a redshift of three lies above the extrapolated $1/z$ value by a full order of magnitude. This is perhaps the best direct observational support for the predictions of Friedmann cosmologies uncontaminated by the evolution of the sample objects, and is also direct evidence that the redshift of galaxies and quasars are really due to the expansion of the universe. The observed angular size—redshift relation can be fit by an Einstein—de Sitter universe with the deceleration parameter, $q_0 = 1/2$.

These results provide direct evidence on the largest observable scales that the mean density of matter in the Universe is close to the critical value, $\Omega = 1$, separating open and closed universes, consistent with the predictions of inflationary cosmologies. We have derived these results using an inhomogeneous data set, although it is unlikely that systematic errors can account for the clear departure of the observed θ - z relation from a linear 1/z law. With the VLBA it will be possible to obtain a much more homogenous sample of angular sizes, perhaps more objectively defined directly by the visibility function. With a larger body of data obtained over a wider range of flux density, it will also be possible to examine separately the dependence of linear size on luminosity. This will eliminate any dependence of angular size on redshift due to a possible linear size—luminosity dependence in a flux limited sample which contains a disproportionate number of high luminosity sources at high redshift. With the close spacings available with the VLBA it will also be possible to a limited extent to make scaled observations at deferent wavelengths corresponding to a fixed emitted wavelength.

References

Cohen, M.H. 1989, in *BL Lac Objects* (ed. Maraschi, L., Maccacaro, T., and Ulrich, M.-H.) 13-21.
Ekhart, A., Witzel, A., Bierman, P., Johnston, K., Simon, R., Schalinski, C., and Kuhr, H. 1987, *Astron. and Astrophys. Suppl. Ser.* **67**, 121.
Gurvits, L.I., Kardashev, N.S., Popov, M.V., Schilizzi, R.T., Barthel, P.D., Pauliny-Toth, I.I.K., and Kellermann, K.I. 1992, *Astron. and Astrophys.* **260**, 82-88.
Hoyle, F. (1959) in *Paris Symposium on Radio Astronomy, IAU Symposium No. 9* (ed. R. Bracewell) 529-532.
Kapahi, V.K. 1987, in *Observational Cosmology, IAU Symposium No. 124* (ed. Hewitt, A. Burbidge, G., and Fang, L. Z.) 251-265.
Kapahi, V.K. 1989, *Astronom. J.* **97**, 1-9.
Kellermann, K.I., Sramek, R., Schmidt, M., Shaffer, D.B. 1989, *Astronom. J.* **98**, 1195-1207.
Legg, T.H. 1970, *Nature* **226**, 65-67.
Miley, G.K. 1970, *Mon. Not. R. ast. Soc.* **152**, 477-489.
Oort, M.J.A., Katgert, P., Steeman, F.W.M., and Windhorst, R.A. 1987, *Astron. and Astrophys.* **179**, 41-59.
Pearson, T.J. and Readhead, A.C.S. 1988, *Astrophys. J.* **328**, 114-142.
Pearson, T. 1990, in *Parsec Scale Radio Jets*, (ed. Zensus, A.) 1-12.
Sandage, A.R. 1988, *Ann. Rev. Astron. and Astrophys.* **26**, 561-630.
Singal, A.K. 1988, *Mon. Not. R. astr. Soc.* **233**, 87-113.
Wardle, J.F.C. and Miley, G., 1974, *Astron. and Astrophys.* **30**, 305-315.
Wehrle, A., Cohen, M.H., Unwin, S., Aller, H.D., Aller, M.F., and Nicholson, G. 1992, *Astrophys. J.* **391**, 589.
Wilkinson, P.N., Polatides, A, Readhead, A.C.S., Xu, W., and Pearson, T.J. 1992, Sub Arc Second Radio Astronomy, ed. R. Spencer, (Cambridge Univ. Press).

Linking the Optical and Radio Reference Frames

R.W.Argyle, L.V.Morrison, J.D.H.Pilkington *   *A.N.Argue* †

Abstract

The current status of a campaign to perform astrometry of radio stars in both the optical and radio is described.

1. Introduction

The radio reference frame is defined by an evenly distributed net of about 400 unresolved extragalactic sources with $<V> \sim 18$ to an accuracy of about 1 mas (see for instance Russell et al, these proceedings) . The current best ground-based optical frame is the FK5 catalogue defined by 1535 stars with V = 7 or brighter having individual errors of about 40 mas at 1990, but with systematic zonal errors reaching 100 mas (Morrison et al [1]) . There is thus a need to improve the optical frame and unify it with the radio frame to correlate structure in extended objects.

2. Why radio stars?

There are two main avenues of approach to linking the optical and radio reference frames; by the use of radio quasars and radio stars. The quasars are strong emitters and sufficiently distant that their proper motions are negligible. However, they are optically faint. Radio stars, on the other hand, are relatively nearby, often binary in nature and optically bright. Corrections are therefore required for parallax, proper motion and orbital motion (Lestrade, these proceedings); but this requires many hours of observation by VLBI. Many radio stars will be well covered by Hipparcos, so in practice a single radio observation should be sufficient to determine the offset (if any) from the optical position.

---

*Royal Greenwich Observatory, Madingley Road, Cambridge, CB3 0EZ, England.
†Institute of Astronomy, Madingley Road, Cambridge, CB3 0HA, England

At present only about 30 radio star positions from VLA or VLBI observations have been published. Twelve radio stars are being observed with MERLIN in collaboration with colleagues from Jodrell Bank. Some of these stars will already have been measured, so these new observations will act as a check. Simultaneously, we will be obtaining photographic plates using the Wide Field Camera on the Jacobus Kapteyn Telescope on La Palma.

3. Optical positions of radio stars

It has been shown by Morrison et al [2] that direct observations of radio stars using the Carlsberg Automatic Meridian Circle on La Palma can yield positions to an accuracy of about 60 mas. Argyle et al [3] have shown that by using photographic astrometry and reducing the plates with reference stars observed by the CAMC, the accuracy can be improved to 40 mas. A further set of radio stars is being reduced (Argyle et al, in preparation).

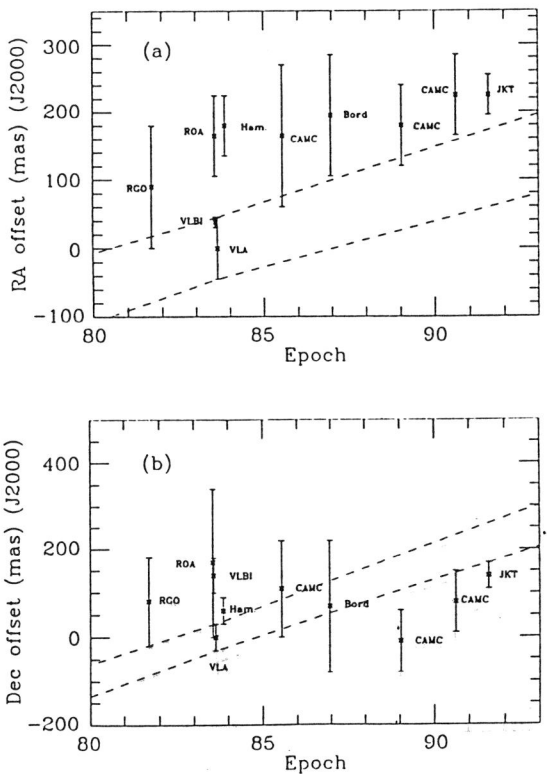

Figure 1. Recent determinations of optical and radio positions for SZ Psc in right ascension (a) and declination (b).

4. Comparison of Optical and Radio

Recent results for the radio star SZ Psc are shown graphically in Figure 1. The optical and VLBI positions are plotted relative to the VLA position. . The positions are from Clements & Argyle (RGO) [4], Florkowski et al (VLA) [5], Muiños (ROA) (private communication), Niell et al (VLBI) [6], Réquième & Mazurier (Bordeaux) [7], de Vegt et al (Hamburg) [8] and Argyle et al (this paper) (JKT). The dashed lines represent the effect of uncertainties in the optical proper motions with time. The proper motions are mean values taken from several Carlsberg Meridian Catalogues. Figure 1a shows what seems to be a significant difference in right ascension between the optical and radio positions, even allowing for the quoted errors in the radio positions. Why should this be so?

Previously, Morrison et al [2] had compared positions obtained from the Bordeaux and La Palma meridian circles with radio positions. They concluded that one radio star (54 Cam) had an anomalous radio position and the correlation between the two sets of data suggested that the radio position errors were somewhat larger than quoted. SZ Psc is relatively low in the sky for the VLA and the northern VLBI network.

The point from which the radio emission emanates should also be considered. SZ Psc is an RS CVn binary with stars of 1.5 and 5.1 solar radii (Popper, 1988) [9]. Mutel & Lestrade (1985) [10] quote a parallax of 10 mas (for which no correction has been made to the positions). The sum of the linear radii thus subtends an apparent angular diameter of 1 mas so the combined effect of distance and binary geometry do not explain the apparent differences between the optical and radio positions.

5. Future Work

A campaign of CCD imaging of the benchmark QSO's and Hipparcos stars within about 10 arc minutes of each other using the prime focus of the 2.5 metre Isaac Newton Telescope on La Palma is being considered. It is anticipated that this will fix the positions of the QSOs relative to the Hipparcos frame to about 10 mas.

References

[1] Morrison, L.V. et al, 1990, A&A,240,173

[2] Morrison, L.V., Argyle, R.W., Réquième, Y., Mazurier, J., 1990, A&A,236,256

[3] Argyle, R.W., Morrison, L.V., Irwin, M.J., Bunclark, P.S., 1991, MNRAS,250,576

[4] Clements, E.D., Argyle, R.W., 1984, MNRAS,209,1

[5] Florkowski, D.R., Johnston, K.J., Wade, C.M., de Vegt, C., 1985, AJ 90,2381.

[6] Niell, A.E, Lestrade, J-F, Preston, R.A., Mutel, R.L., Phillips, R.B., *1987, in Reid, M.J., Moran, J.M., eds., Proc. IAU Symp. 129, The Impact of VLBI on Astrophysics and Geophysics, Kluwer, Dordrecht, p.327*

[7] Réquième, Y., Mazurier, J., *1991, A&A,89,311*

[8] de Vegt,C., Florkowski, D.R., Johnston, K.J., Wade, C.M., *1985, AJ,90,2387*

[9] Popper, D., *1988, AJ,96,1054*

[10] Mutel, R.A., Lestrade, J-F., *1985, AJ,90,493*

THE CURRENT STATUS OF HST ASTROMETRY FOR LINKING THE HIPPARCOS FRAME TO EXTRAGALACTIC OBJECTS

Paul D. Hemenway [*]   *Raynor L. Duncombe* [†]

Abstract

The Fine Guidance Sensors [FGS] of the Hubble Space Telescope are expected to produce relative positions of objects to 17th magnitude at the milliarcsecond level of accuracy. We plan to use this capability to tie the HIPPARCOS reference frame to the VLBI frame. The proposed program and the current status of the FGS Astrometry for this program will be discussed.

1. Introduction

We propose to use the Fine Guidance Sensors (FGS) of the Hubble Space Telescope to tie the HIPPARCOS reference frame to the VLBI frame. The method is to measure the separations and their time derivitives of HIPPARCOS stars with respect to extragalactic objects (EGOs = QSOs, BL Lacs, and AGNs). The program has been described in detail elsewhere ([2] and [3]). The operation of the FGSs for astrometry and their current status is given by [1]. A full discussion of the HIPPARCOS satellite and the input and reduction processes is given in [4]. The first results are described in the May 1992 issue of Astronomy and Astrophysics.

2. HIPPARCOS

HIPPARCOS (HIP) stands for HIgh Precision PARallax COllecting Satellite. The goal is to measure positions, parallaxes and proper motions for 118000 stars to an rms accuracy of 0.002 arcsec in position and parallax, and 0.002 arcsec/yr in proper motion.

---

[*] Dept. of Astron. and Cent. for Space Research, Univ. of Tex., Austin, TX 78712, USA.
[†] Dept. of Aerospace Eng. and Cent. for Space Research, Univ. of Tex., Austin, TX 78712, USA.

HIP measures accurate chords on the celestial sphere. During its lifetime, several million chords are measured, building up a structure like a geodesic dome on the celestial sphere.

The Problem: The entire "rigid body" of the solution has an unknown rotation in inertial space. HIP's limiting magnitude is about 11 with full accuracy and 13 with reduced accuracy. Thus, HIP can observe only one quasar and a limited number of asteroids and natural satellites. The problem reduces to: How to connect the HIP instrumental frame with a dynamical non-rotating frame?

First, consider the question: Why is the system rotation required? a) The rotational OFFSET at some epoch (with respect to the VLBI frame, for example) is required to register radio and optical maps to the milliarcsecond (mas) level of accuracy for purposes including identification and astrophysical interpretation. b) The velocity of rotation is required to have as close to a local inertial frame (non-rotating) as possible so that the physics—dynamics of the galaxy and the solar system—may be understood correctly.

Solutions: 1) Observe with respect to the VLBI reference, assuming extragalactic objects form a quasi-inertial frame at the mas level. 2) Observe solar system objects with HIP and apply dynamics. This is being done.

The question is: How to observe HIP objects in the VLBI frame or with respect to EGOs directly? a) Observe HIP stars with VLBI/MERLIN on the VLBI frame directly. However, most stars observable are RS CVn stars, which are few. b) For the rotational offset, measure HIP stars directly with respect to VLBI optical counterparts (photographically, with CCDs, or with HST). c) To determine the system rotation with time, measure the change in the positions of HIP stars with respect to optical EGOs over time. This only gives the derivitive of the rotational offset matrix, but other objects besides VLBI sources may be included in the solution.

3. The Fine Guidance Sensors

The FGSs can measure accurate separations of objects up to 18 arcminutes apart between 4th and 17th magnitude. The accuracy is expected to be 0.003 arcsec rms for objects closer than 10 arcminutes and within 4 magnitudes of each other, and worse for objects farther apart in angle and magnitude. These observations must be performed in *POS mode* which locks onto the central linear portion of the FGS interferometer transfer function, *fine lock*. Astrometric observations could not begin until the final secondary mirror position was decided in December 1991. We then began our calibration observations.

Coarse track measurements (less accurate than fine lock) of the same star field in December 1990 and December 1991, showed discrepancies which are interpreted as a scale-like change over one year. As a result, all astrometric measurements were held in abeyance until the "anomaly" could be "resolved". From March to July, 1992, we observed a 6 star asterism in the center of FGS3, at about 2 week intervals with fine lock. The data show that FGS3 is a geometrically stable instrument at the milliarcsecond level within a few arcminutes of the center of the pickle. However, a slow scale-like change is occuring, which we can monitor at the milliarcsecond level.

Based on the geometric stability test, we are planning to calibrate the FGSs for milliarcsecond astrometry. The calibrations and the HIP link observations are expected to begin in a few months. With separations of about 80 EGO-HIP star pairs, we expect to determine the HIP–VLBI offset to 0.001 arcsec and the rotation with respect to the EGOs to 0.001 arcsec/year for the HIP Catalogue, and to a factor of 3 better by extending the HST observations over 10 years.

4. Acknowledgements

The HST GO project team comprises the authors plus: N. Argue, D. Jauncey, K. Johnston, J. Kristian, J. Kovalevsky, J.-F. Lestrade, M.A.C. Perryman, R. Preston, B. Tapley, C. Turon, C. deVegt, H. Walter, and G. White. We thank the members of the HST Astrometry Science Team and the HIPPARCOS community for advice and encouragement. Funding by NASA Grant NAGW-1537, is gratefully acknowledged.

References

[1] Benedict G.F., Nelan E., Story D., McArthur B., Whipple A., Jefferys W. H., van Altena W., Hemenway P. D., Shelus P. J., Franz O., Fredrick L. W., Duncombe R. L., and Bradley, A. *Astrometric Performance Characteristics of the* Hubble Space Telescope *Fine Guidance Sensors, Publ. Astron. Soc. Pacific, in press.*

[2] *Hemenway P. D., and Duncombe R. L., in: Proc.129th Symp.IAU, 1988, eds: M. J. Reid and J. M. Moran, pp 335-336, Kluwer: Dordrecht, Boston, London*

[3] *Jefferys W. H., Benedict G. F., Duncombe R. L., Franz O G., Fredrick L W., Gerard T., Hemenway P D, McArthur B, McCartney J, Nelan E., Shelus P J, Story D, van Altena W, Wasserman L., Whipple A., Whitney, J., in: Proc. 127th Col.IAU, 1991, eds: J. A. Hughes, C. A. Smith, and G. H. Kaplan , pp 68-76, USNO, Washington, D.C.*

[4] *Perryman M.A.C. (ESA Project Scientist), Hassan H. (ESA Project Manager), et al. June 1989, THE HIPPARCOS MISSION Pre-Launch Status (3 volumes), ESA Special Publication SP-1111, European Space Agency, Paris*

The Radio/Optical Reference Frame: Progress and Plans

J. L. Russell [a]   A. L. Fey [b]   D. L. Jauncey [c]
K. J. Johnston [b]   N. Kawaguchi [d]   A. Kemball [e]
E. A. King [f]   C. Ma [g]   G. MacLeod [e]   D. F. Malin [h]
P. M. McCulloch [f]   G. Nicolson [e]   J. E. Reynolds [c]
D. Shaffer [i]   Y. Takahashi [d]   C. de Vegt [j]   G. L. White [k]
N. Zacharias [j]

Abstract

In 1987 we began a program to establish a global radio reference frame of about 400 compact and flat spectrum radio sources with optical counterparts, and thus also provide the link to the optical reference frame. So far we have considered about 700 radio sources as part of this program and have compiled a preliminary list of about 400 which have been observed in the radio and have known optical counterparts. These are nearly uniformly globally distributed except for the galactic plane. We will show the progress in the continuing observations and data reductions for the reference frame program in both the northern and southern hemispheres and will discuss the plans for its maintenance.

---

[a] Applied Research Corp, Landover MD, USA. mailing address: Code 7210, NRL, Washington DC 20375-5351, USA.
[b] Code 7210, NRL, Washington DC 20375-5351, USA.
[c] Australia Telescope National Facility, CSIRO, Epping, NSW 2121, Australia.
[d] Kashima Space Research Center/CRL, 893-1 Hirai, Kashima-mach1, Ibaraki-ken, 314, Japan.
[e] Hartebeesthoek Radio Astronomy Observatory, P.O. Box 443, Krugersdorp, South Africa.
[f] Univ. of Tasmania, GPS 252C, Hobart, Tasmania, 7001, Australia.
[g] Goddard Space Flight Center, Code 921, Greenbelt MD 20771, USA.
[h] Anglo-Australian Observatory, Epping, NSW 2121, Australia.
[i] Interferometrics, Inc, Vienna VA 22182-2799, USA
[j] Hamburger Sternwarte, Universität Hamburg, Gojenbergsweg, Hamburg, Germany.
[k] Univ. of Western Sydney, Nepean, NSW, Australia.

1. Introduction

In 1987 we began a program to establish a link between the radio and optical reference frames based on 400 extragalactic radio sources (Johnston et al. 1988). The sources were to be chosen as compact, i.e. quasars or compact galaxies, flat spectrum, strong enough to be measured with VLBI (>0.5 Jy) and from prime focus optical plates ($m \leq 19$), and evenly distributed about the sky.

One example of the importance of such a link is shown in the current work on SN1987A, comparing the HST map with the AT radio map, where the size of many of the image features is smaller than the accuracy with which the coordinate systems can be linked (Jauncey, private communication).

An example of the radio/optical offsets which are common is in a series of papers by de Vegt et al. (1985), Florkowski et al. (1985) and Johnston et al. (1985). They measured the offsets between the radio and optical positions of 22 radio stars. They illustrated the need for a higher density program and thus the plan for the 400 source program described here.

2. Program Status

Using VLBI we have so far obtained observations of 417 sources and have published positions for 287 of them (Ma et al. 1990, Russell et al. 1991, Russell et al. 1992, Fey et al. 1992). The next 3 publications are in progress and will add 117 new sources, as well as improved positions for 122 of those already published. The distribution of all of these sources is shown on an Aitoff projection plot in Figure 1. These results do not include the 1992 observations which, at the time of this meeting, are still in various stages of data analysis.

Optical observations are more difficult to obtain, but have progressed. We have published optical positions of 40 sources in the previous lists of publications. Positions for several dozen additional sources are also in various stages of data analysis.

3. Source Priorities

All in all we have progressed far enough with our program to recommend the list of extraglactic sources which should replace the original radio/optical list of the IAU working group (Argue et al. 1984). The latter catalog was compiled from 9 radio catalogs and included 233 sources, 145 northern and only 6 sources south of -45 deg. We have gone through the list of 703 sources considered for the program and assigned each a priority. The priorities are 1, 2, 3, or 4. Respectively, these stand for good sources, probably good sources, probably bad sources and bad sources. An additional set were given priority 0, i.e. sources not considered at this time. The breakdown

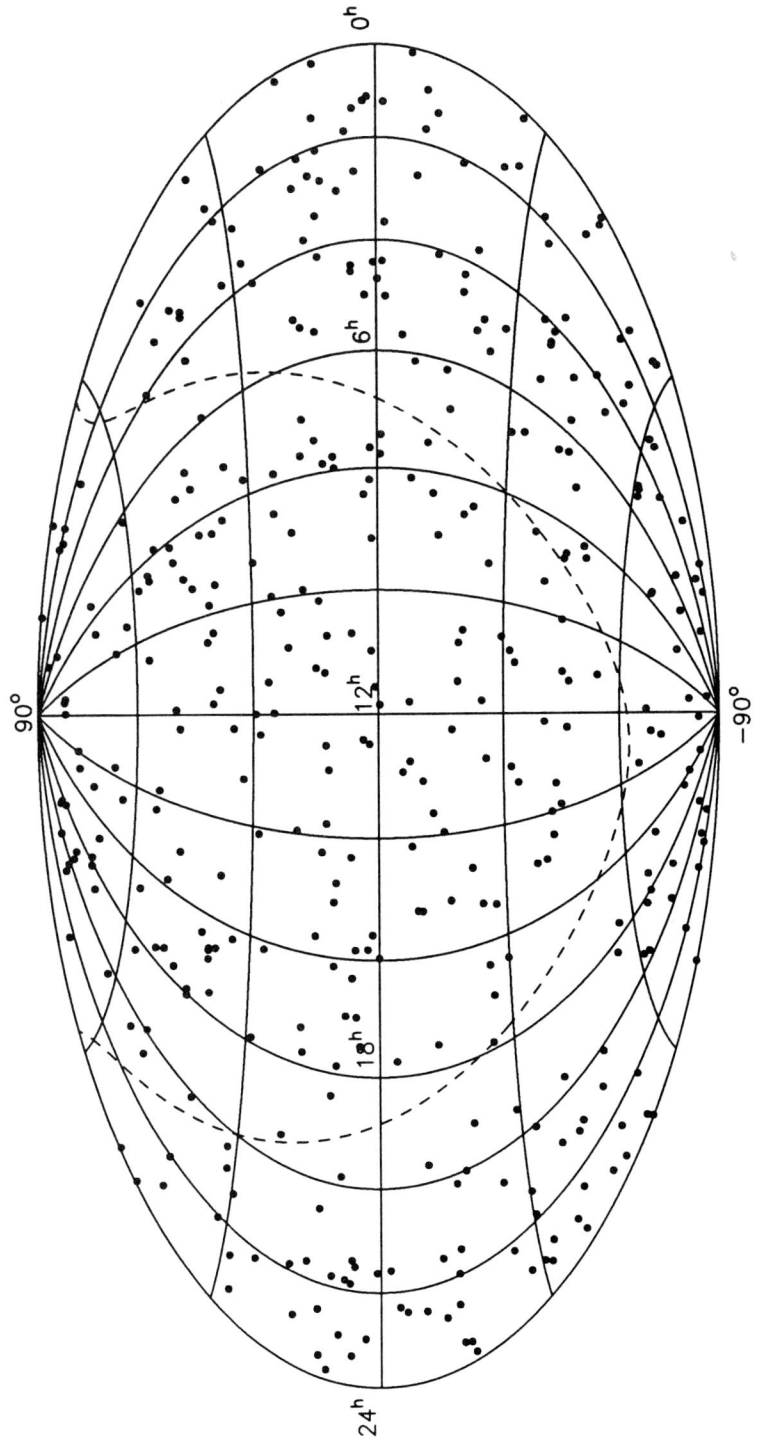

Fig. 1. Distribution of Sources

of number of sources in each class (and the number in each Northern and Southern hemispheres, respectively) is as follows:

PRIORITY		TOTAL	N	S
1	Good	300	181	119
2	Probably Good	106	36	70
3	Probably Not Good	20	11	9
4	Not Good	26	16	10
0	Not Observed	253		
	Total	705		

We note that the total of sources with priorities 1 and 2 is 406. While priorities for several sources will be revised, we expect the total number of sources in the top two priorities not to change significantly. As additional data become available, we expect the list to become much more polarized, i.e. most of the priority 2 sources to become 1's and most of the priority 3 sources to become 4's.

The priorities were assigned by looking at the radio and optical information on hand for each source individually. In order to illustrate what was defined as a "good" source, the breakdown of the 300 priority 1 sources is given below:

N	DESCRIPTION	TOTAL
187	Quasars, m < 20	187
17	Quasars, 23 > m > 20	204
40	BL Lac; 39: m < 20	244
17	Galaxies; 15: m < 20	261
9	Empty fields	270
30	Misc	300

The 187 quasars with optical counterparts brighter than 20th magnitude are the ideal sources for linking the radio and optical reference frames. The other sources in the list are exceptions and additions to the list to fill in where ideal sources are not available. In some cases we are taking advantage of geodetic sources with extensive VLBI observations to strengthen the radio reference frame. The miscellaneous sources are those which have incomplete or conflicting information in our data base and we are working complete or correct data for those entries.

4. Source Summaries

Of the 404 positions which have been published or are in preparation for publication, 308 of them have errors less than 1 mas in right ascension and 308 in declination.

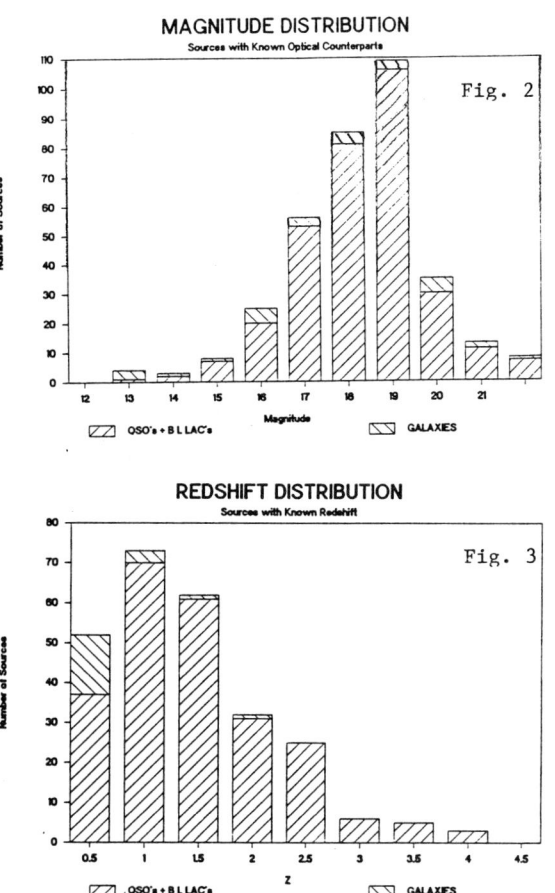

Fig. 2

Fig. 3

Fifty-two sources have errors in either coordinate greater than 4 mas. All but 6 of the 52 have been observed on short baselines only. We expect their position errors to approach those of the other sources with new observations on longer baselines, which have already been scheduled.

The magnitude and redshift distributions of the sources for which we have that data are shown in Figures 2 and 3, respectively.

5. Maintenance

The sources will be further checked, corrections made and recommendations made to the current IAU working group. We expect to add fewer than 50 additional sources.

We have been awarded time on the VLBA for early 1993. We hope to transfer most of the program for sources north of -30 deg to the VLBA, which will not only provide us with good positions, but also with visibility plots and with some image information. This source information will be used to eliminate any sources with structure which would introduce errors into the radio reference frame and will warn us of possible problems in the radio/optical reference frame alignment.

For sources south of -30 deg, we will continue to monitor the visibility plots and survey as much for structure as possible with the limited resources available in the southern hemisphere.

We also plan a long term archive of VLBI data for use to the reference frame project and maintenance. This of course will begin with the data base from the former Crustal Dynamics Project at NASA, will include all of our data obtained specifically for reference frame work and also will include any additional geodetic or astronomical data which includes large numbers of our 400 sources.

6. References

Argue, A.N., de Vegt, C., Elsmore, B., Fanselow, J., Harrington, R., Hemenway, P., Johnston, K.J., Kühr, H., Kumkova, I., Niell, A.E., Walter, H., Witzel, A., *1984, Astron. Astrophys. 130, pp. 191-.*

de Vegt, C., Florkowski, D. R., Johnston, K. J., Wade, C. M., *1985, Astron. J. 90, pp. 2387-.*

Fey, A., Russell, J. L., Ma, C., Johnston, K. J., Archinal, B. A., Carter, M. S., Holdenreid, E., Yao, Z., de Vegt, C., Zacharias, N., *1992, Astron. J. 104, pp. 891-896.*

Florkowski, D. R., Johnston, K. J., Wade, C. M., de Vegt, C., *1985, Astron. J. 90, 2381.*

Johnston, K. J., de Vegt, C., Florkowski, D. R., Wade, C. M., *1985, Astron. J. 90, 2390.*

Johnston, K.J., Russell, J., de Vegt, C., Hughes, J., Jauncey, D., White, G., Nicolson, G., *Proc. of IAU Symp. 129 "The Impact of VLBI on Astrophysics and Geophysics," M. Reid and J. Moran, eds., Reidel, Dordrecht, pp. 317-, 1988.*

Ma, C., Shaffer, D. B., de Vegt, C., Johnston, K. J., Russell, J. L., *1990, Astron. J. 99, pp. 1284-.*

Russell, J. L., Johnston, K. J., Ma, C., Shaffer, D. B., de Vegt, C., *1991, Astron. J., 101, pp. 2266-2273.*

Russell, J. L., Jauncey, D. L., Reynolds, J. E., White, G., Nothnagel, A., Nicolson, G., Ma, C., Kingham, K., Johnston, K. J., Malin, D., *1992, Astron. J. 103, pp. 2090-2098.*

Astrometry with optical interferometers

C.A. Hummel [*]

1. The MK3 interferometer on Mt Wilson

This instrument [SHAO], jointly operated by the Center for Advanced Space Sensing of the Naval Research Laboratory and by the U.S. Naval Observatory since 1988, features two astrometric baselines 12 m in length and oriented north-south and east-south (a variable baseline with lengths ranging from 3 to 31.5 m is used to measure stellar diameters and orbits of spectroscopic binaries). High-precision siderostats, mounted on massive concrete piers located in air-conditioned huts, pick up star light and feed it into vacuum pipes leading to a temperature controlled laboratory. Here, the geometrical delay d is compensated for by delay lines (which fold the ligh path using a movable retroreflector in vacuum, in order to minimize internal dispersion), and the two beams are combined at a beam splitter. The fringe pattern is scanned by modulating the delay with a triangle wave, using piezo-activated mirrors. At the same time, these mirrors are used in a closed loop to track the fringe motion – induced by turbulence in the atmosphere – in a wide band channel. Three additional 25 to 40 nm narrowband channels are centered on 500, 550 and 800 nm wavelength.

In a typical night on a given baseline **B**, some 100 to 200 scans are obtained on a list of about 12 stars, which are distributed uniformly over the sky. The complex visibility is coherently averaged in 200 ms intervals for each scan; only delays (i.e. the product of the fringe phase and the filter wavelength) of data points obtained while tracking the zero-order fringe are then averaged over the length of the scan (75 s). A dispersion-corrected delay $d = d_{red} - D(d_{blue} - d_{red})$ is calculated (with the dry-air dispersion $D = (n_{red} - 1)/(n_{blue} - n_{red})$; n=spectral index).

The fundamental equation relating the measured delays to the star positions $\mathbf{S_i}$, $d_i = \mathbf{B}\mathbf{S_i} + C$, C being the delay constant (i.e. the difference in length between the north (or east) arm and the south arm of the interferometer), is solved in two steps. Firstly, the coordinates X, Y, Z, and C of the *a priori* baseline vector are fitted to the delay data, assuming that the time dependence of the first three is negligible (requiring a baseline of high mechanical stability). By injecting a white light beam into the system, which is reflected off of cornercubes attached to the rear of the

---

[*]NRL/USNO Optical Interferometer Project and Universities Space Research Association

siderostat mirrors, any time dependence of C can be accounted for by monitoring the position of the white light fringe pattern. Secondly, residual delay variations are modelled with offsets of the star positions from the FK5 position.

Preliminary results from five oberving sessions between 1988 and 1992 (each session lasting 4 to 12 days) indicate formal accuracies in the determination of stellar positions of about 5 mas in declination and 10 mas in right ascension. The following figure shows the star positions relative to the FK5 position for each session. Stars are identified by their FK5 catalog number. The error box shows the FK5 positional accuracy of about 50 mas.

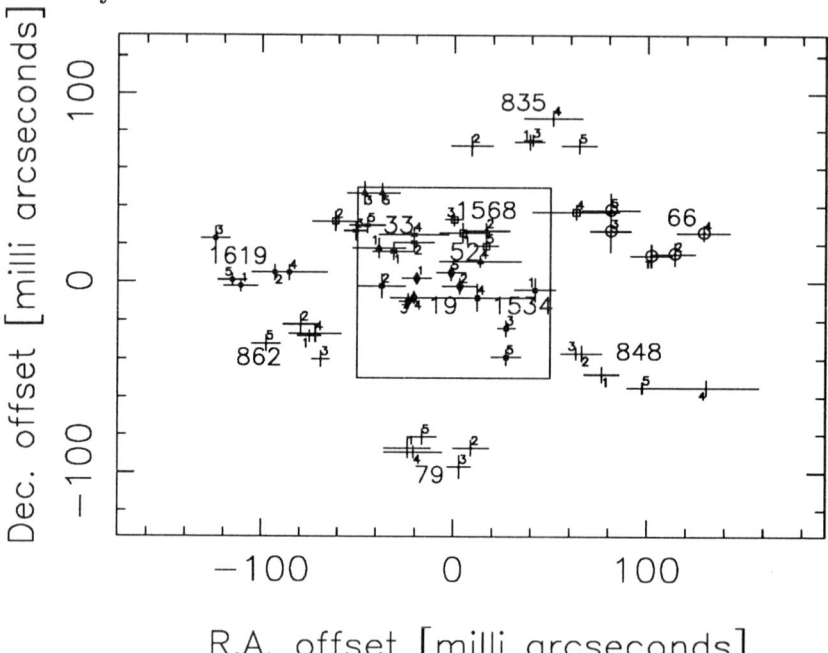

An error analysis, as shown in the following histograms of the normalized deviations of the star positions from a weighted average position, indicates that systematic errors are not completely under control. In fact, no C-monitoring was available in July 1989. However, most of the deviations are less than 2 standard error bars and a factor of about 1.8 applied to the formal errors leads to a global reduced χ^2 of unity.

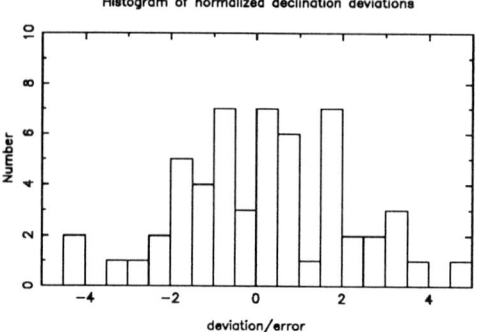
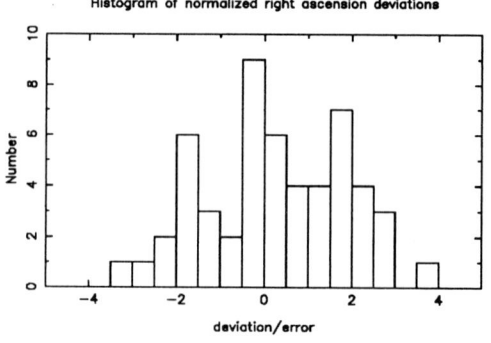

Using these scaled errors, the following figure plots the mean stellar positions using all four sessions. Accuracies are now in the range from 5 to 10 mas. (Errors in right ascension are larger since the north-south baseline is more sensitive for declination offsets and the east-south baseline was observed less frequently.)

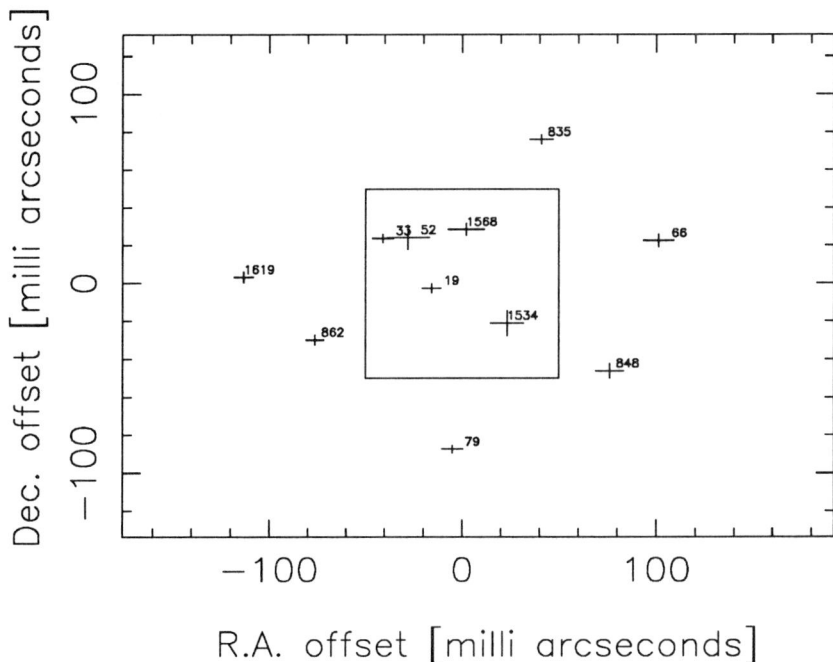

2. The Astrometric Optical Interferometer

NRL and USNO are building a new astrometric interferometer on Anderson Mesa, Arizona. The AOI will have 4 siderostats in a Y-shaped configuration with arms 20 m long, one siderostat being in the array center. It will feature full laser metrology of the array geometry in order to monitor the baseline coordinates, dispersed fringe tracking for real time zero-order fringe identification and a limiting magnitude for star and fringe tracking of about 10^m. It will enable us to determine proper motions of stars and thereby to maintain the accuracy of the HIPPARCOS reference frame for the next decades.

References

[SHAO] Shao M., Colavita M.M., Hines B.E., Staelin D.H., Hutter D.J., Johnston K.J., Mozurkewich D., Simon R.S., Hershey J.L., Hughes J.A., Kaplan G.H., *The MK3 stellar interferometer*. Astron. Astrophys. Vol. 192 (1988), pp. 357.

Astrometry and Structure of SiO Masers

Alain Baudry *

Abstract

The IRAM interferometer was used to measure the absolute positions of the 86 GHz SiO masers toward several late-type stars, and to discuss their spatial structure. The optical positions measured with the Bordeaux automatic meridian circle were compared with the radio positions of the SiO sources. A preliminary analysis of these data indicates that the 'radio-optical' differences lie in the range 40-80 milliarc sec. Our spot maps of the SiO emission reveal that the individual features are clustered within 0.1 arc sec in general; some stars however, show complex spatial and velocity-position structures.

1. Introduction

The $v=1$, $J=2-1$ and 1-0 SiO masers observed at 86 and 43 GHz in the envelopes of many late-type stars have distinctive characteristics: (i) the apparent sizes of the individual hot spots are extremely compact (about 1-6 milliarc sec); (ii) the SiO emission is distributed within 0.1 arc sec in general; (iii) the SiO molecule must be excited close to the stellar photosphere because of the high energy required to populate the $v=1$ state. Therefore, the SiO masers are potentially useful for astrometry and to probe the innermost circumstellar layers of Miras or semi-regular variables. Absolute radio positions of the 86 GHz SiO masers can now be obtained with 0.1 arc sec accuracy with the IRAM interferometer on Plateau de Bure. (In fact the positions are relative to the quasars used to calibrate the baselines.) These masers lie in the direction of variable stars which are often bright enough to be observed with the Bordeaux automatic meridian circle with 0.05 arc sec accuracy. One can then combine the radio and optical positions to align the VLBI (quasars) and FK5 (stars) reference frames at the 0.1 arc sec level. In addition, the interferometer may also be used to derive highly accurate relative positions of the maser spots; these spot maps give access to the overall structure of the SiO masing region.

---

*Observatoire de Bordeaux, BP 89, 33270 Floirac, France. This paper is a preliminary account of a work which will be presented elsewhere. The author is indebted to several colleagues at the IRAM and Bordeaux observatories for their help during the observations.

2. Observations

The radio observations were made in April-May 1991 and in January 1992 with three 15-m antennas of the IRAM array located on the most distant stations. The synthesized beam was 1.5-2 arc sec. The selected stars were: o Ceti, S Per, TX Cam, U Ori, R Cnc, R Leo, RX Boo, RU Her, VX Sgr, NML Cyg, mu Cep, R Peg, R Cas and Orion. They were observed in the v=1, J=2-1 transition of SiO at 86 GHz. For each stellar field, a minimum of three nearby quasars were used to perform accurate complex gain calibrations. The J2000 coordinates of the bandpass and phase calibrators were taken from the IERS compilation of VLBI catalogs; the internal accuracy is better than 1 milliarc sec. The baselines were carefully calibrated, and the resulting phase error was around 3-10 degrees for an average distance of 20 degrees between source and calibrator. The baseline stability was regularly checked with inclinometers installed in each antenna. All of the SiO features in all stars were mapped, and absolute and relative positions were obtained.

The optical positions of the visible radio stars studied here, as well as several other star positions, were determined with respect to the FK5 stars with the Bordeaux meridian circle (Baudry et al., 1990). The accuracy of the measurements, 0.05 arc sec, is the best one can achieve with ground-based instruments, and further improvement will require the Hipparcos satellite data.

3. The Link to Extragalactic Sources

The question of the link of the optical reference frame to the extragalactic radio frame by means of the maser stars was recently discussed in Baudry et al. (1990) and Baudry and Requieme (1991). The unification of the VLBI and Hipparcos reference frames requires a final accuracy of 1 milliarc sec. This seems possible now with a few radio continuum stars (Lestrade et al. 1992). In practice, one needs to observe several stars distributed over the celestial sphere. In this regard the SiO masers may also contribute to the unification because several tens of stars are accessible to connected interferometers, to VLBI and to optical instruments. However, the coincidence of the SiO and optical centers is crucial for the link and one may have difficulties in aligning the reference systems at the level of a few milliarc sec (Baudry et al. 1990). Nevertheless, the observations of several stars selected for their compact SiO emission might (statistically) get round the problem of the 'radio-optical' differences due to structural effects.

We have compared our radio positions measured in 1991, and the positions of three stars observed with the Hat Creek array (Wright et al. 1990), with our optical positions. (The latter were corrected for proper motions.) For a total of 12 stars the means of the differences between the radio-optical positions are 80 and 40 milliarc sec in right ascension and declination respectively. For both coordinates the mean square error is 0.1 arc sec.

4. Structure of the SiO Emitting Region

With maximum baselines of about 300 m the synthesized beam of the IRAM interferometer is too large to measure the sizes of the individual SiO maser spots. However, the interferometer can measure high quality relative positions of the hot spots distributed toward the star because this is not limited by uncalibrated baseline effects and by atmospheric phase effects. When good signal to noise ratio is achieved, our present relative position accuracy is of order 10-20 milliarc sec, i.e. it is comparable to the typical photospheric sizes of Miras. (This accuracy will improve with self-calibration.) Our SiO results show that most stars exhibit some spatial structure. This is the case in R Cnc, R Leo, VX Sgr, R Cas, RX Boo, R Peg or TX Cam. On the other hand our NML Cyg and mu Cep spot maps show little or no structure. In R Leo the positive and negative velocity features are separated by about 50 milliarc sec, and the SiO cloud seems to be rotating. The spatial structure in R Cas and TX Cam is complex and extends for the latter source over 100 milliarc sec. In R Cas, our map of the $v=1$, $J=2-1$ emission shows an overall extent (about 80x80 milliarc sec) comparable to that determined in VLBI by Lane (1984) at 43 GHz. However, the velocity-position diagrams differ at 86 and 43 GHz. This perhaps results from time-variable physical conditions (related to turbulence or shocks?) since the epochs of the observations differ by nearly ten years. It is also worth noting that the $v=1$, SiO spatial extents are much narrower than those determined with the IRAM interferometer from the thermal $v=0$ SiO emission toward several late-type stars (sizes around 1 arc sec). The thermal emission rather traces the outer layers where the grains form.

References

[1] Baudry A., Mazurier J.M., Perie J.P., Requieme Y., Rousseau J.M., *Optical positions of late-type stars associated with microwave line emission*. Astron. Astrophys. 232 (1990) 258-266.

[2] Baudry A., Requieme Y., *The link to extragalactic sources: observations of radio stars*. ASP Conference Series Vol.19, (1991), pp. 330-333.

[3] Lane A.P., *The spatial structure of silicon masers*. IAU Symposium 110, eds. R.Fanti et al., Reidel, Dordrecht (1984), pp. 329-330.

[4] Lestrade J.F., Phillips R.B., Preston R.A., Gabuzda D.C., *High precision VLBI astrometry of the radio emitting star sig CrB- a step in linking the Hipparcos and extragalactic reference frames*. Astron. Astrophys. 258 (1992) 112-115.

[5] Wright M.C.H., Carlstrom J.E., Plambeck R.L., Welch W.J., *Absolute positions of SiO masers*. Astron. Journal 99 (1990) 1299-1308.

VLBA Phase-Referencing Tests

J.E. Conway [*] *and R.C. Vermeulen* [†]

Abstract

Two phase-referencing tests were conducted with the partially complete VLBA using target-reference separations of up to 3.7°. In the first, we found quasi-sinusoidal phase residuals, which we model as being due to a Travelling Ionospheric Disturbance. Such effects may occasionally (perhaps 10% of the time) significantly limit the dynamic range of phase reference maps at frequencies of 1.6GHz and below. The second experiment only showed slowly varying phase residuals due to inaccuracies in modelling the zenith path delay through the troposphere and ionosphere. We show that under such conditions appropriate combinations of the phase-referencing and self-calibration techniques allow noise-limited images to be made of sources of any flux density.

1. Observations

Traditionally, VLBI is limited to imaging relatively bright sources ($\sim > 20$mJy for global MkIIIA VLBI at cm wavelengths). Using fringe-finding and self-calibration methods, the source must be bright enough to find fringes in delay, rate and phase space unambiguously within a relatively short phase-coherence time (5–15 minutes at cm wavelengths). Much weaker sources may be imaged by making frequent observations of a nearby bright, compact calibrator. This phase referencing technique seeks to produce phase stable target source data, by applying the propagation delay, rate and phase, determined for the calibrator, to the target.

We have carried out two test observations on bright sources using the partially complete VLBA at 5GHz with the MKII recording system (for a description see Vermeulen & Conway 1991, Radio Interferometry, IAU Colloq 131 p 350, full results will be given by Conway & Vermeulen 1993, in prep). We particularly wanted to test whether phase-referencing imaging can work over target-reference separations of 2°–3° degrees, in which case there is a sufficient density of catalogued, bright (> 200mJy), compact flat spectrum sources (e.g. Patnaik, Browne, Wilkinson and Wrobel. 1992

---

[*]NRAO, P.O. Box O, Socorro, NM 87801, USA
[†]California Institute of Technology, Pasadena, CA 91125, USA. This work was supported in part by the NSF under grant AST-9117100.

MNRAS 254,655) that any part of the sky is accessible to the technique. After correlation of the experiments on the Block II Caltech-JPL correlator, software corrections were applied to take account of geometrical effects such as improved antenna and source positions, polar motion, estimated tropospheric/ionospheric delays and antenna geometry. Fringe-fitting on the calibrator data was carried out inside AIPS and the results were transferred to the target source data. Typical phase-calibrated data for the largest target-reference separation are shown in Fig 1a,b for the first and second experiments respectively. While the first experiment shows quasi-sinusoidal phase variations with period approximately 20 minutes (whose amplitude depends on the target-reference separation) the second experiment only has long period phase residuals which correlate with source elevation.

2. Effects of TID's

We have fitted the residual phase variations seen in the first experiment with a Travelling Ionospheric Disturbance (TID) model. TIDs are generated by gravity waves in the upper atmosphere which cause variations in the electron density and hence the phase delay. A good fit to the many constraints provided by the data is obtained with a simple model, which consists of a sinusoidal phase screen with wavelength 220km whose amplitude of electron column density variations ΔTEC is 4×10^{15} cm^{-2} (i.e. 2% of the total daytime column density), moving at 180 m s^{-1} in a roughly NE direction at a height of 350km. Such parameters are compatible with TIDs observed by van Velthoven (1990, Phd Thesis, Univ of Eindhoven), who estimates that TIDs of this magnitude (ΔTEC) or greater occur perhaps 10% of the time. Simulations of full track 10 station VLBA observations in which the 5 SW stations are affected by such a TID, show that for a 3° target-reference separation, an image of dynamic range 50:1 is still achievable at 5GHz, but this is reduced to 10:1 at 1.6GHz.

3. Slowly Varying Phase Residuals.

The slowly varying phase-residuals in our second experiment (Fig. 1b) can be traced to inaccuracies in the assumed (i.e. model) zenith delays due to the troposphere and ionosphere. These affect the differential phase because of the differing elevations of the target and reference sources. In general, the zenith path lengths can only be estimated with limited accuracy (see Conway & Vermeulen 1993), which in turn limits the dynamic range achievable by 'direct imaging' (i.e. simple Fourier inversion and CLEANing). For instance, ground meteorological measurements only allow tropospheric delay estimates to an r.m.s. accuracy of ~2cm (e.g. Elgered 1982, IEEE Tran Antennas and Propagation AP30,502), which limits the dynamic range at 8GHz to less than 10:1 for a 3° target-reference separation. As a consequence it will not easy to make, via direct imaging, noise-limited full track VLBA images of sources in the flux range from 20mJy (above which fringe-fitting methods can comfortably be employed) down to about 2mJy (below which thermal noise precludes a higher dynamic range anyway). Fortunately, for sources in this intermediate flux-density range, phase-referencing sufficiently lengthens the phase coherence time that noise-limited images can be made via self-calibration methods using long (e.g. 1 hour) solution intervals (details will be given by Conway and Vermeulen 1993). Figure 2 summarizes the regimes of flux density and r.m.s residual zenith path error in which pure 'direct imaging' and 'self-calibration' techniques may respectively be employed without dynamic range limits. At 5–8GHz the crossover point between the two techniques occurs at about 2mJy. For target-reference separations of 2°–3°, we find that sources above this critical flux density can be self-calibrated, while direct imaging is adequate for weaker sources, provided that the zenith delay can be calibrated to about a wavelength. Hence it appears that, at least at 5 and 8.4GHz, techniques exist to obtain noise-limited images of sources of any flux-density.

Fig. 1 – a) 1st expt; large quasi-sinusoidal phase residuals over a period of 40 Minutes due to a passing TID. b) 2nd expt; slowly varying phase residuals over a period of 9 hours due to inaccuracies in the assumed ionospheric and tropospheric zenith delays. As in a) target-reference separation is 3.7°.

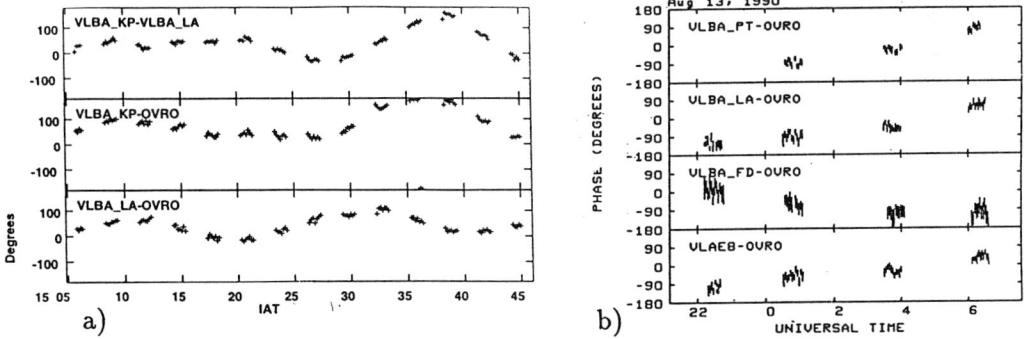

Fig. 2 – Different phase-reference imaging regimes as a function of source flux and r.m.s. errors in the estimated zenith delay assuming a target-reference separation of 2°.

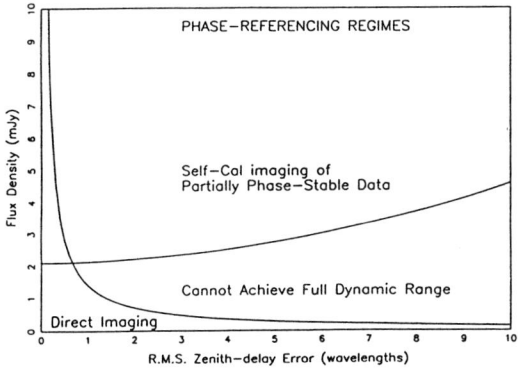

VLBI Phase-Referencing at 5° Separation

P. Elósegui [*]   J. M. Marcaide [†]   A. Alberdi [‡]   M. I. Ratner [*]

I. I. Shapiro [*]   A. Quirrenbach [§]   A. Witzel [¶]   F. Mantovani [||]   A. Rius [**]

Abstract

We observed the quasar 1928+738 and the BL Lac object 2007+777 on 1988 October 28-29 with a dual-wavelength, five-station, VLBI array, and determined their ~5° relative separation to within 0.1 mas using a phase-reference technique previously applied only to sources with smaller separations. This demonstration shows that the technique can be used to constrain the kinematics of the cores and inner parts of the jets of compact radio sources in a manner not possible with closure-phase and hybrid-mapping techniques alone.

1. Introduction

The apparent superluminal expansion of compact radio sources has been explained in the standard relativistic jet model [BK79] as motion of certain components with respect to a core, assumed to be stationary. The use of only closure phases and amplitudes in hybrid mapping and the difficulty of using phase-reference techniques for sources with large angular separations have prevented the gathering of statistical evidence on the core stationarity and hence on the validity of the standard model. We show here the feasibility of estimating with submilliarcsecond uncertainty the relative sky position of two sources about 5° apart; the fractional precision of this determination is at least comparable to that for smaller source separations (*e.g.,* 1038+528 A,B (~0.01°), 3C345-NRAO512 (~0.5°)): 10^{-8} to 10^{-7}. Moreover, because a suitable, strong, compact source (≥ 0.5 Jy) can usually be found within ~5° from any given source, this technique should be applicable to most or even all strong, compact sources.

---

[*]Harvard-Smithsonian Center for Astrophysics, Cambridge, MA 02138, USA.
[†]Universitat de València, E-46100 Burjassot, Spain.
[‡]Instituto de Astrofísica de Andalucía, E-18080 Granada, Spain.
[§]Naval Research Laboratory, Washington, DC 20392, USA.
[¶]Max-Planck-Institut für Radioastronomie, W-5300 Bonn 1, Germany.
[||]Istituto di Radioastronomia, I-40126 Bologna, Italy.
[**]Instituto de Astronomía y Geodesia, E-28040 Madrid, Spain.

2. Observations and Data Reduction

Using the Mark III VLBI system, on 1988 October 28-29, we observed alternately the sources 1928+738 and 2007+777, with a cycle time of 6 minutes, for ~12 hours. The VLBI array was composed of the following telescopes: Onsala (T, 20-m, Sweden), Medicina (L, 32-m, Italy), Effelsberg (B, 100-m, Germany), Madrid (M, 70-m, Spain), and Fort Davis (F, 26-m, Texas). The observations were made simultaneously at λ=3.6 and 13 cm at each telescope, except at Effelsberg, where only a λ=3.6 cm receiver was available. The recorded bandwidths at the two wavelengths were 16 and 12 MHz, respectively.

Each source has a total flux density higher than 1 Jy at both wavelengths and, because each is compact, can be detected with high signal-to-noise ratio on all baselines in a few seconds. After correlating the recorded data, we constructed phase delays, *i.e.* interferometer delays consistent with the observed interferometer phases. Since initially each phase has a 2π ambiguity, each delay is likewise ambiguous. However, we used astrometric analysis of the phase delays together with the unambiguous delay rates to eliminate these ambiguities, except for a single unknown offset for all delays for any particular source, wavelength, and telescope. We used a hybrid-mapping technique to make images of the two sources at the two wavelengths to remove structure effects; defined a reference position in each core at each wavelength; estimated the structure contributions to the phase delays for each of these positions; and subtracted them from the total phase-delays. We used these unambiguous, "structure-free" phase-delays, obtained from both the λ=3.6 and 13 cm data, to estimate and then remove the ionospheric phase-delay contributions.

We formed the differenced phase-delay observable by subtracting from the phase-delay value for each 1928+738 observation the corresponding value for the following 2007+777 observation. (The order of the sources in the pairing of the observations was chosen to minimize the time difference between the paired observations; more importantly, we have shown that making the opposite choice has no significant effect on the astrometric results.) We then obtained a consistent set of Earth orientation parameters and station and source coordinates from GSFC (Solution GLB718; IERS Annual Report 1991) and used the current version of our VLBI-3 program [Rob75], to analyze the differenced phase delays and the undifferenced phase delays of 2007+777 to produce a weighted-least-squares estimate in the J2000 system of the sky position of the core of 1928+738 relative to that of 2007+777. The details of this analysis will be published elsewhere [EMA92].

3. Data Quality

The resulting postfit residuals for the differenced phase delays still have visible systematic trends, but their amplitudes are quite small compared to the ambiguity interval of 128 ps, as shown in Figure 1. The weighted-root-mean scatter of the postfit residuals about their weighted mean shows that we have determined the position of the core of 1928+738 relative to that of 2007+777 with a statistical standard error of 40 μas in right ascension and 45 μas in declination, each under three parts in 10^9 of the 4.6° arc between the two sources.

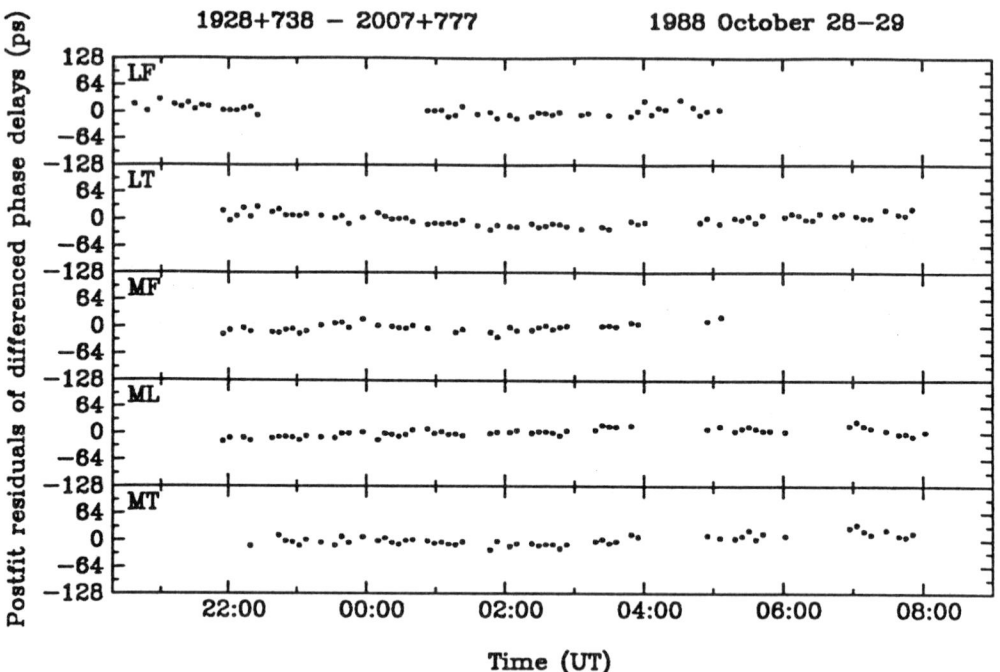

Figure 1. *Postfit residuals of the differenced phase-delay observable versus UT. In the "ionosphere-free" delays, a change of one cycle of phase in the previously constructed λ=3.6 cm phase delay causes a change of 128 ps.*

References

[BK79] Blandford R.D., Königl A., *Astrophys. J.*, Vol. 232 (1979), pp. 34-48.

[EMA92] Elósegui P., Marcaide J.M., Alberdi A., Ratner M.I., Shapiro I.I., Quirrenbach A., Witzel A., Mantovani F., Rius A. *Astron. J.*, (1992), (in preparation).

[Rob75] Robertson D.S., (1975), *Ph.D. Thesis*. MIT.

High-Precision VLBI Trigonometric Parallaxes of Three Radio-Emitting Stars σ CrB, UX Ari and HR5110

Jean-François Lestrade [*]   Robert B. Phillips [†]
Dayton L. Jones [‡]   Robert A. Preston [‡]   Denise C. Gabuzda [§]

Abstract

VLBI phase-referenced observations have yielded trigonometric parallaxes with submilliarcsecond formal precision for three radio-emitting stars whose variable flux densities were typically between 2 and 30 milliJansky during our astrometric monitoring program.

1. Introduction

The future Hipparcos optical reference frame must be linked to the stable VLBI extragalactic reference frame at the milliarcsecond level in order to 1) *unify the optical and radio coordinates systems*. This will enable registration of high-angular resolution images made at both wavelengths to identify counterparts for astrophysical purposes; 2) *stop the global rotation of the Hipparcos reference frame*. This will enable kinematical and dynamical stellar studies in an inertial system. Since 1982, we have conducted a VLBI astrometric program on 11 optically bright radio-emitting stars which are objects common to both the VLBI and Hipparcos frames and are a means to link them at the milliarcsecond level.

2. Phase-referenced VLBI observations

Radio-emitting stars are very weak with typical flux densities smaller than 30 milliJanskies. We employ phase-referenced VLBI observations by alternately switching

---

[*] DERAD, Observatoire de Meudon, 92195 Meudon Principal Cedex, France.
[†] Haystack Observatory-MIT, off route 40, Westford, Mass, 01886
[‡] Jet Propulsion Laboratory-California Institute of Technology, 4800 Oak Grove Drive, Pasadena, CA, 91109
[§] University of Calgary, Department of Physics and Astronomy, 2500 Universty Drive NW, Calgary, Alberta, T2N 1N4, Canada

VLBI array antennas every 2-3 minutes between a star and an angularly nearby stronger extragalactic source for 6 to 10 hours. The coherent integration of all these data provides a large enough signal-to-noise ratio to detect 2 milliJansky radio stars and to perform high-accuracy astrometry. The prime observable used is the VLBI phase. The typical VLBI array we use consists of one high-sensitivity antenna (i.e. the 70m at Goldstone or the 100m at Bonn or the phased-VLA (130m)) and 2 to 4 sensitive antennas (e.g., VLBA, OVRO, Greenbank, Bologna, Noto, Haystack). The recorded VLBI data are correlated at the Haystack Observatory Mark III Processor. The software FRNGX of Haystack was used to detect the low-SNR fringes of the weak star with the constraint of the residual delay and delay-rate of the reference source. The software SPRINT we developped was then used to determine the position of the star relative to the extragalactic reference source by performing a 2-D Fourier inversion of the differenced observed visibilities [LRWNPP90].

3. Astrometric results

We present the trigonometric parallaxes for σ CrB, UX Ari and HR5110 in Table 1. They are extracted from the complete solutions of all 5 astrometric parameters (2 coordinates, 2 proper motion components and parallax) estimated by a least squares fit of the coordinates measured at all the epochs of our VLBI astrometric program for that star (see numbers of observations for each star in Table 1). The formal uncertainties of the 5 parameters are derived by adjusting the measured VLBI coordinate uncertainties (α, δ) to make the normalized χ^2 of the fit unity. These formal uncertainties are $\leq$ 0.5 milliarcsecond in position, annual proper motion and parallax. The complete astrometric solution of σ CrB is fully discussed in [LPPG92]. The comparison with the best optical trigonometric parallaxes shown in Table 1 is satisfactory when the relatively large optical uncertainties are considered. The coordinate post-fit residuals are presented in Figure 1 for the star σ CrB which has the highest number of degrees of freedom (N.d.f) since 16 coordinates were measured by VLBI at 8 epochs. The post-fit rms of these residuals in α (= R in Fig. 1) and δ (= D in Fig. 1) is 0.42 milliarcsecond.

Table 1 : VLBI trigonometric parallaxes π of σ CrB, UX Ari and HR5110

	π_{VLBI} (mas)	π_{opt} (mas)	Number of observations	N.d.f	Refer.-star separation (°)
UX Ari	18.22 ± 0.41	20± ?	12	7	3.0
HR5110	14.24 ± 0.53	19 ± 6	8	3	5.0
σ CrB	43.97 ± 0.20	47 ± 6	16	11	0.5

These are the first trigonometric parallaxes of stars measured by VLBI.

The other interesting astrometric comparison is for the only FK5 star, HR5110, that we observe. The FK5 is regarded as the best optical astrometric catalog based on ground-based observations. The comparison summarised below shows that the differences FK5 minus VLBI for HR5110 in right-ascension α, declination δ and proper motion μ_α and μ_δ are smaller than or equal to the FK5 uncertainties $\sigma_\alpha(FK5)$, $\sigma_\delta(FK5)$, $\sigma_{\mu_\alpha}(FK5)$, $\sigma_{\mu_\delta}(FK5)$:

$|\Delta\alpha(FK5 - VLBI)| = 0.65$ millisec $= 1/2 \times \sigma_\alpha(FK5)$

$|\Delta\delta(FK5 - VLBI)| = 19.7$ milliarcsec $= 1 \times \sigma_\delta(FK5)$

$|\Delta\mu_\alpha(FK5 - VLBI)| = 0.000017$ millisec/year $= 1/2 \times \sigma_{\mu_\alpha}(FK5)$

$|\Delta\mu_\delta(FK5 - VLBI)| = 0.00006$ milliarcsec/year $= 1/2 \times \sigma_{\mu_\delta}(FK5)$

Figure 1 : Post-fit residuals for the complete astrometric solution of σ CrB

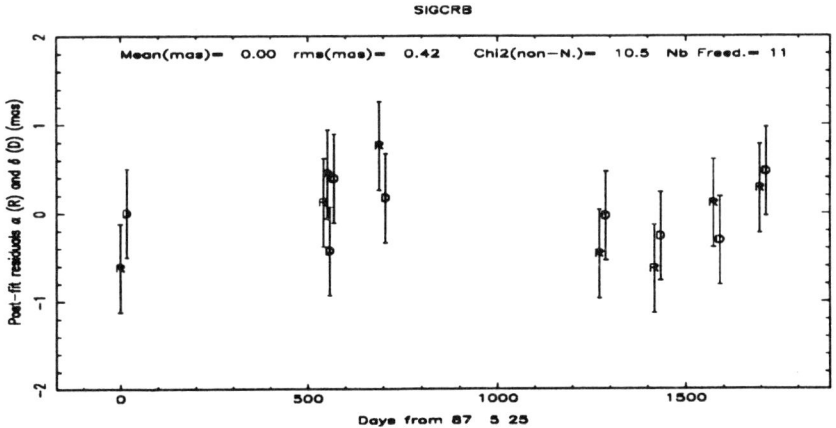

Acknowledgements: We are grateful to the participating VLBI observatories. The research described in this report was carried out, in part, at the Jet Propulsion Laboratory, Caltech, under contract with the National Aeronautics and Space Administration.

References

[LNPM88] Lestrade, J-F, Niell, A.E., Preston, R.A., Mutel, R.L., 1988, *Astron. J.*, 96, 1746-1754.

[LRWNPP90] Lestrade, J-F, Rogers, A.E., Whitney, A.E.E., Niell, A.R., Phillips, R.B., Preston, R.A., 1990, *Astron. J.*, 99, 1663-1673.

[LPPG92] Lestrade, J-F, Phillips, R.B., Preston, R.A., Gabuzda, D.C., 1992, *Astron. Astroph.*, 258, 112-115.

High-accuracy Proper Motions of Millisecond Pulsars from the Nançay Timing Program

I.Cognard [*]   G.Bourgois [*]   D.Aubry [†]   B.Darchy [†]
J.-P.Drouhin [†]   J.-F.Lestrade [*]

Abstract

A timing program of millisecond pulsars has been initiated at Nançay observatory in 1988. The results of the first campaign on the pulsar PSR1937+214 is presented. The timing accuracy is $0.4\mu s$ and is similar to Arecibo. The proper motion has been determined with a formal precision of 0.03 mas/year. However, on a short time span, uncorrected DM variations systematicaly biased this parameter.

1. Introduction

Since December 1988, a high-accuracy timing program of the two millisecond pulsars PSR1937+214 and PSR1821-247 is conducted at Nançay. Observations are made at 21 cm with the meridian-transit radiotelescope characterized by a system temperature of 45 K and by a surface equivalent to a 94 meters dish. Pulsar signals are integrated in four channels (of 8 MHz for PSR1937+214) using the technique of the swept local oscillator. The offset between the Paris Observatory UTC and the Nançay Rubidium UTC is monitored daily through measurements of synchronization TV signal (precision of 40ns). Since January 1992, we are conducting timing observations at 1.4 and 1.7 GHz to derive the Dispersion Measure (DM) variations along the lines of sight of the 2 pulsars. First, we present the results of our timing program on the pulsar PSR1937+214 and a comparison with the Arecibo-Princeton program. Second, we discuss the effect of uncorrected DM variations on the determination of the proper motion of this pulsar.

---

[*]DERAD , Observatoire de Meudon, 92195 MEUDON Principal Cedex, FRANCE.
[†]Nançay , Observatoire de Meudon, 92195 MEUDON Principal Cedex, FRANCE.

2. Results of the Nançay timing observations of 1937+214

From December 1988 to October 1991 we have measured 282 Times Of Arrival (TOA). Each TOA corresponds to 1 hour of coherent integration of pulses at both senses of circular polarization. The analysis of these timing data has been carried out with our software ANTIOPE to solve for the pulsar parameters in Table 1. In this analysis, we have included the DM variations determined by Arecibo for this pulsar and removed 2 months of data around 1989 October 15 where a Refractive Scintillation Event is clearly visible (discussed elsewhere). The Figure below shows our TOA residuals for 3 years which are characterized by a r.m.s. of $0.39\mu s$ similar to the r.m.s. of the Arecibo program. A comparison between the pulsar parameters determined with the 2 programs indicates small discrepancies. The largest discrepancy is 5σ (σ=formal uncertainty) for the period, while it is 2σ for the period derivative, 0.5σ for μ_α, 2σ for μ_δ, 0.5σ for α and δ. The discrepancy in the period is likely to be caused by different Atomic Time Scales used in the 2 analysis (TAI and TA(BIPM)).

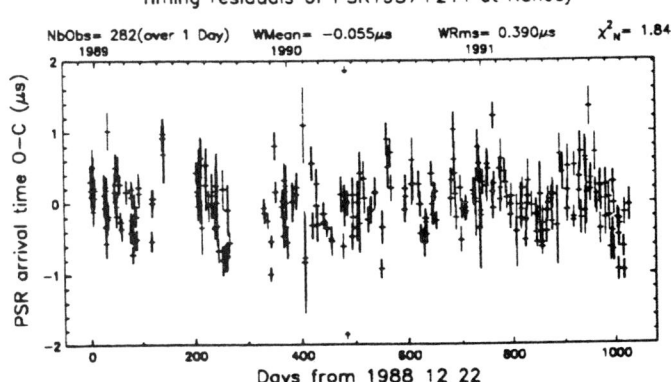

Period=0.001557806472448806(3) s
Pdot=10.51199(2) x 10^{-20} s/s
α(J2000)=19h 39m 38.558721(2)s
δ(J2000)=21d 34' 59.13741(3)"
μ_α(J2000)=-0.34(3) mas/year
μ_δ(J2000)=-0.55(3) mas/year
Adopted π = 0.25714mas
Adopted origin JJ=2447900.0
Adopted ephemeris DE202
TOA post-fit r.m.s. = $0.39\mu s$

Table 1: Parameters of PSR1937+214.

3. Effect of DM variations on proper motion determination

The high-accuracy proper motion of PSR1937+214 measured by timing data can be biased by the variations of DM along the line of sight. We have made two solutions with and without DM corrections for our data from December 1988 to January 1991. Discrepancies between these two analysis are -0.16 mas/year (4σ) in μ_α and 0.08 mas/year (2σ) in μ_δ. Indeed, on such a short time scale the non-linear variations of DM can be similar to the characteristic signature of proper motion in TOA ($t \times \sin(t)$). However, proper motion will separate from DM variations with an increasing span of observations.

Acknowledgments: We are very grateful to F.Biraud of Meudon Observatory and to the Laboratoire Primaire Temps Fréquences (LPTF) of the Observatory of Paris, especially to G.Fréon.

Towards VLBI Determinations of Pulsar Parallaxes and Proper Motions

R. M. Campbell [*]   N. Bartel [†,¶]   I. I. Shapiro [‡]   M. I. Ratner [‡]
R. J. Cappallo [§]   A. R. Whitney [§]   W. H. Cannon [¶,†]
W. T. Petrachenko [‖]   J. Popelar [**]   T. M. Eubanks [††]

Abstract

We have begun a long-term program to measure the parallax distances and proper motions of a set of about ten pulsars. Astrometric VLBI observations of a pulsar and one or more extragalactic phase-reference sources provide a means to determine their differential positions with submilliarcsecond accuracy, corresponding to distance uncertainties for some pulsars as small as 5%.

Interleaved VLBI observations of pulsars and extragalactic sources nearby on the sky can yield trigonometric parallax distances and proper motions of the pulsars, and hence constraints on their spatial distribution. Coupled with pulsar age estimates, these distances and proper motions should also allow useful constraints to be placed on the birth places of pulsars and perhaps on their birth rate. Moreover, distances combined with observed pulsar dispersion measures yield the average electron density along the lines of sight to each pulsar and hence characteristics of the spatial distribution of free electrons in the interstellar medium. Such information for the solar neighborhood provides a local calibration of the dispersion-based distance scale which is complementary to the galactic-scale calibration derived from HI-absorption measurements of more distant pulsars. Furthermore, parallax distances, coupled with angular-diameter determinations and spectrometric scintillation observations, can be used to characterize the turbulence in the interstellar medium. So far, trigonometric parallaxes have been measured for only four pulsars: 0823+16 (2.8 ± 0.6 mas) and 0950+08 (7.9 ± 0.8 mas) [GTW86], 1451-68 (2.2 ± 0.3 mas) [BMK90], and 1929+10 (with discrepant values of 21.5 ± 0.8 mas [SLA79] and < 4 mas [BaS82]).

We have made six-station, 13 cm VLBI observations at two epochs (July 1991 and May 1992) of fifteen pulsars, of which nine were detected (pulsars 0329+54, 0355+54, 0823+16,

---

[*] Department of Astronomy, Harvard University, Cambridge, MA 02138 USA
[†] Department of Physics and Astronomy, York University, North York, ON M3J 1P3 Canada
[‡] Harvard-Smithsonian Center for Astrophysics, Cambridge, MA 02138 USA
[§] Haystack Observatory, Westford, MA 01886 USA
[¶] Institute for Space and Terrestrial Science, North York, ON M3J 1P3 Canada
[‖] Geological Survey of Canada, Ottawa, ON Canada
[**] Geodetic Survey of Canada, Ottawa, ON Canada
[††] U. S. Naval Observatory, Washington, DC 20392 USA

0950+08, 1133+16, 1642-03, 1929+10, 1933+16, and 2021+51). In addition, in earlier three-station VLBI observations (September 1988), we detected eight of these nine pulsars. At each epoch, we observed each pulsar and one or more nearby extragalactic reference source(s) for ~ 6 hr. The resulting u-v coverage should provide sufficient redundancy to allow us to eliminate the ambiguity in the difference fringe phase. Use of two reference sources for each pulsar, where feasible, allows testing of the consistency of our ionospheric corrections by determining whether there appears to be relative proper motion between the reference sources. Analysis of portions of the July 1991 data for PSR 2021+51 and its two reference sources has shown that the phase delays for each of the three sources appear to be of sufficient quality to allow phase connection. A total of about seven epochs of observations of each pulsar well distributed throughout the seasons will allow us to solve for position, proper motion, and parallax with sufficient redundancy to rule out ambiguity errors in the position determinations.

For the first two epochs, the phase reference sources we used were primarily selected from the Preston et al. survey [PMW85]. During the July 1991 observations, we also conducted a search for better-situated reference sources, principally from sources in the Gregory and Condon survey [GrC91] and sources listed as non-detections in the Preston et al. survey. For all but one of the nine detected pulsars, this search resulted in our detecting at least one new reference source located $< 2°$ from the pulsar; for all but three pulsars we have at least one reference source (new or old) located well less than 1° away. Incorporation of these new reference sources into the May 1992 and subsequent epochs should yield more accurate determinations of the trigonometric parallax of the corresponding pulsars and allow distance determinations with standard errors for some of the pulsars as small as 5%. We gratefully acknowledge support from NASA grant NGT-50663 (RMC), NSF grant AST89-02087 (NB), and NSF grants AST91-44439 and AST92-01555 (Haystack VLBI correlator operations).

References

[BaS82] Backer D.C., Sramek R.A., *Apparent Proper Motions of the Galactic Center Compact Radio Source and PSR 1929+10*. Ap. J., Vol. 280 (1982), pp. 512–519.

[BMK90] Bailes M., Manchester R.N., Kesteven M.J., Norris R.P., Reynolds J.E., *The Parallax and Proper Motion of PSR 1451-68*. Nature, Vol. 343 (1990), pp. 240–241.

[GrC91] Gregory P.C., Condon J.J., *The 87GB Catalog of Radio Sources Covering $0° < \delta < 75°$ at 4.85 GHz*. Ap. J. Suppl., Vol. 75 (1991), pp. 1011–1291.

[GTW86] Gwinn C.R., Taylor J.H., Weisberg J.M., Rawley L.A., *Measurement of Pulsar Parallaxes by VLBI*. A. J., Vol. 91 (1986), pp. 338–342.

[PMW85] Preston R.A., Morabito D.D., Williams J.G., Faulkner J., Jauncey D.L., Nicolson G.D., *A VLBI Survey at 2.29 GHz*. A. J., Vol. 90 (1985), pp. 1599–1641.

[SLA79] Salter M.J., Lyne A.G., Anderson B., *Measurements of the Trigonometric Parallax of Pulsars*. Nature, Vol. 280 (1979), pp. 477–478.

MERLIN – Phase 2

P. N. Wilkinson *

The MERLIN (Multi–Element Radio–Linked Interferometer Network), consists of six remotely operated telescopes working in conjunction with either the 76m Lovell or the 26m (equivalent) Mk2 telescopes at Jodrell Bank. A detailed description of the original MERLIN system has been given by [Tom86]. In order to maintain its competiveness throughout the next decade the UK SERC provided funding for major upgrades to the system, including a new high–performance 32m telescope at Cambridge. The "Phase 2" instrument is now in operation at 5 GHz.

1. The Upgrades

The receivers: Cryogenically cooled receivers are available at 5 GHz ($T_{sys} \sim 33K$), L–band ($T_{sys} \sim 37K$), and soon at 22 GHz ($T_{sys} \sim 100K$). The L–band system system covers the ranges 1370–1430 and 1570–1750 MHz (limited by interference) and it is intended to make use of Multi–Frequency–Synthesis or MFS [CCW90] to increase the fidelity of images of large (tens of arcsecs) sources.

The microwave links: In order to increase the instantaneous bandwidth transmitted within the 30 MHz channel width, commercial links have been modified from FM to AM (independent double sideband) operation. Two 15 MHz channels can now be transmitted back from each outstation; usually one for each hand of circular polarisation.

The correlator: A new 2 × 2 bit XF correlator has been built based on a custom VLSI chip. All pairs of circularly polarised inputs from the home and out–stations are correlated, enabling full polarisation information to be obtained. At the full bandwidth data from 32 delay channels per baseline are produced for each of the four polarisation combinations. For line work recirculation will be used to increase the number of channels to 2048 per baseline over a bandwidth of 250 kHz.

The 32m telescope at Cambridge: The telescope (prime contractor MAN GHH) is located at the MRAO site in Cambridge, well to the East of the other telescopes. The new telescope not only increases the maximum baseline to 218 km from 135 km, it also enhances the u, v coverage, particularly at low declinations. In addition the new telescope provides baselines which link the MERLIN with the European VLBI Network (EVN).

---

*University of Manchester, NRAL, Jodrell Bank, Macclesfield, Cheshire, SK11 9DL, UK.

Table 1. The performance of MERLIN at its prime operating frequencies

Frequency (GHz)	Thermal Noise (μJy beam^{-1})	Resolution (milliarcsec)	Jodrell telescope
1.3–1.8	60	150	Mk2(25m)
	30	150	Lovell (76m)
5.0	60	50	Mk2
22.0	250–300	11	Mk2

The optical design was done by NRAL staff. Following the Australia Telescope practice the receivers are mounted on a rotatable carousel in the vertex cabin. Following NRAO practice low frequency (< 1 GHz) operation will be at prime focus with the feeds mounted in front of the sub–reflector. The prime and sub–reflectors are both shaped to optimise the aperture efficiency in Cassegrain operation (for frequencies above 1.3 GHz) and great care was taken to minimise the amount of scattered radiation entering the feed. Measurements at 5 GHz show that the aperture efficiency is $80 \pm 1\%$ (i.e. 0.23 K Jy^{-1}) and that the "on telescope" contribution to T_{sys} (including the contributions of the atmosphere and the CMB) is only 10K. The system noise equivalent flux density at 5 GHz is currently ~ 140 Jy but sub–100 Jy performance can be anticipated in the near future. The pointing accuracy of the telescope has been measured to be better than 5 arcsec rms in precision conditions and the telescope is expected to work well at frequencies up to, and possibly beyond, 50 GHz.

2. The Upgraded Performance

The resolution and the sensitivity of MERLIN for a 12 hour synthesis is given in Table 1. The 5 GHz sensitivity quoted is that actually achieved in maps made with natural weighting and with $\sim 25\%$ of the time spent off source in the phase referencing mode (see section 3). The 1.66 and 22 GHz sensitivities are scaled from that at 5 GHz based on telescope performance data and anticipated system temperatures. As a sparse array MERLIN is best suited for observations of compact sources. It provides higher resolution than the VLA and for imaging OH and H_2O masers this full resolution advantage accrues. For studies of continuum sources MERLIN and the VLA are complementary, providing comparable resolution at widely separated frequencies. MERLIN's sensitivity disadvantage is compensated by the fact that optically thin synchrotron sources are stronger at lower frequencies.

3. Observing with MERLIN

The array is operated at a single frequency for periods of 6 to 12 months; the first observing period was at 5 GHz. External phase calibration or phase referencing is standard. At 5 GHz the target source is observed for 8 mins and the reference source for 2 mins; a shorter cycle time is envisaged at L–band. References sources are drawn from the VLA calibrator list and from the new catalogue of [PBWW91]. At 5 GHz satisfactory phase calibration can usually be achieved as long as the reference source is within 10° of the target (reducing to $\sim 3°$ at L–band) and is stronger than 50 mJy. Most continuum observations are made in "line mode" to allow for wide–field mapping. All Stokes parameters are observed as a matter of routine and the data are principally reduced using the NRAO AIPS software package.

As of July 1992 absolute positions could be determined to ~ 50 mas while relative positions for a 2° target–reference throw could be determined to ~ 2 mas. Phase corrections for UT1–UTC and polar motion variations, obtained from IERS predictions, are applied on–line. A tropospheric model whose input parameters are the local pressures at each site is also used to correct the interferometric phases. Currently no correction for the ionosphere is implemented but at 5 GHz this does not have a major effect on the astrometric performance of the array. We expect to achieve significantly better astrometric accuracies in the future as the baselines and *a priori* phase corrections are refined.

The Phase 2 upgrades have already opened up new areas of study to MERLIN – in particular radio stars, Seyfert galaxies and gravitational lenses – in addition to enhancing its unique role in the study of powerful extragalactic sources and galactic OH and H_2O masers. The following papers in this volume are based on new MERLIN results on continuum sources at 5 GHz: Davis et al. (stars); Muxlow et al. (M82 SNRs); Pedlar et al. (Seyfert galaxies), Patnaik et al. (gravitational lenses), Ludke et al. (polarisation of CSS)

MERLIN is a UK 'National Facility' open to proposals from the world–wide community. Applications for observing time are considered by a Time Allocation Committee linked with the SERC PATT committee structure. Applications should be sent to the Director, NRAL; the deadlines are September 30 and March 30 each year. Throughout most of 1993 MERLIN will observe at L–band. For further information contact Dr. T.W. Muxlow (SPAN 19739::TBM; FAX 44–(0)477-71618.)

References

[Tom86] Thomasson P., *QJRAS 27 (1986) 413–431*.

[PBWW91] Patnaik A. R. , Browne I.W.A., Wilkinson P.N. and Wrobel J.M., *MNRAS, 254, 381-385*.

[CCW90] Conway J.E. , Cornwell T.J. and Wilkinson P.N., *MNRAS, 246, 490–509*.

VLBA CAPABILITY AND STATUS

R. C. Walker *

Abstract

The NRAO Very Long Baseline Array (VLBA) is a major new facility for subarcsecond radio astronomy. It is scheduled for completion around the end of 1992. All 10 antennas have been constructed and 9 are in operation. The correlator has been moved to the Array Operations Center in Socorro, New Mexico and is under intensive testing. It should begin processing astronomy experiments at the end of 1992. The VLBA has supported VLBI Network observations for several years and a number of pure VLBA projects have been done using the Mark II system or the VLBA recorders in their Mark III mode. Early results are very promising for the ultimate performance of the array.

1. The Instrument

The VLBA is a dedicated instrument for Very Long Baseline Interferometry. The 10 antennas are distributed about the United States in a configuration designed to optimize the u-v coverage. Baselines between 200 and 8000 km are covered, which provides resolutions up to 0.2 milliarcseconds at 43 GHz. The shorter baselines, and hence the highest concentration of antennas, are near the VLA for optimal joint observations and to allow for a future project to fill the gap in the range of baselines covered by the two instruments. The antennas are 25 m in diameter and of an advanced design that allows full performance at 43 GHz and useful performance at 86 GHz. In particular, the antenna structures, which will never be replaced, are adequate for full performance at 86 GHz. The antennas are designed for remote operation from the Array Operations Center (AOC) in Socorro, New Mexico. Local intervention is only required for changing tapes, for maintenance, and for fixing problems.

The VLBA is outfitted for observations in 9 frequency bands at approximately factor of two intervals between 327 MHz and 43 GHz. The receivers at 1.4 GHz and above contain cooled HFET amplifiers. The low frequency receiver is a room temperature GaAsFET. The cooled receiver for each band is in a separate dewar mounted directly on the feed to minimize noise contributions from waveguides etc. All receivers

---

*National Radio Astronomy Observatory, Socorro, NM, USA. The NRAO is operated by Associated Universities Inc. under cooperative agreement with the National Science Foundation.

cover both right and left circular polarization. There is a dichroic/ellipsoid system that allows simultaneous observations at 4 and 13 cm, primarily for geodesy and astrometry.

The frequency standard at each VLBA site is a hydrogen maser manufactured by Sigma Tau Corporation. The recording system is based on a Metrum (formerly Honeywell) longitudinal instrumentation tape recorder that has been heavily modified by the Haystack Observatory. The recorder is similar to the one used in the Mark III and future Mark IV VLBI systems. There are two drives at each VLBA station to allow about 24 hours of recording at 128 Mbits s^{-1} between required visits to the station for tape changes. Higher data rates up to 512 Mbits s^{-1} are supported for special cases but cannot be sustained for logistical reasons. The tapes are 16 microns thick with about 5.5 km of tape on a 14 inch (35.6 cm) reel.

The VLBA correlator is now located at the AOC in Socorro. It is able correlate up to 8 channels from each of up to 20 sites. A total of up to 2048 spectral channels can be provided for each baseline. The correlator is of a novel design, pioneered by the Nobeyama Radio Observatory in Japan, in which each bit stream is Fourier transformed to a spectrum before cross correlation (the "FX" architecture). Output data will be archived on DAT tapes while the input tapes are recycled for more observing shortly after correlation. Users will receive their correlated data in FITS format on any of several media including DAT and EXABYTE tapes.

VLBA postprocessing is done in the original AIPS system for now. Considerable software development for VLBI in AIPS is still going on. Astrometric and geodetic processing will be done primarily in the system developed at the Goddard Space Flight Center. Over the next few years, the postprocessing will shift to the AIPS++ system as that system acquires the necessary capabilities. The in-house computing for the VLBA is done mainly on workstations of the SUN IPX and IBM RS/6000-560 classes.

2. Status

The VLBA is nearly complete and will transition from a construction project to an operational user facility in 1993. Currently 9 of the 10 antennas are operational and the correlator is being debugged. Full support will continue to be given to global VLBI Network sessions and pure VLBA science observations will increase rapidly. The ramp-up to full time operation is expected to occur during the first half of the year as the correlator checkout is completed. The last antenna, located on Mauna Kea in Hawaii, is also expected to begin operation during the first few months of the year. By mid 1993, most of the residual construction activities should be complete and the Array should be fully functional.

Proposals for the use of the VLBA are already being accepted and considerable interest is evident. Already at the June 1992 proposal deadline, with full VLBA operation still many months away, proposals for observations involving just the VLBA outnumbered global Network proposals by 60%. Early observations with the partial VLBA have already demonstrated its capabilities in many areas including high dynamic range, ease of use at high frequencies, receiver agility, and scheduling flexibility. Several results of this sort are presented elsewhere in this volume. The very important ability of the Array to serve astronomers who are not specialists in VLBI observational techniques has begun to have effect, although this area still needs much work.

Two especially encouraging VLBA results that were presented at this meeting are those by Wrobel and Conway and by Diamond. Wrobel and Conway (these proceedings) present an image of Mrk 501 that has a dynamic range of 2400 (ratio of peak to off-source rms). This is about as good as is achieved by even the best Network VLBI observations with 15-20 antennas and it was obtained using Mark II data from only 4 VLBA antennas and 1 VLA antenna. The imaging involved only 3 iterations of phase self-calibration and no amplitude self-calibration — far less processing than is typical for Network observations. This demonstrates, as MERLIN has before, the value of good a-priori calibration and holds great promise for what can be done with 10 antennas and the full VLBA bandwidth.

Diamond (in preparation) presents results of spectral line synthesis imaging observations of SiO masers in U Her at 43 GHz. Usually VLBI observations at mm wavelengths involve a tremendous effort and the results are often marginal. The VLBA observations, on the other hand, presented little more difficulty than spectral line observations at other frequencies such as 1.6 and 22 GHz. The promise that the VLBA will make VLBI observations at 43 GHz routine is already being demonstrated.

A major area of technical research on the VLBA is "phase connection". If it is possible to calibrate VLBA phases using calibrator sources, in much the same way as is done on connected interferometers, the range of flux densities that can be observed will be extended by nearly two orders of magnitude. Successful VLBI observations of this sort have been made at optimal wavelengths (*cf.* Lestrade *et al.* A&A, 258, 112) and efforts are being made to determine under what conditions these methods work and to determine optimum observing strategies (*cf.* Conway and Vermeulen, these proceedings).

Finally, the VLBA will be used as a geodetic and astrometric instrument. Early results by NASA's Crustal Dynamics Project include some of the best geodetic data ever. During a period of more than 2 weeks in 1989, geodetic observations were done almost continuously. The day to day rms fluctuations in the resulting position of Pie Town, when holding the positions of the other antennas (not VLBA) fixed, were only 1.5 mm in the horizontal coordinates and 4mm in the vertical. Close cooperation continues with the geodetic community to help insure that the VLBA will be a very capable instrument for high accuracy geometric work.

The Southern Hemisphere VLBI Experiment (SHEVE)

Robert A. Preston and the SHEVE team*

The SHEVE team:

David L. Jauncey, [†]	Robert A. Preston, [*]	John E. Reynolds, [†]
Edward A. King, [‡]	David L. Meier, [*]	Anastasios K. Tzioumis, [†]
Dayton L. Jones, [*]	David W. Murphy, [*]	George D. Nicolson, [††]
Richard H. Ferris, [†]	Marco E. Costa, [§]	Russell G. Gough, [†]
Donald W. Hoard, [*]	Shaun W. Amy, [¶]	David G. Blair, [§]
Duncan Campbell-Wilson, [¶]	Roger W. Clay, [‖]	Phillip Edwards, [‖]
Carl R. Gwinn, [††]	Phillip A. Hamilton, [‡]	Benjamin Johnson, [*]
Paul A. Jones, [**]	Athol J. Kemball, [††]	Earl T. Lobdell, [*]
James E. J. Lovell, [‡]	W. Bruce McAdam, [¶]	Peter M. McCulloch, [‡]
Ray P. Norris, [†]	Eric Perlman, [*]	Malcolm W. Sinclair, [†]
Mary E. St. John, [*]	Lyle Skjerve, [*]	Robin M. Wark, [†]
	Graeme L. White [**]	

Abstract

Radio telescopes at eight sites in Australia and one in South Africa operate as a VLBI array during several periods each year. Several improvements to the array during the last few years have significantly improved its sensitivity, frequency range, and u-v coverage.

---

[*] Jet Propulsion Laboratory, California Institute of Technology, Pasadena, CA 91109, U.S.A
[†] Australia Telescope National Facility, Epping, New South Wales 2121, Australia
[‡] University of Tasmania, Hobart, Tasmania, 7001, Australia
[††] Hartebeesthoek Radio Astronomy Observatory, Johannesburg, South Africa
[§] Department of Physics, University of Western Australia, Nedlands, Australia 6009, Australia
[¶] University of Sydney, Sydney, New South Wales 2006, Australia
[‖] Department of Physics, University of Adelaide, Adelaide, South Australia 5001, Australia
[‡‡] University of California, Santa Barbara, California 93106, USA
[**] University of Western Sydney, Sydney, New South Wales, Australia

The first VLBI imaging experiment in the southern hemisphere took place in 1982 [P89]. In the last few years a significant effort has been undertaken to enhance the southern hemisphere array [J91]. These programs are known as the Southern Hemisphere VLBI Experiment, or SHEVE.

The array now usually operates for one to two week periods every four months. The telescopes in Australia that have frequently participated in these observations include the NASA tracking antennas at Tidbinbilla, the Parkes Radio Telescope, the Mount Pleasant Observatory at Hobart, two individual antennas of the Australia Telescope at Culgoora and Mopra, a Landsat Station at Alice Springs, and an ESA tracking antenna at Perth. The Hartebeesthoek Radio Astronomy Observatory in South Africa usually joins the observations, and northern hemisphere telescopes are sometimes added to improve u-v coverage for the more northerly southern hemisphere sources. Since the early SHEVE experiments in 1982, the sites at Culgoora, Mopra, and Perth have been added to the array and a larger, more versatile telescope has been erected at Hobart. An additional Australian telescope, the Molonglo Synthesis Telescope, has occasionally participated in observations at 0.84 GHz. A map showing the principal telescope locations appears in Figure 1 and an example of the u-v coverage achievable is shown in Figure 2.

The observing frequencies supported by most of the SHEVE array include 1.7, 2.3, 5.0, and 8.4 GHz, although not all telescopes support all frequencies. The present recording systems include Mark II at all sites and Mark III at four sites, with data being correlated at the JPL/Caltech Block II correlator. By 1994, S2 recorders will be implemented in the array and correlation will be done at a new ATNF S2 correlator.

Several new results from the SHEVE array are presented at this conference including observations and images of the new strong Einstein ring 1830-211 (see Jauncey *et al.* and Jones *et al.*), the nucleus of Centaurus A (see Meier *et al.*), southern galaxies and quasars (see Murphy *et al.*), the peculiar radio source MSH04-71 (see Reynolds *et al.*), the first interstellar VLBI speckles (see Gwinn *et al.*), a sample of peaked spectrum sources (see King *et al.*), and methanol masers (Norris).

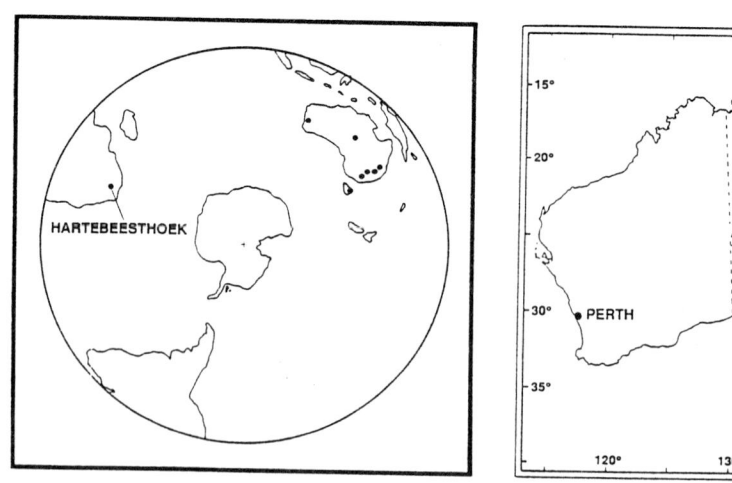

Figure 1.   The SHEVE array

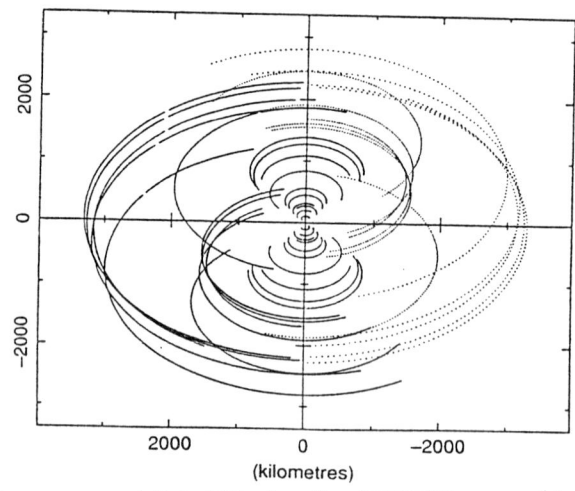

Figure 2.   The UV-coverage at 2.3 GHz for the SHEVE array (Australia only) at $\delta = -42°$

References

[P89] Preston, R. A. et al., *The Southern Hemisphere VLBI Experiment*. Astronomical Journal, Vol. 98 (1989), pp. 1-26.

[J91] Jauncey, D. L., *VLBI in Australia - A Review*. Australian Journal of Physics, Vol. 44 (1991), pp. 785-803.

Millimetre VLBI capability status

Lars B. Bååth *

Abstract

The development of new receiver and data reduction techniques have now made VLBI at mm wavelengths possible. This contribution discusses the capability of the present and future VLBI networks at λ1 and 3mm and compares with radio interferometers at other wavelength regimes.

Very Long Baseline Interferometer observations at mm-wavelengths has now made it possible to increase the resolution from sub-arcsecond to sub-milliarcsecond. This opens up completely new opportunities to study extragalactic objects. Figure 1 demonstrates the significance of these observations by comparing the resolution achieved with the VLA, MERLIN, and global VLBI at λ6cm to global VLBI at λ3mm. Note that λ3mm VLBI compares to λ6cm VLBI as the latter compares to MERLIN. The pioneering work for mm-VLBI is done in a collaboration between groups at Onsala, Nobeyama, Caltech, Berkeley, Haystack, Kitt Peak and FCRAO has produced a new, and effective instrument which already has produced, and will in the future lead to, new discoveries. The resolution we can achieve this VLBI instrument is $40\mu as$ at λ3mm and $15\mu as$ at λ1m. This is *better* than the resolution achievable with the currently proposed satellite VLBI projects, and has the extra advantage of looking into the core region which is presumably optically thick at cm wavelengths.

VLBI at λ3mm started in 1982 and has during these ten years produced a number of papers (e.g. [MR88], [RMM83], [WBC88] and references therein). These earlier experiments and papers concentrated on finding fringes and develop the receiver and recording techniques. In 1988 we tokk the first steps towards a full mapping session, using for the first time a global fringe fitting technique in order to detect weaker fringes. The technique and the results from these first mapping sessions (1988 and 1989) are discussed in [BRI92], a later epoch of 1990 is discussed in [B92].

The antennas used in our network is described in table 1. It is important to note that unlike VLBI observations at cm wavelengths we are at mm wavelengths much more limited by the troposphere. The relevant systemtemperature is in our case the

---

*Onsala Space Observatory, S-439 92 Onsala, Sweden.

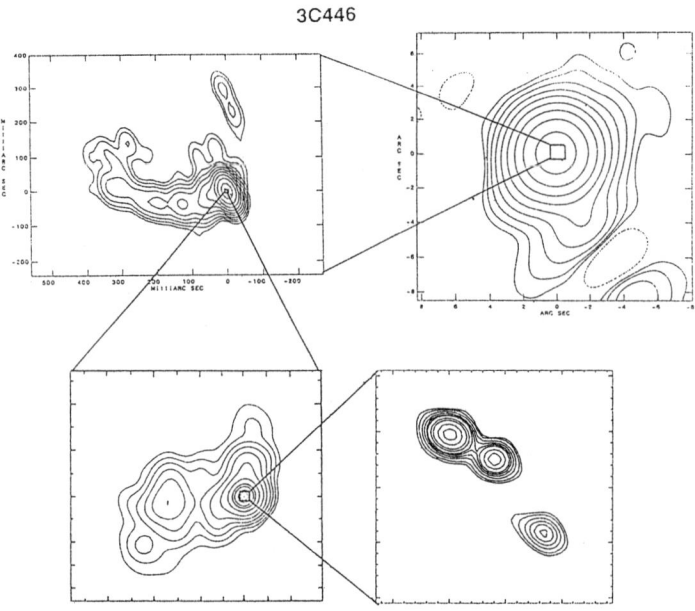

Figure 1. 3C446 observed with interferometers: upper right: VLA at $\lambda 22$cm (tick marks at 2as); upper left: MERLIN at $\lambda 6$cm (tick marks at 100mas); lower left: VLBI at $\lambda 6$cm (from WCU90:tick marks at 1mas); and lower right: VLBI at $\lambda 3$mm (tickmarks at 20μas). (VLA and MERLIN maps are courtesy of C.Akujor)

equivalent systemtemperature *outside* the atmosphere, which will make the *altitude* of a station very important. Note for example that the 15m antenna at SEST has a better sensitivity than the 20m at Onsala because it is situated at an altitude of 2700m compared to 30m. The detection limit of the arrays in the table is 200mJy at $\lambda 3$mm and 800mJy at $\lambda 1$mm using a recording bandwidth of 112MHz. The detection limits refer to a 7 sigma detections in 6.5mins, including expected coherence losses. It is obvious that a large number of Active Galactic Nuclei are available for us to observe with such sensitivities. VLBI at mm wavelengths is still very much a development project but we hope to be able to open $\lambda 3$mm VLBI to external users within the next few years. At that time scale we also hope to develop $\lambda 1$mm VLBI and start test experiments at $\lambda 0.7$mm. Developments are underway to increase the recording bandwidth to 1GHz at Nobeyama (K4) and Haystack (MkIV). Such an increased bandwidth will result in a much needed increase in sensitivity, especially at wavelengths $\leq \lambda 1$mm.

Table 1. Available antennaes for $\lambda 3mm$ VLBI. S_{sys} refer to the systemtemperature in Jansky at the top of the atmosphere.

Antenna	diameter (m)	eff.	$\lambda 3mm$ S_{sys}	experiment	eff.	$\lambda 1mm$ S_{sys}	experiment
Nobeyama	45	0.30	1500	88,89,90	0.10	10000	planned 93
Onsala	20	0.48	5400	88,89,90,92			
Owens Valley	3×10.4	0.60	2700	88,89	0.45	9600	89,90,92
Hat Creek	3×6.1	0.60	13000	88,89,90			
Kitt Peak	12	0.60	8000	88,89,90	0.40	36000	89,90
Quabbin	14	0.50	8700	88			
SEST	15	0.60	3900	90	0.50	12400	90,92
JCMT	15	0.60	3700	planned 93	0.60	8700	92
Pico Veleta	30	0.50	1200	92	0.27	5600	planned 93
Haystack	37	0.15	4200	planned 93			
Effelsberg	60	0.10	2500	92			
Metsähovi	14	0.30	18000				
Quing Hai	14	0.30	15000				

References

[B92] Bååth, L.B., *New millimetric VLBI results.* proceedings of IAP no.7, Paris (1992), in press

[BRI92] Bååth, L.B. et al., *VLBI observations of active galactic nuclei at 3 mm.* A&A, Vol. 257 (1992), pp. 31-46.

[MR88] Moffet, A.T. & Readhead, A.C.S., *Observations of 3C273 at 3mm Wavelength.* in "Superluminal Radio Sources", eds. Zensus, A. & Pearson, T.J., Cambridge Univ. Press (1988), pp. 32-33.

[RMM83] Readhead, A.C.S. et al., *Very long baseline interferometry at a wavelength of 3.4mm.* Nature, Vol. 303 (1983), pp. 504-506.

[WCU90] Wehrle, A.E., Cohen, M.C. & Unwin, S.C., *The Caltech survey of strong, highly variable sources.* in Parsec Scale Radio Jets, eds. Zensus, A. & Pearson, T.J., Cambridge Univ. Press (1990), pp. 49-54.

[WBC88] Wright, M.C.H. et al., *Evolution of the sub-milliarcsecond nucleus of 3C84 at 100GHz.* ApJ, Vol. 329 (1988), pp. L61-L64.

Space VLBI Project VSOP

Makoto INOUE *

Abstract

A space VLBI satellite is planed to be launched in 1995 by the Institute of Space and Astronautical Science (ISAS), Japan. The satellite is the main element of the VLBI Space Observatory Program (VSOP), which is a worldwide collaborative project involving satellite operations, link stations, and ground radio telescopes. The basic parameters and management plan of VSOP project are reviewed.

1. Introduction

The VLBI Space Observatory Program (VSOP) was initiated as a cooperative project between the Institute of Space and Astronautical Science (ISAS) and the Nobeyama Radio Observatory (NRO), Japan, in the late 1980's. The VSOP satellite has an 8-m deployable radio telescope and three radio astronomy receivers, and will be launched in 1995. It is expected to have a lifetime of more than three years. As the VSOP is a space VLBI imaging mission, the worldwide support of link stations and ground observing telescopes is needed. Some detailed descriptions are in the proceedings of the International VSOP Symposium (1992).

2. Orbit

The orbit of the VSOP satellite is elongated (See Table 1) and precesses so that good UV coverage is expected over a wide area of the sky. As the orbit is not synchronous, about 6 hours per orbit, the UV coverage evolves day by day. Since the apogee is not very high, VSOP is good for imaging rather than high resolution. There are, however, several constraints on the satellite. For instance, low declination sources may not have good UV coverage due to the avoidance angle from the Sun. The final situation will depend on the exact time of the launch.

Table 1. Orbit Parameters

Apogee Height	20,000 km
Perigee Height	1,000 km
Period	$\sim$ 6 hrs
Inclination	$\sim 31°$

*Nobeyama Radio Observatory, Nobeyama, Minamisaku, Nagano 384-13, Japan.

3. Radio Telescope

The main reflector is hexagonal in shape. Its surface is shaped by a tension truss structure of cable nets, and six extendible masts which tension the surface. The accuracy of the mesh surface is 0.5 mm rms. The aperture is 50 m^2 and the aperture efficiency is 40%, 56%, and 44% at 1.6, 5, and 22 GHz, respectively.

4. Receivers

The VSOP satellite has three receivers at 1.6, 5, and 22 GHz with uncooled front-ends and left-hand circular polarization feeds. Table 2 summarizes the receiver and system temperatures for each frequency, together with frequency coverage. Two IF channels with 16- or 32-MHz bandwidth are available so that simultaneous observations in any two of the three frequency bands, or at two frequencies within any one band, can be made (See Table 3). The observing frequencies can be set independently for the two channels in 1 MHz steps. The sensitivity ($5\Delta S$) with a VLBA telescope at 5 GHz (T_{sys} = 230 Jy) is, for example, 190 mJy with 32-MHz bandwidth and 2 minutes integration.

Table 2. Receiver and System Temperatures

Band	1.6 GHz	5 GHz	22 GHz
Frequency	1.60-1.73 GHz	4.70-5.00 GHz	21.9-22.3 GHz
T_{RX}	30 K	50 K	150 K
T_{sys}	100 K	120 K	200 K
T_{sys}	14,000 Jy	12,000 Jy	25,000 Jy

5. Link and Observing Mode

The received signal is sampled and sent to the link stations at 128 Mbps at 15 GHz. Each link station also sends a frequency reference signal from its hydrogen maser. A round-trip phase link signal is used to measure the error of the phase transfer. As the uplink and downlink frequencies are high and are very close to each other, the phase fluctuations caused by ionospheric disturbances are greatly reduced.

As a recorder is not installed on board, link stations will have VLBI terminals to record the VSOP satellite data. Consequently, a number of link stations, which are effectively distributed around the world and ground observing telescopes, are needed to get good UV coverage. Five link stations are now planned at Usuda in Japan, Goldstone and Green Bank in the US, Canberra in Australia, and Madrid in Spain.

VSOP observations will have three data parameters: the number of sampling bits, the number of IF channels, and their bandwidths. However, these are not mutually independent and should be chosen so that the total bit rate is 128 Mbps. In Table 3, the available combinations are listed.

Table 3. Observing Modes

Bandwidth	bit	Channel
16 MHz	2	2
32 MHz	1	2
32 MHz	2	1

6. Recorders and correlators

At Usuda link station, a K4 recording terminal is installed which is capable of recording the 32-MHz modes; VLBA recorders will be installed in the other stations. The K4 tapes will be correlated with the VSOP correlator which is planned to be built in Japan. This correlator is designed to cover items special to space VLBI, in particular, having a very wide fringe search window. At the VSOP correlator facility, data translation systems from VLBA to K4, and from S2 to K4 will be installed, to handle all types of recording systems. The S2 system has been developed in Canada and is used in Australia. VLBA tapes will also be correlated using the VLBA correlator.

7. Management

The VSOP International Science Council (VISC) has been established, and will manage all scientific issues relating to VSOP observations. The Announcement of Opportunity (AO) will be released 1.5 years ahead of the launch of the VSOP satellite, and proposals will be screened by peer review. Participation of the ground VLBI networks and telescopes is being negotiated, and a certain amount of their observing time will be devoted to VSOP. The Global VLBI Working Group has been acting as an interface between VSOP and ground networks/telescopes.

References

[1] *Frontiers of VLBI.*, 1992, Proceedings of the International VSOP Symposium and of the mm-VLBI Workshop, eds. H. Hirabayashi, M. Inoue and H. Kobayashi, Universal Academy Press, Inc., Hongo, Tokyo, Japan.

Preliminary Space VLBI Requirements for Observing Time on Ground Radio Telescopes

David L. Meier, David W. Murphy, and Robert A. Preston [*]

Abstract

An initial estimate has been made of the observing time required on ground radio telescopes by the space VLBI missions Radioastron and VSOP. Typical science programs have been adopted for both missions. The missions were assumed to be in orbit simultaneously with each being operated 85% of the time. For the largest telescopes, typically 17% of the hours in a year is required for most telescopes, with notable exceptions being Arecibo (4%), Ooty/GMRT (4%), and Ussurijsk (52%), the latter of which is assumed to be largely dedicated to space VLBI. For the VLBI arrays, typically about 30% of the hours in a year is required for the telescopes of the VLBA and much of the EVN, while 25% is required for a Southern Hemisphere-Pacific VLBI array. Larger time commitments are desired from some telescopes: 52% at Bologna, Noto, Shanghai, Torun, and Urumuqi, and 72% at Algonquin. The scheduling of the space VLBI missions would be most efficient if the missions were able to request time from the ground observatories as needed and return the unused time to the observatories. If a ground observatory wished to set a minimum length of a space VLBI observing period at their telescope, that minimum should be no longer than eight hours. If an observatory wished to set a minimum length on a returned time period, that minimum should also not exceed eight hours.

Radioastron and VSOP, the first two space VLBI observatories, will require joint observations with ground-based radio telescopes (GRTs). In order that the ground observatories can plan their support of these missions, they need to understand the requirements on co-observing time and technical configuration. This paper summarizes the results of a report on the subject to the Interagency Consultative Group, which represents the world's principal space agencies, and the Global VLBI Working Group, which represents the world's GRTs. Further analysis and negotiations will certainly be needed before final allocations can be determined.

---

[*]Jet Propulsion Laboratory, California Institute of Technology, Pasadena, CA. 91109, U.S.A. This research was carried out at the Jet Propulsion Laboratory, California Institute of Technology, under contract with the National Aeronautics and Space Administration.

In our study we have simulated all the major stages of scheduling the actual missions. The mission definition phase, which attempts to understand the details of how science observations would be done by the space observatories, has been studied over the last few years by science simulation teams associated with both missions and support teams at the University of Calgary and at JPL. In the nominal observing plan developed for Radioastron, an observation would last one orbit (28 hours) and require that at least two GRTs form ground-space baselines all the time. For VSOP, an imaging mission, the nominal observing period is 24 hours (four orbits) and would require an array of five or more GRTs throughout the observing period. The proposal and review process was simulated by developing a plausible science program with proposed targets using input from the science simulations teams, the U.S. space VLBI Project Science Group, and other VLBI radio astronomers at large. For Radioastron the resulting program devotes 65% of its observations to high brightness temperature surveys or fine structure modeling, 18% to monitoring a few AGN or QSOs for structural changes, 10% to interstellar scattering studies, 5% to maser spot size measurements, and 2% to radio star observations. For VSOP the sample program devotes 30% of its observations to strong source imaging and 5% to weak source imaging, 50% to AGN/QSO monitoring, 9% to galactic maser studies and 4% to extragalactic masers, and 2% to radio star observations. For each of these subprograms we specify a different number of small (10^2 Jy $< T_{sys} < 10^3$ Jy), large (10 Jy $< T_{sys} < 10^2$ Jy), and giant ($T_{sys} < 10$ Jy) telescopes and the fraction of time each is needed during the observation. To simulate the actual scheduling of observations and telescope time, we developed a new computer program – SVLBSCHED – and generated schedules for periods in the range of 1-12 months. Input to the program included: 1) a list of ground and space radio telescopes, their positions or orbital parameters, their operating frequencies and sensitivities, and their availability and scheduling policies; 2) a list of tracking stations and their policies on tracking each spacecraft; 3) a list of science subprograms, which frequency to observe in each, the number of radio telescopes in the various sensitivity classes required, and specific targets for space VLBI observations (roughly 1000 of the most compact VLBI sources) with different scientific priority. GRTs were assumed to be available continuously, but GRT time which was not needed to satisfy the requirements of the space VLBI science programs was not allocated for space VLBI. A synthesis of our results, along with our additional assumptions, is given in the abstract.

The actual amount of GRT time "charged" to the missions depends strongly on each ground radio observatory's time allocation policies (the minimum allowed observing period and the minimum gap between space VLBI observations) and can be much greater than the hours needed to obtain the science. For example, if the VLBA can make use of gaps between observations as small as 3 hours, then only 30% of the hours in the year is required, whereas 50% is required if the VLBA cannot handle gaps smaller than 24 hours!

32 meter antenna for Toruń Radio Astronomy Observatory

S. Gorgolewski   A. Kus   B. Krygier[*]

Introduction. The 32 m diameter antenna is an altitude-azimuth radiotelescope of the Wheel and Track type. The optics of the telescope can be either Cassegrain or prime focus. The design was planned to meet the requirements of the European VLBI Network. The range of usable frequencies encompasses the 327 MHz and 43 GHz bands. Feeds for 21 cm and shorter wavelengths shall be placed at the vertex of the main paraboloid and shall be selected by an appropriate tilt of the antenna subreflector. At longer wavelengths, the Cassegrain subreflector can be removed for work in the prime focus mode.

Technical specifications. diameter 32.0 m; (F/D) = 0.35; diameter of subreflector = 3.2 m;   diameter of railtrack = 24.0 m; maximum height = 37.2 m.

Design tolerances: reflector surface ¡ 0.35 mm rms; final adjustment 0.2 mm rms; manufactured subreflector : 0.05 mm rms **Steering ranges:** elev. $+2°$ to $+95°$, az.(from South) $\pm 270°$ subreflector axial ± 60 mm and rotation around two axes $\pm 5°$

Speeds: elevation:min. $0.004°$/min; max. $14.7°$/min
           azimuth: min. $0.008°$/min; max. $31.0°$/min

Pointing and tracking max. error: $0.002°$ **Weights:** on railtrack 526.5; on elevation axis 223.5 tons. **Wind velocity:** operational max. 16 m/s; survival: 56 m/s. **Ice deposit:** max. 2 cm; **temperature range** $-25°$ to $+35°C$. **Surface of the Main Reflector** consists of 336 panels arranged in 7 rings. The panels, 2240 mm long and 1200 – 1600 mm wide, are made of 2.5 mm aluminium sheet. They are stretch–formed on a mould and pin–adjusted on an aluminium frame. The measured surface accuracy is better than 0.4 mm rms. The reflector structure is homological and very rigid. Gravitational deformation on the rim: 4.2 mm, gravitational homology imperfections: in zenith 0.11 mm rms, and in horizontal position 0.14 mm rms.

Elevation and Azimuth Drive. Each gearbox is driven by two motors in antibacklash mode. When one drive is operating, the other brakes with a 10 % torque.

Antenna Railtrack. The rail is made of a high–carbon steel 180 mm wide billet. After bending, the rail is machined on both sides. This allows to adjust the rail in level with high accuracy of ± 0.3 mm.

Control System. The antenna control system is designed as a two–level hierarchical system. The managing subsystem, enables the interactive communication with the

---

[*]Toruń Radio Astronomy Observatory, Nicholas Copernicus University, Piwnice, 87–148 Łysomice, Poland, Tel.: 783327 Fax: 11651 Telex: 0552324

operator and external computer systems. The lower level, performs active control of the antenna position, on–line diagnostics and graphic presentation of the control system state. Antenna position is measured by two independent subsystems: absolute (analog converters) and incremental (digital encoders).

Receiver systems shall use frontends covering the 21, 18, 6, 2.8, 1.35 and 0.7 cm bands. All frontends shall use cooled FET and HEMT solid state amplifiers. The IF amplifier system consists of broad band (up to 1 GHz) amplifiers. All LO systems shall be phaselocked to the already ordered EFOS Hydrogen–maser. We plan to use the upgraded MK III VLBI terminal, digital auto correlator spectrograph and a "pulsar machine". The dish is expected to become operational before the end of 1993.

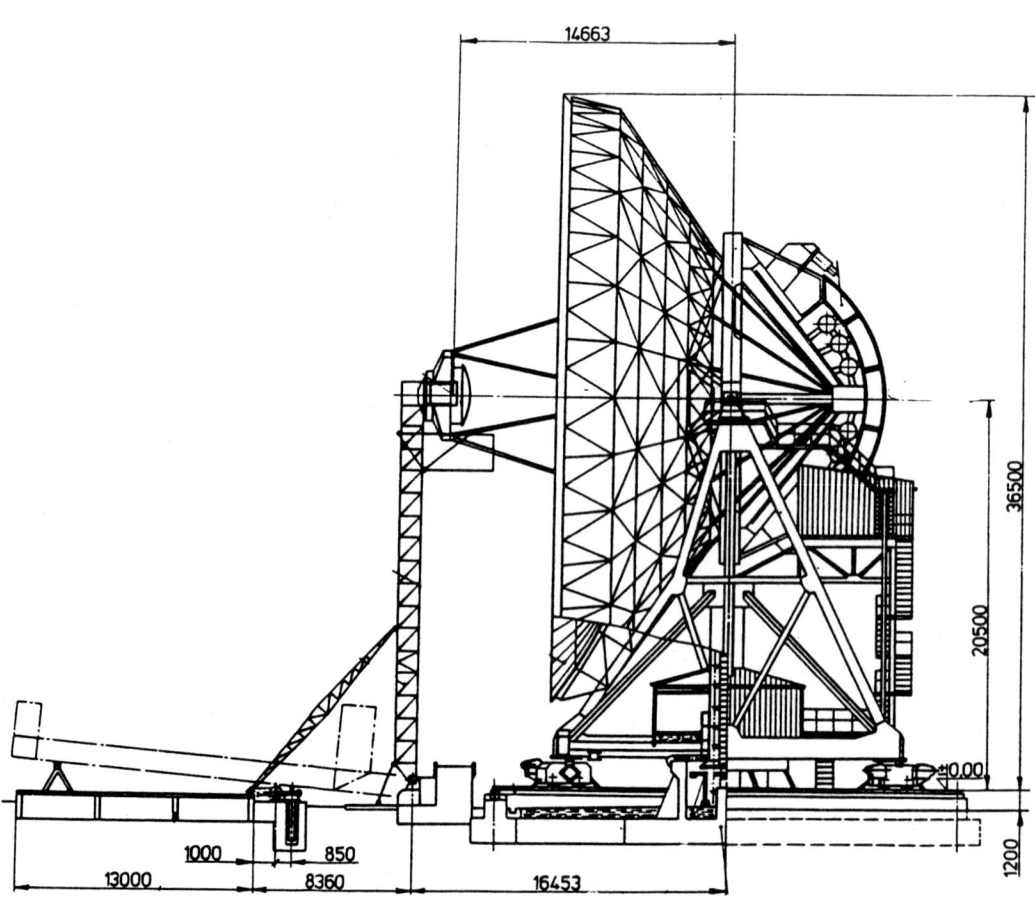

ANTENA D=32m -RT4

Portuguese Radio Interferometry: The MAGRIÇO Project

Maria João Martins [*]   *António A. da Costa* [†]

Abstract

The unique characteristics of Portuguese territory may help bridge the gap in the North Atlantic between the EVN in Europe, and the VLBA in the USA, with radio astronomical facilities which will help the development of Portuguese Astronomy. These are the goals of the MAGRIÇO project.

This paper presents the strategic plan of the MAGRIÇO project as an extension of the EVN, followed by its technical characteristics and requirements. This project provides an interesting and important link with Telecommunications Engineering. A summary of the current stage of development is provided.

Portuguese territory is divided in three different regions in the North Atlantic: the Continent, in European mainland, and the two Autonomous Regions of Azores and Madeira Islands in the Atlantic Ocean. It allows the scattering of telescopes through them, and therefore to make a radio-interferometer network of almost intercontinental dimensions, which might bridge the gap between the EVN in Europe, and the VLBA in the USA. Then, Portugal would take advantage of participating in this cooperation between the two networks. With this idea in mind, the MAGRIÇO (**M**ulti **A**rray **G**rid **R**adio **I**nterferometer **C**ollective **O**peration) project was born.

This VLBI project, a natural extension of EVN, would: allow the development, in Portugal, of Astronomy in general, and Radio Astronomy in particular, in close cooperation with the EVN institutions; allow to work also with VLBA; improve the studies of plate tectonics associated with Azores and Madeira Islands; help to exchange high technology; allow a balanced regional scientific development.

MAGRIÇO must cover the more common bands of radio-astronomy, and obey EVN specifications. Its Phase I will contain 4 telescopes, which will be parabolic antennas. One will be in the Continent (Évora), a reference telescope, with at least 40m. The

---

[*] Centro de Electrodinâmica (INIC), Instituto Superior Técnico, 1096 Lisboa Codex, Portugal.
[†] Centro de Electrodinâmica (INIC), Instituto Superior Técnico, 1096 Lisboa Codex, Portugal; Department of Astronomy, The University, Manchester M13 9PL, England.

other three will have at least 25m diameter, and two will be in Azores (S. Miguel and Flores), and the third in Madeira (Funchal). With this configuration the longest baseline will be of the order of 2 000 km (Évora–Flores) which gives a resolution of 21 milliarcsec, and whose length is the same order of magnitude as the longest EVN baseline. (MAGRIÇO + EVN) has a dramatic improvement in UV-coverage as expected, doubling the size of the present maximum baseline length.

However the construction of MAGRIÇO will start with just one radio-telescope, to be able to join EVN, but keeping open the possibility of full expansion to Phase I. Two choices emerge as the most interesting ones: Flores and Madeira. Flores would allow a better connection between EVN and VLBA. However we prefer Madeira, with better logistic conditions, a choice which would allow EVN expansion to the South.

To start the development of Phase I, we studied the radio-communication links between the several radio-telescopes assuming the whole (Phase I) MAGRIÇO configuration. The radio–communication links consisted of: Continent–Madeira; Continent–Azores; inside Azores. The location of the telescopes was also chosen. Due to the distances involved and the nature of the location, it was necessary to consider microwave, satellite and fiber optical links. We followed MERLIN specifications [Tho86]. All the links were chosen to be digital links. The bandwidth of the signal to be transmitted shows that this is a non trivial project, and it is in the area of HDTV. For the link between the Continent (Sesimbra) and Madeira, we chose the new optical Eurafrica cable. The connection between the Continent (Sintra) and Azores is via satellite. All other connections are through Hertzian links.

The design of the microwave links between Lisbon and Sesimbra, Lisbon and Sintra, and between the islands of Azores, presented some difficulties associated with the large bandwidths to be transmitted, and also the path lengths between the several islands. The project involving the links in Azores was the most difficult one, due to the geographic characteristics of the islands, its scattering in the Atlantic Ocean, and the position of the sattelite station at Terceira.

The problems just summarized show MAGRIÇO as a very interesting project for Telecommunications Engineering. A full account of it may be found in [MdC92].

References

[daC89] da Costa A.A., *Radio Interferometry in Portugal: The MAGRIÇO Project*. Proceedings of the IEEE MELECON'89, Lisbon, 1989.

[MdC92] Martins M.J., da Costa A.A., *Radio Interferometry in Portugal: The MAGRIÇO Project as a Tool for Research and Education*. IEEE Transactions on Education, submitted 1992.

[Tho86] Thomasson P., *MERLIN*. Q. Jl. R. Astr. Soc., Vol. 27 (1986), pp. 413-431.

AUTHOR INDEX

Z. Abraham	186, 191	S.R. Conner	123, 154
C.E. Akujor	225, 263, 265, 273, 275, 277	J.E. Conway	236, 360, 409
		J. Cordes	99
A. Alberdi	50, 184, 241, 267, 355, 412	B. Cordier	47
		B.E. Corey	131
W. Alef	28, 199	W. Cotton	50, 207
P. Alexander	261	S. Curiel	20
H.D. Aller	358	D. Dallacasa	207, 229, 232
M.F. Aller	358	R.M. Danen	34
K. Alvi	131	B. Darchy	418
R.J. Antonucci	331	R.J. Davis	13, 18, 22
A.N. Argue	390	A.A. da Costa	441
R.W. Argyle	390	A.G. de Bruyn	312
D. Aubry	418	B. Dennison	86, 92
D. Axon	312, 314	K.M. Desai	96, 102
W.A. Baan	324	R.J. Dettmar	323
L.B. Bååth	184, 197, 431	P.J. Diamond	99, 102
M. Bang	18	S.M. Dougherty	16
T.M. Bania	84	J-P. Drouhin	418
N. Bartel	107, 420	R.L. Duncombe	394
R. Barvainis	331	W. Eastman	104
A. Baudry	406	P. Elósegui	50, 131, 220, 412
S. Baum	314	R. Estalella	24, 30
J.M. Benson	271	T.M. Eubanks	420
A.O. Benz	28	E.E. Falco	131
W. Benz	375	S.A.E.G. Falle	340
M. Birkinshaw	303	C. Fanti	207, 229, 238
E.E. Bloemhof	34, 76	R. Fanti	107, 207, 229
K.M. Blundell	261	I. Fejes	7
M.F. Bode	18, 36	M. Felli	24
M. Bondi	104, 107, 232	L. Feretti	204, 245
R.S. Booth	74, 184, 222	A.L. Fey	397
G. Bourgois	418	D. Field	68, 74
P.F. Bowers	52	C. Flanagan	96
M. Bowman	378	A. Fletcher	154
A.H. Bridle	258	T. Foley	207
S. Britzen	199	D. Frail	99
I. Browne	137, 144, 267, 269	D.C. Gabuzda	148, 209, 211, 415
B.F. Burke	123, 154	J.A. Garcia-Barreto	323
R. Calder	128	M.A. Garrett	128, 146
J Campbell	199	S.T. Garrington	263, 273, 275
R.M. Campbell	131, 420	M. Gerin	323
W.H. Cannon	420	G. Giovannini	204, 245
Jorge Cantó	20	J.L. Gómez	220, 241, 355
R.J. Cappallo	420	J.I. González-Serrano	290
E.A. Carrara	186, 191		
T.V. Cawthorne	209, 211	M.V. Gorenstein	131
P.L. Cerchiara	232, 269	S. Gorgolewski	439
P. Charlot	218	W.M. Goss	38
G.H. Chen	140	E.M. Gouveia Dal Pino	369, 375
H.S. Chu	249		
D.A. Clarke	258	D.A. Graham	181, 184
I. Cognard	418	M. Gray	68, 74
R.J. Cohen	74	L. Gregorini	104, 107, 321
F. Colomer	59, 184	A. Greve	184
F. Combes	323	M. Grewing	184

443

M. Güdel	28	M.R.W. Masheder	74
J.C. Guirado	220	M. Massi	24
L.I. Gurvits	380	R. Maszkowski	222
C.R. Gwinn	34, 96, 102, 303, 331	R.L. Matveenko	249, 279
P.D. Hemenway	394	P.M. McCulloch	397
L. Herold	154	D.L. Meier	201, 437
J.N. Hewitt	123, 140	K.M. Menten	10, 78
D.H. Hough	193, 195	V. Migenes	74
P.A. Hughes	358	I.F. Mirabel	47
V.A. Hughes	26	J.M. Moran	20, 62, 76, 78
C.A. Hummel	403	R. Morganti	293
M. Inoue	227, 434	L.V. Morrison	390
N. Jackson	267	D.W. Murphy	243, 360, 437
D.L. Jauncey	96, 134, 397	R.L. Mutel	249
K.. Johnson	22	T. Muxlow	137, 144, 146, 241, 252, 269, 273, 312, 314
K.J. Johnston	397		
D.L. Jones	96, 150, 415		
W. Junor	290	S. Neff	331
F.D. Kahn	36	G.D. Nicolson	96, 107, 397
S. Kameno	227	R.P. Norris	81
N. Kawaguchi	397	T.J. O'Brien	36
K.I. Kellerman	386	C.P. O'Dea	314
A. Kemball	397	R. Opher	369
H.T. Kenny	18	L. Padrielli	104, 107, 321
E.A. King	96, 152, 397	N. Panagia	32
L.J. King	128, 137, 144	J.M. Paredes	24, 30
C.S. Kochanek	117	A. Patnaik	137, 144, 146, 277
R.I. Kollgaard	148	J. Paul	47
B. Komberg	373	I.I.K. Pauliny-Toth	234
M. Koribalski	323	P. Pavelin	18, 22
Y.A. Kovalev	371	T.J. Pearson	213, 216, 222, 261
Y. Yu Kovalev	371	A. Pedlar	312, 314
T. Krichbaum	159, 181, 184, 363	I. Pérez-Fournon	290
B. Krygier	439	R.A. Perley	247, 258
M.J. Kukula	314	W.T. Petrachenko	420
A.J. Kus	222, 365, 439	R.B. Phillips	415
R.J. Laing	346	J.D.H. Pilkington	390
L. Lara	50, 241, 245	A.G. Polatidis	213, 216, 225
A. Lazarian	109	J. Popelar	420
F. Lebrun	47	R.W. Porcas	128, 195, 222, 234, 265
J. Lehar	123		
J-F. Lestrade	415, 418	P. Pratap	78
H.M. Lloyd	36	R.A. Preston	96, 415, 428, 437
C.J. Lonsdale	318	A. Quirrenbach	412
E. Ludke	265, 273	P. Rafanelli	321
C. Ma	397	A.C. Raga	352
G. MacLeod	397	F.T. Rantakyrö	197, 232
D.F. Malin	397	M.I. Ratner	412, 420
F. Mantovani	104, 107, 269, 412	A.C.S. Readhead	173, 193, 195, 213, 216, 222
A. Marecki	222		
M. João Martins	441	M.J. Reid	10, 76, 78
J.M. Marcaide	50, 131, 220, 241, 245, 267, 355, 412	Nan Rendong	227
		J. Reynolds	96, 156, 397
A.P. Marscher	84, 297	J.M. Riley	261
J. Martí	30	M.J. Rioja	245
P. Marziani	321	A. Rius	195, 412

AUTHOR INDEX

L.F. Rodríguez	20, 47	K.W. Weiler	32, 107
J. Romney	50, 107	G.L. White	397
A. Rosen	358	A.R. Whitney	420
J.L. Russell	397	P.N. Wilkinson	128, 137, 144, 146, 213, 216, 222, 225, 312, 422
H.S. Sanghera	367		
I. Shapiro	50, 131, 412, 420	T.L. Wilson	78
N. Shapirovskaya	94, 104	A. Witzel	22, 159, 165, 181, 184, 199, 363, 412
Wu Shengyin	234		
SHEVE Team	134, 150, 152, 156, 201, 243, 428	D.M. Worrall	303
		J.M. Wrobel	236
D.L. Shone	265	W. Xu	213, 216
J.H. Simonetti	92	F. Yusef-Zadeh	44
O.B. Slee	94	N.Zacharias	397
K.P. Sokolov	282	J.A. Zensus	181, 186, 191, 195, 363
H.E. Smith	318	Jun-Hui Zhao	38
S. Spangler	104	F.J. Zhang	238, 249
W.B. Sparks	284		
R.E. Spencer	7, 22, 207, 229, 238, 249, 273, 367		
R. Sramek	32		
C. Stanghellini	232		
W. Steffen	159, 363		
H. Steppe	184		
C. Schalinski	22, 181, 184, 199		
R.T. Schilizzi	7, 207, 227, 229, 238		
K.S. Sergey	349		
D. Shaffer	397		
C.N. Tadhunter	293		
H. Takaba	227		
Y. Takahashi	397		
A.R. Taylor	1, 16, 18		
G.B. Taylor	247		
H. Teräsranta	167, 170		
C. Trigilio	232, 245		
G.S. Tsarevsky	94		
J. Ulvestad	331		
G. Umana	22, 232, 245		
S.W. Unger	314		
S.C. Unwin	186, 189, 191		
C.M. Urry	333		
E. Valtaoja	167, 170, 295		
S. van Dyk	32		
H.J. van Langevelde	99		
C. de Vegt	397		
T. Venturi	204, 245		
R.C. Vermeulen	7, 193, 195, 409		
P. de Vicente	28		
S. Wagner	159		
R.C. Walker	271, 425		
D. Walsh	111, 128, 137, 144, 146		
Z. Wang	84		
M.J. Ward	306		
J.F.C. Wardle	209		
P.J. Warner	261		
R. Wegner	159, 165		
A.E. Wehrle	204,		

OBJECT INDEX

Object	Page No.
1E1740.7-2942	47
0106+013	199
0108+388	232
0202+149	107
0212+735	199
0223+341	229
0224+671	107
0229+131	199
0300+470	199
0316+161	229
0316+413	107,181,184
0319+121	229
0333+321	107
0345+337	229
0404+768	213,229
0405-123	107
0420-014	199
0422+044	107
0428+205	229
0454-234	199
0528+134	184,199
0552+398	199
0605-085	107
0607-157	107
0646+600	213
0723-008	107
0727-115	199
0736+017	107
0755+37	204
0851+202	199
0859-140	107
0923+392	199
0954+658	303
0957+561	128,131,234,267
1038+528	220
1042+178	146
1055+018	107
1116+128	107
1127-145	107
1144+35	204
1225+268	229
1226+023	184,186,199,218,284
1308+326	148
1323+321	229
1323+655	263
1334-127	199
1347+539	216
1358+624	229,232
1404+286	199
1413+349	229
1418+546	216
1442+349	229
1504-166	107
1510-089	107
1540+18	263
1600+335	229
1611+343	107,199
1641+399	107,184,189,199,279,363
1642+690	263
1652+398	213
1730-130	107
1741-038	199
1749+096	303
1750+509	144
1803+78	188
1803+784	199,211
1819+39	229
1829+29	229
1857+566	263
1928+738	412
1943+54	213
2007+777	412
2016+112	146
2121+053	199
2134+004	199
2145+067	199
2200+420	107,199
2230+114	229
2234+282	199
2251+158	107,184,199
2255+415	213
2342+821	229
2352+495	173,213
20.7+0.1	62
21.5+0.5	62
26.5+0.6	62
28.7-0.6	62
32.0-0.5	62
32.8-0.3	62
35.6-0.3	62
39.7+1.5	62
40.7+550	312
41.3+596	312
104.9+2.4	62
127.9-0.0	62
3C66B	204
3C67	273
3C84	see 0316+413
3C111	352
3C120	253,271
3C138	207
3C147	249
3C159	269
3C213.1	275
3C216	265

OBJECT INDEX

3C219	258	NGC 315	204
3C264	204,245	NGC 1068	253
3C268	273	NGC 1143/1144	321
3C273	see 1226+023	NGC 1326	323
3C274	204	NGC 2227	32
3C279	184,191	NGC 2748	32
3C286	222,238	NGC 3169	32
3C295	247	NGC 3359	32
3C303	245	NGC 3367	32
3C309.1	222,275,365	NGC 3675	32
3C338	204,245	NGC 3862	284
3C345	see 1641+399	NGC 4045	32
3C346	275	NGC 4151	253,314
3C380	222,225,365	NGC 4254	32
3C395	241,253	NGC 4302	32
3C418	253	NGC 4451	32
3C446	184,431	NGC 4618	32
3C452	245	NGC 4699	32
3C454.3	see 2251+158	NGC 5033	32
3C465	204,245	NGC 5128	32
4C29.30	204,245	NGC 5953/5954	321
4C31.04	245	NGC 6251	303
4C39.25	see 0923+392		
4C74.26	261	NRAO 530	184
AG Pegasi	18	OH 0.319-0.040	99
B0218+35.7	137	OH 0.334-0.181	99
B1422+23.1	137	OH 359.581-0.240	99
B1938+66.6	137		
		OJ 287	see 0851+202
CH Cygni	1		
CTA 102	184,197	OQ 208	see 1404+286
Centaurus A	201,369		
Cepheus A	26,62	Orion	62
DA193	see 0552+398	PKS 0023-263	152
DA406	see 1611+343	PKS 0237-233	243
		PKS 0438-436	243
G331.28-0.19	81	PKS 1151-348	152
		PKS 1245-197	152
HH30	352	PKS 1549-790	243
HM Sge	18	PKS 1830-211	134,150,152
HR 5110	415		
		PSR 1937+214	418
LKH$_\alpha$ 101	34	ψ Persei	16
LSI+61°303	24		
		SN 1983K	32
M87	173,284,253	SN 1984E	32
MG 0414+056	146	SN 1984R	32
MG 1131+0456	140	SN 1985A	32
MG +0-32	32	SN 1985B	32
MSH 04-71	156	SN 1985F	32
		SN 1985G	32
		SN 1985H	32
Mkn 3	253	SN 1985L	32
Mrk 501	236	SN 1986A	32

SN 1986E	32
SN 1986G	32
SN 1986I	32
SN 1986O	32
SN 1987D	32
SS 433	7
SZ Psc	391
Sgr A*	38,50
Sgr A West	38
Sgr B2 M	62
Sgr B2 N	62
Serpens	20
σCrB	415
UX Ari	415
V 1016 Cyg	18
W3(OH)	62,74,76,78
W49N	62,102
W51M	62
W51N	62

SUBJECT INDEX

absorbtion lines
    Galactic   84
accretion disk   333
    corona   250
active galactic nuclei (AGN)
    broad line region   334
    cores   167
    dust enshrouded   318
    IR observtaions   306
    milliarcsecond structure   380
    narrow line region   314, 334
    optical observations   306
    radio observations   167, 170
    relativistic beaming   334
    unified models   249, 333-339
    UV observations   306, 314
antennas
    Torun   439
antenna arrays
    Australia Telescope   81
    IRAM   403
    MERLIN   2, 13, 252, 273, 422-424
    VLA   2
astrometry   390-421
    radio-optical comparison   14, 390-405
atmospheric phase fluctuations   409, 412

barred spiral galaxies   323
black hole   333
blazars   170, 295
BL Lac objects   167, 170, 209, 211, 236, 303, 334, 346
broad line region   249

catalogues
    3CR   229
    Peacock and Wall   229
compact structure   176, 186
compact steep-spectrum sources
    optical observations   293
    polarization   207, 273, 275
    projection effects   367
    radio observation   174, 207, 220, 222, 225, 227, 229, 238, 265
Compton observatory   276
Compton limit   335
core-jet structure
    beaming   175
    complex   265
    counter-jet   258
    curvature   236, 258
    hot spots   263
    knots   271
    kpc structure   258
    limb brightening   230
    motions   271
    numerical simulations   241
    triple   177
cosmology   380-389

distances   62-67
Doppler boosting   162, 222

Faraday rotation   207
FRI sources   201, 204-206, 334, 340, 346, 349
FRII sources   334, 340

Galactic center   38-51
    OH/IR stars   99
Gamma Ray Observatory, GRO   159
gravitational lensing   111-152
    caustics   117
    Einstein ring, quad   121, 123, 134, 150
    lensing galaxy   121
    magnification   120
    structure   128, 131, 134, 137, 140, 146, 150, 156
    surveys   120, 137, 144, 152, 154
    theory   117-122

Herbig Haro objects   20, 26, 353
HIPPARCOS   14, 390, 394, 408, 416
hydrodynamic models   36
Hubble Space Telescope   284, 308, 314, 394

interferometry
    optical   406
interstellar scattering   86-110
    anisotropic
    diffractive   87, 96
    extreme   90, 92
    refractive   89, 94, 104
inverse Compton
    temperature   161
IRAS galaxies   309, 318, 325

jets (see also pc scale jets)
    helical   360, 363, 365
    hypersonic   340

numerical modelling 342, 352, 358, 371, 375, 378
relativistic 347, 349, 355
radio observations 173-180, 204-206, 252, 258
Reynolds number 341
optical/UV observations 284-289, 290
stellar 352

Lorentz factor 223, 335

masers
CH 325
CH$_3$OH 78-83
H$_2$CO 325
H$_2$O 52-56, 64, 102, 324
OH 52-56, 62, 70, 74, 76, 99, 102, 324
SiO 52-56, 59-61
extragalactic 324-330, 331
interstellar 64-67, 68-72, 76, 78, 81, 102
megamasers 325, 326, 331
polarization 68
proper motions 64
saturation 70
scattering 99, 102
stellar 52-56, 59, 62, 99
theory 68-73
variability 62, 76
millimeter wavelength VLBI 59, 167, 181-185
molecular cloud 84

narrow line region (see AGN)
Navier-Stokes equations 341
novae (see radio stars)

parallaxes 390, 415, 418, 420
parsec scale jets 173-180, 201, 204-206
bending 182, 236, 355
interaction with interstellar medium 176
inverse Compton radiatio 178, 300
kinematics 189
knots 355
magnetic fields 300
morphology 174-176, 204-206
polarization 209, 301, 35
relation to kiloparsec jets 216, 360

shocks 301, 356
spectral index 218
superluminal motion 182, 188, 189, 191,199
phase referencing 409-417
polarization 165, 207, 209, 269, 275, 277
pulsars 96
proper motions 418, 420
timing 418

q_0 383
quasars 184, 220, 222, 243, 249, 279, 334
compact steep spectrum 222, 227
core dominated 267
high redshift 381
large sizes 262
lobe dominated 193, 195
morphology 234
polarization 209
spectral index 220

radio galaxies 181, 201, 209, 243, 245, 247, 252, 269, 303
radio sources, extragalactic
angular size-redshift relation 386-389
central engine 177
compact 232
core dominated 360
core-jet 216
double 373
extended 282
filaments 369
flat spectrum 277
flip-flop model 373
hot spots 247
multi-wavelength observations 297-302
polarization 275, 277
steep spectrum 282
(see also compact steep spectrum sources)
structure 199
surveys (see surveys)
X-ray observations 303

radio stars 1-37, 390
Be 16
dMe 2
novae 14, 36
red giant 10-12
RS CVn 1

SUBJECT INDEX

  supernovae 32
  symbiotic 18
  Wolf Rayet 13
  x-ray binary 3, 22, 24, 30
radio variability 159-172
  correlated with UV 159
  correlated with x-ray 159
  correlated with gamma ray 159
  high frequency 159, 167, 170
  intraday 159-166
  low frequency 89, 94, 104, 107, 279
  monitoring 167, 170
Raleigh-Taylor instability 46
recombination lines 39
redshift 383, 386
reference frames
  FK5 390, 403, 406, 416
  J2000 404
  VLBI 13
relativistic beaming 334
ROSAT 159, 302, 303

Seyfert galaxies 252, 309, 314, 321, 325, 331, 334
speckle VLBI 96-98
space VLBI 2, 434, 437
  scheduling 437
spectral index 165, 218, 220
starburst 309, 318
stars (see also radio stars)
  pre-main-sequence 34
  red giant 10
stellar flare 28
stellar wind 37, 44
subluminal motion 203
superluminal motion 182, 188, 189, 191, 193, 197, 199
supernovae
  extragalactic 312
surveys
  Caltech-Jodrell (CJ) 213-215
  3CR 230
  Pearson-Readhead 360
synchrotron emission 368

telescopes
  Hubble Space Telescope 284
  Torun radio telescope 439
turbulence 86, 109, 340-345
  $\kappa-\epsilon$ model 340

visibility 383

VLBI arrays
  European VLBI Network (EVN) 7
  Magrico project 441
  millimetre wavelength 59, 181, 184, 431-433
  SHEVE array 428-430
  VLBA 2, 425-427
VLBI space projects
  VSOP 434, 437
  RADIOASTRON 437
  Scheduling 437

x-ray binaries (see radio stars)